Property of Exxon R & D Laboratories

Composition Modulation of Catalytic Reactors

Topics in Chemical Engineering
A series edited by R. Hughes, University of Salford, UK

Volume 1   HEAT AND MASS TRANSFER IN PACKED BEDS
by N. Wakao and S. Kaguei

Volume 2   THREE-PHASE CATALYTIC REACTORS
by P.A. Ramachandran and R.V. Chaudhari

Volume 3   DRYING: Principles, Applications and Design
by C. Strumillo and T. Kudra

Volume 4   THE ANALYSIS OF CHEMICALLY REACTING SYSTEMS: A Stochastic Approach
by L.K. Doraiswamy and B.K. Kulkarni

Volume 5   CONTROL OF LIQUID–LIQUID EXTRACTION COLUMNS
by K. Najim

Volume 6   CHEMICAL ENGINEERING DESIGN PROJECT: A Case Study Approach
by M.S. Ray and D.W. Johnston

Volume 7   MODELLING, SIMULATION AND OPTIMIZATION OF INDUSTRIAL FIXED BED CATALYTIC REACTORS
by S.S.E.H. Elnashaie and S.S. Elshishini

Volume 8   THREE-PHASE SPARGED REACTORS
edited by K.D.P. Nigam and A. Schumpe

Volume 9   DYNAMIC MODELLING, BIFURCATION AND CHAOTIC BEHAVIOUR OF GAS-SOLID CATALYTIC REACTORS
by S.S.E.H. Elnashaie and S.S. Elshishini

Volume 10   THERMAL PROCESSING OF BIO-MATERIALS
edited by T. Kudra and C. Strumillo

Volume 11   COMPOSITION MODULATION OF CATALYTIC REACTORS
by P.L. Silveston

This book is part of a series. The publisher will accept continuation orders which may be cancelled at any time and which provide for automatic billing and shipping of each title in the series upon publication. Please write for details.

Composition Modulation of Catalytic Reactors

P.L. Silveston

University of Waterloo, Ontario, Canada

GORDON AND BREACH SCIENCE PUBLISHERS
Australia • Canada • China • France • Germany • India
Japan • Luxembourg • Malaysia • The Netherlands • Russia
Singapore • Switzerland • Thailand

Copyright© 1998 OPA (Overseas Publishers Association) N.V. Published by license under the Gordon and Breach Science Publishers imprint.

All rights reserved.

No part of this book may be reproduced or utilized in any form or by any means, electronic or mechanical, including photocopying and recording, or by any information storage or retrieval system, without permission in writing from the publisher. Printed in India

Amsteldijk 166
1st Floor
1079 LH Amsterdam
The Netherlands

British Library Cataloguing in Publication Data

A catalogue record for this book is available from the British Library.

Contents

Introduction to the Series	xi
Bibliography	xiii
Notation	xli

CHAPTER 1   INTRODUCTION 1
- A. Background 1
- B. What is Composition Modulation? 2
- C. Variables in Composition Modulation 4
- D. Modes of Composition Modulation 5
- E. Objectives for Composition Modulation 8
- F. Strategy in Composition Modulation 10
- G. Equipment for Composition Modulation 12
    - Laboratory reactors 13
    - Industrial scale equipment 19
- H. Development of Composition Modulation as a Research Field 22

CHAPTER 2   COMBUSTION 25
- A. Introduction 25
- B. Combustion Modulation 26
- C. Catalytic Combustion 29
    - Application to a combustion system exhibiting hysteresis 30
    - Hydrocarbon catalytic combustion over noble metals 32

CHAPTER 3   CARBON MONOXIDE OXIDATION 38
- A. Introduction 38
- B. Low Temperature Studies over Noble Metal Catalysts 39
    - Experimental 39
    - Modeling 56
    - Forcing of autonomous oscillations 65
- C. Single Crystals of Pt under High Vacuum 68
- D. Noble Metal Catalysts under Higher Temperatures 71
    - Rate enhancement 71
    - Response to forcing in the presence of multiplicity and autonomous oscillations 78
    - Diffusion interference 83

E.	Silver Catalyst	90
F.	Oxide Catalysts	90
	Nickel catalyst	90
	Vanadia catalyst	92
	Copper catalyst	95
	Hopcalite catalyst	97
G.	Summary	98
H.	Future Challenges	101

CHAPTER 4 REDUCTION OF NITROGEN OXIDES — 102

A.	Introduction	102
B.	NO Reduction	102
	Experimental	102
C.	Pt Single Crystals under High Vacuum	121
D.	N_2O Reduction	124
E.	Discussion	130

CHAPTER 5 AUTOMOTIVE EXHAUST CATALYSIS — 132

A.	Background	132
	Engine behavior, driving cycles and exhaust condition	132
	Catalytic converter systems	134
	Oscillation in exhaust gas composition	135
	Objectives of composition modulation research on automotive exhaust	136
B.	Experimental Systems	137
C.	Composition Modulation and Single Reactions in Automotive Exhaust	139
D.	Composition Modulation and Dual Reactions in Automotive Exhaust	141
	Oxidation reactions	141
	Reduction/oxidation reactions	142
E.	Studies on Actual and Simulated Automotive Exhaust	148
	Catalyst development	148
	Catalyst performance with A/F ratio or SN modulation and a simulated exhaust	153
	Catalyst performance under A/F ratio or SN modulation with actual engine exhaust	168
	Effect of A/F ratio modulation on sulpur poisoning	175
F.	A/F Ratio Modulation with Experimental Catalysts without Noble Metals	177
G.	Overview – Does A/F Ratio Modulation Improve Three-way Catalyst Performance?	179

CONTENTS

CHAPTER 6   HYDROGENATION 183
A. Introduction 183
B. Ethylene Hydrogenation 183
C. Aromatics Hydrogenation 189
D. Methanation 190
E. Methanol Synthesis 195
 27:44:29 CuO:ZnO:Al$_2$O$_3$ 196
 30:70 CuO:ZnO and 60:30:10 CuO:ZnO:Al$_2$O$_3$ 199
 Commercial methanol catalysts 204
F. NO$_x$ Reduction with H$_2$ 208
G. Summary 210

CHAPTER 7   AMMONIA SYNTHESIS 211
A. Introduction 211
B. Experimental Methods 211
C. Chromatographic Studies 212
D. Composition Forcing 213
 Iron catalysts 213
 Ruthenium catalyst 226
 Osmium catalyst 231
 Mass transfer interference 235
 Near adiabatic operation 239
E. Analysis 242
 Relaxed steady state 242
 Reactant inhibition model 249
 Reactant storage model 252
 Surface activation/restructuring model 259
F. Overview and Research Challenges 263

CHAPTER 8   SULPHUR DIOXIDE OXIDATION 267
A. Introduction 267
B. Experimental Systems 268
C. Composition Forcing at Low Conversion 268
D. Composition Forcing at High Conversion 271
E. Modelling of SO$_2$ Oxidation under Periodic Operation 280
 Mechanistic model 280
 Application to the final stage of SO$_2$ converter with
  composition forcing 282
 Application to an isothermal backmixed reactor 287
F. Reduction of SO$_3$ by CO over Platinum 292

CHAPTER 9   MISCELLANEOUS REACTIONS 295
A. Ethyl Acetate from the Acid Catalyzed Reaction
 of Ethylene and Acetic Acid 295

B.	Claus Reaction	303
C.	Dehydrogenation of Methanol	305
D.	Deamination and Alcohol Dehydration Reactions	305
	Experimental	305
	Modelling	309
E.	Overview	312

CHAPTER 10   CATALYTIC GAS–SOLID REACTIONS   315

A.	Introduction	315
B.	Non-Catalytic Gas-Solid Reactions	315
	Decomposition of calcium sulphate, gypsum and phosphogypsum	315
	Regeneration of calcium oxide	323
	General comments	323
C.	Iron Catalyzed Gasification	325
	Catalytic coal gasification	325
	Redox mechanism for iron catalyzed gasification	325
	CO_2 gasification of low rank coal	326
	CO_2 gasification with Forestburg coal	328

CHAPTER 11   MULTIPLE REACTIONS AND SELECTIVITY   335

A.	Introduction	335
	Multiple reactions as a target for composition forcing	335
	Selectivity and yield	336
	Classes of multiple reactions	337
B.	Homogeneous Reactions	338
	Concept	338
	Ethyl adipate saponification	339
	Formation of diethanolamine	342
	Assessment	346
C.	Catalytic Consecutive and Competitive Reactions	347
	Catalytic reactions of ammonia and ethylene oxide	347
	Butadiene hydrogenation	350
	Acetylene hydrogenation	353
	Total oxidation	357
	Assessment	362
D.	Competitive Reactions	363
	Methanol synthesis	363
	Ethanol dehydration	364
	Parasitic reaction systems	367
E.	Methane Homologation	369
	Methane homologation at high temperature	369

	Methane homologation at low temperature	373
	Assessment of the cyclic methane homologation processes	382
F.	Oligomerization of Ethene	383

CHAPTER 12 HYDROCARBON DEHYDROGENATION AND PARTIAL OXIDATION — 385

A.	Introduction	385
B.	Oxidative Coupling of Methane	386
	Background	386
	Experimental considerations	387
	Union Carbide and ARCO studies	388
	University of Waterloo studies	393
	Overview and challenges	401
	Partial oxidation to methanol	401
C.	Ethylene and Propylene Oxides from Olefins	402
	Renken's work	402
	Successive pulsing for kinetic studies	406
	Work of Gau at the Institut de Petroleochimie et de Synthese Organique Industrielle	407
	Chinese research	412
D.	Propene Partial Oxidation and Ammoxidation	413
	Sohio research	413
	Early composition forcing experiments	413
	Successive pulsing of the ammoxidation reaction	417
	Acrolein from propene partial oxidation	418
	Acrylic acid from propene partial oxidation	421
	Ketones from olefins	424
E.	Catalytic Dehydrogenation of Propanes, Butanes and Higher Hydrocarbons	426
	Oxidative dehydrogenation of propane on magnesium vanadate catalysts	426
	Oxidative dehydrogenation of butane on MoO_3/MgO catalysts	429
	Dow riser reactor studies	431
F.	Maleic Anhydride from Butane	434
	Investigations at the University of Waterloo	435
	Maleic anhydride from n-butane in a recirculating solids reactor system	440
G.	Maleic and Phthalic Anhydride from Aromatics	448
	Maleic anhydride	448
	Phthalic anhydride from o-xylene	454
H.	Aromatic Nitriles	456
	Lummus isophthalonitrile process	457
	Lummus nicotinonitrile process	459

I. Overview and Assessment 460
    C_1 conversion 460
    Partial oxidation 462

CHAPTER 13   FISCHER-TROPSCH SYNTHESIS 464
A. Introduction 464
B. Experimental Systems Employed 465
C. Experimental Observations 466
    Promoted iron catalyst 466
    Ruthenium supported on Al_2O_3 471
    Molybdenum supported on charcoal 475
    Cobalt supported on kieselguhr or silica 479
D. Comparisons 488
E. Mechanism Conjectures 491
F. Conclusions and Directions for Future Work 495

CHAPTER 14   POLYMERIZATION REACTIONS 498
A. Introduction 498
B. Simulation of Polymerization under Flow
    or Composition Forcing 500
C. Free Radical Polymerization 501
    Simulation 502
    Experimental 505
D. Anionic 'Living' Chain Polymerization 511
    Simulation 511
    Experimental 517
E. Step Growth Polymerization 522
F. Ziegler-Natta Polymerization 524
G. Emulsion Polymerization 527
H. Assessment 528

CHAPTER 15   APPLICATION TO MECHANISTIC STUDIES 530
A. Introduction 530
B. Measurement Applications of Composition Modulation 531
    Qualitative applications 531
    Quantitative applications 532
C. Modulation of Light Intensity 541
D. Application to the Testing of Rival Models 543
E. Overview and Research Challenges 562

AUTHOR INDEX 565

SUBJECT INDEX 575

Introduction to the Series

The subject matter of chemical engineering covers a very wide spectrum of learning and the number of subject areas encompassed in both undergraduate and graduate courses is inevitably increasing each year. This wide variety of subjects makes it difficult to cover the whole subject matter of chemical engineering in a single book. The present series is therefore planned as a number of books covering areas of chemical engineering which, although important, are not treated at any length in graduate and postgraduate standard texts. Additionally, the series will incorporate recent research material which has reached the stage where an overall survey is appropriate, and where sufficient information is available to merit publication in book form for the benefit of the profession as a whole.

Inevitably, with a series such as this, constant revision is necessary if the value of the texts for both teaching and research purposes is to be maintained. I would be grateful to individuals for criticisms and for suggestions for future editions.

R. HUGHES

BIBLIOGRAPHY

Abdul-Kareem, K. H., Silveston, P. L. and Hudgins, R. R. (1980a) "Forced cycling of the catalytic oxidation of CO over a V_2O_5 catalyst: 1 – Concentration cycling", *Chem. Eng. Sci.*, **35**, 2077–2084. [1,3]

Abdul-Kareem, H. K., Jain, A. K., Silveston, P. L. and Hudgins, R. R. (1980b) "Harmonic behavior of the rate of catalytic oxidation of CO under cyclic conditions", *Chem. Eng. Sci.* **35**, 273–282. [3]

Abdul-Kareem, H. K., Silveston, P. L. and Hudgins, R. R. (1981) "Differences in behavior in forced periodic catalytic oxidation of CO", Paper, *2nd World Cong. of Chem. Eng.*, Montreal, October 3–9. [3]

Adesina, A. A., Hudgins, R. R. and Silveston, P. L. (1986) "Feed composition modulation of hydrocarbon synthesis over a cobalt catalyst", *Can. J. Chem. Eng.*, **64**, 447–454. [13]

Adesina, A. A., Hudgins, R. R. and Silveston, P. L. (1988) "Multiple rate resonance phenomena in the forced composition cycling of the Fischer-Tropsch synthesis over a supported cobalt oxide catalyst", *React. Kinet. Catal. Lett.*, **37**, 157–162. [13]

Adesina, A. A., Hudgins, R. R. and Silveston, P. L. (1990) "Effect of ethene addition during the Fischer-Tropsch reaction", *Appl. Catal.*, **62**, 295–308. [13]

Adesina, A. A., Hudgins, R. R. and Silveston, P. L. (1991a) "Influence of concentration waves on the Fischer-Tropsch reaction", *J. Chem. Tech. & Biotech.*, **50**, 535–547. [13]

Adesina, A. A., Hudgins, R. R. and Silveston, P. L. (1991b) "Effect of periodic dosing of ethene in the Fischer-Tropsch synthesis", *Chem. Eng. J.*, **47**, 83–89. [13]

Alassia, L. M., Frontini, G. L., Vega, J. R. and Meira, G. R. (1988a) "The effect of impurities in continuous living anionic polymerizations", *J. Polym. Sci., Part C: Polym. Lett.* **26**(4), 201–206. [14]

Alassia, L. M., Couso, D. A. and Meira, G. R. (1988b) "Molecular weight distrbution control in a semibatch living-anionic polymerization II. Experimental study", *J. Appl. Polym. Sci.*, **36**(3), 481–494. [14]

Al-Taie, A. S. (1977) Ph.D. Thesis, Dept. of Chemical Engineering, University of London, London, U.K. [11]

Al-Taie, A. S. and Kerschenbaum, L. S. (1978) "Effect of periodic operation on the selectivity of catalytic reactions", in Weekman, V. W. Jr. and Luss, D. (Editors), "Chemical Reaction Engineering-Houston", *ACS Symp. Series*, **65**, 512–525.
[11]

Amariglio, H. and Rambeau, G. (1977) "Kinetic manifestations of reversible changes induced by ammonia synthesis upon iron catalyst" in Bond, G. C., Wells, P. B. and Tomplins, F. C. (Editors) *Proc. 6th Int. Congress on Catalysis*, **2**, 1113–1121. [7]

Note: The square bracket [ ] following the reference gives the chapters in which the reference appears.

Amariglio, A., Pareja, P., Belgued, M. and Amariglio, H. (1994) "The possibility of obtaining appreciable yields in methane homologation through a two-step reaction at 250° on a platinum catalyst", *J. Chem. Soc., Chem. Commun.*, 561–562. [11]

Amariglio, H., Pareja, P. and Amariglio, A. (1995a) "Periodic operation of a catalyst as a means of overcoming a thermodynamic constraint. The case of methane homologation on metals", *Catal. Today*, **25**(2), 113–125. [11]

Amariglio, H., Belgued, M., Pareja, P. and Amariglio, A. (1995b) "Carbon monoxide induced desorption of alkanes and alkenes up to C_8 after chemisorption of methane on platinum", *Catal. Let.*, **31**, 19–26. [11]

Amenomiya, Y., Birss, V. I., Goledzinowski, M., Galuszka, J. and Sanger, A. (1990) "Conversion of methane by oxidative coupling", *Catal. Rev.-Sci. Engng.*, **32**, 163–227. [12]

Anderson, R. B. (1984) "The Fischer-Tropsch Synthesis", Academic Press, New York. [13]

Ausikaitis, J. and Engel, A. J. (1974) "Steady state multiplicity and stability in an adiabatic controlled, stirred tank reactor", *AIChEJ.*, **20**(2), 256–265. [1]

Baiker, A. and Richarz, W. (1976) "Verhalten von festbettreaktoren bei periodischen gasdurchsatz-und total druckschwankungen", *Chem. Ing. Tech.*, **48**(12), 1203 (MS 429/76, 24 pages). [6]

Baiker, A., Casanova, R. and Richarz, W. (1980) "Transient behaviour of an adiabatic fixed-bed catalytic reactor in the range of multiple steady states", *German Chem. Eng.*, **3**(2), 112–117. [3]

Baiker, A. and Bergougnan, M. (1985a) "Investigation of a fixed-bed pilot plant reactor by dynamic experimentation. Part 1. Apparatus and experimental results", *Can. J. Chem. Eng.*, **63**, 138–145. [6]

Baiker, A. and Bergougnan, M. (1985b) "Investigation of a fixed-bed pilot plant reactor by dynamic experimentation. Part 2. Simulation of reactor behaviour", *Can. J. Chem. Eng.*, **63**, 146–154. [6]

Bailey, J. E. (1973) "Periodic operation of chemical reactors: a review", *Chem. Eng. Commun.*, **1**, 111–124. [1,2]

Bailey, J. E. (1977) "Periodic phenomena" in Lapidus, L. and Amundson, N. R. (Editors) "Chemical Reactor Theory – A Review", Prentice-Hall, Englewood Cliffs, New Jersey. [1,7]

Bailey, J. E. and Horn, F. J. M. (1968) "An application of the theorem of relaxed control to the problem of increased catalyst selectivity," *J. Opt. Theory Appl.*, **2**, 441–449. [7]

Bailey, J. E. and Horn, F. J. M. (1971) "Improvement of the performance of a fixed bed catalytic reactor by relaxed steady state operation," *AIChEJ.*, **17**, 550–553. [7]

Bailey, J. E. and Horn, F. J. M. (1972) "Cyclic operation of reaction systems: the influence of diffusion on catalyst selectivity", *Chem. Eng. Sci.*, **21**, 109–119. [7]

Bailey, J. E., Horn, F. J. M. and Lin, R. C. (1971) "Cycle operation of reaction systems: effects of heat and mass transfer", *AIChEJ.*, **17**, 818–825. [7]

Balzhinimaev, B. S., Park, D. W. and Gau, G. (1984) "Non steady state oxidation of propylene on supported silver catalysts", *React. Kinet. Catal. Letters*, **24**, 59–64. [12]

BIBLIOGRAPHY

Balzhinimaev, B. S., Belyeava, N. P. and Ivanov, A. A. (1985) "Kinetics of dissolution of inactive crystalline phase in vanadium catalysts for SO_2 oxidation", *React. Kinet. Catal. Letters*, **29**, 465–472. [8]

Balzhinimaev, B. S., Ivanov, A. A., Lapinaa, O. B., Mastikhin, M. and Zamaraev, K. I. (1989) "The mechanism of SO_2 oxidation over supported vanadium catalysts", *Disc. Faraday Chem. Soc.*, **87**(8), 1–15. [8]

Bandermann, F. (1971) "Optimierende reactions fuhrung von polymer reactionen", *Angew. Makromol. Chem.*, **18**, 137–147. [14]

Barshad, Y. and Gulari, E. (1985a) "A dynamic study of CO oxidation on supported platinum", *AIChEJ.*, **31**, 649–658. [1, 3]

Barshad, Y. and Gulari, E. (1985b) "A novel catalytic reactor system for transient response and its use in CO oxidation on Pd/Al_2O_3", *J. Catal.*, **94**, 468–477. [1, 3]

Barshad, Y. and Gulari, E. (1986) "Modification of product distribution through periodic operation: Fischer-Tropsch synthesis over Ru/Al_2O_3", *Chem. Eng. Commun.*, **43**, 39–51. [13]

Barshad, Y., Zhou, X. and Gulari, E. (1985) "CO oxidation under transient conditions: a Fourier-transform infra-red transmission spectroscopy study", *J. Catal.*, **94**, 128–141. [3]

Bartsch, S., Falkowski, J. and Hofmann, H. (1989) "Catalyst development for oxidative methane coupling" *Catal. Today*, **4**, 421–431. [12]

Belgued, M., Pareja, P., Amariglio, A. and Amariglio, H. (1991) "Conversion of methane into higher hydrocarbons on platinum", *Nature*, **352**, 789–790. [11]

Belgued, M., Amariglio, H., Pareja, P., Amariglio, A. and Saint-Just, J. (1992a) "Low temperature catalytic homologation of methane on platinum, ruthenium and cobalt", *Catal. Today*, **13**, 437–445. [11]

Belgued, M., Monteverdi, S., Pareja, P., Amariglio, H., Amariglio, A. and Saint-Just, J. (1992b) "Homologation of methane on metallic surfaces: consideration of reaction pathways", *Preprint, Div. Petrol. Chem., Am. Chem. Soc.*, April, (1992 Meeting, San Francisco, **37**, 324–335. [11]

Bell, A. T. (1981) "Catalytic synthesis of hydrocarbons over group VIII metals. A discussion of the reaction mechanism", *Catal. Rev. -Sci. Eng.*, **23**, 203–232. [13]

Bennett, C. O. (1976) "The transient method and elementary steps in heterogeneous catalysis", *Catal. Rev. -Sci. Eng.*, **13**, 121–148. [15]

Bi, Y.-L., Zhen, K.-J., Jiang, Y.-T., Teng, C.-W. and Yang, X.-G. (1988) "Catalytic oxidative coupling of methane over alkali, alkaline earth and rare earth metal oxides", *Appl. Catal.*, **39**(1–2), 185–190. [12]

Bilimoria, M. R. and Bailey, J. E. (1978) "Dynamic studies of acetylene hydrogenation on nickel catalysts", in Weekman, V. W. Jr. and Luss, D. (Editors) "Chemical Reaction Engineering-Houston", *ACS Symp. Series*, **65**, 526–536. [11]

Boreskov, G. K. and Matros, Yu. Sh. (1983) "Unsteady-state performance of heterogeneous catalytic reactions", *Catal. Rev. -Sci. Eng.*, **25**(4), 551–590. [1, 8]

Boreskov, G. K., Matros, Yu. Sh., Kiselov, O. V. and Bunimovich, G. A. (1977) "Realization of heterogeneous catalytic processes under unsteady-state conditions", *Dokl. Acad. Nauk USSR*, **237**, 160–163. [8]

Brazdil, J. F., Suresh, D. D. and Grasselli, R. K. (1980) "Redox kinetics of bismuth molybdate ammoxidation catalysts", *J. Catal.*, **66**, 347–367. [12, 15]

Briggs, J. P., Hudgins, R. R. and Silveston, P. L. (1976a) "GC measurement of large quantities of sulfur trioxide formed during the catalytic oxidation of sulfur dioxide", *J. Chromatogr. Sci.*, **14**(7), 335–338. [8]

Briggs, J. P., Hudgins, R. R. and Silveston, P. L. (1976b) "Fragmentation of sulfur trioxide in electron impact ionization sources", *Intern. J. Mass Spec. Ion Phys.*, **20**(1), 1–5. [8]

Briggs, J. P., Hudgins, R. R. and Silveston, P. L. (1977) "Composition cycling of an SO_2 oxidation reactor", *Chem. Eng. Sci.*, **32**, 1087–1092. [1, 8]

Briggs, J. P., Kane, D. M., Hudgins, R. R. and Silveston, P. L. (1978) "Reduction of SO_2 emissions from an SO_2 converter by means of feed composition cycling to the final stage" in Moo Young, M. W. and G. Farquhar (Editors)) "Proc. 1st Intern. Waste Treatment and Utilization Conf." (Univ. of Waterloo, Waterloo, Ont., Canada), 521–533. [8]

Budman, H., Kzyonsek, M. and Silveston, P. (1996) "Control of a non-abrabatic packed bed reactor under periodic flow reversal", *Can. J. Chem. Eng.* **74**, 751–759. [12]

Bunimovich, G. A., Strots, V. O. and Goldman, O. V. (1990) "Theory and industrial application of SO_2 oxidation reverse-process for sulphuric acid production" in Matros, Yu. Sh. (Editor)) "Unsteady-state Processes in Catalysis", VNU Science Press, Utrecht-Tokyo, 7–24. [8]

Bunimovich, G. A., Vernikovskaya, N. V., Strots, V. O., Balzhinimaev, B. S. and Matros, Yu. Sh. (1995) "SO_2 oxidation in a reverse-flow reactor: influence of vanadium catalyst dynamic properties", *Chem. Eng. Sci.*, **50**, 565–580. [8]

Callahan, J. L., Grasselli, R. K., Milberger, E. C. and Strecker, H. A. (1970) "Oxidation and ammoxidation of propylene over bismuth molybdate catalyst", *Ind. Eng. Chem. Prod. Res. Devel.*, **9**(2), 135–142. [1, 12]

Capsaskis, S. C. and Kenney, C. N. (1986) "Subharmonic response of a heterogeneous catalytic oscillator, the "Cantabrator", to a periodic input", *J. Phys. Chem.*, **90**(19), 4631–4637. [3]

Casanova, R., Baiker, A. and Richarz, W. (1979) "Dynamisches verhalten eines festbettreaktors im bereich instabiler betriebszustände", *Chem.-Ing. Tech.*, **51**(10), 978–979. [3]

Chakrabarty, T., Hudgins, R. R. and Silveston, P. L. (1982) "Influence of diluent gases on CO oxidation over platinum-alumina catalyst", *J. Catal.*, **77**, 527–538. [3]

Chanchlani, K. G., Hudgins, R. R. and Silveston, P. L. (1992) "Methanol synthesis from H_2, CO and CO_2 over Cu/ZnO catalysts", *J. Catal.*, **136**, 59–75. [1, 6, 11]

Chanchlani, K. G., Hudgins, R. R. and Silveston, P. L. (1994) "Methanol synthesis under periodic operation: an experimental investigation", *Can. J. Chem. Eng.*, **72**, 657–669. [1, 6, 11]

Charon, O., Jouvard, D., Toledo, E. and Genies, B. (1991) "New technique for NO_x emission reduction", Preprint, *1st Intern. Conf. Combust. Technologies for a Clean Environment* (Villamoura, Portugal, Sept., 1991). [2]

Chen, H. T., Kuan, C. N., Setthachayanon, S. and Chartier, P. A. (1980) "Photopolymerization in a continuous stirred-tank reactor-experiment". *AIChE J.*, **26**, 672–675. [14]

Chiao, L. and Rinker, R. G. (1989) "A kinetic study of ammonia synthesis: modelling high-pressure steady-state and forced cycling behaviour," *Chem. Eng. Sci.*, **44**, 9–19.
[7,15]

Chiao, L., Zack, F. K., Thullie, J. and Rinker, R. G. (1987) "Concentration forcing in ammonia synthesis: plug-flow experiments at high temperature and pressure", *Chem. Eng. Comm.*, **49**, 273–289. [1,7,15]

Cho, B. K. (1983) "Dynamic behavior of a single catalyst pellet. I. Symmetric concentration cycling during CO oxidation over Pt/Al_2O_3". *Ind. Eng. Chem. Fundam.*, **22**, 410–420. [3,7]

Cho, B. K. (1994) "Mechanistic importance of intermediate N_2O + CO reaction in overall NO + CO reaction system", *J. Catal.*, **148**, 697–708. [4]

Cho, B. K. and West, L. A. (1986) "Cyclic operation of Pt/Al_2O_3 catalysts for CO oxidation", *Ind. Eng. Chem. Fundam.*, **25**(1), 158–164. [3,4,7]

Cho, B. K., Shanks, B. H. and Bailey, J. E. (1989) "Kinetics of NO reduction by CO over supported rhodium catalysts: isotopic cycling experiments", *J. Catal.*, **115**, 486–499. [4]

Claybaugh, B. E., Griffon, J. R. and Watson, A. T. (1969) "Process for broadening the MWD in polymers", U.S. Patent 3,472,829. [14]

Contractor, R. M. (1987) "Improved vapor phase catalytic oxidation of butane to maleic anhydride", U.S. Patent 4,668,802 (May 26, 1987). [12]

Contractor, R. M. and Sleight, A. W. (1988) "Selective oxidation in a riser-reactor", *Catalysis Today*, **3**, 175–184. [12]

Contractor, R. M., Bergna, H. E., Horowitz, H. S., Blackstone, C. M., Malone, B., Taradi, C. C., Griffiths, B., Chowdhry, U. and Sleight, A. W. (1987) "Butane oxidation to maleic anhydride over vanadium phosphate catalyst", *Catalysis Today*, **1**, 49–58. [1,12]

Contractor, R. M., Bergna, H. E., Horowitz, H. S., Blackstone, C. M., Chowdhry, U. and Sleight, A. W. (1988) "Butane oxidation to maleic anhydride in a recirculating solids reactor", in Ward, J.W. (Editor), "Proceedings of the 10th North American Meeting of the Catalysis Society", Elsevier Science Publishers B. V., Amsterdam, 645–654. [1,12]

Contractor, R. M., Ebner, J. and Mummey, M. J. (1990) "Butane oxidation in a transport bed reactor – redox characteristics of the vanadium phosphorus oxide catalyst" in Centi, G. and Trifiro, F. (Editors) "New Developments in Selective Oxidation", Elsevier Science Publishers B.V., Amsterdam, 553–562. [12]

Cordova, V. R. and Gau, G. (1983) "Redox kinetics of benzene oxidation to maleic anhydride", *Can. J. Chem. Eng.*, **61**, 200–207. [1,12]

Couso, D. A. and Meira, G. R. (1984) "Living anionic polymerizations in a continuous stirred-tank reactor under sustained feed oscillations", *Polymer Eng. Sci.*, **24**(6), 391–397. [14]

Couso, D. A., Alassic, L. M. and Meira, G. R. (1985) "Molecular weight distribution control in a semi batch "living" anionic polymerization" I. Theoretical study" *J. Appl. Polym. Sci.*, **30**(8), 3249–3265 [14]

Creaser, D. (1996) Research Report, Dept. of Reaction Engineering, Chalmers University, Gothenburg, Sweden. [12]

Crone, G. and Renken, A. (1979) *Chem. Ing. Technik* **51**, 842 (Summary). [14]

Cutlip, M. B. (1979) "Concentration forcing of catalytic surface rate processes", *AIChE J.*, **25**, 502–508. [3]

Cutlip, M. B., Hawkins, C. J., Mukesh, D., Morton, W. and Kenney, C. N. (1983) "Modelling of forced periodic oscillations of carbon monoxide over platinum catalyst", *Chem. Eng. Commun.*, **22**, 329–341. [3]

Dath, J.-P., Fink, Th., Imbihl, R. and Ertl, G. (1992) "Periodic and random perturbation of a system exhibiting damped kinetic oscillations – Pt(100)/ NO + CO", *J. Chem. Phys.*, **92**, 1582–1589. [4]

Davis, T. P., O'Driscoll, K. F., Piton, M. C. and Winnik, M. A. (1989a) "Determination of propagation rate constants using a pulsed laser technique", *Macromolecules*, **22**, 2785–2788. [15]

Davis, T. P., O'Driscoll, K. F., Piton, M. C. and Winnik, M. A. (1989b) "Determination of propagation rate constants for the copolymerization of methylmethacrylate and styrene using a pulsed laser technique", *J. Polym. Sci., Part C Polym. Lett*, **27**, 181–185. [15]

Davis, T. P., O'Driscoll, K. F., Piton, M. C. and Winnik, M. A. (1990) "Copolymerization propagation kinetics of styrene with alkyl methacrylates", *Macromolecules*, **23**, 2113–2119. [15]

Davis, T. P., O'Driscoll, K. F., Piton, M. C. and Winnik, M. A. (1991) "Copolymerization propagation kinetics of styrene with alkyl methacrylates", *Polym. Intern.*, **24**, 65–70. [15]

Dettmer, M. and Renken, A. (1983a) "Zur kinetik der ethylacetat-bildung unter stationaeren und instationaeren bedingungen", *Chem. Ing. Tech.*, **55**(2), 146–147. [9]

Dettmer, M. and Renken, A. (1983b) "Kinetic studies on the formation of ethyl acetate under steady-state and non-steady conditions", *Ger. Chem. Eng.*, **6**, 356–365. [9, 11]

Dong, Y. and Lai, H. (1995) "Theoretical studies on flow reversal cycle process of $CaCl_2/CH_3OH$ chemical heat pump", *Chem. Eng. China*, **23**(1), 57–61. [10]

Doroshenko, V. A., Shaporalova, L. P. and Dolya, L. P. (1986) "Preparation and properties of magnesium-molybdenum catalysts for oxidative dehydrogenation of n-butane", *Zhurnal Prikladnoi Khimii*, **59**(5), 1176–1179 Plenum Publ. Corp. (Translation). [12]

Douglas, J. M. (1967) "Periodic reactor operation", *Ind. Eng. Chem. Proc. Design Develop.*, **6**(5), 42–49. [2, 14]

Douglas, J. M. and Rippin, D. W. T. (1966) "Unsteady state process operation", *Chem. Eng. Sci.*, **21**, 305–315. [1, 2, 14]

Dun, J.-W., Ng, K. Y. S. and Gulari, E. (1985) "Fischer-Tropsch synthesis on charcoal supported molybdenum: the effect of preparation conditions and potassium promotion on activity and selectivity", *Appl. Catal.*, **15**, 247–263. [13]

Dun, J.-W. and Gulari, E. (1986) "Rate and selectivity modification in Fischer-Tropsch synthesis over charcoal supported molybdenum by forced concentration cycling", *Can. J. Chem. Eng.*, **64**, 260–266. [13]

Dwyer, D. J. and Somorjai, G. A. (1979) "The role of readsorption in determining the product distribution during CO hydrogenation over Fe single crystals", *J. Catal.*, **56**, 249–257. [13]

Eiswirth, M. and Ertl, G. (1988a) "Forced oscillations of a self-oscillating surface reaction", *Phys. Rev. Let.*, **60**, 1562–1569. [3]

Eiswirth, M., Möller, P., Wetzl, K., Imbihl, R. and Ertl, G. (1989) "Mechanism of spatial self-organization in isothermal kinetic oscillations during catalytic carbon monoxide oxidation on platinum single crystal surfaces", *J. Chem. Phys.*, **90**, 510–521. [3]

El Masry, H. A. (1985) "The Claus reaction: effect of forced feed composition cycling", *Appl. Catal.*, **16**, 301–313. [9]

Elicabe, G. E. and Meira, G. R. (1988) "Estimation and control in polymerization reactors. A review", *Polymer Eng. Sci.*, **28**(3), 121–135. [14]

Elicabe, G. E. and Meira, G. R. (1989) "Model reference adaptive control of a continuous polymerization reactor under periodic operation", *Polymer Eng. Sci.*, **29**(6), 374–382. [14]

Erdöhelyi, A., Cserényi, J. and Solymosi, F. (1992) "Activation of CH_4 and its reaction with CO_2 over supported Rh catalysts", *J. Catal.*, **141**, 287–299. [11]

Ertl, G. (1990) "Oscillatory catalytic reactions at single crystal surfaces", *Advances in Catalysis*, **37**, 213–277. [1,3]

Farhadpour, F. A. and Gibilaro, L. G. (1975) "Continuous unsteady-state operation of a stirred tank reactor", *Chem. Eng. Sci.*, **30**(8), 997–999. [11]

Farhadpour, F. A. and Gibilaro, L. G. (1981) "On the optimal unsteady state operation of a continuous stirred tank reactor", *Chem. Eng. Sci.*, **36**, 143–147. [1]

Feimer, J. L., Hudgins, R. R. and Silveston, P. L. (1984) "Influence of forced cycling on the Fischer-Tropsch synthesis. I. Response to feed concentration step changes", *Can. J. Chem. Eng.*, **62**, 241–248. [13]

Feimer, J. L., Hudgins, R. R. and Silveston, P. L. (1985a) "Influence of forced cycling on the Fischer-Tropsch synthesis. Part II Response to feed concentration square waves", *Can. J. Chem. Eng.*, **63**, 86–92. [1,13]

Feimer, J. L., Hudgins, R. R. and Silveston, P. L. (1985b) "Influence of forced cycling on the Fischer-Tropsch synthesis. III. A model for forced concentration cycling over promoted iron catalyst", *Can. J. Chem. Eng.*, **63**, 481–489. [13]

Fink, Th., Dath, J.-P., Bassett, M. R., Imbihl, R. and Ertl, G. (1990) "A new model for the "explosive" NO + CO reaction on Pt(100)", *Vacuum*, **41**, 301–303. [4]

Fink, Th., Dath, J.-P., Bassett, M. R., Imbihl, R. and Ertl, G. (1991) "The mechanism of the "explosive" NO + CO reaction on Pt(100): experiments and mathematical modelling", *Surf. Sci.*, **245**, 96–110. [4]

Fiolitakis, E. and Hofmann, H. (1983) "Wave front analysis, a specific analysis of transient responses for the investigation of continuously operated distributed heterogeneous reaction systems", *Catal. Rev.-Sci. Engng.*, **24**, 113–157. [12]

Fiolitakis, E., Schmid, M., Hofmann, H. and Silveston, P. L. (1983) "Investigation of the catalytic oxidation of benzene to maleic anhydride by wave front analysis with respect to the application of periodic operation", *Can. J. Chem. Eng.*, **61**, 703–709. [1,12]

Fjeld, M. (1969) "Optimal control of a multivariable periodic process", *Automatica*, **5**(4), 497–506. [2]

Flank, W. H. and Beachell, H. C. (1967) "The geometric factor in ethylene oxidation over gold-silver alloy catalysts", *J. Catal.*, **8**(4), 316–325. [12]

Frontini, G. L., Elicabe, G. E., Couso, D. A. and Meira, G. R. (1986) "Optimal periodic control of a continuous "living" anionic polymerization", *J. Appl. Polymer Sci.*, **31**(4), 1019–1039. [14]

Frontini, G. L., Elicabe, G. E. and Meira, G. R. (1987) "Optimal periodic control of a continuous "living" anionic polymerization II. New Theoretical Results", *J. Appl. Polymer Sci.*, **33**, 2165–2177. [14]

BIBLIOGRAPHY

Gau, G. (1982) "Catalyseurs a l'argent pour la synthese de l'oxyde d'ethylene et de l'oxyde de propylene par un procede regeneratif", Brevet Francaise 80 26624. [12]

Ghazali, S., Park, D. W. and Gau, G. (1983) "Kinetics of ethylene epoxidation on a silver catalyst", *Appl. Catal.*, **6**, 195–208. [12]

Goodman, M. G., Cutlip, M. B., Kenney, C. N., Morton, W. and Mukesh, D. (1982) "Transient studies of CO oxidation over platinum catalyst", *Surf. Sci.*, **120**, L453–L460. [3]

Gordon, D. L. and Weidner, K. R. (1981) "Control of particle size distribution through emulsifier metering based on rate of conversion" in D. R. Bassett and A. E. Hamielec (Editors) "Emulsion Polymers and Emulsion Polymerization", *ACS Symp. Series*, No.165, 515–532. [14]

Gosden, R. G., Auguste, S., Edwards, H. G. M., Johnson, A. F., Meszena, Z. G. and Mohsin, M. A. (1995) "Living polymerization reactors. Part I. Modelling and simulation of flow reactors operated under cyclical-steady-state feed conditions for the control of molecular weight distribution", *Polymer React.Eng.*, **3**(4), 331–359. [14]

Gosden, R. G., Meszena, Z. G., Mohsin, M. A., Auguste, S. and Johnson, A. F. (1996) "Living polymerization reactors. Part II. Theoretical and experimental tests on an algorithm which predicts MWDs from CSTRs with perturbed feeds", *Polymer React. Eng.*, in press. [14]

Grabmueller, H., Hoffmann, U. and Schaedlich, K. (1985) "Prediction of conversion improvements by periodic operation for isothermal plug-flow reactors", *Chem. Eng. Sci.*, **40**, 951–960. [1]

Graham, W. R. C. and Lynch, D. T. (1984) "Model validation through an experimental investigation of resonant behavior for the catalytic oxidation of carbon monoxide on platinum", Eds. Kaliaguine, S. and A. Mahay Catalysis on the Energy Scene, *Stud. Surf. Sci. Catal.*, **19**, Elsevier Science Publ., Amsterdam, pp. 197–204. [3]

Graham, W. R. C. and Lynch, D. T. (1987) "CO oxidation on Pt: model discrimination using experimental bifurcation behavior", *AIChEJ.*, **33**(5) 792–800. [3]

Graham, W. R. C. and Lynch, D. T. (1988) "Forced cycling of catalytic reactors: systems with two inputs", *Stud. Surf. Sci. Catal.*, **38**, Elsevier Science Publ., Amsterdam, 693–701. [3]

Graham, W. R. C. and Lynch, D. T. (1990) "CO oxidation on Pt: variable phasing of inputs during forced composition cycling", *AIChEJ.*, **36**(12), 1796–1806. [3,4]

Gray, P., Nicolis, G., Baras, F., Borckmans, P. and Scott, S. K. (1990) "Spatial inhomogeneities and transient behaviour in chemical kinetics", Manchester University Press, Manchester, U.K. [1]

Greger, M., Ihme, B., Kotter, M. and Riekert, L. (1984) "Non-equilibrium phase transitions with hysteresis on solid catalysts", *Ber. Bunsenges. Phys. Chem.*, **88**, 427–433. [12]

Gugliotta, L. M. and Meira, G. R. (1986) "Control of particle size distribution in lattices through periodic operation of continuous emulsion polymerization reactors", *Macromol. Chem., Macromol. Symp.*, **2**, 209–218. [14]

Gugliotta, L. M. and Meira, G. R. (1988) "Control de la distribution de pesos moleculares y de la distribution de tamanos de particulas en reactures continuous de polimerizacion en emulsion", Preprint 1[er] Simposio Latinoamericano de Polimeros, Porlamar, Venezuela (July, 1988). [14]

Gupta, S. K., Nath, S. and Kumar, A. (1985) "Forced oscillations in continuous flow stirred tank reactors with nonlinear step growth polymerization", *J. Appl. Polymer Sci.*, **30**, 557–569. [14]

Haber, J. (1980) "Dynamics of surface reconstruction and its influence on catalytic properties", *Izvestiya Khimiya*, **13**, 65–75. [12]

Haber, J., Stoch, J. and Wiltowski, T. (1980a) "Surface transformations of copper molybdate catalysts" in Seiyama, T. and Tanabe, K. (Editors)) "New Horizons in Catalysis, Part B", 1402–1403. [12]

Haber, J., Stoch, J. and Wiltowski, T. (1980b) "Surface transformations of copper oxidation catalysts", *React. Kinet. Catal. Letters*, **13**, 161–165. [12]

Halpern, B. and al Mutaz, I. (1986) "A "chopped fast flow" technique for transient surface reaction rate measurements: carbon deposition and oxidation on hot Pt", *Chem. Eng. Sci.*, **41**(6), 899–904. [15]

Hashimoto, S., Kawakami, W. and Akehata, T. (1976a) "A study on control of radiation polymerization by dose rate control", *Ind. Eng. Chem. Process Des. Dev.*, **15**, 244–249. [14]

Hashimoto, S., Kawakami, W. and Akehata, T. (1976b) "Experimental study on control of radiation polymerization by dose rate control", *Ind. Eng. Chem. Process Des. Dev.*, **15**, 549–552. [14]

Hattori, T. and Murakami, Y. (1968) "Study on the pulse reaction technique I. Theoretical study", *J. Catal.*, **10**, 114–122. [1, 15]

Haure, P. M., Hudgins, R. R. and Silveston, P. L. (1989) "Periodic operation of a trickle bed reactor", *AIChEJ.*, **35**, 1437–1444. [1]

Hegedus, L. L., Oh, S. H. and Baron, K. (1977) "Multiple steady states in an isothermal, integral reactor: the catalytic oxidation of carbon monoxide over platinum-alumina", *AIChEJ.*, **23**, 632–639. [3]

Hegedus, L. L., Summers, J. C., Schlatter, J. C. and Baron, K. (1979) "Poison-resistant catalysts for the simultaneous control of hydrocarbon, carbon monoxide, and nitrogen oxide emission", *J. Catal.*, **56**, 321–335. [5]

Hegedus, L. L., Chang, C. C., McEwen, D. J. and Sloan, E. M. (1980) "Response of catalyst surface concentrations to forced concentration oscillations in the gas phase. The NO, CO, O_2 system over Pt-alumina", *Ind. Eng. Chem. Fundam.*, **19**, 367–373. [5]

Helmrich, H., Renken, A. and Schürgerl, K. (1974) "Control of effective rate of heterogeneous catalytic reactions by forced variation of concentration", *Chem.-Ing. Tech.*, **46**(15), 647 (synopsis). [1, 6]

Herman, R. G., Klier, K., Simmons, G. W., Finn, B. P., Bulko J. B. and Kobylinski, T. P. (1979) "Catalytic synthesis of methanol from carbon monoxide/hydrogen, I. Phase composition, electronic properties, and activities of the copper/zinc oxide/M_2O_3", *J. Catal.*, **56**, 407–429. [6, 11]

Herz, R. K. (1981) "Dynamic behavior of automotive catalysts. 1. Catalyst oxidation and reduction", *Ind. Eng. Chem. Prod. Res. Dev.*, **20**, 451–457. [5]

Herz, R. K. (1982) "The dynamic behavior of three-way automotive catalysts" in A. T. Bell and L. L. Hegedus, (Editors), "Catalysis Under Transient Conditions", *ACS Symp. Series*, No.178, American Chemical Society, Washington, D.C., 59–78. [5]

Herz, R. K. (1987) "Dynamic behavior of automotive three-way emmission control systems" in Crucq, A., and A. Frennet Editors, "Catalysis and Automotive Pollution Control", Elsevier Science Publ., Amsterdam, 427–445. [5]

Herz, R. K. and Shinouskis, E. J. (1985) "Dynamic behavior of automotive catalysts. 4. Impact of air/fuel ratio excursions during driving," *Ind. Eng. Chem. Prod. Res. Dev.*, **24**, 385–390. [5]

Herz, R. K. and Sell, J. A. (1985) "The dynamic behavior of automotive catalysts. 3. Transient enhancement of water-gas shift over rhodium", *J. Catal.*, **94**, 166–174. [5]

Herz, R. K., Kleta, J. B. and Sell, J. A. (1983) "The dynamic behavior of automotive catalysts. 2. Carbon monoxide conversion under transient A/F ratio conditions", *Ind. Eng. Chem. Prod. Res. Dev.*, **22**, 387–396. [5]

Hogan, P. (1973) "Ph.D. Thesis", Prague Inst. of Chemical Technology, Prague, Czech Republic. [9]

Horn, F. J. M. (1965) "Attainable and non-attainable regions in chemical reaction technique", *Proc. 3rd European Symp. on Chemical Reaction Engineering*, 292–303, Pergamon Press (Oxford, U.K.) [2]

Horn, F. J. M. and Lin, R. C. (1967) "Periodic processes: a variational approach", *Ind. Eng. Chem. Proc. Des. Devel.*, **6**(1), 21–30. [8]

Horn, F. J. M. and Bailey, J. E. (1968) "An application of the theorem of relaxed control to the problem of improving catalyst selectivity", *J. Opt. Theory Appl.*, **2**(6), 441–449. [1, 11]

Hudgins, R. R. and Silveston, P. L. (1985) "Adsorption/desorption models: How useful for predicting reaction rates under cyclic operation, (Letter)", *Can. J. Chem. Eng.*, **63**, 698–699. [3, 4]

Hudgins, R. R., Silveston, P. L. and Edvinsson, R. (1988) "Claus reaction: effect of forced feed composition cycling", *Appl. Catal.*, **50**, 303–306. [9]

Hugo, A. J. (1985) M. A. Sc. Thesis, Dept. of Chem. Eng., Univ. of Waterloo, Waterloo, Canada. [3]

Hugo, A. J., Jakelski, D. M., Stanitsas, G., Sullivan, G. R., Hudgins, R. R. and Silveston, P. L. (1986) "Long-term transients in a reactor under forced concentration cycling", *Can. J. Chem. Eng.*, **64** 349–351. [1, 3, 4]

Hungenberg, K. D. (1992) "Strategies for PP-homo- and co polymers with broadened MWD in stirred-bed gas phase reactors", DECHEMA - Monographien, No. 127, VCH Verlaggesellschaft (Frankfort, Germany), 501–509. [14]

Irandoust, S. and Andersson, B. (1988) "Mass transfer and liquid phase reaction in a segmented two-phase flow monolith reactor", *Chem. Eng. Sci.*, **43**, 1983–1988. [1]

Ito, T., Wang, J.-X., Lin, C.-H. and Lunsford, J. H. (1985) "Oxidative dimerization of methane over a lithium-promoted magnesium oxide catalyst", *J. Amer. Chem. Soc.*, **107**, 5062–5068. [12]

Ivanov, A. A. and Balzhininckev, B. S. (1987) "New data on kinetics and reaction mechanism for Sulfur dioxde oxidation over vanadium catalysts", *React. Kinet. Catal. Lett.*, **35**(1–2), 413–424. [8]

Jaeger, N. I., Plath, P. J. and Svensson, P. (1990) "Concentration forcing of the oscillating heterogeneous catalytic oxidation of carbon monoxide", in Gray, P., G. Nicolis, Baras, F., Borckmans, P. and Scott, S. K. (Editors), "Spatial inhomogeneities and transient behavior in chemical kinetics", Manchester University Press, Manchester (U.K.), 593–603. [3]

Jagtap, S. B. and Wheelock, T. D. (1996) "Regeneration of sulfided calcium-based sorbents by a cyclic process", *Energy Fuels*, **10**, 821–827. [10]

Jain A. K. (1981) "A study of forced feed composition cycling in the catalytic synthesis of ammonia", Ph.D. Thesis, Dept. of Chem. Eng., University of Waterloo, Waterloo, Ontario, Canada. [7, 15]

Jain, A. K., Hudgins, R. R. and Silveston, P. L. (1981a) "A model to Unni's experiments on cyclic catalytic oxidation of SO_2", *Can. J. Chem. Eng.*, **36**, 231–233. [4]

Jain, A. K., Hudgins, R. R. and Silveston, P. L. (1981b) "A model to Unni's experiments on cyclic catalytic oxidation of SO_2", *Can. J. Chem. Eng,.* **36**, 233–234. [4]

Jain, A. K., Silveston, P. L. and Hudgins, R. R. (1981) "Mechanistic requirements for resonance-like phenomena in forced cycling of heterogeneously catalyzed reactions", Preprint, ACS National Meeting, Las Vegas, NV. [3]

Jain, A. K., Hudgins, R. R. and Silveston, P. L. (1982a) "Reaction rate resonance in concentration cycling of the catalytic oxidation of carbon monoxide", in Bell, A. T. and Hegedus, L. L. (Editors), "Catalysis under transient conditions", *ACS Symp. Ser.*, **178**, ACS (Washington, DC), 267–275. [3, 13]

Jain, A. K., Hudgins, R. R. and Silveston, P. L. (1982b) "Forced composition cycling experiments in a fixed bed ammonia synthesis reactor", Proc. 7th ISCRE, *ACS Symposium Series*, **196**, 97–107. [1, 7]

Jain A. K., Hudgins, R. R. and Silveston, P. L. (1982c) "Evidence for bulk-phase nitrogen dissolution in iron in catalytic ammonia synthesis", *Can. J. Chem. Eng.*, **60**, 809–811. [7, 15]

Jain, A. K., Hudgins, R. R. and Silveston, P. L. (1983a) "Adsorption/desorption models: How useful for predicting reaction rates under cyclic operation", *Can. J. Chem. Eng.* **61**, 46–49. [3]

Jain, A. K., Hudgins, R. R. and Silveston, P. L. (1983b) "Influence of forced feed composition cycling on the rate of ammonia synthesis over an industrial iron catalyst: I – Effect of cycling parameters and mean composition", *Can. J. Chem. Eng.*, **61**, 824–832. [1, 7, 15]

Jain, A. K., Hudgins, R. R. and Silveston, P. L. (1985) "Effectiveness factor under cyclic operation of a reactor", *Can. J. Chem. Eng.*, **63**, 165–169. [7]

Jones, C. A., Leonard, J. J. and Sofranko, J. A. (1985) U.S. Patent 4,560,821. [12]

Jones, C. A., Leonard, J. J. and Sofranko, J. A. (1987a) "The oxidative conversion of methane to higher hydrocarbons over alkali-promoted Mn/SiO_2", *J. Catal.*, **103**, 38–319. [12]

Jones, C. A., Leonard, J. J. and Sofranko, J. A. (1987b) "Fuels for the future: remote gas conversion", *Energy & Fuels*, **1**, 12–16. [12]

Joy, G. J., Lester, G. R. and Molinaro, F. S. (1979) "The influence of sulfur species on the laboratory performance of automotive three component control catalysts", *SAE Tech. Paper Series*, Paper # 790943. [5]

Kane, D. M., Briggs, P., Hudgins, R. R. and Silveston, P. L. (1979) "Reduction of SO_2 emissions from an SO_2 converter by means of feed composition cycling to the final stage", in Moo-Young, M. W. and G. Farquhar, (Editors) "Proc. 1st Intern.

Waste Treatment Utiliz. Conf.", Univ. of Waterloo, Waterloo, Canada, 521–533. [1,8]

Kaneko, Y., Kobayashi, H., Komagoma, R., Hirako, O. and Nakayama, O. (1978) "Effect of the air-fuel ratio modulation on conversion efficiency of three-way catalysts", *SAE Tech. Paper Series*, Paper # 780607, Passenger Car Meeting, Troy (Michigan). [5]

Keil, W. and Wicke, E. (1980) "Über die kinetischen instabilitäten bei der CO-oxidation an platin-katalysatoren", *Ber. Bunsenges. Phys. Chem.*, **84**, 377–383. [3]

Keller, G. E. and Bhasin, M. M. (1982) "Synthesis of ethylene via oxidative coupling of methane", *J. Catal.*, **73**, 9–19. [1,12]

Kellner, C. S. and Bell, A. T. (1982) "Effects of dispersion on the activity and selectivity of alumina supported ruthenium catalysts for carbon monoxide hydrogenation", *J. Catal.*, **75**, 251–261. [13]

Kevrekidis, I. G., Schmidt, L. D. and Aris, R. (1984) "On the dynamics of periodically forced chemical reactors", *Chem. Eng. Commun,*. **30**, 323–330. [3]

Khazai, B., Vrieland, E. G. and Murchison, C. B. (1995) "Process of oxidizing aliphatic hydrocarbons in the presence of a solid heterogeneous catalyst", European Patent 0 409 355 B1. [12]

Khodadadi, A. A., Hudgins, R. R. and Silveston, P. L. (1996a) "Periodic operation of Fischer-Tropsch synthesis on a cobalt catalyst", *Can. J. Chem. Eng.*, **74**, 695–705. [13]

Khodadadi, A. A., Hudgins, R. R. and Silveston, P. L. (1996b) Unpublished Ms, available from authors. [13]

Kilty, P. A. and Sachtler, W. M. H. (1974) "Mechanism of selective oxidation of ethylene to ethylene oxide", *Catal. Rev.-Sci. Engng.*, **10**, 1–35. [12]

King, D. L. (1978) "A Fischer-Tropsch study of supported ruthenium catalysts", *J. Catal.*, **51**, 386–397. [13]

Klahn, J. and Bandermann, F. (1979) "Optimising procedure of reaction of living polymer. Theoretical considerations about molecular weight distribution of living polymers in a periodically stirred tank reactor", *Angew. Makromol. Chem.*, **83**, 1–19. [14]

Kobayashi, M. (1980) "Transient behavior of the oxidation of propylene over a modified silver oxide", *Can. J. Chem. Eng.*, **58**, 588–593. [12]

Kobayashi, M. (1982) "Characterization of transient response curves in heterogeneous catalysis", *Chem. Eng. Sci.*, **37**, 393–401. [15]

Kobayashi, M. and Kobayashi, H. (1974) "Transient response method in heterogeneous catalysis", *Catal. Rev.-Sci. Eng.*, **23**, 1–15. [15]

Koerts, T. and van Santen, R. A. (1991) "A low temperature reaction sequence for methane conversion", *J. Chem. Soc., Chem. Commun.*, 1281–1283. [11]

Koerts, T. and van Santen, R. A. (1992) "A low temperature reaction sequence for methane conversion", *Preprint, Div. Petrol. Chem., Am. Chem. Soc.*, April„ (1992 Meeting, San Francisco, **37**, 336–339. [11]

Koerts, T., Deelen, M. J. A. G. and van Santen, R. A. (1992) "Hydrocarbon formation from methane by a low-temperature, two-step reaction sequence", *J. Catal.*, **138**, 101–114. [11]

Komiyama, H. and Inoue, H. (1968) "Selectivity of gas-solid catalytic reactions", *J. Chem. Eng. Japan*, **1**, 142–148. [11]

Konopnicki, D. and Kuester, J. L. (1974) "Forced oscillations in a non isothermal continuous polymerization reactor", *J. Macromol. Sci.-Chem.*, **A8**, 887–908. [14]

Koubek, J., Pasek, J. and Ruzicka, V. (1980a) "Exploitation of a nonstationary kinetic phenomenon for the elucidation of surface processes in a catalytic reaction", *Proc. 7th Int. Cong. Catal. Tokyo*, **7B**, 852–865. [9, 15]

Koubek, J., Pasek, J. and Ruzicka, V. (1980b) "Stationary and nonstationary deactivation of alumina and zeolites in elimination reactions" in Delmon, B. and Froment G. F. (Editors) "Catalyst Deactivation", Elsevier Science Publ., Amsterdam, 251–260. [9, 15]

Kozyrev, S. V., Balzhinimaev, B. S., Boreskov, G. K., Ivanov, A. A. and Mastikhin, V. M. (1982) "ESR studies of slow relaxations of the rate of SO_2 oxidation on vanadium catalyst", *React. Kinet. Catal. Letters*, **20**, 53–57. [8]

Krischer, K., Lubke, M., Wolf, W., Eiswirth, M. and Ertl, G. (1991) "Chaos and interior crisis in an electrochemical reaction", *Ber. Bunsenges. Phys. Chem.*, **95**, 821–823. [3]

Krylova, A. V. (1990) "Specific features of ammonia synthesis reaction under non-stationary conditions" in Matros, Yu. Sh. (Editor)" Unsteady State Processes in Catalysis", VSP, Zeist, the Netherlands. [8]

Krylova, A. V. and Abdelgani, A. Kh. (1987) "Catalytic synthesis of ammonia under pulse conditions", *Kinet Katal.*, **28**(5), 1126–1131 (in Russian) [8]

Labastida-Bardales, J. R., Hudgins, R. R. and Silveston, P. L. (1989) "Multiplicity in the partial oxidation of propylene over a copper-molybdena catalyst", *Can. J .Chem. Eng.*, **67**, 418–422. [12, 15]

Labastida-Bardales, J. R. (1991) "Oxidation of propylene to acrolein over a copper-molybdena catalyst", Ph.D. Thesis, Univ. of Waterloo, Waterloo, Ontario, Canada. [12]

Lai, H., Zheng, D., Li, C. and Fu, J. (1992) "Chemical heat pump based on $CaCl_2/CH_3OH$: theoretical and practical performance", Proc. 2nd Joint Meeting on Applied Science and Technology, Chemical Industry Press, Beijing. [10]

Lai, H., Li, C., Zheng, D. and Fu, J. (1993) "Modeling of cycling process of $CaCl_2/CH_3OH$ solid-vapor chemical heat pump", Proc. Intern. Absorption Heat Pump Conf., *AES, Am. Soc. Mech. Eng.*, **31**, 419–423. [10]

Lang, X.-S., Hudgins, R. R. and Silveston, P. L. (1989a) "Maleic anhydride formation from 1,3-butadiene oxidation over a vanadia molybdate catalyst", *Can. J. Chem. Eng.*, **67**, 423–431. [12]

Lang, X.-S., Hudgins, R. R. and Silveston, P. L. (1989b) "Application of periodic operation to maleic anhydride production", *Can. J. Chem. Eng.*, **67**, 635–645. [1, 12]

Lang, X.-S., Hudgins, R. R. and Silveston, P. L. (1991) "Selectivity improvement using alternate flushing and reactant cycle steps", *Can. J. Chem. Eng.*, **69**, 1121–1125. [1, 12]

Langner, B. and Bandermann, F. (1978) "Optimising of reaction conditions for living polymerization. Preparation of living polymers with narrow molecular weight distribution in a periodically stirred tank reactor", *Angew. Makromolec. Chem.*, **71**, 101–115. [14]

Laurence, R. L. and Vasudevan, G. (1968) "Performance of a polymerization reactor in periodic operation", *Ind. Eng. Chem. Proc. Des. Devel.*, **7**, 427–432. [14]

Lee, C. K. and Bailey, J. E. (1974) "Influence of mixing on the performance of periodic chemical reactors", *AIChE J.*, **20**, 74–81. [14]

BIBLIOGRAPHY

Lee, C. K. and Bailey, J. E. (1980) "Modification of consecutive-competitive reaction selectivity by periodic operation", *Ind. Eng. Chem. Process Des. Dev.*, **19**, 160–166. [11]

Lee, C. K., Yeung, S. Y. S. and Bailey, J. E. (1980) "Experimental studies of a consecutive-competitive reaction in steady state and forced periodic CSTRs", *Can. J. Chem. Eng.*, **58**, 212–218. [11]

Lefort, L., Amariglio, A. and Amariglio, H. (1994) "Oligomerization of ethylene on platinum by a two-step reaction sequence", *Catal. Let.*, **29**, 125–131. [11]

Leupold, E. I. and Renken, A. (1977) "A new non-stationary process for the production of ethyl acetate", *Chem. Ing. Tech.*, **49**(8), 667 (synopsis). [9]

Leupold, E. I. and Renken, A. (1978) "A new ethyl acetate process using periodic operation", *Ger. Chem. Eng.*, **1**, 218–222. [9, 11]

Leupold, E. I., Arpe, H.-J., Renken, A. and Schlosser, E. G. (1979) "Verfahren zur herstellung von essigsaeureaethylester", Ger. Patent No. DE 2545845C3. [9]

Lewis, W. K., Gilliland, E. R. and Reed, W. A. (1949) "Reaction of methane with copper oxide in a fluidized bed", *Ind. Eng. Chem.*, **41**, 1227–1237. [1, 12]

Li, C.-Y., Hudgins, R. R., Silveston, P. L. (1984) "Modelling ammonia synthesis kinetics over iron catalysts under cyclic operation", in Kaliaguine, S. (Editor), *Stud. in Surf. Sci. and Catal.*, **119**, 229–235 (Elsevier Science Publishers, Amsterdam). [4]

Li, C.-Y., Hudgins, R. R. and Silveston, P. L. (1985a) "Modelling of non-stationary ammonia synthesis kinetics over an iron catalyst. I. Deriving a model and estimating parameters from transient response data", *Can. J. Chem. Eng.*, **63**, 795–802. [7, 15]

Li, C.-Y., Hudgins, R. R. and Silveston, P. L. (1985b) "Modelling of non-stationary ammonia synthesis kinetics over an iron catalyst. II. Modelling results of forced feed composition in periodic operation and for step changes", *Can. J. Chem. Eng.*, **63**, 803–809. [7, 15]

Li, C.-Y., Su, D., Wang, J., Jiang, Q. and Ai, D. (1992a) "Dynamic studies of ethylene selective oxidation on silver catalyst: (I) structure and kinetics of the elementary process sequences", *J. Chem. Industry Engin. (China)*, **43**(3), 256–261. [12]

Li, C.-Y., Su, D., Wang, J., Jiang, Q. and Ai, D. (1992b) "Dynamic studies of ethylene selective oxidation on silver catalyst: (II) the performance under forced composition cycling", *J. Chem. Industry Engin. (China)*, **43**(3), 262–267. [12]

Li, Y.-E., Willcox, D. and Gonzalez, R. D. (1989) "Determination of rate constants by the frequency response method: CO on Pt/SiO_2", *AIChEJ.*, **35**, 423–428. [15]

Liauw, M. A., Plath, P. J. and Jaeger, N. I. (1995) "Periodic perturbation of an oscillating reaction", Preprint, 2nd International Unsteady State Processes in Catalysis Conference, St Louis, Missouri, September. [3]

Lööf, P., Kasemo, B., Björnkvist, L., Andersson, S. and Frestad, A. (1991) "TPD and XPS studies of CO and NO on highly dispersed Pt + Rh automotive exhaust catalysts: evidence for noble metal-ceria interaction", in Crucq, A. (Editor), "Catalysis and Pollution Control II", Elsevier Science Publishers, Amsterdam, 253–273. [5]

Luk'yanenko, V. P., Shapovalova, L. P., Kutnyaya, M. Yu. and Solodkaya, V. S. (1987) "Effects of the diluent on the activities of molybdenum-bearing oxide catalysts in the oxidative dehydrogenation of n-butane to butadiene", *Zhurnal Prikladnoi Khimii*, **605**, 1169–1171 (Plenum Publ. Corp. Translation). [12]

Luu, H. D. (1991) M. A. Sc. Thesis, Dept. of Chem. Eng., Univ. of Waterloo, Waterloo, Ontario, Canada. [1, 3]
Luu, H. D., Hudgins, R. R. and Silveston, P. L. (1992) "Anomalous excitation of CO oxidation over vanadia by periodic concentration cycling", Preprint, Canadian Symp. Catal., Banff, Alberta, Canada. [3]
Luu, H. D. (1993) Internal Seminar, Dept.of Chem. Eng., Univ. of Waterloo, Waterloo, Ontario, Canada. [3]
Lynch, D. T. (1983) "Modelling of resonant behavior during forced cycling of catalytic reactors", *Can. J. Chem. Eng.*, **61**, 183–188. [3, 5]
Lynch, D. T. (1984) "On the use of adsorption/desorption models to describe the forced periodic operation of catalytic reactors", *Chem. Eng. Sci.*, **39**(9), 1325–1328. [3]
Lynch, D. T. (1986) "Forced cycling of catalytic reactors: number of cycles to reach cycle invariance", *Can. J. Chem. Eng.*, **61**, 1035–1038. [4]
Lynch, D. R. and Walters, N. P. (1990) "Frequency response characterization of reaction systems: external recycle reactor with a solid adsorbent", *Chem. Eng. Sci.*, **45**, 1089–1096. [15]
Lynch, D. T., Emig, G. and Wanke, S. E. (1986) "Oscillations during CO oxidation over supported metal catalysts: III. Mathematical modelling of the observed phenomena", *J. Catal.*, **97**, 456–468. [3]
Madon, R. J., Bucker, E. R. and Taylor, W. F. (1977) Final Report, Contract E(46-1)-8008, U.S. Dept. of Energy, U.S. Printing Office, Washington. [13]
Marconi, P. F. and Pagni, R. (1979) "Polimerizzazione di stirene in reattori a regime periodico: influenza sulla distribuzione dei pesi moleculari", *Ing. Chim. Ital.*, **15**(9–10), 93–99. [14]
Marek, M. (1985) "Periodic and aperiodic regimes in forced chemical oscillations", Ed., Rensing, L. and Jaeger, N. I. *"Temporal Order"*, Springer Verlag, Berlin, 105–115. [3]
Marwood, M., Van Vyve, F., Doepper, R. and Renken, A. (1994) "Periodic operation applied to the kinetic study of CO_2 methanation", *Catal. Today*, **20**, 437–448. [6, 15]
Marwood, M., Doepper, R. and Renken, A. (1996) "Modeling of surface intermediates under forced periodic conditions applied to CO_2 methanation", *Can. J. Chem. Eng.*, **74**(5), 660–663. [15]
Mastikhin, V. M., Lapina, O. B., Balzhinimaev, B. S., Simonova, L. G., Karnatovaskaya, L. M. and Ivanov, A. A. (1987) "Catalytically active complexes and influence of SiO_2 on the catalytic properties of the active components for SO_2 oxidation", *J. Catal.*, **103**, 160–169. [8]
Matros, Yu. Sh, (1985) "Unsteady Processes in Catalytic Reactors", Elsevier Science Publishers, Amsterdam. [8, 11, 12]
Matros, Yu. Sh, (1989) "Catalytic Processes under Unsteady-state Conditions", Elsevier Science Publishers, Amsterdam. [8]
Matsumoto, H. and Bennett, C. O. (1978) "The transient methods applied to the methanation and Fischer-Tropsch reactions over fused iron catalyst", *J. Catal.*, **53**, 331–344. [13]
Matsunaga, S.I., Yokota, K. Muraki, H. and Fujitani, Y. (1987) "Improvement of engine emissions over three-way catalysts by periodic operations", *SAE Tech. Paper*

Series, Paper # 872098, Intern. Fuels and Lubricants Meeting, Toronto (Canada), November 2–5. [5]

McCarty, J. G. and Wise, H. (1979) "Hydrogenation of surface carbon on alumina-supported nickel", *J. Catal.*, **57**, 406–416. [11]

McKarnin, M. A., Schmidt, L. D. and Aris, R. (1988) "Forced oscillations of a self-oscillating bimolecular surface reaction model", *Proc. Roy. Soc.*, **A 415**, 363–387. [3]

McNeil, M. A. and Rinker, R. G. (1994) "An experimental study of concentration forcing applied to the methanol synthesis reaction", *Chem. Eng. Commun.*, **127**, 137–149. [6]

McNeil, M. A., Schack, C. J. and Rinker, R. G. (1994) "Methanol synthesis from hydrogen, carbon monoxide and carbon dioxide over a $CuO/ZnO/Al_2O_3$ catalyst. II Development of a phenomenological rate expression", *Appl. Catal.*, **50**(3), 256–265. [6]

Meira, G. R. (1981) "Forced oscillations in continuous polymerization reactors and molecular weight distribution control. A survey", *J. Macromol. Sci.-Rev. Macromol. Chem.*, **C20**(2), 207–241. [1, 14]

Meira, G. and Johnson, A. F. (1981) "Molecular weight distribution in continuous 'living' polymerizations through periodic operation of the monomer feed", *Polym. Eng. Sci.*, **21**, 415–423. [14]

Meira, G. R., Johnson, A. F. and Ramsay, J. (1979) "Molecular weight distribution control in continuous flow reactors", in J. N. Henderson and Bouton, T. C. (editors), *ACS Symposium Series*, No.104, 253–266. [14]

Meshcheryakov, V. D., Boreskov, G. K., Sheplev, V. S. and Ivanov, A. A. (1982) "Influence of phase transitions of vanadium catalysts active components on kinetics of SO_2 oxidation" in Part 2) "Sulfuric Acid Catalysis" (Inst. Catalysis, Novosibirsk), 61–69. [8]

Mielczarski, E., Monteverdi, S., Amariglio, A. and Amariglio, H. (1993) "Direct conversion of methane to higher alkanes by platinum loaded zeolites", *Appl. Catal.*, **A104**, 215–228. [11]

Mikami, J., Satoh, S. and Kobayashi, H. (1970) "Studies on the catalytic oxidation of ethylene by means of the pulse technique", *J. Catal.*, **18**, 265–270. [12]

Monroe, D. R., Krueger, M. H., Beck, D. D. and D'Aniello, Jr., M. J. (1991) "The effects of sulfur on three-way catalysts", in A. Crucq, (Editor), "Catalysis and Pollution Control II", Elsevier Science Publishers, Amsterdam, 593–616. [5]

Moravek, V. (1992) "Steady state and transient kinetics of displacement adsorption and educt inhibition in dehydration of alcohols on alumina", in "Progress in Catalysis", Elsevier Science Publ., Amsterdam. [15]

Morris, C. E., Wheelock, T. D. and Smith, L. L. (1987) "Processing waste gypsum in a two-zone fluidized bed reactor", in Yang, W.-C. (Editor) "New Developments in Fluidization and Fluid-Particle Systems", *AIChE Symp. Series*, **83**(255), 94–104. [15]

Mortazavi, Y. (1990) "Application of periodic operation in oxidative coupling of methane", M. A. Sc. Thesis, Dept. of Chem. Eng., Univ. of Waterloo, Waterloo, Ontario, Canada. [12]

Mortazavi, Y. (1993) "Steady state, transient and periodic operation studies of oxidative coupling of methane", Ph.D. Thesis, Dept. of Chem. Eng., Univ. of Waterloo, Waterloo, Ontario, Canada. [12]

Mortazavi, Y., Hudgins, R. R. and Silveston, P. L. (1992) "Segregated operation of the oxidative coupling of methane on a LiMgO catalysts", Proceedings, 12th Canadian Symp. on Catalysis, Banff, Alberta, May, 1992. [12]
Mortazavi, Y., Hudgins, R. R. and Silveston, P. L. (1996) "Catalytic methane coupling under periodic operation", *Can. J. Chem. Eng.*, **74**, 683–694. [12]
Moulijn, J. A. and Kapteijn, F. (1986) "Kinetics of catayzed and uncatalyzed coal gasification", in Figueiredo, J. L. and Moulijn, J. A. (Editors), Coal and Carbon Gasification: Science and Technology", Martin Nijhoff Publishers (Dordrecht, Netherlands), 181–195. [10]
Mukesh, D., Kenney, C. N. and Morton, W. (1983) "Concentration oscillations of carbon monoxide, oxygen and 1-butene over a platinum supported catalyst", *Chem. Eng. Sci.*, **38**, 69–77. [3]
Müller, E. and Hofmann, H. (1987) "Dynamic modelling of heterogeneous catalytic reactions - I. Theoretical considerations", *Chem. Eng. Sci.*, **42**(7), 1695–1704. [15]
Muraki, H. (1991) "Performance of palladium automotive catalysts", *SAE Tech. Paper Series*, Paper # 910842, Intern. SAE Congress, Detroit (Michigan), February 25-March 1. [5]
Muraki, H. and Fujitani, Y. (1986) "NO reduction over noble metal catalysts under cycled feed streams", *Ind. Eng. Chem. Prod. Res. Dev.*, **25**, 414–419. [4]
Muraki, H., Sobukawa, H. and Fujitani, Y. (1985a) "Periodic operation effects on carbon monoxide oxidation over noble metal catalysts", *Nippon Kagaku Kaishi*, No.2, 176–181. [3,4,5]
Muraki, H., Shinjoh, H. and Fujitani, Y. (1985b) "$NO-CO-O_2$ reaction on Pd/Al_2O_3 catalyst", *Nippon Kagaku Kaishi*, No.9, 1682–1688. [5]
Muraki, H., Shinjoh, H., Sobukawa, H., Yokota, K. and Fujitani, Y. (1985c) "Behavior of automotive noble metal catalysts in cycled feedstreams", *Ind. Eng. Chem. Prod. Res. Dev.*, **24**, 43–49. [5]
Muraki, H., Shinjoh, H. and Fujitani, Y. (1986a) "Reduction of NO by CO over alumina-supported palladium catalyst", *Ind. Eng. Chem. Prod. Res. Dev.*, **25**, 419–424. [4]
Muraki, H., Sobukawa, H. and Fujitani, Y. (1986b) "Dynamic behavior of automotive noble metal catalysts in model gas systems", *Nippon Kagaku Kaishi*, No.4, 532–537. [5]
Muraki, H., Yokota, K. and Fujitani, Y. (1989) "Nitric oxide reduction performance of automotive palladium catalysts", *Appl. Catal.*, **48**, 93–105. [5]
Murchison, C. B. (1982) "The use of molybdenum catalysts for Fischer-Tropsch", 4th Intern. Conf. on the Chem. and Use of Molybdenum, Climax Molybdenum Co. [13]
Murchison, C. B., Vrieland, G. E., Khazai, B. and Weihl, E. (1993) "Anaerobic oxidation of butane to butadiene on magnesium molybdate catalyst", Preprint, 11th No. Amer. Mtg. Catal. Soc., Pittsburgh. [12]
Nam, Y. W., Hudgins, R. R. and Silveston, P. L. (1990) "Storage models for ammonia synthesis over iron catalyst under periodic operation", *Chem. Eng. Sci.*, **45**(10), 3111–3121. [7,15]
Nappi, A., Fabbricino, L., Hudgins, R. R. and Silveston, P. L. (1985) "Influence of forced feed composition cycling on catalytic methanol synthesis", *Can. J. Chem. Eng.*, **63**, 963–970. [1,6,11]

Niwa, M. and Murakami, Y. (1972) "Study on olefin oxidation by periodic pulse technique", *J. Catal.*, **26**, 359–369. [12,15]

Nowobilski, P. J. and Takoudis, C. G. (1986) "Periodic operation of chemical reactor systems: are global improvements attainable", *Chem. Eng. Comm.*, **40**, 249–264. [1,3,9,15]

Oh, S. H. and Hegedus, L. L. (1982) "Dynamics of high-temperature CO chemisorption on Pt/Al_2O_3 by fast response IR spectroscopy", in A. T. Bell and L. L. Hegedus, (Editors) "Catalysis Under Transient Conditions" *ACS Symposium Series*, 178 Am. Chem. Soc. (Washington, D.C.), 79–103. [3]

Oh, S. H., Baron, K., Cavendish, J. C. and Hegedus, L. L. (1978) "Carbon monoxide oxidation in an integral reactor: transient response to concentration pulses in the regime of isothermal multiplicities" *ACS symposium Series*, No. 65 "Chemical Reaction Engineering–Houston", Am. Chem. Soc. (Washington, D.C.) 461–474. [3]

Oh, S. H., Cavendish, J. C. and Hegedus, L. L. (1980) "Mathematical modelling of catalytic converter light off: single-pellet studies", *AIChEJ.*, **26**(6), 935–943. [3]

Ohtsuka, Y., Tamai, Y. and Tomita, A. (1987) "Iron-catalyzed gasification of brown coal at low temperatures", *Energy Fuels*, **1**, 32–35.

Olaj, O. F., Bitai, I. and Gleixner, G. (1985) "The laser flash-initiated polymerization as a tool of evaluating kinetic constants of free radical polymerization. 1 Outline of method and first results", *Makromol. Chem.*, **186**, 2569–2580 [15]

Olaj, O. F., Bitai, I. and Hinkelmann, F. (1987) "The laser flash-initiated polymerization as a tool of evaluating kinetic constants of free radical polymerization. 2 The direct determination of the rate constant of chain propagation", *Makromol. Chem.*, **188**, 1689–1702. [15]

Olsson, P. and Schöön, N.H. (1985) "Transient formation and reduction of sulfur trioxide in the presence of an unsupported platinum catalyst", *Chem. Eng. Sci.*, **40**, 1123–1134. [8]

Olsson, P. and Schöön, N.H. (1986) "Suppression of sulfur trioxide formation in a monolithic catalytic converter for cars due to oscillating reaction conditions", *Ind. Eng. Chem. Process Des. Dev.*, **25**, 528–531. [8]

Onken, H. U. and Wicke, E. (1986) "Statistical fluctuations of temperature and conversion for catalytic CO oxidation in an adiabatic packed bed reactor", *Ber. Bunsenges. Phys. Chem.*, **90**, 976–981. [3]

Ozaki, A., Taylor, H. S. and Boudart, M. (1960) "Kinetics and mechanism of the ammonia synthesis", *Proc. Reg. Soc. London (A)*, **258**, 47–62. [7]

Padeste, L. and Baikers, A. (1994) "Three-way catalysts in hybrid drive system. 1. Experimental study of dynamic behaviour", *Ind. Eng. Chem. Res.*, **33**, 1113–1119. [5]

Pannell, R. B., Kibby, C. L. and Kobylinski, T. P. (1981) in Seiyama, T. and Tanabe, K. (Editors) Proc. 7th *Intern. Congr. Catal.*, Tokyo, Japan. [13]

Park, D. W., Ghazali, S. and Gau, G. (1983) "A cyclic reactor study of ethylene epoxidation", *Appl. Catal.*, **6**, 175–193. [1,12,15]

Park, D. W. and Gau, G. (1986) "Simulation of ethylene epoxidation in a multitubular reactor", *Chem. Eng. Sci.*, **41**(1), 143–150. [1,12]

Penlidis, A., MacGregor, J. F. and Hamielec, A. E. (1985) "Dynamic modelling of emulsion polymerization reactors", *AIChEJ.*, **31**(6), 881–889. [14]

Pierson, W. R. (1976) "Sulfuric acid generation by automotive catalysts", *Chemtech*, **6**, 333–337. [8]
Pitchai, R. and Klier, K. (1986) "Partial oxidation of methane", *Catal. Rev.-Sci. Eng.*, **28**(1), 13–88. [12]
Ponec, V. (1978) "Some aspects of the mechanism of methanation and Fischer-Tropsch synthesis", *Catal. Rev.- Sci. Eng.*, **18**, 151–171. [13]
Prairie, M. R. and Bailey, J. E. (1987) "Experimental and modelling investigations of steady-state and dynamic characteristics of ethylene hydrogenation on Pt/Al_2O_3", *Chem. Eng. Sci.*, **42**(9), 2085–2102. [6]
Prairie, M. K., Cho, B. K., Oh, S. H., Shinouskis, E. J. and Bailey, J. E. (1988) "Steady-state and transient studies of CO oxidation on alumina supported rhodium via transmission IR spectroscopy", *Ind. Eng. Chem. Res.*, **27**, 1396–1407. [3]
Prauser, G., Keimer, H. and Dialer, K. (1976) "Untersuchung zum mechanismus der aethylenoxidation an silberkontakten mit der pulstechnik", *Ber. Bunsen Gesell.*, **80**(1), 49–60. [12]
Prokopowicz, R. A. (1985) M. A. Sc. Thesis, Dept. Chem. Eng., Univ. of Waterloo, Waterloo, Canada. [3]
Prokopowicz, R. A., Silveston, P. L., Baudais, F. L., Irish, D. E. and Hudgins, R. R. (1987) "Design of a transmission cell for the high temperature study of transient catalytic adsorption and reaction in a flow system", *J. Appl. Spectros.*, **42**, 385–389. [1,3]
Prokopowicz, R. A., Silveston, P. L., Hudgins, R. R. and Irish, D. E. (1988) "Oxidation of CO over a Cu II Oxide/Silica Catalyst", *React. Kinet. Catal. Letters*, **37**(1), 63–70. [3]
Putnam, A. A., Belles F. E. and Kentfield J. A. C. (1986) "Pulse Combustion", *Prog. Energy Combust. Sci.*, **12**, 42–79. [2]
Qin, F. and Wolf, E. E. (1995) "Vibrational control of chaotic self-sustained oscillations during CO oxidation on a $Rh\text{-}SiO_2$ catalyst", *Chem. Eng. Sci.*, **50**(1), 117–126. [3]
Rabl, A. and Renken, A. (1986) "Untersuchungen zur produktinhibierung under katalysator-desaktivierung bei der vinyl acetat-bildung", *Chem. Ing. Tech.*, **58**, 434–435 (synopsis). [11]
Rabl, A., Doepper, R. and Renken, A. (1986) "Steady state and transient kinetic studies of an industrial vinyl acetate catalyst", *Proc.III World Congr. Chem. Eng. (Tokyo)*, **IV**, Paper 9c-257, 387–390. [11,15]
Rambeau, G. and Amariglio, H. (1978a) "No. 14 La synthése de l'ammoniac sur le fer en tant que réaction auto catalytique – I.", *J. Chimie Phys.*, **75**(1), 110–115. [7]
Rambeau, G. and Amariglio, H. (1978b) "No. 47 La synthése de l'ammoniac sur le fer en tant que réaction auto catalytique – II.", *J. Chimie Phys.*, **75**(3), 333–340. [7]
Rambeau, G. and Amariglio, H. (1978c) "No. 47 La synthése de l'ammoniac sur le fer en tant que réaction auto catalytique – III.", *J. Chimie Phys.*, **75**(4), 397–405. [7]
Rambeau, G. and Amariglio, H. (1981a) "Ammonia synthesis on Ruthenium powder from 100° to 500°C and hydrogenation of pre-adsorbed nitrogen down to $-70°C$", *J. Catal.*, **72**, 1–11. [7]
Rambeau, G. and Amariglio, H. (1981b) "Improvement of the catalytic performance of a ruthenium powder in ammonia synthesis by the use of a cyclic procedure", *Appl. Catalysis*, **1**, 291–302. [1, 7, 13]

Rambeau, G. and Amariglio, H. (1984) "Reflections upon the conditions required for obtaining rate improvement by periodically changing the feed composition of a catalytic reactor", in Kaliaguine, S. and A. Mahay (Editors), "Catalysis on the energy scene", *Surf. Sci. Catal. Series*, 213–220, Elsevier Sci. Publ. (Amsterdam). [7]

Rambeau, G., Jorti, A. and Amariglio, H. (1982a) "Ammonia synthesis on osmium powder and hydrogenation of pre-adsorbed nitrogen from 100° to 500°C, *J. Catal.*, **74**, 110–120. [7]

Rambeau, G., Jorti, A. and Amariglio, H. (1982b) "Improvement in the catalytic performance of an osmium powder in ammonia synthesis by use of a cyclic procedure", *Appl. Catal.*, **3**, 273–282. [1, 7]

Randall, H., Doepper, R. and Renken, A. (1996) "Model discrimination by unsteady-state operation: application to the reduction of NO with CO on iron oxide", *Can. J. Chem. Eng.*, **74**, 586–593. [15]

Rao, A. S., Prasad, K. B. S. and Rao, M. B. (1992) "An experimental study of cyclic operation of a benzene hydrogenation reactor", *J. Catal.*, **136**, 242–245. [6]

Raupp, G. B. and Delgass, W. N. (1979) "Moessbauer investigation of supported Fe catalysts. II. Carbides formed by Fischer-Tropsch synthesis", *J. Catal.*, **58**, 348–360. [13]

Rawlings, J. B. and Ray, W. H. (1988a) "The modeling of batch and continuous emulsion polymerization reactors. Part I: model formulation and sensitivity to parameters", *Polymer Eng. Sci.*, **28**(5), 237–256. [14]

Rawlings, J. B. and Ray, W. H. (1988b) "The modeling of batch and continuous emulsion polymerization reactors. Part II: comparison with experimental data from continuous stirred tank reactors", *Polymer Eng. Sci.*, **28**(5), 257–274. [14]

Ray, W. H. (1967) "Modelling polymerization reactors with application to optimal design", *Can. J. Chem. Eng.*, **45**, 356–364. [14]

Ray, W. H. (1968) "Periodic operation of polymerization reactors", *Ind. Eng. Chem. Process Des. Dev.*, **7**, 422–433. [1, 14]

Ray, W. H. (1972) "On the mathematical modelling of polymerization reactors" *J. Macromol. Sci.-Revs. Macromol. Chem.*, **C8**, 1–34. [14]

Razon, L. F. and Schmitz, R. A. (1986) "Intrinsically unstable behaviour during the oxidation of CO on Platinum", *Catal. Rev.-Sci. Eng.*, **28**, 89–164. [1]

Razon, L. F. and Schmitz, R. A. (1987) "Multiplicities and instabilities in chemically reacting systems – a review", *Chem. Eng. Sci.*, **42**, 1005–1047. [1]

Renken, A. (1972) "The use of periodic operation to improve the performance of continuous stirred tank reactors", *Chem. Eng. Sci.*, **27**, 1925–1932. [2, 11]

Renken, A. (1974) "Verbesserung von Selektivitaet und Ausbeute durch periodische Prozessfuehrung", *Chemie-Ing.Technik*, **46**(3), 113 (Extended Manuscript). [11]

Renken, A. (1982) "Unsteady-state operation of continuous reactors", *Chem.-Ing. Tech.*, **54**(6), 571–580. [1]

Renken, A. (1984) "Unsteady-state operation of continuous reactors", *Int. Chem. Eng.*, **24**(2), 202–213. [1]

Renken, A. (1990a) "Application of unsteady state processes in modelling heterogeneous catalytic kinetics", in Yu. Sh. Matros (Editor), "Unsteady State Processes in Catalysis", VSP, Utrecht, The Netherlands. [1,15]

Renken, A. (1990b) "Instationäre reaktionsführung zur modellierung heterogen katalytischer reaktionen", *Chem.-Ing.-Tech.*, **62**(9), 724–733. [15]

Renken, A. (1993) "Transient operation for the purpose of modeling heterogeneous catalytic reactions", *Intern. Chem. Eng.*, **33**(1), 61–71. [15]

Renken, A., Helmrich, H. and Lenthe, M. (1974) "Experimental studies on the behaviour of instantaneous isothermal fixed-bed reactors", *Proc. GVC/AIChE Joint Meeting*, Vol.1, Paper A 3-1, Verlag Chemie GmbH, Weinheim, Germany. [11]

Renken, A., Mueller, M. and Wandrey, C. (1976) "Experimental studies of fixed-bed reactors by periodic operation – the catalytic oxidation of ethylene", *Proc. 4th Inter. Symp. Chem. React. Eng.*, Vol. III, Dechema, Frankfurt, Germany, 107–116. [12]

Renken, A., Truffer, M.-A. and Dettmer, M. (1984) "The effect of cyclic operation on heterogeneous catalytic reaction with educt inhibition", *I. Chem. Eng. Symp. Series*, No. 87, 117–133. [9, 11]

Renken, A., Doepper, R. and Schmid, M. (1989) "Application of non-stationary and periodic processes in the study of heterogeneous catalytic kinetics", *Dechema Monographs*, **120**, VCH Verlagsgesellschaft, 273–290. [11,12]

Ross, G. S., Hudgins, R. R. and Silveston, P. L. (1987) "The Fischer-Tropsch synthesis over a ruthenium catalyst under composition cycling", *Can. J. Chem. Eng.*, **65**, 958–965. [13]

Sadhankar, R. R. and Lynch, D. T. (1994) "N_2O reduction by CO over an alumina-supported Pt catalyst: forced composition cycling", *J. Catal.*, **149**, 278–291. [4,15]

Sadhankar, R. R. and Lynch, D. T. (1995) "Slow convergence during forced oscillations of the NO + CO reaction over a platinum catalyst", *Chem. Eng. Sci.*, **51**(10), 2061–2068. [4]

Sadhankar, R. R. and Lynch, D. T. (1996) "Transient and resonant behaviour for NO reduction by CO over a Pt/Al_2O_3 catalyst during forced composition cycling", *Can. J. Chem. Eng.*, in press. [4]

Sadhankar, R. R., Ye, J. and Lynch, D. T. (1994) "N_2O reduction by CO over an alumina-supported Pt catalyst: steady state multiplicity", *J. Catal.*, **149**, 511–522. [4]

Salah-Alhahamed, Y. A., Hudgins, R. R. and Silveston, P. L. (1992) "Periodic operation studies on the partial oxidation of propylene to acrolein and acrylic acid", *Chem. Eng. Sci.*, **47**, 2885–2896. [12]

Salah-Alhahamed, Y. A., Hudgins, R. R. and Silveston, P. L. (1993) "The role of water in the partial oxidation of propylene" in Guczi, L. (Editor)) "Proceeding of the 10th Intern. Congr. Catal. [12]

Sanayei, R. A., O'Driscoll, K. F. and Klumperman, B. (1994) "Pulsed laser copolymerization of styrene and maleic anhydride", *Macromolecules*, **27**, 5577–5582. [15]

Schack, C. J., McNeil, M. A. and Rinker, R. G. (1994) "Methanol synthesis from hydrogen, carbon monoxide and carbon dioxide over a $CuO/ZnO/Al_2O_3$ catalyst. I. Steady state kinetics experiments", *Appl. Catal.*, **50**(3), 247–263. [6]

Schädlich, K., Hoffmann, U. and Hofmann, H. (1983) "Periodical operation of chemical processes and evaluation of conversion improvements", *Chem. Eng. Sci.*, **38**, 1375–1384. [1]

Schlatter, J. C. and Mitchell, P. J. (1980) "Three-way catalyst response to transients", *Ind. Eng. Chem. Prod. Res. Dev.*, **19**, 288–293. [5]

Schlatter, J. C., Sinkevitch, R. M. and Mitchell, P. J. (1983) "Laboratory reactor system for three-way automotive catalyst evaluation", *Ind. Eng. Chem. Prod. Res. Dev.*, **22**, 51–56. [1, 5]

Schüth, F., Henry, B. E. and Schmidt, L. D. (1993) "Oscillatory reactions in heterogeneous catalysis", *Advances in Catalysis*, **39**, 51–127. [1, 3]

Schwankner, R. J., Eiswirth, M., Müller, P., Wetzl, K. and Ertl, G. (1987) "Kinetic oscillations in the catalytic CO oxidation on Pt100–periodic perturbations", *J. Chem. Phys.*, **87**, 742–749. [3]

Schwartz, S. B. and Schmidt, L. D. (1987) "Is there a single mechanism of catalytic rate oscillations on Pt", *Surf. Sci.*, **183**, L269–L278. [4]

Schwartz, S. B. and Schmidt, L. D. (1988) "The NO + CO reaction on clean Pt(100): multiple steady states and oscillations", *Surf. Sci.*, **206**, 169–186. [4]

Schwarz, J. A. and Madix, R. J. (1974) "Modulated beam relaxation spectrometry", *Surface Sci.*, **46**, 316–341. [15]

Schwendeman, M. N., McBride, R., Reuter, D. and Isaacs, M. (1983) "Porous metal filters prevent fluid bed catalyst loss", *Chemical Processing*, July, 100–101. [1, 12]

Severyanin, V. S. (1971) "Prospects for pulsating combustion in power engineering", in Brown, D. J. (editor), *Proc. 1st Intern. Symp. on Pulsating Combustion*, Sheffield, U.K., September. [2]

Severyanin, V. S. (1982) "Applications of pulsating combustion on industrial installations", in Gas Research Institute, *Proc. Symp. on Pulse Combustion Applications*, **1**, GRI Report 82/0009 (NTIS PB82-240060, Washington, D.C.) Atlanta, U.S.A., March. [2]

Shanks, B. H. and Bailey, J. E. (1987) "Modelling of slow dynamics in the oxidation of CO over supported silver", *AIChE J.*, **33**(12), 1971–1976. [3]

Shanks, B. H. and Bailey, J. E. (1988) "Experimental investigations using feedback induced bifurcation: carbon monoxide oxidation over supported silver", *Chem. Eng. Commun.*, **61**, 127–149. [3]

Shapovalova, L. P., Voznyuk, V. I., Lysukho, T. V., Luk'yanenko, V. P., Kutnyaya, M. Yu. Solodkaya V. S. and Dolya, L. P. (1987) "Influence of the calcination conditions on the activity of a catalyst for oxidative dehydrogenation of n-butane", *Zhurnal Prikladnoi Khimii*, **60**(2), 369-373 (Plenum Publ. Corp. Translation). [12]

Sheintuch, M. and Schmitz, R. A. (1977) "Oscillations in catalytic reactions", *Catal. Rev.-Sci. Eng.*, **15**, 107–172. [1]

Shinjoh, H., Muraki, H. and Fujitani, Y. (1987) "Periodic operation effects on automotive noble metal catalysts: Reaction analysis of binary systems", in A. Crucq and A. Frennet, (Editors) "Catalysis and Automotive Pollution Control", Elsevier Science Publishers (Amsterdam). [2, 6]

Shinjoh, H., Muraki, H. and Fujitani, Y. (1989) "Periodic operation effects in propane and propylene oxidation over noble metal catalysts", *Appl. Catal.*, **49**, 195–204. [2, 4]

Silveston, P. L. (1979) "Influence of composition modulation on product yields and selectivity in the catalytic partial oxidation of propylene", Preprint, 6th *Intern. Symp. Chem. Reaction Eng.*, Nice, France. [12]

Silveston, P. L. (1980) "Influence of composition modulation on product yields and selectivity in the partial oxidation of propylene", Preprint, 73rd Ann. Mtg, A.I.Ch.E., Chicago, Nov. 16–20. [12]
Silveston, P. L. (1987) "Periodic operation of chemical reactors – a review of the experimental literature", *Sadhana*, **10**, 220–246. [1, 7]
Silveston, P. L. (1991) "Catalytic oxidation of carbon monoxide under periodic operation", *Can. J. Chem. Eng,*. **69**, 1106–1120. [3]
Silveston, P. L. (1995) "Automotive exhaust catalysis under periodic operation", *Catal. Today*, **25**(2), 175–195. [4, 15]
Silveston, P. L. and Hudgins, R. R. (1981) "Reduction of SO_2 emissions from a H_2SO_4 plant by means of feed modulation", *Environ. Sci. Technol.*, **15**, 419–422. [8]
Silveston, P. L. and Forrissier, M. (1985) "Influence of composition modulation on product yields and selectivity in the partial oxidation of propylene over an antimony-tin oxide catalyst", *Ind. Eng. Chem. Prod. Res. Dev.*, **24**, 320–325. [1, 12]
Silveston, P. L. and Hudgins, R. R. (1987) "Periodic operation of chemical systems", *Chem. Eng. Comm.*, **51**, 333–335. [3]
Silveston, P. L., Hudgins, R. R., Abdul-Kareem, H. K., Jain, A. K., Fabbricino, L., Roussel, J., Wagler, R. and Huber, S. L. (1983) "Forced composition modulation of CO oxidation over various catalysts", Paper, AIChE National Meeting, Detroit, Michigan. [3]
Silveston, P. L., Hudgins, R. R., Adesina, A. A., Ross, G. S. and Feimer, J. L. (1986) "Activity and selectivity control through periodic composition forcing over Fischer-Tropsch catalysts", *Chem. Eng. Sci.*, **41**, 923–928. [13]
Silveston, P. L., Hudgins, R. R., Bogdashev, S., Vernijakovskaja, N. and Matros, Yu. Sh. (1990) "Modelling of a periodically operated packed bed SO_2 oxidation reactor at high conversion", in Yu. Sh. Matros (Editor)) "Unsteady State Processes in Catalysis", VSP (Utrecht, Netherlands). [8]
Slin'ko M. M. and Slin'ko, M. G. (1977) "Self-oscillations of heterogeneous catalytic rates", *Catal. Rev.-Sci. Eng.* **17**, 119–153. [1]
Slin'ko, M. M. and Jaeger, N. I. (1994) "Oscillating Heterogeneous Catalytic Systems", Elsevier Science Publishers, Amsterdam. [1]
Snyder, J. D. and Subramaniam, B. (1993) "Numerical simulation of a periodic flow reversal reactor for sulfur dioxide production", *Chem. Eng. Sci.*, **48**(24), 4051–4064. [8]
Sofranko, J. A., Leonard, J. J. and Jones, C. A. (1987) "The Oxidative conversion of methane to higher hydrocarbons", *J. Catal.*, **103**, 302–310. [8, 12]
Solymosi, F. and Cserényi, J. (1994) "Decomposition of CH_4 over supported Ir catalysts", *Catal. Today*, **21**, 561–569. [11]
Solymosi, F. and Cserényi, J. (1995) "Enhanced formation of ethane in the conversion of methane over Cu-Rh/SiO_2", *Catal. Letters*, **34**, 343–350. [11]
Solymosi, F., Erdöhelyi, A. and Cserényi, J. (1992) "A comparative study on the activation and reactions of CH_4 on supported metals", *Catal. Letters*, **16**, 399–405. [11]
Solymosi, F., Erdöhelyi, A., Cserényi, J. and Felvgi, A. (1994) "Decomposition of CH_4 over supported Pd catalysts", *J. Catal.*, **147**, 272–278. [11]

Solymosi, F., Erdöhelyi, A. and Szke, A. (1995) "Dehydrogenation of methane on supported molybdenum oxides. Formation of benzene from methane", *Catal. Letters*, **32**, 43–53. [11]

Spitz, J. J., Laurence, R. L. and Chappelear, D. C. (1976) "An experimental study of polymerization reactor in periodic operation", *AIChE Symp. Ser.*, **160**(72), 86–101. [14]

Stamicarbon B. V. (1974) "Procede pour la preparation de cetone saturees et d'un catalyseur pour la mise en oeuvre du procede", Brevet Francaise No. 2,187,747. [12]

Stanitsas, G. (1986) M. A. Sc. Thesis, Dept. Chem. Eng., Univ. of Waterloo, Waterloo, Canada. [3]

Stegenga, S., Dekker, N., Bijsterbosch, J., Kapteijn, F., Moulijn, J., Belot, G. and Roche, R. (1991) "Catalytic automotive pollution control without noble metals", in Crucq, A. (Editor), "Catalysis and Pollution Control II", Elsevier Science Publishers, Amsterdam, 353–369. [5]

Sterman, L. E. and Ydstie, B. E. (1989) "The steady-state process with periodic perturbations", *Chem. Eng. Sci.*, **45**(3), 721–736. [1]

Strots, V. O., Matros, Yu. Sh. and Bunimovich, G. A. (1992) "Periodically forced SO_2 oxidation in CSTR", *Chem. Eng. Sci.*, **47**(9–11), 2701–2706. [8]

Stuchlỳ, V. and Klusáček, K. (1990) "Global activity improvement by feed composition cycling in $CO/H_2 - Ni/SiO_2$ methanation reaction system", in Matros, Yu. Sh. (Editor), "Unsteady State Processes in Catalysis" VSP (Utrecht, Netherlands), 422–428. [6]

Su, E. C. and Rothschild, W. G. (1986) "Dynamic behavior of three-way catalysts", *J. Catal.*, **99**, 506–510. [5]

Su, S., Prairie, M. R. and Renken, A. (1992) "Reaction mechanism of methanol dehydrogenation on a sodium carbonate catalyst", *Appl. Catal. A*, **91**, 131–139. [9]

Su, S., Zasa, Ph. and Renken, A. (1994) "Catalytic dehydrogenation of methanol to water-free formaldehyde", *Chem. Eng. Technol.*, **17**(1), 34–40. [9]

Suzuki, T., Inoue, K. and Watanabe, Y. (1988) "Temperature-programmed desorption and CO_2-pulsed gasification of sodium or iron loaded Yallourn coal char", *Energy Fuels*, **2**, 673–679. [1,10]

Suzuki, T., Chouchi, H., Naito, K. and Watanabe, Y. (1989) "Cyclic feed CO_2 gasification of iron-loaded coal char. Approach to activate iron catalyst in coal gasification", *Energy Fuels*, **3**, 535–536. [10]

Suzuki, T., Sakashita, Y., Ohme, H. and Watanabe, Y. (1990) "Reactivities of various coal chars to carbon dioxide and oxygen measured with pulse reaction technique", *Energy Fuels*, **4**, 593–598. [10]

Svensson, P., Jaeger, N. I. and Plath, P. J. (1988) "Phase and frequency relations with forced oscillations of the heterogeneous catalytic oxidation of CO on supported Pd", *J. Chem. Phys.*, **92**, 1882–1888. [3]

Swift, W. M. and Wheelock, T. D. (1975) "Decomposition of calcium sulfate in a two-zone reactor", *Ind. Eng. Chem. Process Des.Dev.*, **14**(3), 323–327. [10]

Sze, M. C. and Gelbein, A. P. (1975) "Synthesis of aromatic nitriles", Preprint, 1st Chemical Congress of the North American Continent, Mexico City, Mexico.
[1, 12]

BIBLIOGRAPHY

Sze, M. C. and Gelbein, A. P. (1976) "Make aromatic nitriles this way", *Hydrocarbon Processing*, 103–106. [1, 12]

Tan, H., Dai, Y.-C. and Yuan, W.-K. (1994) "A novel kinetic model discrimination method: eigenfunction distribution via forced concentration oscillation", *Chem. Eng. Sci.*, **49**(24b), 5563–5569. [15]

Taylor, K. C. (1984) "Automobile catalytic converters" in Anderson, J. R. and M. Boudart (editors) "Catalysis: science and technology", Springer Verlag, Berlin, 119–170. [5]

Taylor, K. C. and Sinkevitch, R. M. (1983) "Behaviour of automobile exhaust catalysts with cycled feed streams", *Ind. Eng. Chem. Prod. Res. Dev.*, **22**, 45–50. [1, 5]

Thiele, R. (1984) "Dynamic behaviour and stability of continuous reactors for bulk polymerization", *Inst. of Chem. Engrs. Symp. Ser.*, No. 87, "Chemical Reaction Engineering", Pergamon Press. [14]

Thullie, J. and Renken, A. (1990) "Transient behaviour of catalytic reactions with stop-effect", *Chem. Eng. Comm.*, **96**, 193–204. [9]

Thullie, J. and Renken, A. (1991) "Forced concentration oscillations for catalytic reactions with stop-effect", *Chem. Eng. Sci.*, **46**(4), 1082–1088. [9, 15]

Thullie, J. and Renken, A. (1993) "Model discrimination for reactions with a stop-effect", *Chem. Eng. Sci.*, **48**(23), 3921–3925. [9, 15]

Thullie, J., Chiao, L. and Rinker, R. G. (1986) "Analysis of concentration forcing in heterogeneous catalysis", *Chem. Eng. Comm.*, **48**, 191–205. [1, 7]

Thullie, J., Chiao, L. and Rinker, R. G. (1987) "Generalized treatment of concentration forcing in fixed-bed plug-flow reactors", *Chem. Eng. Sci.*, **42**(5), 1095–1101. [7]

Truffer, M. A. and Renken, A. (1986) "Transient behaviour of heterogeneous catalyst reactions with educt inhibition", *AIChEJ*, **32**(10), 1612–1621. [9, 11, 15]

Unger, B. D. and Rinker, R. G. (1976) "Ammonia synthesis reaction in the chromatographic regime", *Ind. Eng. Chem. Fundam.*, **15**(3), 225–227. [7]

Unni, M. P., Hudgins, R. R. and Silveston, P. L. (1973) "Influence of cycling on the rate of oxidation of SO_2 over a vanadium catalyst", *Can. J. Chem. Eng.*, **31**, 623–629. [1, 7, 8, 15]

Vamling, L. (1987) "On the use of periodic operation for improving chemical processes", Technical Paper and Ph.D. Thesis, Chalmers University of Technology, Gothenburg, Sweden. [1, 11]

Van Doesburg, H. and De Jong, W. A. (1978) " Dynamic behaviour of an adiabatic fixed-bed Methanator", in Weekman, V. M. Jr. and D. Luss (Editors), "Chemical Reaction Engineering – Houston", *ACS Symp. Series*, **65**, 489–503. [6]

Vannice, M. A. (1976) "The catalytic synthesis of hydrocarbons from CO and H_2", *Catal. Rev.-Sci. Eng.*, **14**, 153–191. [13]

Vaporciyan, G. G. and Kadlec, R. H. (1989) "Periodic separating reactors: experiments and theory", *AIChEJ*, **35**, 831–844. [1]

Vaporciyan, G., Annapragada, A. and Gulari, E. (1988) "Rate enhancement and quasi-periodic dynamics during forced concentration cycling of CO and O_2 over supported $Pt-SnO_2$", *Chem. Eng. Sci.*, **43**(11), 2957–2966. [3]

Vega, J. R., Frontini, G. L. and Meira, G. R. (1991) "Optimal periodic control of a continuous "living" anionic polymerization. III. Presence of impurities in the monomer feed", *J. Apl. Poly. Sci.*, **42**, 3181–3193. [14]

Verma, A. and Kaliaguine, S. (1973) "Estimation of rate coefficients from pulsed microcatalytic reactors, *J. Catal.*, **30**, 430–437. [12]

Visser, J. B. M., Stankiewicz, A., van Dierendonck, L. L., Manna, L., Sicardi, S. and Baldi, G. (1994) "Dynamic operation of a three-phase upflow reactor for the hydrogenation of phenylacetylene", *Catal. Today*, **20**, 485–500. [6]

Vrieland, G. E. and Murchison, C. B. (1995) "Anaerobic oxidation of butane to butadiene over magnesium molybdate catalysts. I. Magnesia supported catalysts", *Appl. Catal.*, **134**, 101–121. [12]

Vrieland, G. E., Khazai, B. and Murchison, C. B. (1995) "Anaerobic oxidation of butane to butadiene over magnesium molybdate catalysts. II. Magnesia alumina supported catalysts", *Appl. Catal.*, **134**, 123–145. [12]

Wainwright, M. S. and Hoffman, T. W. (1974) "The oxidation of o-xylene in a transported bed reactor", in Hulburt, H. M. (Editor) "Chemical Reaction Engineering II, *ACS Advances in Chemistry Series*, No. 113, Am. Chem. Soc., Washington. [12]

Walker, B. and Renken, A. (1983) "Direkte mikrogravimetric messung der sorptionskinetik under reaktions bedingungen", *Chem. Ing. Tech.*, **55**, 806–807 (synopsis). [9]

Walker, B. and Renken, A. (1984) "Direct microgravimetric studies of sorption kinetics under reaction conditions", *Ger. Chem. Eng.*, **7**, 227–232. [9]

Wandrey, C. and Renken, A. (1973) "Zur Beeinflussung der Produktverteilung durch periodische Konzentrationsschwankungen bei der Oxidation von Kohlenwasserstoffen", *Chem.-Ing. Tech.*, **45**(12), 854–859. [1, 11]

Wandrey, C. and Renken, A. (1974) "On improvement of autothermal operation by enforced concentration variations", *Proc. GVC/AIChE Joint Meeting*, Vol.1, Paper A 3-3, Verlag Chemie GmbH, Weinheim, Germany. [11]

Wandrey, C. and Renken, A. (1977) "Ignition and extinction of autothermal reactors at periodic operation", *Chem. Eng. Sci.*, **32**, 448–451. [2]

Wang, L., Tao, L., Xie, M., Xu, G., Huang, J. and Xu, Y. (1993) "Dehydrogenation and aromatization of methane under non-oxidizing conditions", *Catal. Letters*, **21**, 35–41. [11]

Watanabe, N., Onogi, K. and Matsubara, M. (1981) "Periodic control of continuous stirred tank reactors - I The pi criterion and its application to isothermal cases", *Chem. Eng. Sci.*, **36**, 809–818. [1]

Weibel, B., Garin, F., Bernhardt, P., Maire, G. and Prigent, M. (1991) "Influence of water on the activity of catalytic coverters" in Crucq, A. (Editor), "Catalysis and Automotive Pollution Control II, Elsevier Science Publishers (Amsterdam). [5]

Wheelock, T. D. and Boylan, D. R. (1968) "Sulfuric acid from calcium sulfate", *Chem. Eng. Prog.*, **64**(11), 87–92. [10]

Wheelock, T. D. and Morris, C. E. (1986) "Recovery of sulfur dioxide and lime from waste gypsum", *TIZ-Fachberichte*, **110**, 37–46. [10]

Wheelock, T. D. and Riel, T. (1991) "Cyclic operation of a fluidized bed reactor for decomposing calcium sulfate", *Chem. Eng. Comm.*, **109**, 155–168. [10]

Wheelock, T. D., Fan, C. W. and Floy, K. R. (1988) "Utilization of phosphogypsum for the production of sulfuric acid", *Proc. 2nd Int. Symp. phosphogypsum* **2**, 3–24, Florida Inst. Phosphate Res., Bartow, Florida, U.S.A. [10]

Wilson, H. D. and Rinker, R. G. (1982) "Concentration forcing in ammonia synthesis – I. Controlled cyclic operation", *Chem. Eng. Sci.*, **37**. 43–355. [7]

Winslow, P. and Bell, A. T. (1984) "Application of transient response techniques for quantitative determination of adsorbed carbon monoxide and carbon present on the surface of a ruthenium catalyst during the Fischer-Tropsch synthesis", *J. Catal.*, **86**, 158–172. [11, 13]

Winslow, P. and Bell, A. T. (1985) "Studies of the surface coverage of unsupported ruthenium by carbon- and hydrogen-containing adspecies during CO hydrogenation", *J. Catal.*, **91**, 142–154. [11, 13]

Yadav, R. and Rinker, R. G. (1989) "The efficacy of concentration forcing", *Chem. Eng. Sci.*, **44**(10), 2191–2195. [3]

Yeramian, A., Silveston, P. L. and Hudgins, R. R. (1970) "Influence of inert gases on the catalytic oxidation of SO_2", *Can. J. Chemistry*, **48**(8), 1175–1182. [8]

Yokota, Y., Muraki, H. and Fujitani, Y. (1985) "Rh-free three-way catalysts for automotive exhaust control", *SAE Tech. Paper Series*, No. 850129, Society of Automotive Engineers. [5]

Yu, F. C. L. (1969) "Periodic operation of a non-isothermal polymerization reactor", M. Sc. Thesis, University of Massachusets, Amherst, Massachusetts. [14]

Zasa, Ph., Randall, H., Doepper, R. and A. Renken, (1994) "Dynamic kinetics of catalytic dehydrogenation of methanol to formaldehyde" in Klusácek, K. (Editor)) "Dynamics of Catalytic Systems", *Catal. Today*, **20**(3), 233–239. [9]

Zhang, Z.-G., Scott, D. S. and Silveston, P. L. (1994a) "Steady state gasification of an Alberta sub bitumimnous coal in a micro fluidized bed", *Energy Fuels*, **8**, 637–642. [10]

Zhang, Z.-G., Scott, D. S. and Silveston, P. L. (1994b) "Successive pulsing of an iron loaded sub-bituminous coal char", *Energy Fuels*, **8**, 943–946. [10]

Zhang, Z.-G., Scott, D. S. and Silveston, P. L. (1995) "Iron-catalyzed gasification of coal char under composition modulation", *Energy Fuels*, **9**, 479–483. [10]

Zhou, X. and Gulari, E. (1986a) "CO oxidation on Pt/Al_2O_3 and Pd/Al_2O_3 transient response and concentration cycling studies", *Chem. Eng. Sci.* **41**, 883–890. [3, 15]

Zhou, X. and Gulari, E. (1986b) "CO adsorption and oxidation on Pd/Al_2O_3 under transient conditions", *Langmuir*, **2**, 709–715. [3]

Zhou, X. and Gulari, E. (1987) "CO hydrogenation on Ru/Al_2O_3: selectivity under transient conditions", *J. Catal.*, **105**, 499–510. [13]

Zhou, X., Barshad, Y. and Gulari, E. (1986) "CO oxidation on Pd/Al_2O_3: transient response and rate enhancement through forced concentration cycling", *Chem. Eng. Sci.*, **41**(5), 1277–1284. [3, 13]

Notation

Special notation is indicated by a bracketed number (or numbers) on the far right. The number(s) refers to the chapter(s) in which the notation is used. If there are no bracketed numbers the notation is used throughout the monograph.

A	amplitude (various units)	
	reactant species	
	functional group	[14]
	heat transfer surface area in CSTR (m^2)	[14]
A/F	air/fuel	[5]
$AIBN$	azobisisobutyronitrile	[14]
ASF	Anderson-Schulz-Flory	[13]
$A\text{-}s$	adsorbed species A	[9]
a	volume specific interfacial area (m^{-1})	[12]
a_s	specific surface area (m^2/kg)	[4]
B	reactant species	
Bz	benzene	[12,15]
b_i	coefficient in Elovich isotherm for species i ( )	[7,15]
C	product species	
	concentration (mol/m^3)	
	carbon	[10]
C	matrix formed from the deviation matrix V	[15]
C_i	concentration of species i (mol/m^3)	
	concentration of species i in the gas phase (mol/m^3)	[15]
C_i^s	intraparticle concentration of species i (mol/m^3)	[3,4]
C_i^g	bulk gas concentration of species i (mol/m^3)	[3,4]
$(C_i)^{sat}$	saturated concentration of i (mol/m^3)	[9]
$\bar{C}_i$	time average concentration of species i (mol/m^3)	[3,4]
C_i'	normalized concentration	[4]
C_k	concentration of the kth species in the gas phase (mol/m^3)	[15]
C_N	concentration of N atoms in catalyst bulk (g atoms/g)	[7,15]
C_o	concentration (mol/cm^3)	[8]
C_P	heat capacity or specific heat (kJ/mol/°C)	[14]
C_T	total gas phase concentration (moles/m^3)	[7]
C_V	total gas in melt (mol/cm^3)	[8]
C_{2+}	concentration of hydrocarbons with carbon numbers 2 or greater	[11]
$CSTR$	continuous stirred tank reactor	
c_p	specific heat (kcal/kg °C)	

NOTATION

D	product species	
	dissymmetry factor defined by Eqn. (8.16)	[8]
	polydispersity	[14]
	dimensionless group	[15]
D	matrix formed from the deviation matrix V	[15]
DEA	diethanolamine	[11]
D_{eff}	effective diffusivity (m^2/s)	[3]
d_p	catalyst particle diameter (mm)	[7]
E	product species	
	activation energy (kJ/mol)	
	dimensionless group	[15]
$(E_{\text{ad}})_i$	activation energy for adsorption for the ith species	[15]
$(E_{\text{de}})_i$	activation energy for desorption for the ith species	[15]
Et_2O	ethylene oxide	[11]
F	product species	
	dimensionless group	[15]
F_t	cross section for mass transfer (cm^2)	[7]
FID	flame ionization detector	[1,11]
FTS	Fischer-Tropsch synthesis	[13]
f	fraction of surface flux transmitted from active to inactive surface sites	[7]
	cycle frequency (s^{-1})	[14]
	number of functional groups on monomer	[14]
	fraction of spillover species not consumed by reaction	[15]
G	dimensionless group	[15]
$\Delta G°$	Gibbs free energy for reaction (kJ/mol)	[11]
H_i	Henry's law constant for species i (various units)	[8]
H_k	Heat of reaction for the kth reaction	[8]
H_{in}	heat effect on active to inactive transformation (kcal/mol)	[8]
H^{dis}	heat of solution (kcal/mol)	[8]
HC	hydrocarbon	
$(\Delta H)_i$	heat of the reaction for the ith species (single reaction) (J/mol)	
$(\Delta H_k)_i$	heat of reaction for the ith species in the kth reaction	
$\Delta H_{r \times n}$	heat of reaction (kJ/mol)	[14]
h	heat transfer coefficient (kcal/m^3 °C s)	
	number of terms in an expansion	[15]
h_p	fluid-particle heat transfer coefficient (kcal/m^3 °C s)	[8]
h_w	bed-to-wall heat transfer coefficient (kcal/cm^3 °C s)	[8]
I	initiator species or initiator concentration (mol/L)	[14]
IK	deactivated initiator	[14]
i	species index	
i.d.	internal diameter	
J	surface flux (moles/min cm)	[7]
K	equilibrium constant (various units)	
	adsorption equilibrium constant (various units)	[4]
	impurity, impurity concentration	[14]

NOTATION

K_i	adsorption equilibrium constant for species i (various units)	[11]
$(K_A)_i$	adsorption equilibrium constant for species i (various units)	[4]
$(K_{ad})_i$	adsorption equilibrium constant for species i (various units)	[11]
K_H	solubility constant (mol/cm^3 bar)	[8]
$(K_L)_k$	equilibrium constant in the melt for the kth reaction (various units)	[8]
k	reaction rate constant, mass transfer coefficient (various units)	
	reaction index	
	species index	[7]
	shift variable	[14]
k_a	adsorption rate constant (units depend on stoichiometry, but usually mol/m^3 gas·s)	[3,4,7]
k_{ad}	adsorption rate constant (various units)	
$(k_{ad})_i$	adsorption rate constant for species i (various units)	
(k_D)	apparent mass transfer coefficient for N diffusion into/out of catalyst bulk (min^{-1})	[7,15]
k_d	desorption rate constant (units depend on stoichiometry, but usually s^{-1})	
k_{de}	desorption rate constant	[15]
k_{des}	desorption rate constant (various units)	[7]
k_f	rate constant for forward reaction (various units)	[7]
k_i	adsorption or rate of reaction constant for species i (various units)	[11]
	initiation rate constant	[14]
k_k	rate constant for the kth forward reaction (various units)	
k_{-k}	rate constant for the kth backward reaction (various units)	
	rate constants for reaction and/or adsorption, desorption	[9]
k_n	rate constant for the nth reaction (various units)	[7]
k_{-n}	rate constant for reverse of the nth reaction (various units)	[7]
k_r	rate constant for the reverse reaction	[7]
k_p	propagation rate constant	[14,15]
	fluid-particle mass transfer coefficient (m/s)	[8]
k_s	gas-solid mass transfer coefficient (m/s)	[3,4]
	apparent mass transfer coefficient for surface diffusion between active and inactive sites (mol/min g)	[7]
k_t	termination rate constant	[14,15]
L	length of catalyst bed (cm, m)	
L_A	adsorption capacity (moles/m^2)	[4]
LO	localized, adsorbed oxygen atom	[15]
M	monomer species or monomer concentration (mol/L)	[14,15]
M_c	catalyst mass (g)	[7]
M_n	number of average molecular weight (g/mol)	[14]

NOTATION

M_W	weight average molecular weight (g/mol)	[14]
MA	maleic anhydride	[12,15]
MEA	monoethanolamine	[11]
MWD	molecular weight distribution	[14]
m	feed composition or time average feed composition expressed as mole fraction H_2	[7]
	mass of catalyst (kg)	
	number of monomer units in chain, chain length index	[14]
	time intervals in a time series	[15]
m_{cat}	mass of catalyst (kg)	
mss	maximum steady state	[4,13]
NaEtA	sodium ethyl adipate	[11]
N_{CO}	CO adsorption exclusion factor	[4]
N_{cyc}	number of cycles needed to achieve an invariant cycling state	[4]
N_i	site density, concentration of active sites for i^{th} species or in i^{th} catalyst state (mol/g)	[7]
	concentration of ith species on spillover or support surface	[15]
N_N^*	site density, concentration of inactive sites for N chemisorption (moles/g)	[7]
n	integer, cycle index, time step index	
	reaction index	[7]
	number of data points	
	index for number of measurements, data points	[7]
	number of monomer units in chain chain length index	[14]
	measurement point	[15]
n_i	molar flow rate or number of moles of species i	
P	pressure (Pa, bar, atm)	
P_i	partial pressure of ith species (Pa, bar, atm)	[7]
P_n	'dead' polymer species or living chain polymer species of chain length n, chain length or degree of polymerization or dead polymer concentration (gmol/L)	[14]
P_{O_2}	oxygen partial pressure (kPa)	[5]
PFR	plug flow reactor	
PSD	particle size distribution	[14]
PVC	polyvinyl chloride	[14]
PVAc	polyvinyl acetate	[14]
p	pressure (kPa)	[4]
p_i	partial pressure of species i (kPa)	
Q	volumetric flow rate (m³/s)	
	dimensionless group	[15]
Q_{is}	surface capacity in the ith model for species s (gmol/gm)	[15]
Q_n	'dead' polymer of chain length n	[14]
Q_o	volumetric feed rate at reactor inlet (cm³/min)	
QSS	quasi steady state	
q	shrinkage factor	[3,4]
qss	quasi steady state	[13]

NOTATION

R	gas constant (various units)	
	pellet radius (m)	
	rate (usually mol/cm^2 s)	[15]
$(R_{ad})_i$	rate of adsorption for the ith species	[15]
R_c	free radical initiator, concentration of free radical initiator (mol/L)	[14]
R_{ik}	rate in the ith model for a fluid species k	[15]
R_{is}	rate in the ith model for a surface species s	[15]
R_n	'live' polymer species of chain length n (free radical mechanism), 'live' polymer concentration (mol/L)	[14]
$(R_{re})_i$	rate of reaction of the ith species	[15]
$\bar{R}$	z transform of function R	[14]
R^*	free radical concentration (mol/L)	[15]
RA_f	monomer with f functional groups of type A	[14]
r	rate of reaction, product formation, reactant disappearance (mol/s g cat)	
	radial position (m)	
	time average rate	
r_{ad}	adsorption rate (mol/min g cat)	[7]
r_i	rate of disappearance or formation of species i (mol/s g cat)	
r_k	rate of the kth reaction (mol/m^3 s or mol/kg cat s)	[11]
r_{mss}	maximum steady state reaction rate at T, P of reactor system (mol/min g cat)	[7]
r_p	rate of polymerization	[15]
r^s	rate on catalyst support	[3,4]
rss	relaxed steady state	[13]
S	specific surface area (various units)	
	surface vacancy, vacant site	[7,15]
	solvent species or solvent concentration (mol/L)	[14]
S_i	surface concentration (mol/m^2)	[3,4]
$S_{i/j}$	overall selectivity to product 'i' in terms of reactant 'j'	[11]
S_p	specific surface (m^2 catalyst/m^3 cat. volume)	[3,4]
S_t	site density (mol/m^2)	[3,4]
S_v	volumetric specific surface area (m^{-1})	[8]
SN	stoichiometric number (various definitions)	[2,3,4,5]
SS	steady state	
SV	space velocity, usually STP, $= Q/V_{cat}(h^{-1})$	
STP	standard temperature and pressure (25°C, 1 bar)	
s	cycle split, duty fraction	
	active site, surface reaction site in mechanistic equation	
	solid	[10]
$s_{i/j}$	instantaneous selectivity to product 'i' in terms of reactant 'j'	[11]
s_m	surface site for adsorption on catalyst (metal)	[15]
s_s	surface site for adsorption on catalyst support	[15]
s_1	acidic sites on an alumina surface	[15]

NOTATION

s_2	basic sites on an alumina surface	[15]
s^v	complex density (mol/cm^3)	[8]
ss	steady state	[14]
T	temperature (°C, °K)	
	transfer agent species or transfer agent concentration (mol/L)	[14]
ΔT	temperature difference (°K)	
TEA	triethanolamine	[11]
TCD	thermal conductivity detector	
THF	Tetrahydrofuran	
TWC	three-way catalyst	[2,3,5]
t	time (s, min)	
	age (s, min)	[14]
$\bar{t}$	space time (s)	
t'	dimensionless time	[3,4]
t_c	characteristic or relaxation time (s, min)	
t_f	time duration between light flashes (s)	[15]
t_i	duration of exposure to species (reactant) i (s, min)	[7]
$t_{1/2}$	initiator half life (s)	[14]
U	mean, overall heat transfer coefficient in CSTR (kJ/m^2 °C/s)	
u	superficial velocity (m/s)	
V	reactor volume (m^3)	
	void volume of reactor (cm^3)	[7]
$V(z,t)$	deviation from a time-average concentration at position z and time t	[15]
$V, V(z,t)$	deviation matrix	[15]
V_R	reactor volume (cm^3)	[7]
v	oxygen vacancy on an oxide surface	[15]
v_T	volumetric flow rate (L/s)	[14]
W	weight of catalyst (g)	[7]
	distribution parameter	[14]
W_{cat}	weight of catalyst (g)	[4]
WGS	water-gas-shift reaction	[5]
X	conversion	[5]
X_j	conversion of reactant 'j'	
X_m	monomer conversion	[14]
x	stoichiometric number in the dissociative adsorption of CH_4	[11]
x_i	mole fraction of i in the melt phase	[8]
$Y_{i/j}$	yield of product 'i' in terms of key reactant 'j'	[11]
y	mole fraction	
y_i	mole fraction of ith species in the gas phase	
y_j	mole or mass fraction of complex j in melt phase	[8]
z	axial position (cm, m)	
	dimensionless axial position in reactor	[7]
	transform variable	[14]

Greek

α	Henry's law constant for N_2 dissolution in Fe catalyst, exponent in Temkin Eqn. (Eqn. (7.1))	[7]
	chain growth probability in the Flory model for polymerization	[13]
β	constant in the Elovich equation for adsorption	[15]
β'	constant in the Elovich equation for desorption	[15]
Γ	material balance expression (Table 7.4)	[7]
Δ'	mol shrinkage term defined in Table 7.4	[7]
δ	depth of catalyst layer (μm)	[7]
ε	void fraction	
ε_B	reactor or bed void fraction	
ε_m	fraction occupied by the melt phase	[8]
ε_p	particle void fraction	[3,4]
ζ_i	parameter $= \bar{t} N_i$	[15]
η	effectiveness factor	
η^*	apparent effectiveness factor $= r/r_{\text{fine particle}}$	[7]
θ	fractional occupancy of surface sites	
	dimensionless time	[9]
	space time (s)	[14]
	dimensionless time	[14]
θ_i	fraction of surface sites occupied by species i	
θ_i^S	fraction of support sites occupied by species i	[3,4]
θ_i^*	fraction of sites occupied by species i on the spillover or support surface	[15]
θ_m	fraction of unoccupied sites on the metal surface	[15]
θ_N^*	fractional occupancy of inactive N chemisorption sites	[7]
θ_S	fraction of unoccupied sites on the catalyst surface	[15]
θ_s	fraction of unoccupied sites on the support material	[15]
	fraction of sites occupied by adsorbed species s	[15]
θ_1, θ_2	fraction of surface sites of type 1, type 2	[9]
λ_i	ith moment of the chain number (or molecular weight) distribution of the living polymer	[14]
	coefficient in an eigenfunction expansion	[15]
μ_i	ith moment of the chain number (or molecular weight) distribution of the dead polymer	[14]
μ_{jk}	stoichiometric coefficient for component j in the melt for the kth reaction	[8]
μ_p	number chain length of polymer	[15]
v	stoichiometric coefficient	
v_{ad}	stoichiometric coefficient in adsorption (sites/mol adsorbate)	[3,4]
v_{ik}	stoichiometric coefficient for species i in the kth reaction	
ξ	normalized radius	[3,4]
	constant $= \tau_0/N_i$	[7]
	dimensionless time, $t/\bar{t}$	[15]

NOTATION

ξ_i^T	term of an eigenfunction of the matrix $\mathbf{D}$	[15]
ρ	density (kg cat/m^3)	
ρ_B	bulk density of the bed (kg cat/m^3)	[15]
ρ_{cat}	catalyst density (kg cat/m^3)	[4,8]
ρ_g	gas density (kg/m^3)	
σ	relative temperature standard deviation (Eqn. (7.12))	[7]
σ_i	ith moment of the chain number (or molecular weight) distribution of the total polymer	[14]
ς	selectivity ratio	[11]
	gelation parameter $(= k_p/2\,(f-2)^2[P_1]^\circ \bar{t})$	[14]
ς_{ik}	selectivity ratio for desired product 'i' in terms of undesirable product 'k'	[11]
τ	cycle or forcing period, cycle time (s, min)	
τ_A	adsorption response time (s)	[4]
τ_D	diffusional response time (s)	[3,4]
τ_L	relaxation time (min)	[7]
τ_R	reaction response time (s)	[3,4]
τ_p	dimensionless time, cycle period (min)	[7]
τ_o	space time $= V/Q_0$ (min)	[7]
τ_{opt}	optimal cycle period	
τ_s	space time (s, min)	[8]
τ'	space time (1/SV) (s)	[3,4]
ϕ	phase lag or lead (°, rad)	
	constant $= \tau_0\, W/VC_T$	[7,15]
ϕ_i	term of an eigenfunction of the matrix $\mathbf{C}$	[15]
χ	spillover flux	[15]
Ψ	enhancement or improvement factor, usually r/r_{ss}	
Ψ_S	selectivity enhancement	[1,11]
Ψ_Y	yield enhancement	[1,11]
Ψ^*	global enhancement or improvement factor, usually r/r_{mss}	
Ψ_S^*	global selectivity enhancement	[11]
Ψ_Y^*	global yield enhancement	[11]
ω	cycling frequency (Hz)	
ω_o	natural frequency of a self oscillating system (Hz)	[3,4]
ω_{opt}	optimal cycle frequency	

Subscript

A	reactant, usually key reactant	
a	adsorption	
ad	adsorption	[7,15]
Bz	benzene	[15]
C	product, usually desirable product	[11]
cat	catalyst	[3]
D	product, usually undesired product	[11]
	desorption	[7]
fm	chain transfer to monomer	[14]

fs	chain transfer to solvent	[14]
fT	chain transfer to transfer agent	[14]
g	gas	
H_2, H	hydrogen	[7]
H	solubility	[8]
I	initiator	[14]
i	species (reactant or product)	
	state of catalyst	[7]
	initiator, initiation	[14]
	model index	[15]
	index for a term in an expansion	[15]
in	inactive	[8]
	feed or start up condition	[11]
	intermediate index	[15]
i/j	reference relation for selectivity or yield	[11]
j	melt component	[8]
	species, usually a reactant species	[11]
	jacket	[14]
k	reaction index or number	
	species, usually a product species	[11]
	gas species index	[15]
	moment index	[14]
-k	index of reverse reaction or kinetic step	[9]
M	monomer, molar	[14]
MA	maleic anhydride	[15]
m	monomer, chain number	[14]
	metal or catalyst	[15]
mss	maximum steady state	
N	nitrogen	
N_2	nitrogen	[7]
NH_3	ammonia	[7]
n	number index	
	chain number, number variable	[14]
o	inlet, feed, initial, natural, entrance	
opt	optimal	
out	outlet or final condition	[11]
p	particle, constant pressure	
	propagation	[14]
RSS	relaxed steady state	[15]
rss	relaxed steady state	
rxn	reaction	
qss	quasi steady state	
SM	solvent monomer chain transfer	[14]
SS	steady state	[15]
s	surface	
	support, surface species	[15]
	solvent	[14]
ss	steady state	

sup	support	[3]
T	total	[7]
t	termination by combination	[14]
tw	termination by disproportionation	[14]
W	water	[11]
w	weight variable	[14]
0,1,2	moments of chain number/molecular weight distribution	[14]
1,2,3	reaction index	
	species index	[15]
1,2,3..	reactants, intermediate and product species in the hydrogenation of	[15]
	CO_2 over ruthenium steps in initiation	[14]
	chain species	[14]
1	CO	[3]
2	O_2	[3]
	N_2O	[4]
3	CO_2	[3]
α	carbidic carbon	[11]
β	refractive carbon	[11]
γ	graphitic carbon	[11]
θ	surface vacancy	[7]

Superscript

g	gas	[8]
o	feed, initial	[7]
	intercept or base value	[15]
p	particle	[8]
m	melt	[8]
in	inactive	[8]
s	support	[3,4]
sat	saturation	[9]
T	vector or matrix transpose	
$-$	average or mean quantity	
$*$	inactive	[7]
$\cdot$	free radical	[14]
$+$	just after time or position zero, time t	
$-$	before time or position zero, time t	

CHAPTER 1
Introduction[1]

A. BACKGROUND

This monograph discusses a single weapon within the arsenal for engineering chemical reactors: composition modulation. Composition modulation is an operation in which the feed to a reactor is switched periodically between two or more compositions to prevent the reactor from ever attaining a steady state. It is just one of several types of periodic operation that can be adopted to improve reactor performance.

Composition modulation is a contacting technique that is temporal rather than spatial. Choice of contacting pattern (back mixing, spatial distribution of feed) and, where applicable, the level of mixing are well known ways of influencing reactor performance. We will demonstrate in this mongraph that effects on reactor performance for composition modulation can be of the same order of magnitude as those observed with spatial contacting. The peculiarity of this contacting technique is that it forces a reaction to proceed under transient rather than steady state conditions. Under this type of operation, production becomes a time-average quantity.

Periodic operation is hardly new. Any heterogeneous, catalytic reactor in which the catalyst is regenerated *in situ* operates periodically. Even those reactors whose catalyst charge is removed for regeneration or replaced periodically share many similarities with periodic operation. Normally, the periods encountered for systems involving catalyst regeneration are measured in days, in months or even in years. In such cases, the reactor operates at steady state on the scale set by the residence time of reactants in the system. This is an important difference between the systems we consider in this monograph and periodic catalyst regeneration. When we discuss composition modulation, the catalytic reaction is transient at all times.

Not all catalytic systems employing regeneration operate with long periods. One of the largest regenerative systems in terms of volume of production, catalytic cracking of gas oils, employs periods of the order of minutes. Nevertheless, catalytic cracking is still treated as if a pseudo steady state exists. Composition modulation differs significantly from periodic catalyst regeneration in purpose. Regeneration is utilized to

---

[1] A version of this chapter appeared in *Catalysis Today*, **25**(2), 1995

restore performance whereas modulation means to achieve performance that is either difficult or impossible to attain under steady state operation.

Why consider composition modulation? This is the argument of this monograph. Simply stated it is that large improvements in reactor performance are possible. For ammonia synthesis using a ruthenium catalyst, a 1000-fold increase in the synthesis rate has been demonstrated in an operation which switches the reactor feed every several seconds between H_2 and N_2 (Rambeau and Amariglio, 1981b). Or, the yield of acrolein in the partial oxidation of propene over a Sb—SnO catalyst can be doubled by switching between air and propene mixtures of different composition (Silveston and Forrissier, 1985). Moreover, composition modulation allows investigation of the true kinetics of a catalytic process and permits separation of model parameters that are lumped when determined from steady state measurements (Renken, 1990a, b).

B. WHAT IS COMPOSITION MODULATION?

A terse definition was given at the beginning of this chapter. Notwithstanding, further discussion will make the operation more easily understood. Composition modulation of a reactor is shown schematically in Figure 1.1. Here, two inputs, the volumetric flow rates of reactants "A" and "B", are switched periodically between two settings so as to generate a chain of step changes representing a square wave variation of reactant partial pressure or concentration in the reactor feed. In most of the systems studied in the

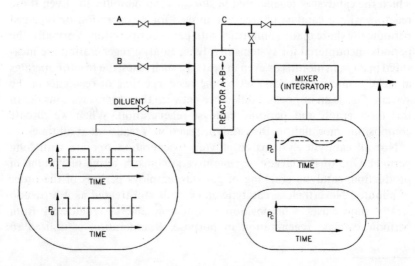

FIGURE 1.1 Schematic of a periodically operating reactor and mixer-integrator with two feed streams

INTRODUCTION

laboratory, the flow rate variations are matched so that the space velocity remains constant, but this is not a necessary condition for composition forcing. When the variation of reactants are unequal, perhaps because of reaction stoichiometry, the flow rate of a diluent can be adjusted so as to maintain a constant space velocity. Provision for diluent adjustment is shown in the figure.

Product "C" is formed in the reactor, but because of mixing in the reactor and transients on the catalyst surface variation of the product partial pressure or concentration is often sinusoidal as illustrated in Figure 1.1. If product variations are undesirable, the flow from the reactor can be directed to a mixer of suitable size to damp them out. This is the purpose of the mixer shown in the figure.

The terms composition modulation, forcing or cycling will be used interchangeably reflecting the wide range of terminology in the current literature. All of these terms refer to a periodic operation in which one or more reactant concentrations fed to a reactor vary with time, but in such a way that each "state" of an input is revisited after a time τ corresponding to the period.

Figure 1.2 compares composition modulation to steady state operation. Composition modulation appears on the right hand side of the figure, while steady state is shown on the left hand side. Input variations result in a time

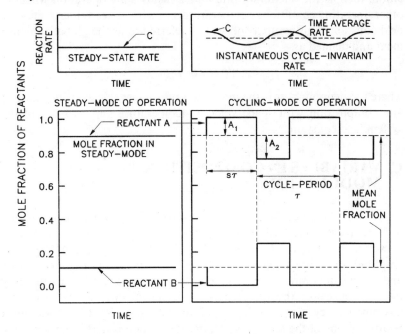

FIGURE 1.2 Comparison of steady state (left side) and periodic (right side) operation showing definition of the cycling variables: cycle period (frequency), τ, cycle split (duty fraction), s, amplitudes, A1, A2

varying output shown in the upper portion of the right hand side. These are referred to as instantaneous concentrations, yields or rates. When integrated over time, they provide the mean or time-average production rate, yield or product concentration. Many studies have found that these time-averages are greater than steady state values under identical time-average feed conditions. This is illustrated in Figure 1.2.

To compare cycling with steady state operation, a ratio of performances will be used and referred to as the enhancement under periodic operation:

$$\Psi \equiv r/r_{ss} \qquad (1\text{-}1)$$

where r is the time-average rate of a forced reaction and r_{ss} is the rate at steady state for a reactant composition corresponding to the time-average composition under cycling. Enhancements can also be defined for selectivity or for product concentration or for yield.

Enhancement with respect to the steady state rate at the time average feed composition is often not the significant comparison. It can be argued that if the feed composition can be changed (as it is in cycling), it should be possible to find a composition which will improve reactor performance in steady state operation. The meaningful comparison for composition modulation, then, is the maximum rate under steady state operation when all other operating conditions (temperature, pressure, space velocity, etc.) are held constant. This comparison will be referred to as the global enhancement and defined as

$$\Psi^* \equiv r/(r_{ss})_{max} \qquad (1\text{-}2)$$

where the denominator is the maximum rate achievable at steady state at any composition with space velocity, temperature and pressure of the system held constant.

C. VARIABLES IN COMPOSITION MODULATION

Figure 1.2 illustrates most of the variables that arise in composition modulation: period (τ) – the time between repetitions of a change in an input condition, split (s) – the duration of one part of a cycle relative to the period, amplitude (A) – the change in the value of an input condition from its mean, and the mean composition. The split, sometimes referred to as the duty fraction, measures the symmetry of a cycle. A value $s = 0.5$ indicates a symmetrical cycle with both parts of equal duration. Split must be defined relative to one of the reactants and the convention is that it measures the relative duration of the portion of the cycle when that reactant is at its highest concentration. Amplitude takes on just a single value for symmetrical forcing, but, if $s \neq 0.5$, two amplitudes must be given, one for each portion of the cycle. Mean composition is also a variable for composition forcing, just as it is when operating at steady state.

A further variable not illustrated in Figure 1.2 is the phase lag. The composition changes shown in the figure are 180° or π radians out of phase. Other phase lags could be used. In a pulsing operation, the phase lag is zero.

The part cycles shown in the figure are possible when just one reactant is modulated or, if two rectants are modulated, only when the phase lag is 0 or 180°. Other phase lags, multiple ampitudes or variation of a further feed component lead to 3 or more part cycles.

Cycle period or its inverse, frequency, is a variable in almost all cycling applications, but this cannot be said of the other variables shown in Figure 1.2. Even when a variable can be used, another variable might be a better choice. For example, split is not the appropriate variable in the pulse that we consider in the next section. The variable is pulse length or duration. Similarly, amplitude should be replaced by pulse magnitude; for example, a flow rate, or the reactant concentration in the pulse. Phase lag becomes a variable only when more than one input is changing with time. Time-average feed composition is a variable only in the periodic mode; it is not a variable in a pulse mode. Even cycle period becomes a restricted variable in the relaxed and quasi steady state modes; period needs to be determined to established the mode, but since performance is independent of period, once the mode has been found, it ceases to be variable.

When there are two components and the phase lag is not held at 0 or 180°, a cycle will have 3 or more parts. Stated differently, composition forcing can have two cycle periods and two cycle splits, one for each component when this occurs. Even when the phase lag is 180° out of phase, cycling must be described by two amplitudes if $s \neq 0.50$. When the number of components becomes greater than two, multi-part cycles become the general rule and the number of variables which must be considered increases rapidly. There have been neither simulation or experimental investigations of systems having more than three time changing inputs.

There are also other choices in composition modulation, such as the choice of reactant(s) to be modulated. A distinction can be made between strategy and operating variables. Strategy refers to the choice of input(s) or mode of variation or even cycle structure. Operating variables are cycle period, split, phase lag and amplitude.

D. MODES OF COMPOSITION MODULATION

It is useful to distinguish different types of cycles. The simplest differentiation is cycle structure. On one hand there is a more or less uniform division of a cycle as illustrated in Figure 1.2. Both parts of the cycle have the same duration. This is the cycle structure that is most frequently investigated. The other extreme is when one part of the cycle is very short. This will be referred to as a pulse mode. Note that the cycle structure may be described by the split variable.

Distinction between modes can also be made with respect to the characteristic response or relaxation time of the catalytic reaction or reactor that is being periodically forced. One extreme are cycles with long periods with respect to the relaxation time (t_c), that is, $\tau \gg t_c$. At this extreme, the reactor remains overwhelmingly at steady state. The transient portion of a cycle, roughly 2 t_c, is a negligible portion of the cycle period. Referring to Figure 1.2, r becomes the time-average of the steady state rates for the two feed concentrations shown in the figure. Even though the operation is periodic, the time-average reaction rate, r, is uniquely determined by the steady state behaviour of the reaction or reactor. The term 'quasi-steady state' (qss) is commonly used for this mode.

At the other extreme, very short cycle periods, two situations can arise. If mixing occurs, as it does in all real situations, and the characteristic time of mixing is shorter than the cycle period, time variations in an input, such as a reactant concentration, will be smoothed out. In this extreme, the reaction or reactor behaves as though it is at steady state at a condition represented by the time average of the forced input. This situation arises with recycle reactors or reactors in which large scale back mixing takes place.

If mixing is negligible, that is for systems which approach the plug flow limit, and $t_c \gg \tau$, the "relaxed steady state" mode arises. The catalyst and transport to the catalyst surface are no longer able to follow the composition variation. The state of the catalyst surface and product concentration then become time invariant, but the state is often different from that seen with a continuous steady feed. The relaxed steady state is widely used in analytical treatments of composition forcing (see, for example, Bailey (1973, 1977)), but is reported only rarely in experimental studies. Reaction rate at the relaxed steady state limit will be represented by $(r)_{rss}$. Both $(r)_{qss}$ and $(r)_{rss}$ are asymptotic conditions for periodic forcing.

Between the fast and slow cycling extremes lies the region where $\tau \approx t_c$. It is on this region or for this mode that past and current research focus. Even for this cycling mode, several different types of operations can be distinguished. Consider the common situation in catalytic reactors where there are two vastly different characteristic response times: one associated with response of the surface (adsorbate concentrations, surface state, temperature) to changes in reactant concentration or reactor temperature and the other associated with catalyst deactivation. Response times of the first sort are typically measured in seconds or at most in minutes, whereas those associated with deactivation are measured in weeks or months of on stream time. One of the operations in the $\tau \approx t_c$ region is a process with significant decay of catalyst activity. Once fouling or surface poisoning has brought

INTRODUCTION

catalyst activity down to a low fraction of its initial level, the catalyst is regenerated or replaced and the cycle begins again. Clearly, period, that is, the time between regeneration or replacement, is about equal to the characteristic response time for deactivation. Bailey (1973) has labeled such cases "process life cycle" operations. With respect to the first characteristic response time, the process life cycle is very much longer than this response time so that at any instant, if temperature or feed composition are undisturbed, the reaction system can be treated as though it is operating at steady state. The steady state assumption is standard for models of catalytic reactors. Process life cycle systems will not be discussed further in this monograph.

The two modes of periodic operation that arise in the $\tau \approx t_c$ region have been introduced already, but in a somewhat different manner. The first of these is shown in Figure 1.2. In this mode, the changes in reactant concentration (or volumetric flow, if Figure 1.1 is considered) are 180° out of phase. While reactant "A" increases in concentration, reactant "B" decreases. This is the "standard" cyclic mode. Phase lags other than 180° are possible as has been mentioned. Several investigators have used a mode in which the concentration or flow changes have a phase angle of zero. Usually the duration of reactant flow to the reactor is very short compared to the cycle period and this is referred to as a pulse operation or as periodic pulsing. If the parts of the cycle are about equal, it is called a stop flow mode or operation. A typical operation is illustrated in Figure 1.3 for the "A" + "B" → "C" reaction system.

The pulse mode has attracted considerable research attention because it offers mechanistic insight when used before the time when the cycles become invariant (see, for example, Hattori and Murakami (1968)). The

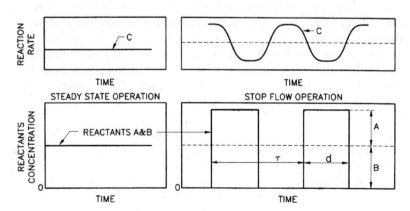

FIGURE 1.3 Comparison of steady state (left side) and pulse (right side) operation showing definition of variables: period (frequency), τ, pulse duration, d, and amplitudes, A, B

mode is also interesting because chromatographic effects can arise leading to reactant and product separation and significant yield and/or selectivity improvement when equilibrium limitations exist.

E. OBJECTIVES FOR COMPOSITION MODULATION

Increased conversion, catalyst activity or rate of reaction are often cited as reasons to explore modulation. However, these are reasonable goals for industrial processes only in unusual circumstances. Composition modulation is a more complicated operation than steady state and therefore more expensive. Usually, more than one reactor vessel would be used so that the capital cost for the same throughput could be greater unless large saving in recycle or separation costs are achieved. Consequently, if increased conversion is the goal, it usually can be achieved more cheaply by increasing reactor size or catalyst charge. Under these circumstances, changing from steady state to composition forcing can be justified only by an increase in catalyst activity on the order of 100%.

What then are the circumstances in which increased conversion is a reasonable goal? One of these is when a reaction is carried out at extreme temperature and/or pressure, or when an expensive catalyst is used. Saving in vessel size or catalyst charge from higher catalyst activity could offset the higher cost of two reactor vessels. Another circumstance for which composition modulation could be attractive is when it is desirable to increase capacity of an existing reactor. Finally, composition modulation may be an attractive alternative when conversion per pass is limited, often by equilibrium, so that reactants must be separated downstream and recycled. Kane et al. (1979) claim SO_2 conversion in excess of the equilibrium steady state conversion in their study of periodically flushing the final stage of an SO_2 converter with air.

Current studies of periodic composition forcing are aimed at improved selectivity or yield. This is always a reasonable objective because selectivity often cannot be altered significantly by increasing the catalyst charge or even by changing pressure or temperature.

Another situation in which composition forcing might be beneficial is when hysteresis is present in a reaction system. Unstable operating states often exist for catalytic systems exhibiting hysteresis. If it is advantageous for the reactor to operate at an unstable state, simulation and experiment have demonstrated that this is possible through composition modulation (Ausikaitis & Engel, 1974). Indeed, there is evidence that composition forcing may be a means of permitting reactors to operate safely in regions of high parametric sensitivity. Jain et al. (1982b) demonstrated that composition forcing smooths reactor temperature distribution and reduces peak temperatures while maintaining conversion achieved under steady state.

INTRODUCTION

Control of catalyst deactivation could be an objective too. There is evidence (Chanchlani *et al.*, 1994) that composition modulation can decrease the rate of catalyst deactivation. Figure 1.4 compares deactivation data for methanol synthesis over a $Cu/ZnO/Al_2O_3$ catalyst under periodic forcing of the H_2—CO_2 feed mixture and operating at steady state. Loss of activity is clearly slower for composition forcing. The $Cu/ZnO/Al_2O_3$ catalyst is similar in composition and performance to the widely used ICI low pressure methanol catalyst.

There are also scientific objectives which can be achieved by using modulation. The rate controlling step in a reaction or the mechanism by which products are formed or, occasionally, how the catalyst deactivates can be identified (Renken, 1990 a). A research application for composition modulation should not be surprising. Pulse experiments, in which a catalyst sample is exposed to a series of reactant pulses, is a standard research tool in catalysis.

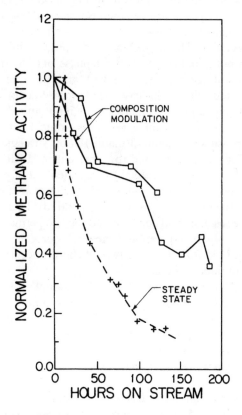

FIGURE 1.4 Loss of catalyst activity of a $Cu/ZnO/Al_2O_3$ catalyst for methanol synthesis with time on stream under steady state and composition modulation of the CO_2 feed mixture (Adapted from Chanchlani *et al.* (1994))

F. STRATEGY IN COMPOSITION MODULATION

There are many ways of modulating a reactor periodically. It is this richness in opportunity that usually makes it possible to find at least one way that will achieve one of the objectives just mentioned. On the other hand, this very richness presents a challenge: namely, which of the various modes should be used and how can a researcher efficiently find that mode. The terms strategy and variables were introduced earlier. By strategy we mean the method or means of periodic forcing, while variables are those variations that can be made in the operation without changing the mode.

A reactor input is a manipulated variable and almost all inputs can be forced periodically. Choice of input is part of the strategy. Reactant, promoter or diluent are candidates for manipulation in time. Usually, more than one input will be modulated. Often these will be the concentrations of two reactants. In this situation there are two strategy options. Figure 1.5, taken from a study of the application of composition modulation to methanol synthesis (Chanchlani et al., 1994), shows these in 1). They are (a) vary both reactants, in this case the H_2 and CO mole fractions, simultaneously but out of phase, (b) vary both simultaneously but in phase, or (c) hold the concentration of one reactant constant, such as CO, and vary the other reactant concentration periodically, in this case the H_2 mole fraction. There are two version of this manipulation depending on the input forced.

If the reaction system has three reactants, some 13 different periodic operations are possible, reflecting 3 manipulation strategies. In 2), "three reactant cycles," operations (a) and (b) represent modulation of all three reactants in phase or with one reactant modulated out of phase. (b) shows only one of three possible combinations. Two reactants are forced and one is held constant in the operations shown by (c) and (d). There are 6 variants depending on the reactant held constant and the two permissable phase lags of 0 and 180°. The third manipulation strategy is to force the concentration of one reactant and hold the remaining two reactant concentrations constant (see (e)). There are three variants of this strategy.

In several of the operations in Figure 1.5, for example, (b) in "two reactant cycles" and (b) or (e) in "three reactant cycles," both flow rate through the reactor and feed composition must vary together. If flow rate variation is to be avoided, a further component, an inert, must be introduced and must vary periodically with time. This expands the strategies available, although forcing the flow rate through the reactor to be constant limits the possible operations considerably.

Further strategies can arise when more than two components are modulated. Although only two part cycles are shown in Figure 1.5, multi-part cycles are possible. There are circumstances when a multi-part cycle is useful, for example, higher selectivity might be obtained by flushing the

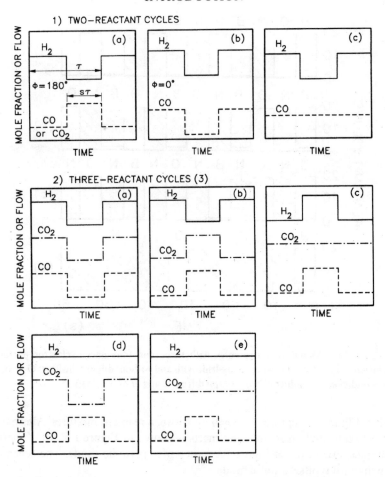

FIGURE 1.5 Different possible composition forcing operations with two and three reactants for methanol synthesis (Adapted from Chanchlani *et al.* (1994))

catalyst surface to desorb reactant or product before exposing the surface to a second reactant. Park *et al.* (1983) and Lang *et al.* (1989b, 1991) have employed multi-part cycles in their investigations of the partial oxidation of C_4 hydrocarbons. Figure 1.6, taken from Lang *et al.* (1991), illustrates 3 part cycles in operations (a) and (b), while the bottom two operations in the figure show 4 part cycles. Note that the location of the N_2 flush differs in (a) and (b). There is diluent flow after each exposure to a reactant in (c) and (d). The latter illustrates a pulse mode: there is a pulse of a reactant followed by a long duration of diluent flow before the next reactant pulse.

The manipulation strategies just discussed can be viewed as modes of operation, just like the relaxed and quasi steady state modes introduced

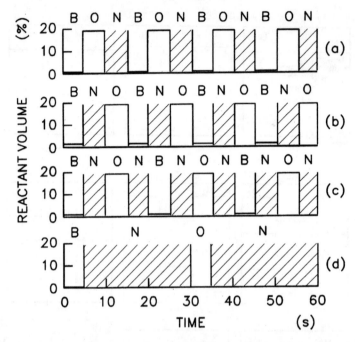

FIGURE 1.6 Modulation strategies with three and four part cycles for partial oxidation of butadiene to maleic anhydride with and without diluent flushing: O = air, B = butadiene, N = diluent (N_2) (Adapted from Lang et al. (1991))

earlier. Figure 1.5 represents switching between reactant mixtures. Alternating between single reactants and inerts is shown in Figure 1.6. All of these examples, except (d) in Figure 1.6 represent a periodic mode. The scheme shown in (d) typifies a pulse mode.

Selecting a forcing strategy requires choosing the variables to be manipulated and the mode of operation. There is some ambiguity in strictly separating strategy and operating variables. Are relaxed, periodic and quasi steady state really strategies? The difference in these strategies is only cycle period. Similarly, the difference in switching between mixtures and a single reactant is the amplitude, which again is a variable. We speak of strategy in this context because the behaviour of the reaction system may differ dramatically when the strategy is changed.

G. EQUIPMENT FOR COMPOSITION MODULATION

Periodically altering the reactant composition seen by catalyst held in a reactor can be undertaken by changing the reactant concentrations or by using two reactors each having a constant feed and transporting the catalyst from one reactor to the other. Composition switching using a single reactor is simple to build and operate. It is the system widely used in laboratory investigations. A two reactor design with a circulating catalyst is the design favored for industrial scale.

Laboratory Reactors

Almost all studies of composition modulation reported in the literature have used a single reactor vessel and a valving system to periodically change the composition of the feed flowing to the reactor. Modulation equipment of this type is easily fabricated and is inexpensive. Microreactors can be used which require just milligrams of catalyst and provide differential performance. Figure 1.7 shows a microreactor that has been frequently used in investigations of composition forcing in the author's laboratory (see, for example, Feimer et al., 1985a). This reactor is made from 1 cm (o.d.) steel tubing. The catalyst is supported on a steel screen held in place by the fitting at the bottom of the tube. A screen is often pressed fitted on top of the bed to prevent disturbance of the bed when the reactor is moved. Prior to pressing the top screen in place, the tube is rapped to settle the catalyst bed to its maximum, random packed density. The figure shows a separate coil for preheating reactants to within 3 to 5°C of the reaction temperature. Extending the reactor tube and filling the tube above the catalyst with ceramic or steel beads can also be used. However, this can result in reactor-preheater lengths greater than 15 cm which may exceed the depth of the heating bath or furnace. Generally, a coil preheater provides a more compact design as the coil can be wound around the reactor. Two thermocouples are shown in Figure 1.7. The upper is fixed above the bed and measures the temperature reached in preheating. The second thermocouple penetrates the bed from the bottom and passes through the supporting screen. It measures the bed temperature. Catalyst bed depths in these microreactors range from 1 to 6 cm. Depth depends on catalyst activity at the reaction temperature.

Use of microreactors allows observation of reaction dynamics by attaching a continuous analyzer, such as an IR or UV spectrophotometer, to the reactor outlet. Indeed, if the packed, tubular reactor in Figure 1.7 is replaced by an optical cell containing the catalyst pressed into an IR transparent wafer, it is possible to observe the dynamics of the catalyst surface. Figure 1.8 shows a cell designed and built by Prokopowicz et al. (1987) to investigate the modulation dynamics of CO oxidation on a copper oxide catalyst. Copper oxide is opaque in the IR spectra so the wafer was formed from KCl with some copper oxide admixed. The same cell was also used to study suface dynamics using Raman spectroscopy.

Integral reactors must be used if high conversion or equilibrium limited systems are to be investigated. Reactor design is similar to Figure 1.7 only

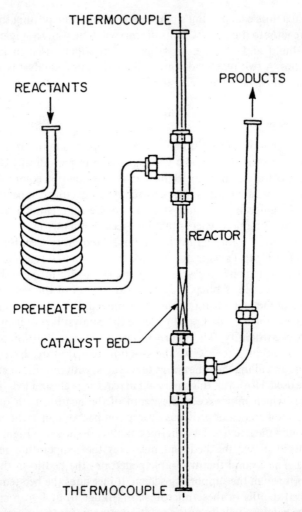

FIGURE 1.7 Schematic of a microreactor for use in composition forcing studies (Figure adapted from Chanchlani *et al.* (1992) with permission, © 1992, Elsevier Science Publishers)

reactor length may be greater. Isothermality is often important. A 1 cm tube diameter is usually adequate to make the bed isothermal provided $L/d > 10$ and the reactor is immersed in a constant temperature bath or furnace. Larger diameter tubing with heavy insulation and counter heating can be used to obtain near adiabatic behaviour. Figure 1.9 shows an integral reactor design used by Vamling (1987) in a study of the production of ethanolamines from ethylene oxide and ammonia over an ion exchange resin. Reactor i.d. was 24 mm and its length was 1000 mm with about 830 mm of this length filled with catalyst.

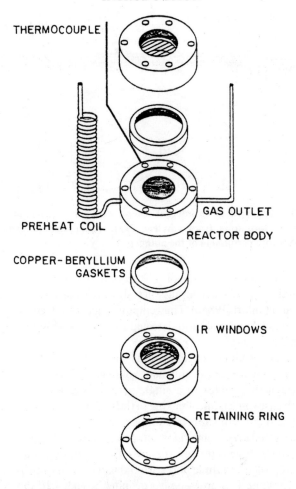

FIGURE 1.8 Schematic of high temperature, pressurized cell for transmission IR, Raman or UV spectroscopy used to study CO oxidation over a copper oxide catalyst under composition forcing (Figure adapted from Prokopowicz *et al.* (1987) with permission, © 1987, Journal of Applied Spectroscopy)

Integral reactors often have non-uniform temperature distributions making data more difficult to interpet. Several investigators (Abdul-Kareem *et al.*, 1980a; Jain *et al.* 1983b; and Nappi *et al.*, 1985) have overcome this problem by using an internal circulation reactor in place of a fixed bed. All internal recirculation reactors have a limit on the shortest cycle period which can be used because of mixing within the reactor bed. Mixing distorts the square wave composition input and limits thereby cycle periods to 30 to 60 seconds. Use of capillary tubing with short connections welded or brazed, and carefully chosen valves permits shorter periods. Luu (1991) describes

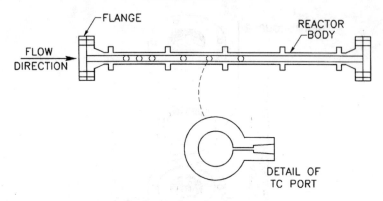

FIGURE 1.9 Schematic of an integral reactor used for composition modulation of ethanolamine formation from ethylene oxide and ammonia (Adapted from Vamling (1987) with permission of the author)

a design which allows the use of 1/2 second periods with just a small distortion of the square wave input. Such short periods have been also used by Barshad and Gulari (1985b). They employed a catalyst cast as a monolith of uniform, parallel channels and detected products by passing an IR beam down a channel.

Time average product concentrations and rates of formation can be obtained by placing a mixer of sufficient volume, ca. 3 to 4 times the volume passing through the reactor in a single period, after the reactor (see Figure 1.1). With gases, mechanical mixing of the integrator is not necessary. The design must only prevent short circuiting.

The simplest feed arrangement for composition forcing with two reactants involves separate reservoirs for the two feed compositions and a flow switching valve, such as a three port solenoid valve. In laboratory scale work, a reservoir would be a compressed gas cylinder. Figure 1.10 shows just such a two reservoir system with an on-off solenoid valve. This system periodically adds ethylene oxide to a continuous flow of ammonia. It was used by Vamling (1987) to study the modulation of ethanolamine formation. Hugo et al. (1986) used a similar arrangement but with a three port solenoid valve and the reservoirs were gas cylinders each containing a different feed composition. Figure 1.11 shows a three component system consisting of O_2, N_2 cylinders, a benzene-air reservoir and timer operated, three port solenoid valves that was used by Park and Gau (1986) to produce 3 and 4 part cycles.

Flexibilty of the equipment increases if feed compositions are prepared by mixing reactants. Needle valves in two separate lines leading from each gas cylinder set different flow rates for each reactant. Solenoid valves switch the flows between the lines. Thus, opening a specific combination of lines blends a feed composition and sends it to the reactor. Closing the first combination and opening another formulates the feed composition for the second part of

INTRODUCTION 17

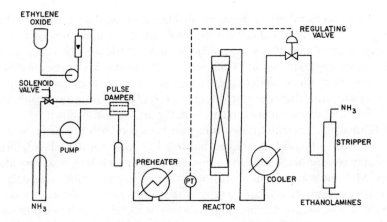

FIGURE 1.10 Two reservoir system with a flow switching valve for composition forcing (Adapted from Vamling (1987) with permission of the author)

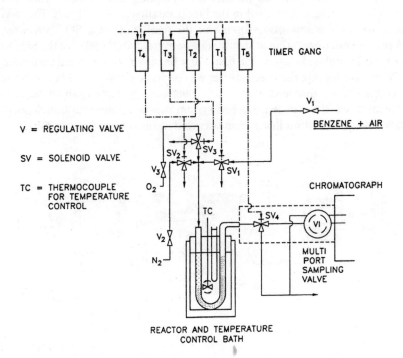

FIGURE 1.11 Flow switching system for composition forcing with three and four part cycles (Adapted from Cordova and Gau (1983))

the cycle. A timer controls the switching. Compositions of the feeds can be changed from one experiment to another by adjusting the needle valves. With this arrangement, composition forcing is accomplished by flow control.

Switching and blending can be automated by replacing the needle valves by mass flow controllers and driving the flow controllers and solenoid valves through a computer. This arrangement is shown in Figure 1.12. It is a schematic of the apparatus used by Chiao *et al.* (1987) in their high pressure study of ammonia synthesis under composition modulation. These investigators were able to switch between reactors, measure feed composition and calibrate their flow and detection instruments with the arrangement shown.

Multiple reactors with catalyst transport back and forth between reactors can be used in laboratory studies, but this has been not done heretofore, probably because of the difficulty of controlling catalyst transport at very low rates. Multiple reactors can be used with feed switching to simulate catalyst transport systems. In principle, a separate reactor is required for each part of a cycle. If composition forcing employs a two part cycle so that two different feed streams arise, at least two reactors will be required. If a cycle has three parts, three reactors are needed. A four part cycle requires four beds or vessels. Only one two bed periodically operating reactor employing feed stream switching is described in the literature (Briggs *et al.*, 1977). This was used to study composition forcing of the final stage of a SO_2 converter. A preconverter operating at steady state converted about 90% of the SO_2 to SO_3 and simulated thereby the first 3 or 4 beds in a conventional converter. The stream leaving the preconverter was periodically switched between the two beds. The other feed, also switched between the beds, is an air flushing stream. Flow through the two bed system was continuous, although small flow and composition fluctuation occurred at the moment of switching.

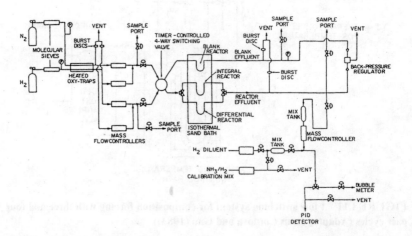

FIGURE 1.12 Computer driven apparatus for composition modulation by flow rate control used for ammonia synthesis over an iron catalyst (Figure adapted from Chiao *et al.* (1987) with permission, © 1987, Gordon and Breach Science Publishers)

Industrial Scale Equipment

The single reactor system for periodic operation seems poorly suited for large scale utilization because of the problem of rapidly opening and then tightly shutting large valves. On the downstream side of the reactor a hold up vessel would be needed in most situations to smooth out composition variation forced on the reactor. Two reactors simplify the design. Feed streams could then be periodically switched between the reactors so the problem of tight shut off and quick opening is avoided. Blending the streams leaving the reactors substantially reduces the time variation of the product gas so that a smaller hold up vessel becomes practical.

Circulating catalyst between the two reactors each using a different feed composition and, if desirable, a different flow rate provides composition modulation for the catalyst, but allows segregating the discharge streams. This could be advantageous for product recovery. Consider partial oxidation using this system. One reactor would be fed a hydrocarbon stream with little or no air, while the second reactor would be fed just air for catalyst regeneration. Only the hydrocarbon stream contains product and so it alone passes to the product recovery train. The product is at higher concentration than in the comparable discharge stream from a continuous, steady state operation and should be separable from the discharge at lower cost. Separation advantages were recognized many years ago, well before the composition forcing concept appeared in the literature. Figure 1.13 shows a system of two fluidized beds and catalyst transfer via a pneumatic lift proposed by Lewis et al. (1949). A strong resemblence to early versions of catalytic cracking can be seen. One of the beds is fluidized by a hydrocarbon which is selectively oxidized while the catalyst is reduced. The second bed is fluidized with air which re-oxidizes the catalyst. The catalyst, thus, functions as the O_2 carrier.

The advantages of this system are obvious. Molecular oxygen is absent during oxidation of the hydrocarbon reducing the extent of total oxidation. Higher hydrocarbon concentrations are possible in the reactor because there is no longer a detonation hazard. This reduces the cost of downstream separation. Air can be used as the oxidant in place of oxygen because the oxidant is separated from the hydrocarbon so that N_2 does not become a load on the downstream separation. Furthermore, catalyst oxidation and reduction steps can be carried out at different temperatures to minimize contact time because each step occurs in a different reactor. A less obvious advantage is that the heat release for the highly exothermic partial oxidation reaction is split between two vessels reducing thereby the cost of cooling.

The two fluidized reactors, circulating catalyst system was first studied experimentally in the late 1960's (Callahan et al., 1970). It was not adopted for commercial use because, according to Callahan et al. (1970), the advantages just mentioned are offset or at best partially offset by the disadvantage of

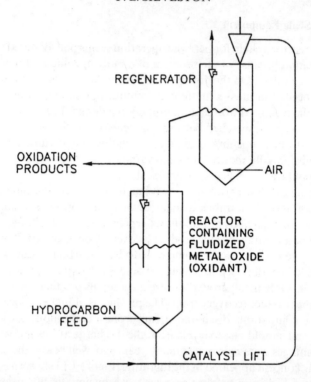

FIGURE 1.13 Conceptual system for composition forcing of partial oxidation reactions using fluidized beds with catalyst transfer between beds (Figure adapted from Lewis *et al.* (1949) with permission, © 1949, the American Chemical Society)

circulating large amounts of catalyst. Because the redox catalysts are only seldom reducible reversibly to the metal, a large weight of catalyst must be circulated per weight of product produced. Furthermore, subjecting a catalyst to transport, often at high speed, and alternating environments leads to attrition and catalyst loss.

The balance between advantages and disadvantages can be shifted if composition modulation provides higher selectivities. This is the case for the partial oxidation of butane to maleic anhydride as long as short butane – catalyst contact times are used. To achieve the short contact time, a riser reactor or a fast fluidized bed replaces one of the conventional fluidized beds shown in Figure 1.13. The Du Pont Company has announced the construction of a maleic anhydride facility using the two bed – circulating catalyst scheme. Figure 1.14 shows a recirculating reactor system employing a riser and a fluidized bed that has been proposed for the partial oxidation of butane to maleic anhydride (Contracter *et al.*, 1987a). Re-oxidation of the catalyst is

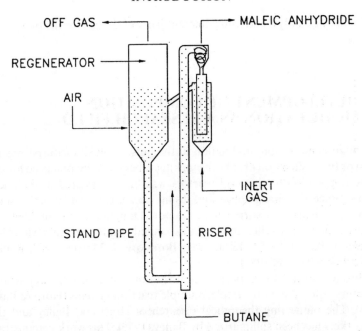

FIGURE 1.14 Schematic of a riser reactor – fluid bed reactor recirculating solids proposal for partial oxidation under periodic operation (Figure adapted from Contractor *et al.* (1987a) with permission, © 1987, Elsevier Science Publishers)

not a critical step for high maleic anhydride yield so this step is performed in the fluidized bed. Although multi-reactor systems using valving to initiate periodic composition changes require more than two reactors if the cycle has more than two parts, this may not be necessary with circulating catalyst. For partial oxidation, multi-part cycles arise when the catalyst is stripped with a gaseous inert between contact with the reactants. With recirculating solids in fast fluidized beds, the catalyst separation and transport steps provide the opportunity for stripping so a further vessel is not needed. This can be seen in Figure 1.14. Stripping of the hydrocarbons and oxygenates from the catalyst is accomplished in the accumulator following the riser cyclone and in the line transferring the catalyst to the fluidized bed regenerator.

Du Pont was not the first company to commercialize the two bed, circulating catalyst system. The technology was available some 20 years ago (Sze & Gelbein, 1975; 1976) and apparently a commercial unit exists (Schwendeman *et al.*, 1983).

It is possible to circulate catalyst between zones in a single bed. A design for such a reactor to be used for hydrocarbon partial oxidation was proposed by Park and Gau (1986) but has not been built. A spouted bed is another means of circulating catalyst and using different parts of the bed for the two parts of a cycle.

H. DEVELOPMENT OF COMPOSITION MODULATION AS A RESEARCH FIELD

Although composition modulation in the guise of in situ catalyst regeneration or the gasification of coke in cracking tubes has been practiced for sixty years, application to chemical reactors was first considered in the middle 1960's. Since the first publications using numerical simulation of simple reaction systems in a stirred tank reactor (Douglas & Rippin, 1966), the research effort and the literature have expanded rapidly. Several reviews have appeared (Bailey, 1973; Meira, 1981; Boreskov & Matros, 1983; Renken, 1982;1984; Silveston, 1987).

Chronologically, the important contributions have been the applications of optimization theory to simple, multiple reaction systems (Horn & Bailey, 1968). The pioneering theoretical research of Horn and Bailey and their co-workers has been summarized by Bailey (1973). This work demonstrated that improved reactor performance was attainable.

Experiments in which catalytic reactors were operated with periodic composition forcing were published almost simultaneously in Germany and Canada (Wandrey & Renken, 1973; Helmrich et al., 1974; Unni et al., 1973). This early experimental work is discussed by Renken (1982; 1984). Renken's review addressed the question of the proper comparison of periodic and steady state operations. This question was discussed earlier in this chapter. Renken also noted that models which predict improved performance under cycling are not supported by experimental data, whereas experimental systems for which improvement has been found do not have adequate models. This regrettable circumstance still persists.

Experimental studies of periodic composition forcing of important industrial catalytic reactions continued into the 1980's. Very large rate enhancement through cycling were demonstrated for CO oxidation over noble metal catalysts (Barshad & Gulari, 1985a) and for the ammonia synthesis over ruthenium and osmium (Rambeau & Amariglio, 1981b; Rambeau et al., 1982b). The presence of unintentional periodic operation in catalytic mufflers used for automotive exhausts was identified and studied with conflicting results on the benefits of this mode (Schlatter et al.,1983; Taylor & Sinkevitch, 1983). During the 1980's, the emphasis in experimental studies began to shift towards complex reactions and questions of selectivity. Several papers were published on the partial oxidation of hydrocarbons (Fiolitakis et al., 1983; Silveston & Forrissier, 1985; Lang et al., 1989b) and on the FischerTropsch synthesis (Feimer et al., 1985a; Adesina et al., 1986).

The analytical literature of the 1980's concentrated on three questions: 1) can composition modulation improve the performance of a chemical reaction system ?; 2) can performance under modulation exceed the best possible performance under steady state ?; 3) what properties of a catalytic system result in improved performance under forcing ? The first question was extensively discussed in the '70's where the issue was the criteria that could be used for determining optimal forcing operations. The discussion continued into the next decade (Watanabe et al., 1981; Farhadpour & Gibilaro, 1981; Sterman & Ydstie, 1989). Several contributions by Nowobilski and Takoudis (1986), Thullie et al. (1986) have dealt with the second question, while Schädlich et al. (1983) and Grabmueller et al. (1985), among others, consider both of the final questions.

Both modelling and experimental work on the application of composition forcing to polymerization reactions was summarized by Meira (1981) who concluded that periodic variation of initiator, monomer or transfer agent concentration increased operating flexibility with respect to molecular weight distribution, however the improvements were modest. Research on this application appears to have languished since the 1980's.

A physical explanation and a conceptual foundation for composition modulation was provided in the early 1980's by Boreskov and Matros (1983). In addition to effects caused by shifts in adsorbate concentrations on catalyst surfaces, Boreskov and Matros point out that improvements in reactor performance under composition forcing result from favourable changes in catalyst activity through a variety of mechanisms, often a change in the composition of the catalyst surface. When the catalyst state is altered in response to adsorbate composition and/or temperature, both the activation energy and the frequency factor change. Consequently, as input variables are cycled between two or more conditions, reaction rates vary because of both adsorbate changes and changes in the activation energy and the frequency factor. If the surface or even bulk phase transitions are slow relative to the forcing period, the catalyst will operate in a transient, i.e., periodic, state which can lead to improved rates, selectivity or yields. Bailey (1973) commented on the difficulty of analyzing reaction systems in which the switching period is the same order as the relaxation time. Considering systems now in which catalyst properties depend on the gas phase and on temperature makes the analysis even more challenging. Not only must the states of the catalyst corresponding to different composition and temperature environments be known, the dynamics of the transition from state to state must be determined.

Late 1980 discoveries that modulation is applicable to gas-solid reactions (Suzuki et al., 1988) and to three phase reactors (Haure et al., 1989; Irandoust and Andersson, 1988) seems likely to furnish new research themes in the 1990 decade. A further theme will be the coupling of catalytic reactions and separation. A 1989 paper by Vaporciyan and Kadlec discusses catalytic CO oxidation in a specially modified pressure swing adsorber.

Another development stream parallels the growth of the research effort, but commenced almost twenty years earlier. In the late 1940's, Lewis *et al.* (1949) proposed that fluidized bed technology could be used for selective oxidation of hydrocarbons by letting the catalyst transport the oxygen needed for the oxidation reaction just as it transport heat for catalytic cracking. Sohio (now BP) in the United States pursued this two bed concept and tested the reactor part (Callahan *et al.*, 1970), but rejected it because of the weight of catalyst which would have to be circulated. Research on a two bed system using the catalyst to transport one of the reactants, usually oxygen, continued into the 1970's and 1980's (Park and Gau, 1986; Keller and Bhasin, 1982; Cordova and Gau, 1983). The Lummus Company developed technology for the two bed system (Sze and Gelbein, 1975; 1976) and an application of this technology to the ammoxidation of m-xylene into isophthalonitrile has been mentioned (Schwendeman *et al.*,1983). Laboratory research and pilot scale application of a circulating fluidized bed arrangement to the selective oxidation of butane to maleic anhydride continued to the end of the 1980 decade. A full scale plant is under construction. It has been only recently realized that the two bed operation with circulating catalyst is a version of periodic feed composition forcing.

Modulation causes concentration or partial pressure oscillations. Oscillations in reaction systems are possible, however, without external forcing. This is a subject that has attracted a good deal of research attention, even though they appear to be unusual in industrial processes. There have been several comprehensive reviews of spontaneous oscillations (Sheintuch and Schmitz, 1977; Slin'ko and Slin'ko, 1977; Razon and Schmitz, 1986; 1987; Gray *et al.*,1990; Ertl, 1990; Schüth *et al.*,1993; Slin'ko and Jaeger, 1994).[2]

---

[2]Notation and Bibliography for this chapter will be found following the Table of Contents.

CHAPTER 2
Combustion

A. INTRODUCTION

Combustion is a homogeneous reaction and thus appears to be an inappropriate topic for a monograph on catalytic reactions. There are several reasons to consider this class of reactions as our first subject. The first of these is that the far-reaching research on composition modulation we discuss in the chapters which follow began with several simulation studies of homogeneous reactions. These considered an arbitrary (A + B → Products), isothermal, homogeneous reactions in a CSTR with 2nd order kinetics. Douglas (1967), Douglas and Rippin (1966), and Renken (1972) demonstrated through simulation that modulation of a reactant concentration increased the time average conversion of the limiting reactant over that achieved in steady state operation. A resonance with cycling frequency was not observed and quasi steady state operation gave the highest conversion. Fjeld (1969) recognized that the limiting reactant concentration is a concave function of cycle period and observed that a periodic operation will always be better than steady state when such a functionality is present. Horn (1965) introduced the concept of attainable sets to handle this type of problem (Bailey, 1973). The importance of this early work is that it demonstrated that periodic manipulation of an input was a means of controlling reactor operation.

There appears to be just three experimental studies of the periodic forcing of a homogeneous reactions, all of them liquid phase systems. Lee *et al.* (1980) examined the saponification of ethyl adipate, Renken (1972) investigated the formation of di- and triethanolamines from ethylene oxide and monoethanolamine, and Vamling (1987) studied the reaction of ethylene oxide and ammonia to yield various ethanolamines. Each of these investigations dealt with the effect of modulation on selectivity. Because we will discuss selectivity and product distribution beginning in Chapter 10, our examination of these studies will be postponed until that chapter.

The second reason for beginning with combustion, like the first, is one of sequence. One of the earliest, large scale systems with periodic variations of composition in a reactor is the internal combustion engine. It is a bit far fetched to portray these engines as examples of composition modulation, but their are similarities, in peculiar with a class of modulation that is referred to as stop flow. In stop flow, reactants do not flow continuously to a reactor. Their flow is restricted to one part of the cycle. In the other part, diluents flush out the reactor. This is essentially what happens in an internal combustion engine. Of course, the internal combustion engine cycle has 6 parts rather than the two we will consider throughout most of this and the succeeding chapters, but this is not the reason cyclic engines are ignored in this monograph. The periodic operation is not imposed on engines because it offers a better way of conducting combustion. It is rather a requirement for withdrawing work from the burning of a fuel.

Pulsed combustors, on the other hand, are periodically operated systems that are much closer to composition modulation. This classification covers a range of devices from the engine that operated the V-1 rockets deployed in the last years of World War II to more or less conventional burners operated under a de facto composition forcing. Of these devices the mechanically valved or aerovalved pulse combustors resemble the internal combustion engine in that they do not use composition forcing, but oscillate instead at a natural frequency which depends on the design, the fuel type and feed rate, and the fuel/air ratio. It is the oscillatory flow and pressure that have attracted interest in pulsed combustors. Frequencies are very high compared to those we will encounter in this monograph. Generally pulsed combustors operate at frequencies between 15 and 230 Hz. Oscillatory flow and pressure result in high heat transfer rates and rapid mixing in the combustor. Because of the rapid mixing, both the mechanically and aerovalved combustors operate at high combustion efficiencies and have lower NO_x emissions than steady state burners. Putnam *et al.* (1986) provide a comprehensive survey of pulse combustors and some performance data.

B. COMBUSTION MODULATION

Research on pulsed combustion appears to have spun off in the 1950's an interest in modulated combustion. All of the early work was done in the Soviet Union and reports are difficult to obtain. They are mentioned by Putnam *et al.* (1986) in their survey of pulse combustion and in reviews published by Severyanin (1971, 1982). In modulated combustion, the fuel supply to the burner or combustion chamber is switched on and off so it is a composition forcing with large amplitudes. Motor driven rotary valves as well as fluidic devices are used for flow switching so that a wide range of forcing frequencies are possible. Applications require frequencies between 1 and 20 Hz. These are well below the natural or acoustic frequencies that pulse combustors operate at.

The attractions of modulated combustion are first of all that flow and pressure oscillations are generated in the burner tube or combustion chamber. Such oscillation increase mixing and rates of heat transfer from the hot combustion gas to heat exchanger surfaces. Modulation with a properly designed burner tube provides waves, namely detonation waves, with large, sharp pressure increases. Modulated combustion (and pulsed combustors) significantly lower NO_x formation. Corliss et al. (1984) have reviewed this subject. The consensus is that it is the rapid flow mixing generated in pulse devices or in modulation that are responsible for reduced NO_x generation.

Current activity in modulated combustion focuses on NO_x generation. NO_x formation from nitrogen in the fuel or in the air charge proceeds rapidly in certain, high temperature industrial furnaces. Air Liquide (France) has presented a paper describing experiments on modulating the combustion of natural gas in oxygen (Charon et al, 1991). This mixture is used to fire glass remelt furnaces as well as other processes where temperatures in excess of 1200°C are needed. Charon et al. mention that NO_x levels can reach 20 vol.% in the stack gas from reheating furnaces if natural gas containing a high level of N_2 is used. Usual NO_x concentrations range from 0.5 to 4 vol.%.

In their paper, Charon et al. observed that NO_x formation is at a maximum for stoichiometric or excess oxygen in the oxygen-fuel mixture and investigated modulating the mixture to reduce formation of this pollutant for these conditions. Figure 2.1 is a schematic of the Air Liquide experimental unit. Two furnaces were used: 1) a 0.5 m (i.d.) × 1.5 m unit lined with an alumina refractory wool and fired at a heating rate of 20 kW, and 2) a larger 0.6 m (i.d.) × 2 m unit lined with alumina brick fired at 60 kW. The smaller

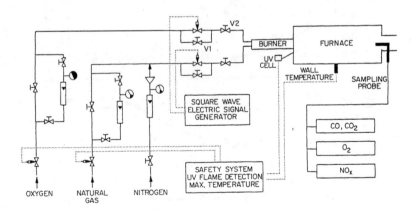

FIGURE 2.1 Schematic of the experimental unit used for modulation of natural gas-oxygen combustion. V1 is either a solenoid or a rotary plug valve, while V2 is a needle valve for flow adjustment. CO and CO_2 are monitored by IR, NO_x by chemiluminescence (Figure adapted from Charon et al. (1991) with the permission of the authors)

furnace had to be held below 1350°C, whereas the second unit could operate at temperatures up to 1800°C. In the experiments, a 99.99% O_2 and a natural gas stream containing 7 to 15 vol.% N_2 were used. Symmetrical forcing, $s = 0.5$, was employed in all experiments and tests were made varying fuel or oxygen alone, holding the other component constant, or varying both simultaneously. The time average oxygen volume % in the feed was set to give 6 vol.% O_2 in the furnace off gas. As indicated in the figure, CO, CO_2, and O_2 were monitored in addition to NO_x.

Composition forcing of both natural gas and O_2 appears to sharply reduce NO_x formation in the lower frequency portion of the range studied. This is evident in Figure 2.2(a) which plots the ratio of the time average

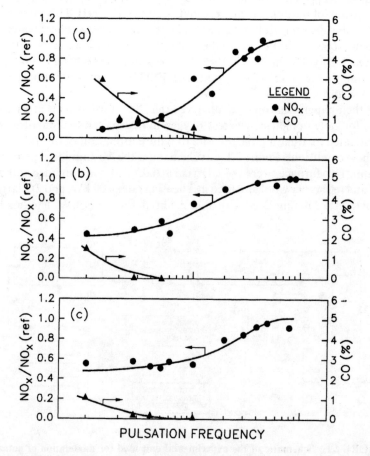

FIGURE 2.2 Normalized NO_x emission versus frequency: (a) composition modulation of both natural gas and oxygen, (b) natural gas alone, (c) oxygen alone. Amplitude of the modulation in (b) and (c) is 1/2 that in (a) (Figure adapted from Charon *et al.* (1991) with the permission of the authors)

NO_x under forcing to the NO_x emerging from the furnace at steady state for the same mean air to fuel ratio against the forcing frequency. Unfortunately, Charon et al. (1991) do not give the frequency and amplitudes used, but it is possible to estimate what these are based on the equipment used and the author's comment that the gases emerging from the furnace are not mixed. It is likely that the logarithmic scale in Figure 2.2 covers two decades: probably from 0.1 to 10 Hz. The probable amplitude is 0.2 SN where SN = stoichiometric number defined as the moles of oxygen per mole of methane divided by the moles of oxygen needed to completely oxidize methane. The figure shows that periodically lowering the air-fuel ratio increases the formation of CO, although Charon et al. claim that if the emerging gas was completely mixed, CO would be oxidized to CO_2 at the flue gas temperature of 1300°C. The forcing experiments shown in a) were repeated in the larger furnace using a rotary plug valve which distorted the air-fuel ratio square wave. In these experiments a 40% reduction in NO_x was measured. CO formation was also lower suggesting more mixing occurred when the larger furnace was used. Nitrogen in the feed was 7 vol.% and this could also have contributed to the smaller NO_x reduction observed.

Figure 2.2(b) and (c) show the effect of forcing the natural gas flow rate and the oxygen flow rate respectively. The amplitude of the forcing is half that in a) and just a 45 to 60% reduction in the NO_x emission is realized. CO formation is also lower. A comparison of the curves in Figure 2.2 shows that the effect of frequency on NO_x formation is the same for all three forcing experiments. This indicates that amplitude has a strong effect on performance when composition forcing is employed. Charon et al. explored the influence of amplitude and confirmed its importance.

C. CATALYTIC COMBUSTION

The combustion of propylene in air has been studied by two groups of investigators each with quite different objectives. Wandrey and Renken (1977) thought to demonstrate that periodic operation could be used to operate a reaction system exhibiting hysteresis in a region of reactant concentration and feed temperature where the system would be unstable at steady state. Hysteresis resulting from backmixing can result in multiple steady states at a single operating condition, one of which is unstable. Without control or composition forcing, the system cannot operate at this state.

The objective of the Toyota Research Laboratory team was to determine the behaviour under cycling of the individual reactions occurring in automotive exhaust catalysis for the different noble metal components of three-way catalysts (Shinjoh et al. 1987, 1989). Ultimately, the team hoped to improve the performance of the automotive emission control system through catalyst formulation or control of the cycling frequency and/or amplitude. They also wanted to discover why rates increase under composition forcing.

In neither of these studies was composition modulation considered as a technique for increasing throughput or conversion; although improved rate or conversion is usually the objective when the question of product distribution is not important.

Application To A Combustion System Exhibiting Hysteresis

For their demonstration, Wandrey and Renken (1977) switched between a feed containing 1.8 vol.% C_3H_6 and 8.1 vol.% O_2 with the balance helium and a stream containing helium only. In place of heat exchange between exhaust and feed, the feed was brought to a constant inlet temperature in a heated fluidized bed. The reactor was a layer of platinum gauze. The experimental system is quite different from those discussed in Chapter 1 so it is shown in Figure 2.3. Feed preparation is not shown but it consisted of solenoid valves which switched the flow between the combustion mixture and the helium diluent. A multi-loop gas sampling valve, coupled to a gas chromatograph, permitted measurement of the time variation of the exhaust gas composition. With this system, Wandrey and Renken investigated the influence of inlet feed temperature and the cycle period on the time-average conversion and bed temperature using symmetrical forcing. The system shown in the figure was operated at steady state prior to initiating concentration forcing so that the hysteresis loop could be delineated.

Figure 2.4 shows the temperatures measured, either steady state or time averaged, on the platinum gauze as a function of the inlet temperature. These are the solid circular points connected by the dashed or solid line. At inlet temperatures of 70 °C and 200 °C, the different points represent different forcing frequencies. At 200 °C, these frequencies correspond to cycle periods between 12 and 20 seconds. As the period increases, the gauze temperature increases. The cross hatched zone is the region of multiplicity so the figure shows that the strategy of alternating between diluent and a reaction mixture allows the system to operate at a state between "ignition" (the upper steady state line) and "extinction" (the steady state line at the bottom left hand corner). It is the time average behaviour, not the instantaneous one that rests inside this region. Choice of period controls where within the region the system operates.

The success of this experiment is easily explained. The platinum gauze fluctuates between ignition when a reactant mixture flows across it and extinction when it sees only the helium diluent. This can be seen in Figure 2.5 which shows the instantaneous temperature of the gauze as a function of

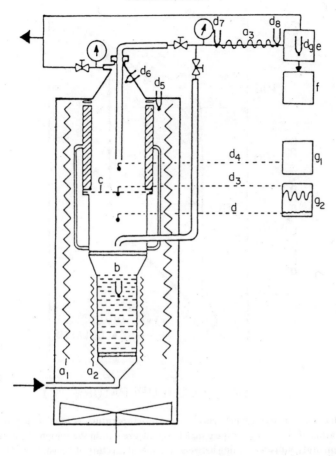

FIGURE 2.3 Schematic of the experimental unit used for the unstable state study. Legend: (a) heating filament for the preheater(2) and the reactor enclosure(1), (b) fluidized bed used to preheat feed, (c) platinum gauze layers, (d) thermocouple locations: preheat, (3) gauze, (4) exhaust sample, (5) reactor enclosure, (6) top of reactor, (7,8) inlet to sampling valve, (e) multi-loop sampling valve, (f) gas chromatograph, (g) recorders. (Figure reprinted from Wandrey & Renken (1977) with permission, © 1977, Elsevier Science Publishers)

time for different cycle periods. The maximum gauze temperature increases with cycle period and this causes an increase in the time average temperature. Wandrey and Renken demonstrate that the period effect arises from a gauze temperature lag with respect to ignition caused by the thermal inertia of the gauze and the reactor walls. It can be seen in the figure that at a period of 12 seconds, forcing causes a small oscillation around either the ignition or extinction stable state depending on which steady state occurred prior to the initiation of cycling. A small increase of the period to 13.6

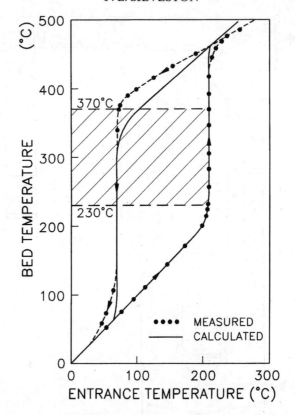

FIGURE 2.4 Hysteresis behaviour of gauze temperature vs inlet temperature and stabilization of the time average gauze temperature within the region of multiplicity. Experimental results for cycling between a reactant mixture of 1.8 vol.% C_3H_6 and 8.1 vol.% O_2 and pure He diluant (Figure reprinted from Wandrey & Renken (1977) with permission, © 1977, Elsevier Science Publishers)

s brings the system into the multiplicity region. Above 20 s, the system moves out of the unstable region and it oscillates with a large amplitude around the high temperature stable state. Wandrey and Renken measured the steady state kinetics of propylene combustion and assuming the gauze behaves like a continuous stirred tank reactor, they simulated the behaviour just described. Their simulation is given as the solid line in Figure 2.4. It represents the experimental results closely.

Hydrocarbon Catalytic Combustion Over Noble Metals

Both propylene and propane combustion under forcing of the air/fuel ratio have been studied by a Toyota Central Research and Development Lab-

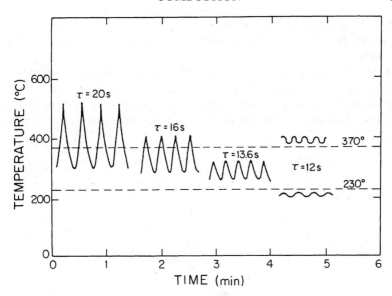

FIGURE 2.5 Effect of cycle period on the variation of the platinum gauze temperature with time inside of the multiplicity region for an inlet temperature of 200 °C. Other experimental conditions given in the caption of the previous figure (Figure reprinted from Wandrey & Renken (1977) with permission, © 1977, Elsevier Science Publishers)

oratories team as part of a comprehensive program concerned with understanding the effect of composition modulation on automotive emission catalysis. This program is discussed in Chapter 5. It is also mentioned in conjunction with CO oxidation in Chapter 3. Propylene has been widely assumed to typify all hydrocarbons in automotive emission catalysis.

Conversions were examined for a fixed space velocity under both steady state and air-fuel ratio cycling. Measurements were repeated for each of the noble metal catalysts used in commercial three-way catalytic mufflers now so widely used. Palladium received special attention as it was being considered then as a replacement for the much more expensive rhodium in three-way-catalysts (TWC). The support used was α-Al_2O_3, an inert alumina. This support was chosen to minimize support-catalyst interactions so that it is not typical of those employed commercially. The metal loading on the support was approximately the total metal loading used in TWCs.

For their experimental program, Shinjoh et al. (1989) used platinum (Pt) and palladium (Pd) at 0.05 g/L of catalyst or 0.006 wt.% metal, while rhodium (Rh) was used at 2.0 g/L. The space velocity was 30,000 h^{-1}. Temperature and cycle period were the independent variables. Feed compositions were switched between 0.01 and 0.4 vol.% hydrocarbon and

0.1 and 2.0 vol.% O_2. These typify concentrations in rich and lean fuel mixtures. Both hydrocarbon and O_2 were varied in a symmetrical cycle ($s = 0.5$).

Composition forcing of the hydrocarbon-O_2 concentrations is capable of increasing conversion at a constant throughput. This is illustrated in Figure 2.6 for two of the three catalyst studied. Only Pd is not activated by cycling. Figure 2.6(a) shows that for the Pt catalyst, cycling at 1 Hz increases

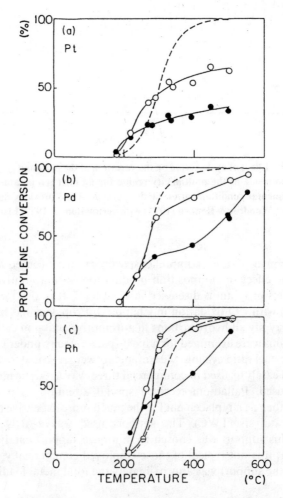

FIGURE 2.6 Propylene conversion as function of temperature at a constant space velocity (30,000 h^{-1}) for steady state and periodic operation: (a) Pt catalyst, (b) Pd catalyst, (c) Rh catalyst. Legend: ○ = 1 s period, ● = 2 s, ⊖ = 0.65 s, dashed line = steady state (Figure reprinted from Shinjoh et al. (1989) with permission, © 1989, Elsevier Science Publishers B.V.)

conversion up to about 260 °C. Thereafter, conversion is depressed. The explanation of this observation is that the optimum cycle period depends on temperature. This can be seen in Figure 2.7 which plots time average conversion *vs.* cycle period for different temperatures. Steady state data for the time average feed composition is given by the point at $\tau = 0$. Consequently, at 330 °C, the cycle period would have to be less than 1 second for an improvement to be seen. The smallest period used in Figure 2.6(a) was 1 second. Although the authors did not include results for rhodium, it is likely that the cycle frequency resonance seen in Figure 2.7 for Pt will also be observed for Rh.

Measurements for propane oxidation of Shinjoh *et al* .(1989) for the Pt catalyst are given in Figure 2.8 and are similar to Figure 2.6(a) except that conversions are lower by a factor of 2 to 3 × and steady state provides higher conversions than modulated operation above 480 °C rather than 260 °C. It seems likely that oxidative dehydrogenation to propylene is the first step in propane oxidation over Pt. Figure 2.8 suggests that C_3H_8 is less strongly adsorbed than C_3H_6 so that at 160 °C adsorbed O_2 or oxygen adatoms dominate the surface.

Examination of the transient concentrations of reactant and product species after a composition switch (Figure 2.9) show a complex and relatively slow rise in O_2 leaving a catalyst bed after a switch from a C_3H_6/He to an O_2/He feed, whereas it disappears from the outlet

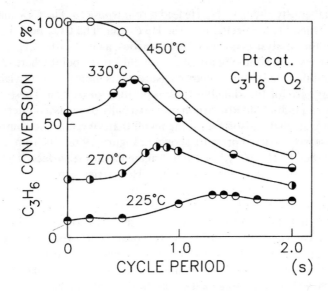

FIGURE 2.7 Propylene conversion as a function of cycle period at different temperatures for the Pt catalyst with a space velocity of 30,000 h^{-1} (Figure reprinted from Shinjoh *et al.* (1989) with permission, © 1989, Elsevier Science Publishers B.V.)

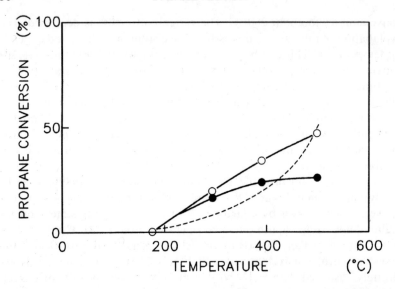

FIGURE 2.8 Propane conversion over a Pt/α-Al$_2$O$_3$ catalyst as function of temperature at a constant space velocity (30,000 h^{-1}) for steady state and periodic operation: Legend: ○ = 1 s period, ● = 2 s, dashed line = steady state. (Figure reprinted from Shinjoh *et al.* (1989) with permission, © 1989, Elsevier Science Publishers B.V.)

instantaneously after a C$_3$H$_6$/He feed is re-introduced. The CO$_2$ maximum occurs when O$_2$ is injected into the He carrier. The CO$_2$ peak in the gas leaving the catalyst bed is not instantaneous, as it is with C$_3$H$_6$ injection, but reaches a maximum in about 2 s, roughly at the point where O$_2$ shows a minimum. Steady state kinetic measurements indicate a propylene order less than zero for Pt, while the order is positive for oxygen. These observations mean that the hydrocarbon or its partially oxygenated intermediates are more strongly adsorbed on the metal than oxygen. The complex and slow rise in O$_2$ partial pressure shown in Figure 2.9 results from oxidation of adsorbed C$_3$H$_6$ or hydrocarbon fragments on the surface. The smaller CO$_2$ peak on C$_3$H$_6$ injection in the He carrier can be explained by stoichiometry, but it may also suggest that propylene molecules cannot react with adsorbed O$_2$ but must adsorb first.

The 2 s maxima for CO$_2$ evolution corresponds, according to the authors, with a maxima at a period of 2 s in C$_3$H$_6$ conversion at 160 °C. Indeed, this is suggested by Figure 2.7. Thus, the optimal cycle period indicated by the figure may come about from the time needed for O$_2$ to scavenge adsorbed hydrocarbons or their fragments from the catalyst surface.

Shinjoh *et al.* (1989) conclude that under steady state operation the noble metal catalysts are reactant inhibited so that the mechanism for the improvement through periodic operation is to balance the adsorbate concen-

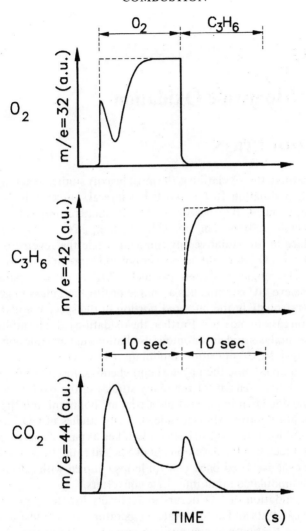

FIGURE 2.9 Transients for O_2, C_3H_6, and CO_2 for bang-bang switching between oxygen and propylene injection into a He stream flowing over a Pt/α-Al_2O_3 catalyst at 160 °C (Figure reprinted from Shinjoh et al. (1989) with permission, © 1989, Elsevier Science Publishers B.V.)

trations on the surface. For propylene, the inhibiting species is the hydrocarbon, whereas for propane, Shinjoh et al. conclude it is oxygen. In an earlier paper, Shinjoh et al. (1987) observe that when ever the product of the reaction orders is negative, composition forcing activates the noble metal catalyst provided an optimum or near optimum cycling frequency is invoked. A negative product requires inhibition by one of the reactants.

CHAPTER 3

Carbon Monoxide Oxidation[1]

A. INTRODUCTION

Without question, this oxidation is the most heavily studied reaction under composition modulation. Its popularity has several sources. It is a reaction that is easily studied. It proceeds readily at temperatures below 100°C on noble metal catalysts. The one side reaction, $2CO \rightarrow CO_2 + C$, does not take place in an oxidative environment; while the reverse reaction, $CO_2 \rightarrow CO + 1/2\, O_2$, does not become important for temperatures studied heretofore. The single oxidation product, CO_2, is easily separated in a molecular sieve GC column, has a unique positive ion mass finger print and absorbs strongly in the infra red portion of the spectrum making its concentration easy to measure. Further, the oxidation exhibits multiplicity over various catalysts and autonomous oscillations under some conditions over noble metals. Composition modulation has been used by several researchers to investigate the physical and chemical mechanisms of these phenomena. A final reason for the many studies of the oxidation is that carbon monoxide is an important industrial air pollutant and the major pollutant in automotive exhaust. Indeed, many studies of CO oxidation under composition forcing have been undertaken as part of a large international effort directed at the development of efficient catalytic converters for automotive exhaust. Feed back control loops coupled with sensor dynamics cause a modulated input into these converters.

Catalytic oxidation of CO occurs at relatively low temperature over noble metal catalysts and at moderate temperatures over transition metal oxide catalysts. Response to composition forcing is not the same for these two catalyst groups. It is convenient, therefore, to organize this chapter around the noble metals and metal oxides. The former will be subdivided into separate sections on low temperature studies, on temperatures found in automotive catalytic converters, and on single crystals because of the large literature that exists. CO oxidation with NO has also been studied under periodic composition forcing. This work will be examined in the chapter.

---

[1]An early version of this chapter appeared as a review paper: Silveston, P. L. (1991), "Catalytic oxidation of carbon monoxide under periodic operation" *Can. J. Chem. Eng.*, **69**, 1106–1120.

B. LOW TEMPERATURE STUDIES OVER NOBLE METAL CATALYSTS

Experimental

With the proper choice of cycle split and period, modulation of the CO and O_2 partial pressures in the reactor feed can result in a large increase in the CO oxidation rate. One of the first studies to demonstrate this enhancement was made by Cutlip (1979) using a 0.5 wt.% platinum (Pt) on γ alumina as the catalyst. Cutlip's study was directed at the mechanism of activity improvement under periodic forcing so an exhaustive investigation of the cycling variables was not undertaken. Cutlip reasoned from studies on spontaneous oscillations of CO oxidation that the coverage of the catalyst surface by adsorbed reactant was implicated so his primary variables were the duration of CO exposure, measured as cycle split, and temperature. He employed a recycle reactor and used alternating exposure of the catalyst charge to 2 vol.% CO and 3 vol.% oxygen with argon the diluent for both feeds. These concentrations and the temperatures used (28 and 60°C) meant that Cutlip's experiments were performed in the region of steady state multiplicity.

Cutlip (1979) observed a sharp resonance in the oxidation rate with cycle period as illustrated by Figure 3.1 drawn from his results. The cycle split, s, in the figure is 0.67. Note that for this split, the oxidation rate drops to

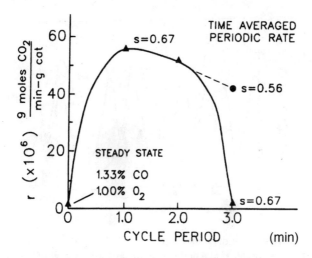

FIGURE 3.1 Comparison of steady state and periodic operation for the oxidation of carbon monoxide over a 0.5 wt.% Pt/γ alumina catalyst at 60°C and atmospheric pressure (Redrawn from Cutlip (1979) and reproduced by permission of the American Institute of Chemical Engineers, © 1979 *AIChE*. All rights reserved)

nearly zero at $\tau = 3$ min. However, if $s = 0.56$, there is just a small decrease in the time average rate. The steady state rate at the time average feed composition is shown by the point at $\tau = 0$. The time average rate under periodic operation does not change with period in the range 1 to 3 minutes if $s = 0.5$. This rate, about the maximum rate in Figure 3.1, appears to equal the rate at the higher steady state in the multiplicity region.

Figure 3.2 a plots reactant and product mole fractions leaving the reactor for $\tau = 3$ min and $s = 0.56$. Because CO is limiting and conversion is high, little CO was detected. The double oxygen peak may be explained by a surface saturated with CO when the feed is switched to 3% O_2. Due to

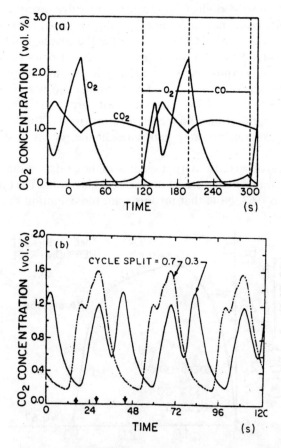

FIGURE 3.2 Reactor outlet mole fractions for CO oxidation over a 0.5 wt.% Pt/γ alumina catalyst at: (a) 60°C at a period of 3 min and a cycle split of 0.56 (b) 120°C at a period of 1 min and a cycle split of 0.3 and 0.7 [(a) (Reprinted from Cutlip (1979) and (b) reprinted from Barshad and Gulari (1985a). Both figures reproduced by permission of the American Institute of Chemical Engineers, © 1979, 1985 AIChE. All rights reserved)]

mixing in the recycle reactor, the O_2 partial pressure does not step up instantaneously. Some of the oxygen reacts with adsorbed CO freeing adsorption sites for O_2. The build up of oxygen on the surface and continued reaction between the adsorbates increase CO_2 formation and drive down the O_2 partial pressure. As the surface fills with O_2 adatoms and CO adatoms are consumed, the O_2 partial pressure rises again, while the CO_2 partial pressure drops. The persistence of O_2 in the CO portion of the cycle suggests equilibrium adsorption of O_2, while double CO_2 peaks each occurring when a reactant was detected at low concentration indicates strong adsorption of the reactants. Thus, Figure 3.2a supports Cutlip's contention that shifts in the coverage of the surface by each reactant during composition modulation is the source of the enhanced activity.

The largest enhancements of the CO oxidation rate have been reported by Gulari and co-workers (Barshad and Gulari, 1985a; Barshad et al., 1985; Zhou and Gulari, 1986a,b, Zhou et al., 1986) who took up Cutlip's investigation of CO oxidation over Pt/Al_2O_3 and extended the study to Pd catalysts on the same support. A feature of the Gulari work was fast switching which enabled the investigators to examine periods as short as 2 seconds. In their first published study, a bed packed with 3.2 mm particles of 0.5 wt.% Pt on Al_2O_3, non-uniformly distributed, was used (Barshad and Gulari 1985a). About 95 wt.% of the Pt was deposited within 160 μm of the particle surface. Fast switching was accomplished by solenoid on-off valves with flow trimmed by electronically driven control valves operated by mass flow meters with unusually quick response. The experimental unit was run differentially at 90 to 120°C. A remarkably strong resonance effect was observed as may be seen in Figure 3.3 drawn from data presented by Barshad and Gulari (1985a). Bang-bang switching between CO and O_2 results in a 10 × increase in the CO oxidation rate at a time-average CO/O_2 ratio in the feed $= 2$, T $= 120°C$ and $s = 0.3$ (based on the CO exposure). A rate increase greater than 15 × can be seen at the same split, but at 90°C and a time-average CO/O_2 ratio in the feed $= 1$.

The maximum conversion or rate was not always observed at $s = 0.3$. Indeed, Barshad and Gulari (1985a) found that split was an important variable as Figure 3.4 illustrates. Lines of equal conversion are plotted for experiments in which split varied from 0.035 to 0.85, based on the duration of the CO + diluent exposure, and period ranged from 5 to 35 s. Rates along the vertical line at $s = 0.3$ appear in the previous figure. The importance of split confirms the comments above on Cutlip's (1979) results. Figure 3.2b shows a double CO_2 peak at a period of 60 s and a cycle split of 0.7 that is even more pronounced than that observed by Cutlip (Figure 3.2a). However, when the split is reduced to 0.3, the double peak disappears. As the period decreases, the effect of split changes and a double peak is observed only at $s = 0.3$ when $\tau = 20$ s. Barshad and Gulari (1985a) speculate that the time of exposure to CO may be the important variable and not split or period.

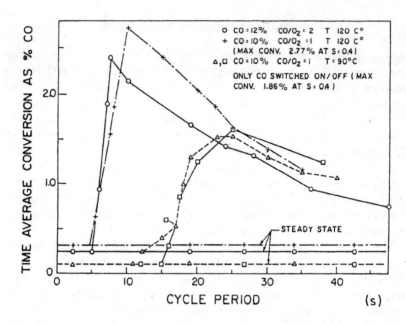

FIGURE 3.3 CO conversion to CO_2 as % as a function of cycle period for a split of 0.3 based on the CO pulse. Catalyst: 0.5 wt.% Pt on Al_2O_3

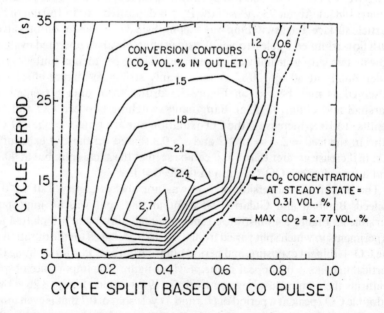

FIGURE 3.4 Conversion contours (lines of equal conversion) as functions of cycle period and split, based on duration of the CO pulse, for CO mole % = 10, $CO/O_2 = 1.0$, 0.5 wt.% Pt on Al_2O_3, 1 bar and 120 °C with an N_2 diluent. (Reprinted from Barshad and Gulari (1985a) with permission of the American Institute of Chemical Engineers, © 1985 *AIChE*. All rights reserved)

Barshad and Gulari discovered also that periodically switching CO between 0 and 10 mole% while maintaining a constant mole% O_2 leads to a large increase in the CO oxidation rate. A maximum of 19 × was found as seen in Figure 3.3. The longer period associated with the maximum arises from slower adsorption rates and thus slower formation of surface regions in which either CO or O_2 predominate as the temperature decreases from 120°C to 90°C.

Figure 3.5 plots the optimal cycle splits and periods for the Pt/Al_2O_3 experiments as well as for other catalysts considered further on vs reciprocal bed temperature, $1/T$. With the exception of the Pt catalyst the optimal cycle period depends strongly on temperature, decreasing as the temperature rises; optimal cycle split on the other hand, increases with rising reactor temperature. Explanations for these temperature effects appear to be straight forward. The increase in the optimal split reflects a lower adsorption equilibrium constant for CO. Probably the magnitudes of CO and O_2 adsorption constants are approaching one another. Reduction in the optimal period indicates reaction or displacement of CO by O_2 and vice versa, when the opposite change is made, proceeds more rapidly as temperature increases.

From these experiments as well as their step change observations, the authors conclude that the reaction proceeds via a surface reaction between adsorbed CO and adsorbed oxygen. They speculate that the adatoms are not randomly distributed over the surface, but instead occur as "islands" in which one adatom predominates or even in which the other adatom is absent.

Additional studies were undertaken by Gulari and co-workers to understand the behaviour under periodic operation as well as to confirm the mechanism for the large increase in CO conversion proposed. The packed bed reactor was replaced by a monolith of parallel, rectangular channels with a characteristic dimension of 1.2 mm and with 50.5 channels/cm² of monolith face. Thickness of the walls separating the channels was 0.3 mm and channel length was 34 mm. To provide catalytic activity, the support was given a wash coat using an Al_2O_3 slurry, calcined, and then impregnated with a Pt solution by successive dipping. The resulting "green" catalyst was calcined at 673°K and reduced in flowing H_2. Use of a monolith permitted an IR beam to pass through the parallel channels and on to a detector. Thus, gas phase concentrations in the immediate neighborhood of the catalyst surface could be observed. With this system, concentration transients could be followed with a time resolution of 0.15 s. This clever design is shown in Figure 3.6. Further details of the equipment and its application are given by Barshad and Gulari (1985b). These investigators found a more complex structure than the double bursts of CO_2 generation shown in Figure 3.2b. The peak on switching from O_2 to CO or on introducing CO to a constant O_2 stream occurred with a lag of just seconds

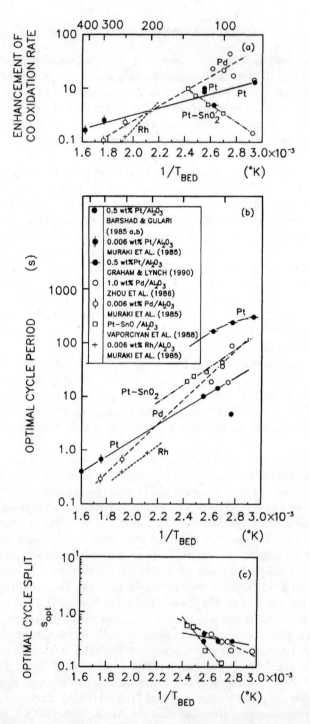

FIGURE 3.5 Temperature dependence of enhancement, optimal cycle periods and split for the maximum time average rate of oxidation on the reactor temperature

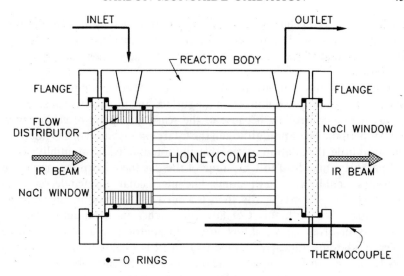

FIGURE 3.6 Experimental monolith reactor for rapid determination of surface response to concentration change (Reprinted from Barshad and Gulari (1985b) with permission, © 1985 Pergamon Press PLC)

and did not coincide with the rise in CO concentration thus establishing that the reaction proceeds through adsorbates and not via gaseous CO reacting with adsorbed oxygen. Production quickly fell to zero indicating that oxygen was rapidly swept from the surface after CO is introduced. On the switch from CO to O_2 or when CO is removed from the gas stream, a moderately long lag occurred before the CO_2 peak appeared. The lag as well as the peak height were temperature dependent. After the peak, CO_2 production continued and in some experiments a second peak appeared before CO_2 generation vanished. This result suggests two separate mechanisms for CO_2 formation when O_2 is contacted with preadsorbed CO.

Mechanism was investigated by Barshad et al. (1985) using transmission IR with a flow cell consisting of a 0.25 mm thick wafer formed from 6.2 wt.% Pt on Al_2O_3 with gas flow on both sides. Measurements were performed at 413 K. Use of an FTIR permitted time resolution of 0.6 s in transient experiments. Spectra associated with various adsorbate states were identified. Under steady state operation with a CO—O_2-diluent feed, linearly bonded CO on Pt or on PtO sites dominates. For step changes between CO-diluent and O_2-diluent, or in periodic operation, the linearly bonded CO on reduced Pt and a CO species sharing oxygen with Pt disappears or reappears depending on the direction of the composition shift. There is a significant lag on the CO to O_2 switch, but almost no lag for the reversed direction. Peak heights for CO linearly bonded to PtO and for bridge-bonded CO change little in composition switching and for bursts of CO_2 production so they are probably not active intermediates. Curiously, the oxide bound CO is apparently the dominant species under periodic operation.

From their observations, Barshad *et al.* (1985) conclude that at steady state CO adatoms predominate on the surface and dissociative adsorption of O_2 is rate controlling because of the scarcity of pairs of vacant sites. Under periodic operation, although the total concentration of CO adatoms changed little, the adatom species distribution differed substantially. Several pathways may lead to CO_2. One of these is the adsorption of CO on oxygen shared by surface Pt atoms. The intermediate formed decomposes quickly to CO_2. Consequently, there is a burst of CO_2 production when the peak associated with this CO adatom reaches its maximum. Another pathway involves surface diffusion of CO adatoms to a site containing oxygen adatoms. Linearly bonded CO on Pt is weakly held and diffuses easily on the surface. Barshad *et al.* suggest this may be the species participating in the diffusion pathway. Composition modulation allows oxygen to adsorb on the surface creating oxygen rich sites for CO adsorption. Island formation may be taking place, but this was not tested by the investigators. Local temperature surges were studied, but they were not found to be the source of the bursts in CO_2 production.

Unlike platinum, palladium crystallites are difficult to oxidize at low temperature. Furthermore, the adsorption of CO and O_2 as islands of predominately CO or oxygen adatoms is well established. These observations led Gulari's group to extend their study to Pd/Al_2O_3. The monolith support discussed above was used and the catalyst was formed just as with Pt, but impregnation was carried out with a $PdCl_2$ solution to yield a catalyst with 1.0 wt.% Pd. Surface area (B.E.T.) was 15 m^2/g and the Pd dispersion was 21%. Under periodic operation, surges in CO_2 production occurred after composition switches that lagged the switch by several seconds going from O_2 to CO and by ca. 10 seconds from CO to O_2. The latter surge vanished in 10 to 15 seconds, whereas CO_2 production dropped slowly after the peak which occurs when CO is re-introduced. Zhou *et al.* (1986) explain their observations by postulating that CO exposure saturates the Pd surface with CO adatoms. The 10 s lag seen on O_2 re-introduction allows for dissociative adsorption of oxygen. This is followed by a rapid CO—O reaction on the surface, eventually leaving the surface covered by O adatoms. The absence of a lag before CO_2 appears on the switch to CO-diluent and the slow decay of the production rate indicate oxygen islands form among adsorbed CO and that the O and CO adatoms are consumed in a surface reaction on the island periphery. At one point, the island abruptly changes to mixed CO and O adatoms and reaction continues. Figure 3.7 illustrates the island model and contrasts conditions under steady state and composition modulation.

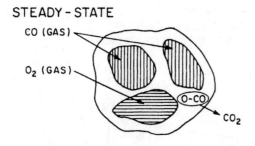
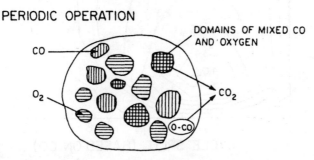

FIGURE 3.7 Schematic description of CO and O adatom coverage on the Pd surface under steady state and periodic operation (Reprinted from Zhou et al., (1986) with permission, © 1986, Pergamon Press PLC)

The island mechanism results in a dramatic increase in the CO_2 production rate under composition forcing according to measurements of Zhou et al. (1986) for cycle splits (based on the CO exposure) from 0.01 to 0.8 and cycle periods from 10 to 210 seconds. Figure 3.8 summarizes measurements made at 366 K with switching between 15 vol% O_2 in N_2 and 10 vol% CO in N_2. The maximum time average rate of CO oxidation occurs at a split of 0.3 and a 20 second period. As may be seen from the figure, the time average CO oxidation rate is 40 × the steady state rate. This latter rate is the maximum steady state rate at 344 K and one atmosphere for any CO/O_2 ratio. The global enhancement, thus, is 40 × In the experiments of Zhou et al. (1986), the rate increases ranged from 14 to 44 ×. These are much higher than the improvements found with the Pt catalyst and are among the largest ever observed for composition modulation. Indeed, CO conversion reached 70%, so depletion of reactants in the monolith catalyst was significant. Under differential operation of the reactor, the rate increases would have been even larger.

At low τ, split has little effect on rate, whereas at $s < 0.1$ or $s > 0.5$, period does not influence the rate. This suggests a minimum length of catalyst exposure to oxygen is needed to gain improvement under composition modulation. This exposure time appears to be about 10 seconds at 366 K in

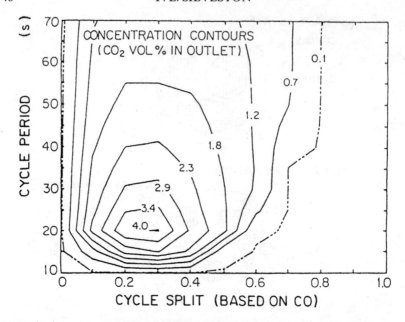

FIGURE 3.8 Rates of CO oxidation at 366 K under periodic switching between 15 vol % O_2 and 10 vol % CO both in N_2. Solid lines are lines of equal rates of oxidation. The dashed line is the maximum steady state rate of oxidation at 10 vol % CO and 15 vol % O_2 and gives the cycling conditions needed to achieve this rate (Reprinted from Zhou et al. (1986) with permission, © 1986, Pergamon Press PLC)

agreement with the time lags seen before surges in CO_2 production after a switch to O_2. Decreasing the temperature, decreases the improvement under periodic operation for supported Pd in contrast to Pt. The optimum period and optimum time of exposure to oxygen increase, however, as can be seen in Figure 3.5.

Vaporciyan et al. (1988) examined a platinum based bi-metallic catalyst, Pt-SnO_2, using once again the monolith support discussed above. Chloroplatinic acid in HCl was used to coat tin oxide. The precipitate was calcined and suspended in acetone. The acetone slurry was applied as a wash coat to the monolithic support. Prior to use, the coated monolith was reduced in flowing hydrogen for 3 h. Rates under steady state and periodic composition forcing were measured in the temperature range of 82 to 142°C. A wide range of cycle splits and periods were employed for the experiments. Lags or induction times were observed on switching and were similar to those seen for the Pt and Pd catalysts. The significant lag for the three catalysts followed a switch from CO + diluent to an O_2 + diluent mixture and varied from about 5 to 15 s depending on temperature. This temperature dependence suggests that an activated process must be occurring and rules out an Eley-Rideal mechanism involving gaseous oxygen.

CARBON MONOXIDE OXIDATION

Variation of cycle split and period leads to a three dimensional representation of the time averaged rate of CO oxidation that is shown in Figure 3.9. The maximum time average rate is about 7 × the optimal rate at steady state at 117°C. Similar diagrams could have been shown for the other temperatures investigated. Optimal time average and steady states rates are compared in Figure 3.10, while cycle splits and period which are optimal are plotted in Figure 3.5. Like the Pd catalyst, the optimal cycle split and enhancement achieved increase with temperature; the temperature influence, however is more pronounced. Vaporciyan et al. (1988) explain the cycle split effect in terms of strong CO adsorption which tends to saturate the surface with CO thereby sharply reducing the oxidation rate.

Vaporciyan et al. (1988) discovered in the course of varying cycle split and period a region of quasi-periodic behaviour shown by the blackened region in Figure 3.9. CO_2 production in this region became irregular as shown in Figure 3.11. Auto-correlation analysis disclosed the presence of two lower frequency variations: the first appearing every 12 cycles of the forcing period, while the second arose about every 175 cycles. The authors link the quasi-periodic phenomenon to storage of potentially reactive

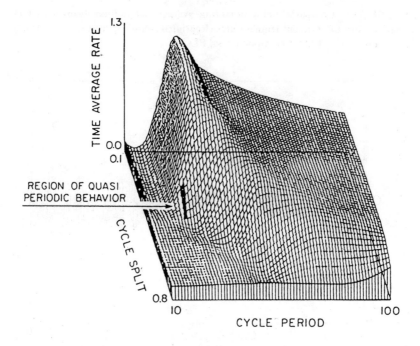

FIGURE 3.9 Time average rate of CO oxidation as a function of cycle split and period for switching between 10 vol % CO in N_2 and 10 vol % O_2 in N_2 over a Pt-SnO_2 catalyst at 117°C (Reprinted from Vaporciyan et al. (1988) with permission, © 1988, Pergamon Press, PLC)

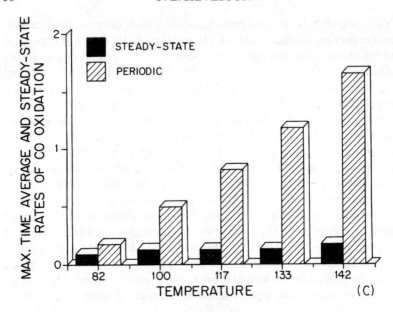

FIGURE 3.10 Comparison of optimal time average and optimal steady state CO oxidation rates for different temperatures (Reprinted from Vaporciyan *et al.* (1988) with permission, © 1988, Pergamon Press, PLC)

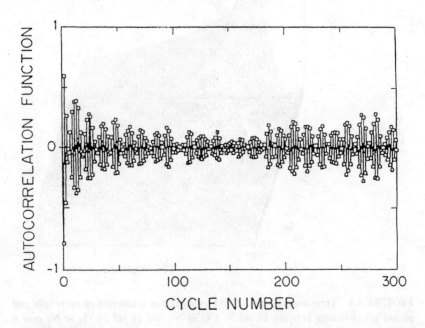

FIGURE 3.11 Results of autocorrelation analysis of quasi-periodic behaviour over a Pt-SnO$_2$ catalyst at 117°C (Reprinted from Vaporciyan *et al.* (1988) with permission, © 1988, Pergamon Press, PLC)

intermediates on the catalyst surface and suggest the storage may be associated with an Sn^0-Sn^{4+} transition for the 175 cycle periodicity. The 12 cycle periodicity was thought to have several possible explanations, among them Sn–Pt overlay and island formation. Experiments at other temperatures disclosed quasi-periodic behaviour in the same cycle split-period region. The behaviour was also seen with the 0.5 wt% Pt/Al_2O_3 catalyst but only the 12 cycle periodicity was observed.

Quasi-periodic response to concentration forcing arises when the compositions spanned by the forcing include a composition region where autonomous (or spontaneous) oscillation under steady state operation are observed. Vaporciyan et al. did not report autonomous oscillation for any of the steady state data they collected. Quasi-periodic response has been predicted through simulation (Kevrekidis et al., 1984; McKarnin et al., 1988) and observed for many systems including higher temperature CO oxidation over noble metal catalysts. It is just one of several responses that arise when systems exhibiting steady state multiplicity or autonomous oscillations are forced. This forcing situation will be discussed in some detail in a companion volume to this monograph. There is some further discussion several pages further on in this chapter.

Forcing of the autonomous oscillating CO oxidation on a Rh/SiO_2 catalyst was considered by Qin and Wolf (1995). These investigators were interested in stabilizing the spontaneous oscillations rather than observing the response. Experiments were performed at 117°C on a wafer prepared from the catalyst and mounted inside a controlled temperature IR cell. This arrangement permitted visualization of temperatures at points on the surface and from their observations Qin and Wolf concluded that autonomous oscillations result from an irregular re-ignition and quenching of a mobile hot spot on the wafer surface. When a sinusoidal variation of the O_2 and CO feed rates π radians out of phase is imposed, the quenching and the amplitude of the CO_2 variation are suppressed provided the forcing amplitude is large enough and a period of under 50 s is used. Forcing enhanced substantially the time average oxidation rate. Enhancements up to 30% were measured. Modulation appeared to reduce the turning point temperatures for extinction and ignition for the catalyst.

A comprehensive examination of the modulated oxidation of CO over a 0.5 wt. % Pt/Al_2O_3 is reported in a series of papers published by Graham and Lynch (1984, 1987, 1988 and 1990). A packed bed of catalyst was used with sufficient external recycle that the reactor behaved as a continuous stirred tank reactor (CSTR). Their reactor was operated isothermally at a total pressure of 0.1 MPa; temperatures were between 90 and 110°C. In their 1988 and 1990 papers, Graham and Lynch were concerned with the forcing strategy. In the first contribution, they compared a strategy manipulating just the CO concentration (between 0 and 2 vol%) to one switching both O_2 and CO concentrations. They observed that although the strategy of manipulating just CO changed the time behaviour of CO_2 production, it had just a small effect on the time average rate of CO_2 oxidation.

In their 1990 paper, Graham and Lynch examined the symmetrical cycling of both CO and O_2 at a phase lag of π radian over a range of cycle frequencies. Switching was undertaken between feeds with $CO:O_2 = 0:1$ and $2:0$ by volume. Double CO_2 production peaks within a cycle such as in Figure 3.2 were seen. Their shape and indeed the occurrence of the doublet depended on the cycle frequency. Results appear in Figure 3.12. The frequency of 0.00167 Hz in (a) allows a qualitative deconvolution of the peak pattern into regions. Increasing the frequency by about 3 × in (b) gives a time pattern of CO_2 production comparable to Figures 3.2, a further 50% increase in frequency in (c) drops the conversion to about 5% and the time variation vanishes. Distinct regions of the cycle in (a) are numbered consecutively beginning at 1 with the switch to a feed containing CO + diluent. The Pt surface has a high, but incomplete O atom coverage. CO entering the reactor adsorbs and quickly reacts with oxygen. As the reactor is well mixed, the surface is replenished with O_2 from the gas phase. As the

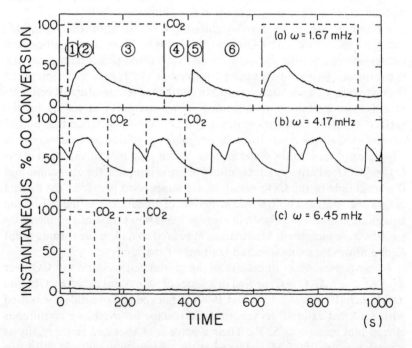

FIGURE 3.12 Experimental time variations of CO conversion or CO_2 production within a cycle for symmetrical cycling between $CO:O_2$ mixtures of 2:0 and 0:1 at 90°C over a 0.5 wt.% Pt/Al_2O_3 catalyst in a recycle reactor: (a) $\omega = 1.67$ mHz, (b) $\omega = 4.17$ mHz, (c) $\omega = 6.45$ mHz (Adapted from Graham and Lynch, (1990) with permission of the American Institute of Chemical Engineers, © 1990, *AIChE*)

CO concentration increases, the rate of CO_2 production rises rapidly. The change in slope at 2 signals the consumption of all the gas phase O_2. The rate continues to rise as the CO adatoms and O atoms achieve equal densities on the surface. Dropping CO_2 formation in 3 occurs as O adatoms are scavenged from the Pt surface which quickly becomes saturated with CO. Graham and Lynch note that CO adsorption is fast and that the CO_2 formation on the surface goes to zero shortly after the maximum. The apparent slow drop seen in 3 results from CO_2 desorption from the Al_2O_3 support. This support does not adsorb significant quantities of CO or O_2, but it has a large capacity for CO_2. The CO feed is turned off at the end of the 3 region and O_2 + diluent is re-admitted at the beginning of the 4 region in the figure. Dynamic CO adsorption-desorption makes sites available for O_2 adsorption, but this occurs to only a limited extent because two neighbouring sites are required. Nevertheless, some O_2 adsorbs and reacts. This is sufficient to end the drop in CO_2 production. CO remains in the gas phase due to mixing. As CO is flushed from the reactor, CO desorbs freeing sites and O_2 begins to adsorb in larger amounts. O atom islands form and grow explosively by reacting with the neighboring CO adatoms. This is the phenomenon indicated by 5. CO is scavenged from the Pt surface by reaction and, in 6, the rate of CO_2 formation goes to zero. The surge in CO_2 production in region 5 has recharged the Al_2O_3 support, so that desorption from the support accounts for the tailing off of the apparent CO_2 production in region 6.

Increasing the frequency does not affect the phenomena in regions 1, 2 and 5, but regions 3 and 4 are combined and shortened. Region 6 is also shortened. CO_2 storage on the support smooths out the rate of formation variations. In Figure 3.12(b), the shortening of the periods of zero CO_2 formation increases the time average rate of CO conversion. Further increases of frequency now interfere with replacement of O atoms by the CO adsorbate and vice-versa causing the rate to drop. At a critical frequency the adsorbate replacement process fails and the systems fall into steady state. This is a consequence of the recycle reactor used for the experiment. The lapse into steady state would be delayed to a much higher frequency if a plug flow system had been employed. An interesting result for this system is that the relaxed steady state is identical to the steady state at the time average feed composition.

The experimental behaviour with frequency is given in Figure 3.13. In this figure, the influence of the mass of catalyst, that is the ratio of the gas phase to catalyst surface storage capacity is examined because the feed rate is increased with the mass so that the space velocity per mass of catalyst is constant. It can be seen that increasing the catalyst to gas phase ratio has a large effect on conversion, but fails to alter the basic frequency behaviour which shows a strong resonance and a critical frequency after which the system falls back to steady state. Figure 13.13a show that the mass to gas

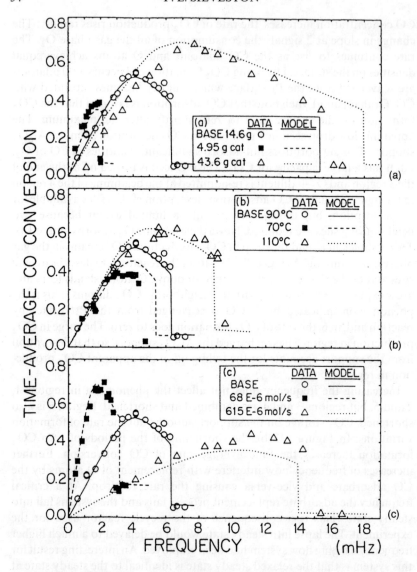

FIGURE 3.13 Effect of cycle frequency on time average CO conversion for symmetrical cycling between $CO:O_2$ mixtures of 2:0 and 0:1 over a 0.5 wt.% Pt/Al_2O_3 catalyst in a recycle reactor: (a) effect of m_{cat}/V ratio at 90°C, constant SV based on catalyst mass, (b) effect of temperature, (c) effect of SV with m_{cat} constant at 90°C (Adapted from Graham and Lynch (1990) with permission of the American Institute of Chemical Engineers, © 1990, *AIChE*)

phase ratio has a large influence on the critical frequency. This frequency increases as the catalyst mass to gas volume increases.

The effect of a 40°C temperature change can be seen in Figure 13.13b. Increasing temperature raises both the critical frequency and the frequency corresponding to the resonance maximum. Conversion is unaffected by temperature at low frequency, but the maximum conversion is temperature dependent. Maximum conversion, as enhancement above steady state, and the optimal frequency have been plotted in Figure 3.5. The Graham and Lynch optimal frequency data fall about an order of magnitude higher than the Barshad and Gulari (1985a) and the Muraki et al. (1985a) results, but show the same temperature trend. There is not enough data available from the Barshad and Gulari work to compare enhancement. It is probably significant that the temperature effect on enhancement is just the opposite to that observed by Gulari and co-workers for $Pt-SnO_2$ and Pd catalysts.

Figure 3.13c shows that flow rate or space velocity has a large effect on the conversion-frequency behaviour. Higher space velocity increases the optimal frequency and lower conversions. Space velocity and/or catalyst mass to gas volume ratio may account for the differences in optimal frequency just discussed. The lowest SV gave a maximum conversion of 71% and $\Psi = 4.7$. Graham and Lynch point out that the conversion is below conversion corresponding to the upper branch in the multiplicity region for the temperature, pressure and space velocity used. It is also below the best steady state conversion achievable with multiple reactors at the conditions used so that $\Psi^* < 1$.

The collapse to the steady state is predictable, according to Graham and Lynch, from the bifurcation map for the 0.5 wt% Pt/Al_2O_3 $CO—O_2—N_2$ oxidation system at 90°C (Graham and Lynch, 1987). The time average feed composition, $CO:O_2 = 1:0.5$ as vol%, is in the low rate region of the map as is one of the feed concentrations ($CO:O_2 = 2:0$); the other feed concentration (0:1) is in the high rate region. At low cycling frequencies, the reactor composition will approach the high rate feed mixture and this stabilizes operation in the high rate branch. Because the reactor is well mixed, as frequency increases, mixing reduces the range of compositions attainable. At the critical frequency, the composition no longer intrudes into the multiplicity region for a duration sufficient to lift the oxidation into the higher rate region.

Graham and Lynch (1990) also reported, for the first time, the effect of phase lag on performance with the periodic forcing of two reactant concentrations. They chose to describe the effect in terms of O_2 phase lead. This is defined as the ratio of the time the introduction of the O_2 rich feed leads the introduction of the CO rich feed divided by the duration of the O_2 cycle. For symmetrical forcing of two reactants, this ratio is always < 1. The definition of O_2 phase lead is illustrated by Figure 3.14. Multiplying the ratio by 360° converts the definition into a phase lag. Thus, if the CO

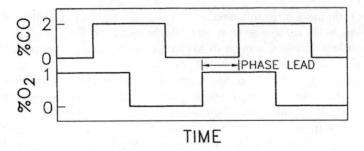

FIGURE 3.14 Definition of O_2 phase lead (Adapted from Graham and Lynch (1990) with permission of the American Institute of Chemical Engineers, © 1990, AIChE)

concentrations steps down as the O_2 concentration steps up, the ratio is 1/2 and the O_2 phase lead, ϕ, is 180°. This is the forcing mode that is considered in most of this chapter. Graham and Lynch (1990) carried out their phase lag experiments by switching a mechanically actuated 4 way valve between reservoirs of different $CO:O_2$ volume ratios, namely 2:0, 2:1, 0:1, and 0:0. Figure 3.15 demonstrates that phase lag can significantly affect CO conversion. Figure 3.15d shows that at 90°C and w = 0.00833 Hz the usual cycling mode with $\phi = 180°$ leads to a time average conversion similar to that observed in steady state operation. A like performance would be observed for $\phi = 30°$. Between 45 and 150°, conversion greater than 60% can be achieved and these reach almost 90% between 45° and 90°. The advantage of a small ϕ persists as the frequency is lowered. Figure 3.15 shows that for the $CO-O_2-N_2$ system with a 0.5 wt.% Pt/Al_2O_3 catalyst, $\phi > 180$ is never attractive; $\phi = 180°$ is always better than $\phi = 0$ and that for the experiments pursued by Graham and Lynch the optimal phase lead is between 45 and 90°. The optimal phase lead (or lag) varies only slowly with the forcing frequency.

Modeling

There have been several attempts to model the response to forcing discussed in the previous subsection. These efforts have led to experimental studies whose purpose has been to test the models proposed.

Modeling generally assumes that the oxidation proceeds via a sequence of elementary, two body reactions such as shown below as Equation (3.1). One approach has been to assume a single elementary step controls and that other steps are so fast that they can be assumed to be at equilibrium. Through often tedious algebra, a rate expression can be derived which is then introduced into the heat and material balances for the reactor under consideration. Langmuir adsorption/desorption expressions are inevitably used and models obtained in this way are loosely referred to as Langmuir-

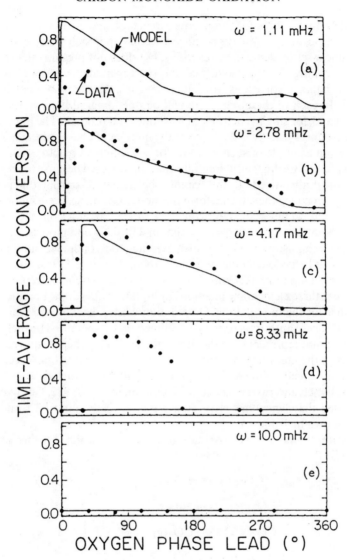

FIGURE 3.15 Effect of O_2 phase lead on time average CO conversion for various cycle frequencies with a time average feed composition $CO:O_2:N_2 = 1:0:0.5:98.5$ over a 0.5 wt.% Pt/Al_2O_3 catalyst at 90°C: (a) $\omega = 1.11$ mHz, (b) $= 2.78$ mHz, (c) $= 4.17$ mHz, (d) $= 8.33$ mHz, (e) $= 10$ mHz (Adapted from Graham and Lynch (1990) with permission of the American Institute of Chemical Engineers, © 1990, AIChE)

Hinshelwood models. These models are satisfactory for steady state systems, but are often unable to describe transient data. They do not predict the experimentally observed resonance of time average CO oxidation rate as the forcing frequency varies.

Models that successfully reproduce reactant and product behaviour under composition modulation utilize a second approach which continues to assume a set of elementary reactions, but does not presume that one is rate controlling. As a consequence, the reactor model consists of sets of material balance equations, usually, a set for each phase present in the reactor. In each of these sets there will be one material balance for each reactant and, for single reactions, at least one for a product. The number of energy balances usually is the same as the number of phases present. Table 3.1 gives an example of such a model for a backmixed reactor with three phases present. Ordinary differential equations are used because the reactor is spatially fully mixed. If this cannot be assumed, partial differential equations make up the sets that form the model. Of course, these models are more difficult to integrate. Unfortunately, models arising from the second approach are also referred to as Langmuir-Hinshelwood models.

Some mechanistic complexity must exist if models of the second type are to represent behaviour under periodic forcing. Jain *et al.* (1983a) observed that simple Langmuir adsorption-surface reaction systems as represented by the first three elementary reactions in Equation (3.1) do not predict rate resonance for the modulated CO oxidation when a back mixed reactor is assumed. These authors erroneously claimed that Langmuir-Hinshelwood models cannot represent modulated reactors, invoking thereby quite a controversy in the literature. At that time, one modeling contribution (Lynch, 1983) had already shown that resonance can be predicted from these models. Lynch and two earlier studies (Goodman *et al.*, 1982; Cutlip *et al.*, 1983) used the elementary step sequence given in Equation 3.1 and the model shown in Table 3.1.

Most low temperature CO oxidation models assume the sequence shown below (Graham and Lynch, 1984)

$$CO(g) + s \rightleftharpoons CO \bullet s$$
$$O_2(g) + 2s \rightarrow 2O \bullet s$$
$$CO\, s + O \bullet s \rightarrow CO_2(g) + 2s \qquad (3.1)$$
$$CO_2(g) + S \rightleftharpoons CO_2 \bullet S$$

where s is a surface site on a noble metal capable of bonding CO or O_2, while an S site, assumed to be on the support, can adsorb CO_2 only. Some investigators, such as Cho (1983), neglect adsorption of CO_2 on the catalyst support. Shanks and Bailey (1987) left out the CO adsorption step and assume reaction between gaseous CO and adsorbed oxygen. However, they considered a silver catalyst. Recent versions of the sequence allow for a change in surface activity as CO adsorption changes. This change is attributed to surface restructuring. The Langmuir adsorption assumption also has been dropped (Lynch *et al.*, 1986; Graham and Lynch, 1990).

CARBON MONOXIDE OXIDATION

Table 3.1 Kinetic Model for CO Oxidation as a Finely Divided Pt/Al$_2$O$_3$ Catalyst Mounted in a Fully Back Mixed Reactor (Lynch, 1983, 1984; Lynch et al., 1986; Graham and Lynch, 1984, 1987, 1990)

Gas Phase:

$$V\frac{dC_i}{dt} = Q_o(C_i)_o - qQ_oC_i - \left(\frac{S_pm}{\rho}\right)_{cat} r_i - \left(\frac{S_pm}{\rho}\right)_{Sup} r_i^s \quad i = 1, 2, 3$$

Catalyst Surface:

$$\frac{d\theta_i}{dt} = (v_{ad})_i r_i - r_3 \quad i = 1, 2$$

Support Surface:

$$\frac{d\theta_i^s}{dt} = (v_{ad})_i^s r_i^s \quad i = 3$$

Gas Shrinkage:

$$q = 1 - y_1 r_1 - y_2 r_2 + y_1 r_3 - y_1 r_3^s$$

Initial Conditions:

$$t = 0: \quad \theta_1 = 1, \theta_2 = 0, \theta_3^s = 0$$
$$C_1 = (C_1)_o, \quad C_2 = C_3 = 0$$

$r_i, r_1, r_2, r_3, r_i^s$ depend on the kinetics assumed. Lynch (1983) employed a classic Langmuir-Hinshelwood Model:

CO: $r_1 = S_t(k_a)_1 C_1(1 - \theta_1 - \theta_2) - S_t(k_d)_1 \theta_1; \quad r_1^S = 0$

O$_2$: $r_2 = S_t^2(k_a)_2 C_2(1 - \theta_1 - \theta_2)^2 - S_t(k_d)_2 \theta_2; \quad r_2^s = 0$

CO$_2$: $r_3 = S_t^2 k \theta_1 \theta_2; \quad r_3^s = (k_a^S)_3 C_3(1 - \theta_3^s) - (k_d^S)_3 \theta_3^S$

Goodman et al. (1982), Cutlip et al. (1983), and Graham and Lynch (1984) applied the set of elementary reactions given by Equation (3.1) to a fully back-mixed, isothermal reactor containing a finely divided, suspended catalyst. Dynamic response was provided in the model by including time derivatives for gas concentrations and fractional surface coverage of the adsorbed species. This model is presented in Table 3.1 in the version given by Graham and Lynch (1984). It contains six ordinary differential equations and an algebraic equation. Depending on the parameters chosen, the dynamic model can be stiff.

Using the dynamic model, Goodman et al. (1982) and later Cutlip et al. (1983) were able to predict experimental CO$_2$ partial pressure variations for CO oxidation at 100°C over 0.5 wt.% Pt on alumina under symmetrical bang-bang forcing of CO and O$_2$ in Ar mixtures. Prediction was unsuccesful if Langmuir-Hinshelwood rate models were employed, eventhough both the elementary step model (Eqn. (3.1) and Tab. 3.1) and the Langmuir-Hinshelwood models using a steady state rate expression successfully reproduced steady state rate measurements. Goodman *et al.* obtained parameters in the model from step chance observations.

Figure 3.16 shows the predictions of the model in Table 3.1 using arbitrarily chosen rate parameters (Lynch, 1983). Resonance is evident in (a) and the rate reaches roughly 10 × the steady state rate at the time average reactant composition. The CO oxidation rate approaches the quasi-steady state limit as the cycling frequency diminishes and steady state rate at the high frequency limit. A comparison of Figures 3.16a and 3.3 demonstrates that the dynamic model reproduces the shape of the oxidation rate resonance quite well after allowance is made for the difference in the abscissa for each figure. Period and frequency are inversely related to each other. The map of constant oxidation rate in Figure 3.16b for symmetrical forcing show that there are optimal cycling frequencies and amplitudes. The importance of amplitude has not been studied experimentally; cycle split, however, has been shown to be important. Optimal splits have been identified. It is reasonable, therefore, to compare Figure 3.16b with Figures 3.4, 3.8 and 3.9. The predicted confinement of the optimum to a narrow region of the parameter is borne out by the experimental observations.

Graham and Lynch (1984) describe cycling experiments performed to verify the dynamic model given in Table 3.1. A commercial, shell impregnated 0.5 wt.% Pt on γ-Al_2O_3 catalyst was employed in a packed bed with

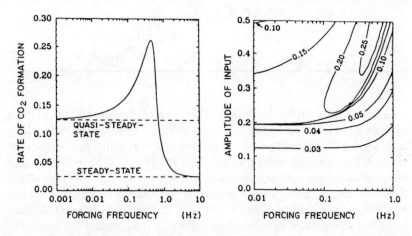

FIGURE 3.16 Predicted rates of CO oxidation as a function of cycle frequency and amplitude from Lynch's model using arbitrarily chosen parameters: a) predictions for a time average equimolar feed mixture of CO and O_2 and switching between $y_{CO} = 0.1$ and $y_{CO} = 0.9$; (b) lines of constant oxidation rate for the same time average feed mixture (Reprinted from Lynch (1983) with permission)

an estimated 28% dispersion of the platinum. Both steady state and concentration forcing experiments were performed at 90°C. The former disclosed a region of steady state multiplicity extending from 0.1 to 1.0 mole % O_2 and 0.1 mole %CO to 1.2 mole % CO. Cycling experiments were carried out just outside of this region by employing time average CO and O_2 mole % of 1.0 and 0.5, respectively. Cycles were symmetrical and switching took place between a feed of 2.0 mole % CO in N_2 and a CO free feed of 1.0 mole % O_2 in N_2. Conversion was followed using a dedicated IR set to the CO_2 absorption line at 2,355 cm^{-1}. Resonance was observed as the cycling frequency varied. At a normalized frequency of about one, the time average CO oxidation rates were close to the rates measured in the lower branch of the multiple steady state region. Decreasing the frequency raised the time average rate quickly and maxima were observed at normalized frequencies between 0.4 and 0.8 depending on the amount of catalyst placed in the reactor. The time average rate maxima were between 10 and 22 × greater than the steady state rates at the corresponding time average reactant composition. As the frequency was reduced further, the time average CO oxidation rates declined gradually towards the quasi-steady state limit of zero for the CO + diluent to O_2 + diluent switching experiment performed.

Parameters for the Lynch model were evaluated from the steady state measurements and step change experiments with a CO_2 and N_2 stream. The composition modulation data were not exploited. With parameters evaluated in this way, the model represented steady state results well. Predictions for the modulated reactor were less satisfactory. Experimental and predicted variations of the CO_2 concentration in the reactor outlet are compared in Figure 3.17 for normalized frequencies between 0.28 and 0.76. At the later frequency, the time average CO oxidation rate was about 12x the steady state rate at the corresponding time average feed composition. In general, the simulation captured the shape of the CO_2 variation with time, but details of the variations were not reproduced. Similarly, the shape of the time average rate of CO oxidation *vs* cycling frequency was predicted by the model, but the maximum rates were just 60 to 70% of the measured values.

The periodic switching modeled by Lynch (1983) used CO concentrations which were inside the region of multiplicity predicted by the model. Hudgins and Silveston (1985) re-worked the controversial arguments of Jain *et al.* (1983a) and claimed that models based on an elementary adatom reaction and Langmuir adsorption expressions, namely Langmuir-Hinshelwood models, cannot predict rates which exceed the maximum steady state rate at the optimal composition of the reactants provided temperature and total pressure are not changed. Whether or not the rate increases for CO oxidation under composition modulation can be explained by compelling the reaction system to operate at the high rate when multiple steady

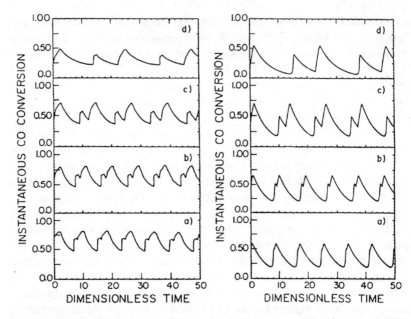

FIGURE 3.17 Experimental and predicted instantaneous CO conversions at different normalized frequency for switching at 90°C between 2.0 mole % CO in N_2 and 1.0 mole % O_2 in N_2 for a 0.5 wt.% Pt/Al_2O_3 catalyst. Model parameters and experimental conditions are given by Graham and Lynch (1984): (a) $\omega = 0.76$; (b) $\omega = 0.69$; (c) $\omega = 0.51$; (d) $\omega = 0.28$. (Reprinted from Graham and Lynch (1984) with permission, © 1984, Elsevier Science Publishers)

states are possible is a serious question. It seems likely that this is not the explanation for the results of Zhou et al. (1986) mentioned in the preceding subsection, but the dynamic model shown in Table 3.1 implies that this is the explanation.

Several years before the controversy on the suitability of Langmuir-Hinshelwood models for predicting reactor performance under composition modulation, Oh et al. (1978) demonstrated experimentally using an egg-shell impregnated Pt/Al_2O_3 catalyst pellet (1.7 cm diameter) "aged" for 1000 h in automotive exhaust and a feed of 0.3 vol%CO, 2 vol% O_2 (balance N_2) that abruptly dropping the CO vol% pulse-wise for various durations raised conversion to a higher level. The level depended on the amplitude of the negative CO pulse and its duration. For a large enough amplitude and a duration of at least 3 s, the change in conversion corresponded to a switch between the higher and lower rate branches in the region of steady state multiplicity identified experimentally by Hegedus et al. (1977). This is shown in Figure 3.18. After the negative pulse, conversion remains at the higher level.

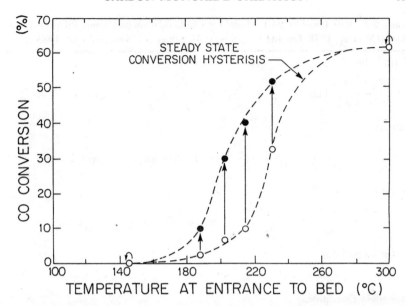

FIGURE 3.18 Steady state CO conversion prior to and following a negative CO pulse of 3 s duration from $y_{CO} = 0.316$ to $y_{CO} = 0.108$ and return for an automotive exhaust aged 0.06 wt % Pt/Al$_2$O$_3$ catalyst [Reprinted from Oh et al. (1978) with permission, © 1978, American Chemical Society]

To interpret their experimental results, Oh et al. constructed a model of a porous catalyst particle located within a CSTR. This model neglected CO_2 adsorption on the catalyst support. Because of this, only material balances for the reactants are needed and there are just two phases. Table 3-2 presents the model. Concentrations in the particle vary with depth so a distributed parameter model must be used which because of time variations is in the form of a partial differential equation. This model has also been adopted by Oh et al. (1980), Oh and Hegedus (1982) and Cho (1983).

Employing the dynamic model given in Table 3.2, and assuming inhibition kinetics with equilibrium adsorption of O_2 and CO to obtain the rate term,

$$r_{CO_2} = \frac{k(T) C_{CO}^s C_{O_2}^s}{(1 + K(T) C_{CO}^s)^2} \tag{3.2}$$

Oh et al. (1978) were able to reproduce their experimental results. The model shows isothermal multiplicity in some ranges of CO and O_2 concentrations. If a pulse represents square wave composition forcing, some of the cycling results over the noble metals can be explained by stabilization of the

Table 3.2 Reaction-Diffusion Model for CO Oxidation on Porous Pt/Al$_2$O$_3$ Catalysts (Oh et al., 1978; Oh and Hegedus, 1982; Oh et al., 1980, and Cho, 1983).

Intra Solid:[1]

$$\varepsilon_p \frac{\partial C_i^s}{\partial t} = \frac{1}{r^2}\frac{\partial}{\partial r}\left((D_{\text{eff}})_i r^2 \frac{\partial C_i^s}{\partial r}\right) - \frac{S_p}{i}[(k_a)_i C_i^s (S_t - S_1 - S_2)^i - (k_d)_i S_i] \qquad i = 1, 2$$

Surface:

$$\frac{\partial S_i}{\partial t} = (k_a)_i C_i^s (S_t - S_1 - S_2)^i - (k_d)_i S_i^i - k S_1 S_2 \qquad i = 1, 2$$

Reactor:[2]

$$V\varepsilon_B \frac{dC_i^g}{dt} = Q((C_i^g)_o - C_i^g) - 4\pi R^2 (k_s)_i (C_i^g - (C_i^s)_{r=R}) \qquad i = 1, 2$$

Initial Conditions:

$$C_i^g = C_i^s = (C_i)_o \quad S_i = K(C_i)_o \qquad i = 1, 2 \quad (0 \leq r \leq R)$$

Boundary Conditions:

$$\frac{\partial C_i^s}{\partial r} = 0 \quad (D_{\text{eff}})_i \left(\frac{\partial C_i^s}{\partial r}\right)_{r=R} = (k_s)_i (C_i^g - (C_i^s)_{r=R})$$

$$\text{for } t \geq 0, \quad i = 1, 2$$

Feed Condition:

$$(C_i^g)_o = (\bar{C}_i^g)_o \pm A\,\omega(t)$$

---

[1] Spherical geometry; Oh and Hegedus (1978) treat a plate geometry; Oh et al (1980) treat spherical particles with catalyst deposition in bands.
[2] Single spherical particle suspended in a well mixed reactor of volume V; Oh et al. (1978) treat an integral flow reactor.

---

oxidation rate at a level corresponding to the high rate branch if multiplicity is possible at one of the compositions visited during cycling. In this case the rate or conversion improvement cannot not exceed the maximum rate under steady state operation.

Use of the model given in Table 3.1 with rate expressions assuming the reactions in Equation (3.1) are elementary were not able to reproduce the peak shape and timing in Figure 3.12 or the time average conversions in Figure 3.13. Modification of the model to allow for next neighbour interaction improved representation of the data. This modification can be written as an exclusion factor which reduces O$_2$ and CO adsorption. For CO the rate equation in Table 3.1 becomes

$$r_2 = S_t (k_a)_1 C_1 (1 - \theta_1 - \theta_2)\left(\frac{1 - \gamma_1 \theta_1}{1 - \theta_1}\right) - S_t(k_d)_1 \theta_1 \tag{3.3}$$

CARBON MONOXIDE OXIDATION

This type of a model has good support in the literature, according to Graham and Lynch (1987), but the modification improved the fit to the data only marginally for $1.0 < \gamma_1 < 1.2$. A surface reconstruction modification was also explored. It has been observed that the Pt (100) face changes from a quasi hexagonal structure (CO free surface) with a low sticking probability for O_2 to a 1×1 structure (normally found) when the CO coverage reaches a critical value. The 1×1 phase exhibits much higher O_2 sticking coefficients. It was also assumed that the reconstruction affects the reaction rate to the same extent. The surface reconstruction modification was handled by replacing (k_a), in the CO_2 rate equation, Table 3.1, by $(k_a) \delta$, and k in the CO rate equation by $k\delta$. The sticking coefficient δ took on two values depending on θ_1. With $\gamma_1 = 1.004$, $\delta = 1.250$, while for $\theta_1 = 0.95$, $\delta = 0.10$. The modified model provides the dotted, dashed and solid curves in Figure 3.13. This is a remarkably good fit of the experimental data. The dynamic model was also applied by Graham and Lynch (1990) to their phase lag experiments. The solid lines in Figure 3.15 show predictions of the model just discussed. They are quite good except at the extremes. The model predicts no effect of ϕ at $\omega = 0.00833$ Hz, while at $\omega = 0.00111$ Hz, the ϕ effect is over predicted.

In summary, it appears that a reasonably good model is available for describing composition modulation of low temperature CO oxidation over Pt/Al_2O_3 catalysts. The model does not assume island formation, but it employs a non-linear adsorption isotherm and allows for surface modification that is expressed as two levels of adsorption and rate constants. The success of the model implies that rate enhancement through modulation can be explained in terms of the high rate branch when multiplicity is present in the kinetic or reactor system.

It remains to be seen if the dynamic model is successful with oxidation over Pd catalysts. Island formation appear to apply for this and perhaps other catalysts.

Forcing of Autonomous Oscillations

Experiments of Qin and Wolf (1995) have been mentioned earlier. This is only one of a number of studies of composition modulation of CO oxidation over noble metal catalysts under condition where the reaction rate oscillates spontaneously. We have separated our discussion of these studies because Qin and Wolf reported rate measurements, while the remainder of the contributions on this subject are concerned with the excitation diagram, that is, the response of the rate of oxidation or of CO conversion as the forcing amplitude and frequency change.

The information conveyed by an excitation diagram and the wide range of response possible when an autonomously oscillating reaction system is forced is illustrated by Figure 3.19. In the diagram, ω is the forcing

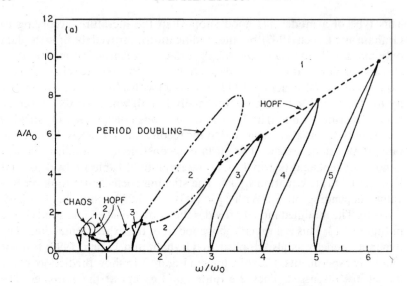

FIGURE 3.19 (a) Excitation diagram for an autonomously oscillating surface reaction under sinsusoidal forcing of reactant partial pressure. Numbers, n, indicate regions of a stable response of period $n\tau_o$ (Figure adapted from McKarnin et al. (1988) with the permission of Elsevier Science Publishers and the Royal Society, the copyright holders, © 1988)

frequency, while ω_o is the natural frequency of the oscillating system. A represents amplitude and the subscript o on A refers to the mean amplitude of the autonomous oscillations. The diagram shows regions extending upwards from the axis, called entrainment horns or tongues, where the period of the response is an integer multiple of the natural frequency. In between these regions, the response is quasi or aperiodic, that is, more than the forcing period is present in the response. The diagram shows that if the forcing amplitude is large enough, the natural frequency is swamped out. Small horns can also be seen at frequency ratios which are rational numbers, such as 1.5. Chaotic response is also possible, but only in very restricted regions of the amplitude-frequency space. The diagram shown is taken from McKarnin et al. (1988) and describes the forcing of a surface reaction consisting, essentially, of the first three reactions in Equation (3.1). Most experimentally determined excitation diagrams resemble Figure 3.19, although the entrainment horns can be swept toward the left as well as toward the right and regions of chaos can have different locations.

Low temperature forcing of CO oxidation over Pt/Al_2O_3 was investigated by Capsaskis and Kenney (1986), however, propene was introduced to facilitate spontaneous oscillation. 3 mm × 3 mm particles of 0.5 wt.% Pt on Al_2O_3 were used as a shallow bed in a well mixed micro-reactor. Argon served as the diluent and oxygen was maintained at 3 vol.%. Autonomous oscillations were observed at 136.5°C for a feed containing 4 vol.% CO and 1% C_3H_6. The natural period was about 180 s. Decreasing the propene concentration to 0.5 vol.% destroyed the oscillations. However, raising the reactor temperature to 150°, reinstated the oscillations for 0.5 vol.% C_3H_6 and 2% CO, but the period fell to 27 s. Temperature fluctuations were also observed with 1 vol.% C_3H_6, but were irregular and the period, between 7 and 14 minutes, was sensitive to the gas residence time (Cutlip et al., 1983; Capsaskis & Kenney, 1986).

Periodic forcing of the $CO-O_2-C_3H_6-Ar$ system was carried out at both 135 and 150° by switching symmetrically between 0.5 and 1 vol.% C_3H_6 or between 2 and 4 vol.% CO. At 135°, for propene switching and CO at 4%, three forcing periods were used: 50 s, 30 s and 20 s. The width of the autonomous oscillations region is about 0.5 vol.% C_3H_6 so the amplitude ratio, A/A_o, for forcing is about unity. For the first forcing period, corresponding to $\omega/\omega_o = 3.5$, the response resembled the natural oscillation in shape and was periodic at the forcing period. Entrainment occurred. Decreasing the forcing period to 30 s, or using $\omega/\omega_o = 6$, led to period doubling, while at $\tau = 20$ s or $\omega/\omega_o = 9$, the oscillations were damped out.

Capsaskis and Kenney repeated their experiments at 150°C. The natural frequency drops to 27 s at the higher temperature so the range of ω/ω_o becomes 0.54 to 2.7. Entrainment occurred at all the frequencies used so the response oscillated at the forcing period. When the forcing period was increased to 60 s, the response became quasi-periodic or chaotic. Cycling at 150°, under the conditions used, crossed out of the limit cycle bounds so that perhaps the forcing was near a Hopf bifurcation.

A second set of forcing experiments was conducted at 135° holding C_3H_6 constant and symmetrically switching the CO concentration. Cycle periods ranged from 20 to 40 s. Only period one response (entrainment at the forcing frequency) was observed even though switching took the system out of the steady state limit cycle region. The third set of experiments allowed both C_3H_6 and CO to vary. Once again, this type of switching brought the system outside of the limit cycle region for part of the cycle. Only a response at the forcing period was observed at both 135 and 150°C. In one experiment, the forcing period was lowered to 10 s. This relatively high frequency obliterated the oscillatory response.

An isothermal model formulated by Mukesh et al. (1983) was applied to the system studied by Capsaskis and Kenney. It made use of the elementary reactions given as Equation (3.1) and added propene balances. Rate terms in these balances assumed the reversible competitive adsorption of propene on two adjacent sites and the irreversible complete oxidation of this adsorbate by O adatoms. This model was fitted using steady state and step change measurements (Cutlip et al., 1983). Only qualitative agreement was found between the model and the experiments: period one entrainment, period doubling, quasi-periodic and disappearance of oscillations at high frequency were all predicted but not always under the conditions at which they were observed experimentally. The model predicted period tripling and quadrupling, but these were not observed. Comparing the observed responses to the excitation diagram shown in Figure 3.19, the experimental entrainment horn near 1/1 seems to be much wider than the horn predicted by the surface model. This difference could be due to the different dimensionality of the experimental system, apparently dimension 4 (4 dependent variables), and the 2 dimensional model used by McKarnin et al. (1988).

The studies mentioned above and continued in the next section sound a caution to those wishing to apply composition modulation. If multiplicity is possible in the reaction system or in the reactor, attention should be paid to the time behaviour of the reaction products. Complex periodicity can be observed. Limit cycles or autonomous oscillations are often associated with multiplicity.

C. SINGLE CRYSTALS OF Pt UNDER HIGH VACUUM

Studies of composition modulation of CO oxidation on specific cuts or faces of Pt single crystals appear primarily in the chemical physics or physical chemistry literature. It is a large literature that has been subjected to a thorough review (Ertl, 1990). Unlike the viewpoint in this mongraph, the focus of these studies has been preparation of excitation diagrams, mechanisms resulting in spontaneous oscillation, or synchronizing the millions of active sites on a face, each of which can function as a single oscillator.

Eiswirth and Ertl (1988a) prepared an excitation diagram at 525 to 530°K over Pt(110) by holding p_{CO} at 2.1 to 2.8 × 10^{-3} Pa and modulating p_{O_2} around a mean of 4 or 5.5 × 10^{-3} Pa. This diagram appears as Figure 3.20. Note that in this diagram, the abscissa is the ratio of the forcing period to the natural period, that is the inverse of Figure 3.19. The amplitude is expressed as a percentage of the time average partial pressure rather than as a ratio with respect to range of the variable for steady state limit cycle behaviour. Resonance horns or tongues of the frequency entrainment regions are evident; these are separated by regions of quasi-periodic response. The response, variation of CO formation, in the 1/1 horn oscillates at the forcing frequency. In the transition from the natural frequency under self oscillation to the forcing frequency in this horn, there is a substantial increase in amplitude of the response.

Figure 3.21 shows the observed response in the horns at 1/2 and 2/1 and in the quasi-periodic region to the right of the 2/1 horn in the previous figure after the stationary cycling state is reached. The forced oscillation of

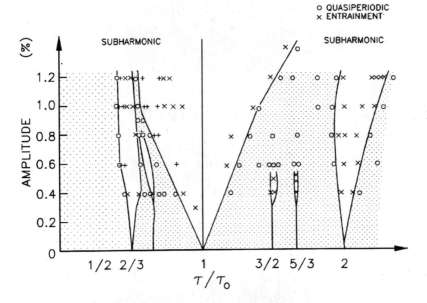

FIGURE 3.20 Excitation diagram for sinusoidal forcing of oxygen partial pressure for CO oxidation over a Pt(110) single crystal under conditions of autonomous oscillation (Figure from Eiswirth and Ertl (1988a), adapted with permission, © 1988, Academic Press)

the oxygen partial pressure appears above each response. Period doubling is evident when $\tau/\tau_o = 1/2$, while in Figure 3.21b at $\tau/\tau_o = 2/1$, there are two peaks in the response for each peak in the forcing although the peaks have different magnitudes. There are apparently two periods in the quasiperiodic region (Figure 3.21c). There is a superharmonic response of varying magnitude with about 25 peaks or cycles for 12 cycles in the forcing partial pressure. Superimposed on this is a subharmonic oscillation of low amplitude in the response magnitude with a period about 12 × the period of the forcing function. The irregularity of this subharmonic may signal the presence of an additional subharmonic having even a larger period.

The experimental excitation diagram in Figure 3.20 for CO oxidation resembles the low amplitude region of Figure 3.19 which models a surface reaction between two adsorbate molecules. Better agreement between experimental and model excitation diagrams has been achieved using a model for CO oxidation on the Pt(110) face (Ertl, 1990; Krischer *et al.*, 1991). This model incorporates the first three elementary reactions in Equation (3.1). Like the Graham and Lynch model, surface reconstruction is assumed to occur. In the Ertl model, this begins when the CO coverage exceeds 20% and is complete when 50% is reached. The mathematical model for this

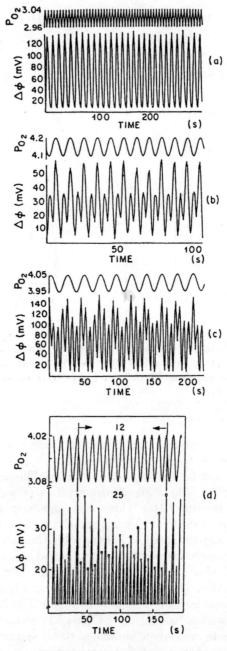

FIGURE 3.21 Stationary cycling response to oxygen partial pressure forcing of CO oxidation over the Pt(110) face in different regions of the excitation diagram: (a) in the 1/2 horn with $\tau/\tau_o = 1/2$, (b) in the 2/1 horn with $\tau/\tau_o = 2.275$, (c) in the quasi-periodic region with $\tau/\tau_o = 2.135$. Upper part of the figure shows 2 × the amplitude as the p_{O_2} variation with time (Figure adapted from Eiswirth *et al.* (1989) with permission, © 1989, *Surface Science*)

mechanism assumes modified Langmuir isotherms in which adsorption depends on the ratio of the fractional coverage to the fractional coverage on saturation by the adsorbate. Adsorption is non-dissociative for CO but requires the concerted action of three to four adjacent vacant sites, whereas it is dissociative for O_2 requiring two adjacent vacant sites. Reconstruction is kinetically controlled and proceeds at a rate proportional to the fractional CO coverage. It is reversible. Oxygen has different sticking coefficients on the original and on the reconstructed surfaces. Modelling of the mechanism has also been undertaken using a cellular automata approach (Eiswirth et al., 1989).

A variety of mechanisms have been proposed to explain self oscillations of CO oxidation on Pt single crystals. Schüth et al. (1993) discuss these critically. They note that the surface reconstruction mechanism has to date the most experimental support.

Forcing of autonomous oscillations on a Pt(100) single crystal have also been undertaken (Schwankner et al., 1987). This face exhibits irregular oscillations when not forced. Under modulation of the O_2 partial pressure, phase locking occurs and a remarkably regular oscillation is established. The effect of forcing frequency and amplitude were investigated by Schwankner et al. using work function and LEED measurements, the latter to follow surface transformations. Fourier transforms of the time series obtained from the work function produced a power spectrum which was used to characterize the dynamic response. Entrainment at 1/1, period doubling and quasi-periodic behaviour was seen at low amplitude as the modulation frequency increased. Phase shift between the response and the O_2 modulation increased with the forcing frequency. In an interesting experiment, Schwankner et al. distorted the forcing sinusoid and concluded that wave shape influences response. Temperature modulation, feasible with a single crystal, was also used and responses similar to those found with O_2 modulation were recorded.

D. NOBLE METAL CATALYSTS UNDER HIGHER TEMPERATURES

Rate Enhancement

Composition modulation studies of CO oxidation above 250°C over the noble metal catalysts have been undertaken in several cases because this oxidation is one of the important reactions in automotive exhaust catalysis. The CO content of the exhaust fluctuates with a dominant frequency of about 1 Hz. This modulation is not intentional, but comes about because of the control system used to maintain the exhaust composition within the optimal operation window for the noble metals based three way catalysts.

In this section we consider measurements and modelling of the periodic composition forcing of the CO—O_2-diluent feed to an oxidation reactor. The next chapter deals with automotive exhaust catalysis under composition modulation and continues the discussion begun here; however, the chapter examines measurements made with artificial or actual exhaust mixtures so that other reactions besides CO oxidation occur.

In their studies of composition modulation, Abdul Kareem et al. (1981) and Silveston et al. (1983) were interested in the influence of the catalyst on the behaviour of catalytic reactions to periodic composition forcing rather than in automotive exhaust catalysis. Pt was just one of several catalysts investigated. Experiments were performed on a commercial 0.1 wt% Pt on γ-Al_2O_3 catalyst. Concentrations of CO and O_2 were varied so as to create two time average ratios with respect to volume: 0.36 and 0.86. Switching was undertaken between two CO/O_2 ratios, such as 0.12 and 0.60, for the smaller time average ratio (0.36). The same amplitude, 0.24, was also used for the higher time average ratio. Experiments were performed at 247°C and cycling was symmetrical. The enhancement vs. cycle period measurements are shown in Figure 3.22. Data collected at a time average ratio of

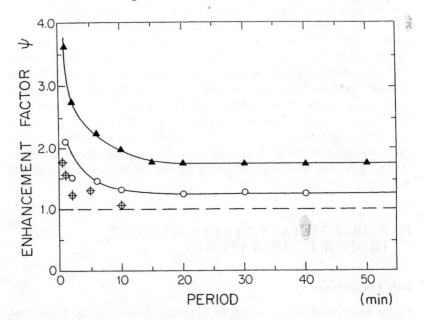

FIGURE 3.22 Effect of cycle period on the time average rate of CO oxidation over a 0.1 wt.% Pt on an alumina catalyst at two time average CO/O_2 ratios and 247°C, with switching between CO/O_2 ratios of 0.12 and 0.6, and 0.62 and 1.1. Rates have been normalized with the steady state rates at the time average CO/O_2 ratios of 0.36 and 0.86 to give the enhancement factor. ▲, ○ = Data of Abdul-Kareem et al. (1981); ⊕ = Data of Roussel (Silveston et al., 1983) (Reprinted from Silveston (1991))

0.86 were obtained by different researchers working about a year apart (Abdul-Kareem et al., 1981; Silveston et al., 1983). Enhancements are smaller than those seen for 0.5 wt% Pt/Al_2O_3 at lower temperatures in Figures 3.3 and 3.4. However, Figure 3.22 indicates cycle periods required for large rate increases are under 60 s, in agreement with Figure 3.5.

Abdul-Kareem et al. (1981) point out that the time average rate increase when cycling around a time average ratio of 0.36 can be explained by multiplicity considerations, that is, the operation periodically brings the system to ignition. High rates at the ignition composition spill over into the regions of lower rates where the reaction is CO inhibited to give a mean behaviour well above that observed under steady state or even quasi-steady state at the time average feed composition. Quasi-steady state seems to be attained when the period exceeds 20 minutes and the rates observed under this state could be predicted from steady state data reported by Chakrabarty et al.(1982). However, multiplicity cannot account for the significant rate increase found cycling around a time average ratio of 0.86. In this case switching was carried out for ratios well outside of the multiplicity region.

The influence of the catalyst on performance under periodic composition forcing was also addressed by Muraki et al. (1985a), but as part of a broad study of the individual reactions occurring in automotive exhaust catalysis. Muraki et al. (1985a) investigated composition modulation of CO oxidation over 5 different noble metal catalysts supported on a low surface area (11 m^2/g) α-Al_2O_3. Most experiments were performed on catalysts containing 0.006 wt.% metal. Particle size was 2 mm, but the low surface area probably limited diffusional interference. Figure 3.23 plots the time average CO conversion against cycle period for different temperatures. Because conversion at $\tau = 0$ is the steady state conversion at the time average feed composition (1.33 vol.% CO and 0.665 vol.% O_2), the figure also compares steady state and periodic operation. The cycle periods correspond to forcing frequencies of 0.5 to 5 Hz. Composition forcing was symmetric ($s = 0.5$) and the stoichiometric number (SN) at the time average feed composition was one. This stoichiometric number represents a stoichiometric CO, O_2 mixture. Amplitude was 0.665 vol.% for CO and 0.3325 vol.% for O_2; the space velocity was held constant at 30,000 h^{-1}.

At the lowest temperature used and with the exception of Iridium (Ir), periodic forcing caused a 2 × to 2.5 × conversion enhancement. Increasing the temperature by 50°C sharply decreased the enhancement. For Palladium (Pd), enhancement all but disappears. On the other hand, the increase in conversion over steady state for Ir is about 125%, roughly the same as was observed at the lower temperature. A further increase of 50°C eliminates the conversion advantage of modulation for Pt, Pd and rhodium (Rh), although modest conversion enhancements remain for ruthenium (Ru) and Ir. These are just 10%. Cycle periods corresponding to the maximum conversion enhancement, and thus, the maximum rate of oxidation de-

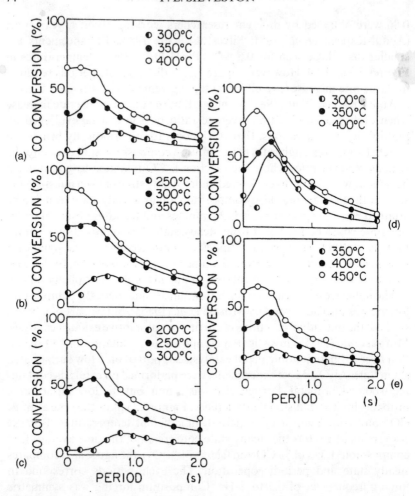

FIGURE 3.23 Effect of cycle period and temperature on CO conversion over (a) Pt, (b) Pd, (c) Rh, (d) Ru, (e) Ir catalysts with cycling at SN = 1 (Reprinted from Muraki *et al.* (1985a) with permission, © 1985, Nippon Kagaku Kaishi)

crease with temperature. Data points from Figure 3.23 are plotted in Figure 3.5 and it can be seen that the trends of the low and higher temperature data sets are similar.

Muraki *et al.*(1985a) investigated the effect of the stoichiometric number (SN) and CO concentration on conversion under composition forcing for the Pd catalyst. Their results are shown in Figure 3.24. As SN increases and the mixture becomes oxidizing, conversion improvement through modulation disappears. Lower SN increases the enhancement through forcing.

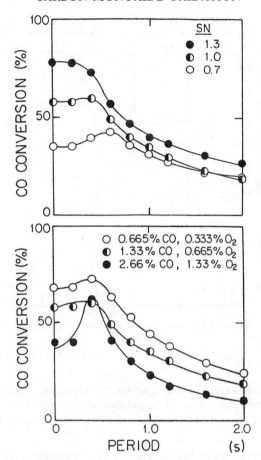

FIGURE 3.24 Effect cycle period on CO conversion of a 0.006 wt.% Pd on Al_2O_3 catalyst at 300°C (a) influence of stoichiometric number, (b) influence of CO mol percent at SN = 1 (Reprinted from Muraki *et al.* (1985a) with permission, © 1985, Nippon Kagaku Kaishi)

Raising the time average CO concentration in the feed from 0.665 vol.% to 2.66 vol.% sharpens the resonance as may be seen in Figure 3.24. The time average SN was 1.0 for this experiment. Bed temperature was 300°C and SV = 30,000 h^{-1}.

CO reacted rapidly with adsorbed oxygen, but Muraki and co-workers observed an induction period when oxygen was re-introduced. This is shown in Figure 3.25 for the Pd catalyst, a cycle period of 20 s, a bed temperature of 85°C, and for switching between CO and O_2 plus diluent mixtures around SN = 1. Higher frequency forcing leads to the CO_2 evolution pattern shown in Figure 3.12a. The induction period depends strongly on

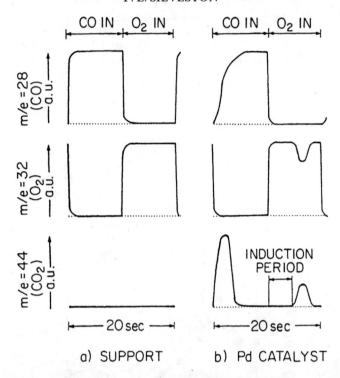

FIGURE 3.25 Mass spectrometer signal for CO, O_2 and O_2 during composition forcing for a 0.006 wt.% Pd on Al_2O_3 catalyst at 85°C with SN = 1 (Reprinted from Muraki et al. (1985a) with permission, © 1985, Nippon Kagaku Kaishi)

catalyst temperature and falls to less than 1 s as the temperature approaches 200°C. It becomes greater than 5 s at 70–80°C.

Muraki et al. comment that their results may be explained by CO inhibition of the oxidation at 200–300°C for the noble metal catalyst. Composition modulation increases O_2 adsorption by periodically decreasing CO coverage of the metal surface. They support their argument by adsorption measurements on clean metal surfaces using heavily loaded catalysts (0.54 wt.% metal). At room temperature the amount of CO adsorbed is 20 to 100% greater than the amount of O_2 adsorbed, whereas at 400°C CO adsorption is much reduced and the amount of oxygen adsorbed becomes larger than the amount of CO. Once adsorption of O_2 matches CO adsorption, the oxidation rate is high. Periodic composition forcing in this situation reduces CO and O_2 adsorbed and thus suppresses the rate.

The importance of CO inhibition in the kinetics of CO oxidation on the noble metals indicates that this oxidation is controlled by the reaction between adsorbed species and suggests these species must respond rapidly to gas phase fluctuations if concentration forcing at about 1 Hz effects rate and CO conversion. This was tested by Oh and Hegedus (1982) using fast response IR spectroscopy with a wafer made of 1 wt.% Pt/Al_2O_3 powder mounted in a transmission cell heated to 450°C. Oh and Hegedus employed the 2070 cm^{-1} line, corresponding to Pt-CO stretching of linearly adsorbed CO, and found from step change forcing and monitoring CO concentration that the transmittance at 2070 cm^{-1} followed CO conversion closely. Transmittance response to cycling between 1 vol.% CO and 1.03 vol.% O_2 (oxidizing) and 1 vol.% CO and 0.23 vol.% O_2 (reducing) at 1 Hz appear in Figure 3.26. These measurements show that adsorbed CO closely tracks the gas phase modulation. The two curves depict the behaviour starting with surfaces initially exposed to an oxidizing or a reducing gas phase. The initial condition dies away after just 3–4 cycles. Similar results are shown in the next chapter for this line, but using an automotive exhaust. Oh and Hegedus comment that adsorption/desorption is rapid enough that equilibrium can be assumed at 450°C, however, diffusion into the wafer or powder particles can significantly influence the response of the surface to transients.

Catalytic muffler considerations led to a study of CO oxidation on alumina supported rhodium that used transmission FTIR to explore the oxidation mechanism and refine the kinetic model for CO oxidation over this catalyst (Prairie *et al.*, 1988). Some composition modulation experiments were performed as part of this study at 300°C by switching between

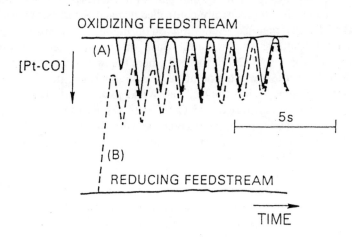

FIGURE 3.26 IR transmittance signal at 2070 cm^{-1} for composition forcing at 1 Hz, 450°C over a 1.0 wt.% Pt/Al_2O_3 catalyst. Curve A starting from a surface exposed to an oxidizing mixture; curve B starting from a surface exposed to a reducing mixture (Reprinted from Oh and Hegedus (1982) with permission, © 1982, American Chemical Society)

a lean stream (0.25 vol.% CO and 0.25 vol.% O_2) and a rich stream containing 1.4 vol.% CO and 0.25 vol.% O_2. The catalyst was 0.3-0.5 wt.% Rh supported on a fumed alumina. Impregnation used a $RhCl_3$ solution and the catalyst was calcined thereafter at 500°C. Prior to use, the catalyst was first oxidized and then reduced at 450°C. H_2 chemisorption gave a Rh dispersion of 49%. Bursts of CO_2 production were seen after switching, similar to the pattern shown in Figure 3.2b. With the rhodium catalyst, the peaks lagged the switching point by 1-2 s and did not appear to depend on the direction of the switch. This suggests that both CO and oxygen are adsorbed strongly by supported rhodium. Modulation using cycle periods of 2, 4 and 8 s gave time average CO conversions below the steady state conversions at the time average feed composition except at the 2 s period where conversions were identical. Consequently, forcing at 1 Hz or greater are needed to enhance the CO oxidation rate for Rh under the conditions employed.

Response to Forcing in the Presence of Multiplicity and Autonomous Oscillations

Forcing of CO oxidation in these situations has also been studied for the same reasons considered earlier. Keil and Wicke (1980) describe 5 experiments on concentration forcing of CO oxidation over a 0.3 wt.% Pt/γ Al_2O_3 catalyst. Cylindrical 3 mm ϕ × 3 mm catalyst particles were mounted on a glass frit support in a 20 mm ϕ tube for these experiments. The tube was placed inside a temperature-controlled furnace. Steady state measurements disclosed a region of oscillations on the higher temperature or ignited branch for multiple steady states which disappeared on reaching the turning point in the % CO in the feed where the rate and the temperature fall abruptly to the lower temperature or extinguished state.

Figure 3.27 summarizes the Keil and Wicke experiments and the results obtained for a temperature of 275°C, SV = 2.5 s/L g catalyst and air as the oxidant. Percent CO_2 in the off gas is shown versus % CO in the feed to the bed. The solid lines show the steady state behaviour, the vertical dashed lines show the region where autonomous oscillation was observed, while the 5 horizontal lines with arrows give the % CO in the feed used for the modulation experiments. All of these used a symmetrical cycle with a period of 8 seconds. Crosses show the CO_2 conversion measured at each feed composition in the cycle, while the circle is an estimate of the time-average conversion. Amplitudes in the excitation diagram of Figure 3.19 were made dimensionless by the width of the region for autonomous oscillations, while frequencies were rendered dimensionless with the natural frequency of the oscillator. Keil and Wicke did not give these parameters but they can be crudely estimated from results in their paper. The width of the oscillatory region was about 1.5% CO, while the period of

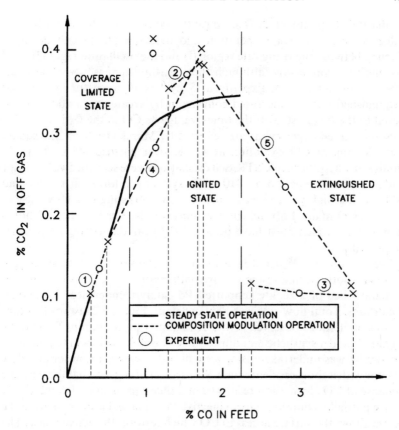

FIGURE 3.27 Square wave forcing of CO oxidation over Pt/γ-Al$_2$O$_3$ at 275°C with $\tau = 8\,s$ between CO concentrations as vol.% shown by the vertical dashed lines in the figure. Time average percent conversion to CO$_2$ given by the open data points, while the cross points show the % conversion in each half cycle. The solid curve shows % conversion at steady state as a function of % CO in the feed. Circled numbers indicate the experiment. Vertical lines extending from the axis separate the ignition state I (stable node) from the limit cycle region II (unstable focus) and the quenched state III (stable node)(Figure adapted from Keil and Wicke (1980) with permission, © 1980, VCH Verlagsgesellschaft mbH)

the natural oscillations was about 1 s. Amplitude ratios for experiments 1 and 2 were less than 0.5, whereas for 3 the ratio was about 1 and for 4 and 5 the ratio exceeded one. The frequency ratio was about 1/8 which puts the experiments near the left axis in Figure 3.19. For these conditions it seems that the response would be τ periodic and may explain why Keil and Wicke did not comment on this question. The interesting result is the ca. 15 % increase in conversion for cycling inside the region of autonomous oscillation (Experiment 2). This puts the conversion above the highest achieved in steady state operation and indicates global enhancement. Cycling between the rising rate region (I) and the oscillation region (II) did not increase conversion, although there was an increase over the quasi steady state conversion. Experiment 5, cycling between the ignited and extinguished regions, increased conversion by greater than 100 % compared to the steady state at the time-average % CO in the feed.

Some limited experiments are reported by Marek (1985) for the square wave forcing of CO oxidation at 428°K over particles of a Pt/Al_2O_3 catalyst in an isothermal CSTR. Self oscillations were seen for 5 vol.% in air and had a natural period of 1410 s. The system was forced at $\tau = 400$ and 600 s with a highly asymmetric square wave consisting of a 60 s pulse of 7 vol.% CO in air. Entrainment was observed on the 3/1 and 2/1 horns (Figure 3.19) which must have been located near the forcing frequencies employed.

Svensson et al. (1988) and Jaeger et al. (1990) investigated the sinusoidal and saw-tooth forcing of the CO partial pressure for CO oxidation over a shallow bed of a zeolite supported Pd catalyst under near isothermal conditions. Total flow rate was kept constant. Figure 3.28a shows autonomous oscillations observed at 483°K in a mixture containing 9.84 vol.% CO in N_2 and a synthetic air adjusted so CO was 0.5 vol.% in the mixture. Oxidation was undertaken using a small bed of catalyst powder in a flow reactor with a short residence time. Svensson et al suggest that the periodic collapse of CO_2 formation results from a phase transition of catalytically active partially oxidized Pd to an inactive PdO phase. Desorption from the phase allows the surface access to CO which restores the active phase. The experimental observations are in the form of a time series which can be subjected to a Fourier transformation. This technique can be used to analyze aperiodic or quasi-periodic unforced or forced behaviour. The power spectrum for the autonomous oscillation appears in Figure 3.28b. The principal frequency is 1.75 min^{-1}. There seems to be incommensurable frequencies at about 1.3 and 1.5 min^{-1}. Harmonics of the principal frequency are evident at 3.5 and 5.25 min^{-1}; other peaks appear to be harmonics of the other major frequencies. Disturbing the bed changed the principal frequency. Svensson et al. report the principal frequency becomes 1.09 min^{-1} under the same conditions, but the general appearance of the power spectrum was unchanged.

Forcing the CO vol.% around 0.5 vol.% with an amplitude of 0.075 % or 15 % at $\tau = 15$ s gives the response shown in Figure 3.28c and the spectrum given immediately below. There is a principal oscillation at the forcing period and peaks of periodically changing amplitude which suggest period doubling. The small doublet indicates a superharmonic response. Peak shapes seems sinusoidal. The spectrum shows almost complete entrainment with a primary frequency of 4 min^{-1}, period doubling and a weak peak at

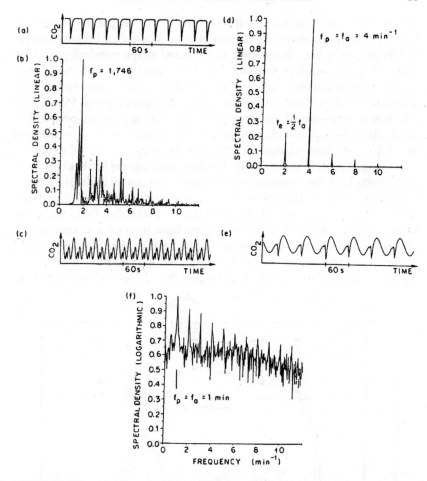

FIGURE 3.28 (a) Time variation of CO_2 concentration leaving a microreactor containing particles of 4.6 wt.% Pd on a zeolite carrier and operating at 483°K with a feed containing 0.5 vol. % CO in an air-N_2 mixture. Concentrations variations correspond to conversion changes from 95 to 75 %. (b) Power (Fourier transform) spectrum of the time series in (a). (c) Response to forcing the system in (a) at $\omega = 4$ min^{-1} with an amplitude of 0.075 vol.% CO. (d) Power (Fourier transform) spectrum of the time series in (c) showing full entrainment of the oscillations of the individual catalyst particles. (e) Response to forcing the system in (a) at $\omega = 1$ min^{-1} with an amplitude of 0.1 vol.% CO. (f) Power (Fourier transform) spectrum of the time series in (e) showing the loss of entrainment (Figure adapted from Jaeger *et al.* (1990) with permission, © 1990, Manchester University Press and the authors)

$2/3\,\tau$. Entrainment, in this case synchronization of the oscillators, continues as the forcing period increases, but is lost when the forcing frequency approaches the natural or autonomous frequency. This is illustrated in

Figure 3.28e. The response now displays the sharp collapse in CO_2 formation seen for spontaneous oscillation, it has taken on an irregular shape. The power spectrum retains some entrainment as the principal peak is at 1 min^{-1} and harmonics at about 2, 3, 4 etc. show $\omega_o = 1.75$ min^{-1}, but many other frequencies appear. Conversion is higher, apparently, when a high degree of synchronization is achieved. This can be seen in Figure 3.29 which plots time-average conversion of CO versus the forcing period. The results, from Svensson et al., exhibited $\omega_o = 1.09$ min^{-1} so $\omega/\omega_o = 1$ occurs at $\tau = 60$ s. Multiple resonances as τ varies are suggested by the figure.

Forcing of zeolite embedded Pd has been investigated further by Liauw et al. (1995). In the autonomous oscillations of this system under the reactor conditions used, the Pd cluster in the zeolite mixture appears to undergo oxidation and reduction as the surface takes on a CO adsorbate layer or an O adatom one. Change of Pd oxidation state triggers a change in the adsorbate layer. Liauw et al. explain improvement under forcing by interference with the triggering phenomenon. With sinusoidal forcing, the excursion into the low oxidation rate region (Fig. 3.28a) still occurs, but the

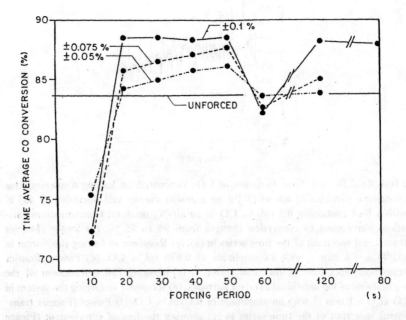

FIGURE 3.29 Effect of the forcing cycle period and amplitude on the time average CO conversion from a microreactor containing particles of 4.6 wt.% Pd on a zeolite carrier and operating under conditions for autonomous oscillations, namely 483°K with a feed containing 0.5 vol. % CO in an air-N_2 mixture. The horizontal line gives the conversion of the autonomous oscillating system (Figure adapted from Svensson et al. (1988) with permission, © 1988, the American Chemical Society)

change happens while the CO/O_2 ratio is increasing and this leads to high instantaneous rates which raise the time average CO oxidation rate. As the frequency decreases, the response changes leading to the type of complex response seen in Fig. 3.28e. Use of a triangular rather than a sinusoidal wave form with different ascending and descending slopes causes the excursion into the low rate region to occur when the CO/O_2 ratio is small and this leads to even greater rate enhancement. Liauw *et al.* find that a pulse mode with a brief interruption of CO flow to the reactor synchronizes the Pd clusters in the zeolite particles and yields, apparently, the highest time average rates.

Unusual experiments on periodic forcing of CO oxidation using an adiabatic 90 mm deep packed bed reactor were carried out by Onken and Wicke (1986) using 0.3 wt.% Pt on γ-Al_2O_3. Using an array of axially located thermocouples, Onken and Wicke show that spontaneous oscillations initiated at the inlet of the catalyst bed are amplified as they move axially through the bed. Measured temperatures were quasi-periodic. A power spectrum manifests a principal frequency of 0.01 s^{-1} ($\tau_o = 630 \text{ s}$) at a feed temperature of 440° and 1·vol.% CO concentration.

When the system at 1 vol.% CO and $T_o = 440°$ was symmetrically forced at $\tau = 200 \text{ s}$ with a 5 % variation in the CO vol.%, the natural frequency continues to dominate and the response is quasi-periodic. However, increasing the amplitude to 10 % largely synchronizes the response and gives the time record seen in Figure 3.30c. The autonomous oscillations are entrained at the forcing frequency, but a multiple period response, including period doubling, is superimposed. The quasi-periodic behaviour has been washed out. This is evident in the figure, as are the multiple periods, and is clearly demonstrated in the power spectrum of the time record. The principal frequency is the forcing frequency. The doublet at about $\omega = 0.015$ indicates period doubling. Raising the forcing period to 360 s (to give $\omega/\omega_o = 1.75$) at the same amplitude brings the bed to a high level of synchronization and results in a period doubling response.

Diffusion Interference

The role of intraparticle diffusion in CO oxidation on Pt/Al_2O_3 under composition modulation has been investigated by Cho (1983) using a single catalyst pellet model and later by Cho and West (1986) experimentally. Cho's model is given in Table 3.2. It was briefly discussed earlier. An isothermal system is assumed with the pellet located in a well stirred gas phase. Competitive adsorption is assumed and the sequence of rate steps is given by the first three stoichiometric equations in Equation (3.1). Initial and boundary conditions are given in the table.

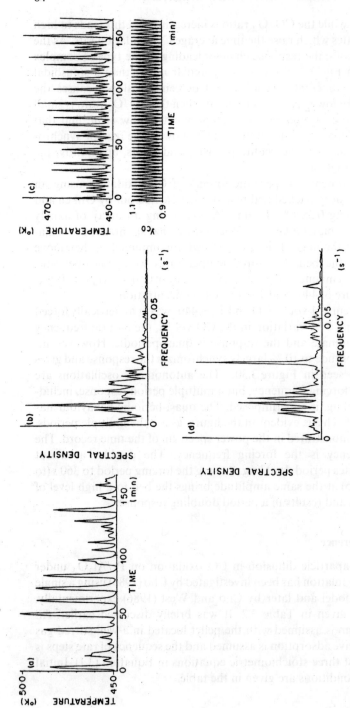

FIGURE 3.30 (a) Time series generated by sinusoidally forcing an adiabatic packed bed of 3 mm × 3mm cylindrical 0.3 wt % Pt/γ-Al$_2$O$_3$ catalyst pellets at 440°, $\omega = 0.031$ s^{-1} and with an amplitude of 5 % of the inlet CO; (b) Power (Fourier transform) spectrum of the time series in (a). (c) as in (a) but at an amplitude of 10 % of the inlet CO. The forcing variations are shown beneath the time series. (d) Power (Fourier transform) spectrum of the time series in (c) showing entrainment as well as multiple period response (Figure adapted from Onken & Wicke (1986) with permission, © 1986, VCH Verlagsgesellschaft mbH)

CARBON MONOXIDE OXIDATION

Normalizing time in the model can be done in several ways. Each leads to a characteristic time:

$\tau = 1/f$ - period of composition modulation
$\tau_R = 1/S_t k$ - response time for the surface reaction
$\tau_D = \varepsilon_p R^2/(D_{eff})_i$ - response time for diffusion for species i
$\tau' = \varepsilon_B V/Q$ - space time

For typical operating conditions in automotive catalytic converters: $\tau = 1$ s, $\tau' = 10^{-3}$ s, $\tau_R = 10^{-2}$ s and $\tau_D = 1$ s. The large τ/τ' means the gas phase composition surrounding the particle follows the feed fluctuations closely. Since $\tau \approx \tau_D$, a significant influence of intraparticle diffusion on behaviour under cycling is possible. It is this consideration which makes the work of Cho (1983) and Cho and West (1986) important.

Steady state solution of the model in Table 3.2 shows that the Pt surface is largely saturated with CO. The solution with sine wave forcing of both CO and O_2 concentrations indicates that at times in a cycle there is a large increase in oxygen stored on the catalyst surface. This provides a better balance of the adsorbate species and thus higher rates of the surface reaction. Consequently, CO conversion increases. The effect of phase lag between the sinusoidal forcing functions was explored by Cho for a pellet with an uniform radial Pt distribution. CO conversion was twice that calculated at steady state for the time average feed composition when the phase lag was 180°, the mode of forcing most usually employed. At a zero phase lag, where CO reaches a maximum in the feed concurrently with O_2, is still much better than steady state operation; indeed, conversion is only about 7% less than that obtained with a 180° lag.

Under periodic composition forcing with $\phi = 180°$ not all the CO diffusing into the catalyst particle can be consumed by adsorbed O_2 and likewise for O_2 diffusion in the opposite half cycle. Some of the CO (or O_2) must be stored. The storage at the end of the rich (CO excess) and lean (O_2 excess) half cycles can be seen in Figure 3.31. Most of the storage occurs on the Pt surface between $0.95 < \xi < 0.65$ where ξ is the dimensionless radial position. Further within the particle some of CO stored earlier in the rich half cycle is given off by the surface in the region $0.65 < \xi < 0.45$. The center of the particles, $\xi < 0.45$, is unaffected by cycling. A similar situation can be seen for O_2 in the bottom half of the figure. Cycling has no effect on adsorbed O_2 for $\xi < 0.6$, indeed there is no adsorbed O_2. The striking conclusion from Figure 3.31 is that the cycling effects are constrained to the outer portion of the catalyst pellet ($\xi > 0.4$ for CO and $\xi > 0.6$ for O_2). Cho (1983) also examined the behaviour of band or egg shell impregnated catalysts and noted that conversion under composition forcing was as much as 2 × greater than for pellets with an uniform radial distribution of metal.

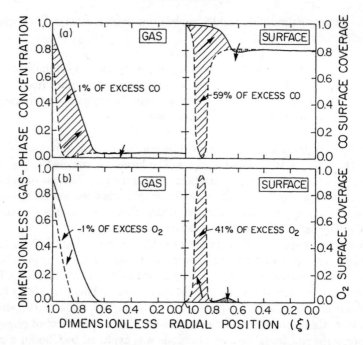

FIGURE 3.31 Storage and consumption of (a) CO and (b) O_2 during composition forcing. Simulation results (Reprinted from Oh and Hegedus (1982) with permission, © 1982, American Chemical Society)

If at equal gas phase concentrations, the adsorption of one reactant is favored, chromatographic displacement when reactants are introduced leads to adsorption of the favored reactant on the outer part of the pellet and adsorption of the least favored at the pellet center if any significant adsorption occurs at all. Periodic forcing of composition counteracts this segregation and results in a large increase of reaction rate because it provides access of the less favored reactant to the surface at the outer portion of the catalyst particle. On the other hand, if concentrations are adjusted so that adsorption between reactants is balanced, composition forcing causes imbalance and suppresses the reaction rate. From these considerations, Cho reasoned, and demonstrated by simulation, that forcing increases CO conversion for $SN < 1.2$ and for $SN > 5$ where O_2 adsorption dominates, whereas in between these regions steady state is the best mode of operation when adsorption rate constants are equal. CO adsorption on Pt is favored at temperatures below 150°C, but above 300°C, O_2 adsorption becomes favored. Consequently, the benefits achievable through composition modulation will depend on temperature.

Some of the consequences of Cho's simulation were examined experimentally. Cho and West (1986) worked with 0.1 wt.% Pt on low density, high surface area γ-Al_2O_3 supports which were impregnated to different depths and by different methods so as to have Pt dispersions from 50 to 100%. Feed was cycled at 0.5 Hz between a "lean" stream consisting of 0.4 vol.% O_2 and 0.2 vol.% CO and a "rich" stream with 0.2 vol.% O_2 and 1 vol.% CO. A 200–550°C temperature range was explored.

For steady state operation, "shell" impregnation was found to be advantageous only at intermediate temperatures. At temperatures below 300°C, CO oxidation over the Pt catalyst is kinetically controlled whereas above 450°C the oxidation is controlled by mass transfer to the catalyst particle. Depth of impregnation is inconsequential in both these cases. Under periodic composition forcing, however, the effect of impregnation depth is complicated. This is indicated in Figure 3.32. Below 300°C, neither impregnation or dispersion affects CO conversion (catalyst "A" has half the Pt dispersion of catalysts "B" and "C"). Shell impregnation is useful between 300 and 450°C, but at higher temperatures a uniformly impregnated catalyst performs better. Maximum conversion at an intermediate tempera-

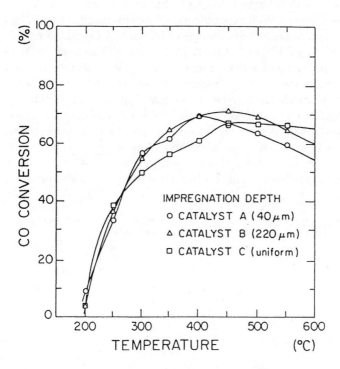

FIGURE 3.32 Influence of Pt impregnation depth and dispersion for CO oxidation under periodic operation on a 0.1 wt.% Pt/Al_2O_3 catalyst as a function of temperature. Switching between feeds of 0.2 vol.% CO and 0.4 vol.% O_2, and 1.0 vol.% CO and 0.2 vol.% O_2 (Reprinted from Cho and West (1986) with permission, © 1986, American Chemical Society)

ture is notable and explained by Cho and West (1986) as the result of the growing desorption rate of reactants from the catalyst surface as temperature increases. Cho and West argue that reactant storage on the catalyst surface causes CO oxidation rates to be higher under periodic than under steady state operation. As a consequence of this argument, diffusional resistance for desorption will be more important for the uniformly impregnated than for the "shell" impregnated catalysts. Adsorbate coverage by the reactants will be greater and therefore CO oxidation rates should be higher. This is what one sees in Figure 3.32 as the temperature approaches 600°C.

These investigators also attempted to unravel the complex influence of the oxidizing/reducing nature of the reactants and temperature on the performance under periodic and steady state. For this, they introduced a stoichiometric number, defined for CO oxidation as 2 × the time average concentration of O_2 divided by the time average concentration of CO. It is evaluated for the feed condition. A stoichiometric mixture of CO and O_2 would have an SN of unity. For mixture with excess O_2, SN would exceed one. Thus, with equal moles of CO and O_2 in the feed, SN = 2.

In the Cho and West experiments, SN was fixed at 0.5, 1 and 2 by cycling around time average feed mixtures of 1.2 vol.% CO and 0.3 vol.% O_2, 0.6 vol.% CO and 0.3 vol.% O_2, and 0.6 vol.% CO and 0.6 vol.% O_2. The results are summarized in Figure 3.33. Under reducing conditions, in Figure 3.33a, there is little difference between modulated and steady state operation up to 250°C. Above this temperature, steady state operation offers higher CO conversion for the uniformly impregnated catalyst. Cho and West (1986) explain this by noting that at temperatures below about 250°C, CO is strongly adsorbed and saturates the catalyst surface. CO sorption weakens above this temperature so that CO and O_2 achieve a more even balance on the surface. Oxidation occurs via sorbate reaction

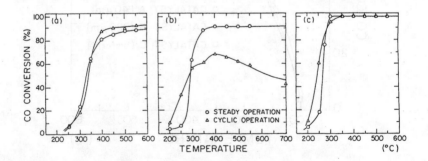

FIGURE 3.33 Influence of stoichiometric number and temperature on CO oxidation under periodic and steady state operation with a uniformly impregnated, 0.1 wt.% Pt/Al_2O_3 catalyst: (a) SN = 0.5, (b) SN = 1, (c) SN = 2 (Reprinted from Cho and West (1986) with permission, © 1986, American Chemical Society)

so that the higher surface concentration of oxygen adatoms increases the oxidation rate much more than the rise due to more rapid surface diffusion of the sorbate species. Both of course contribute to the higher oxidation rate under steady state. Composition modulation improves the O_2 coverage only slightly above 250°C, so this change is outweighed by lower CO and O adsorbates on the surface during the alternating reducing and oxidizing portions of a cycle. Above 350°C however, sorption equilibria reduces O_2 surface coverage, but this reduction is less for periodic forcing because of diffusional resistances. These considerations apply for the net reducing condition of SN = 0.5.

The temperature behaviour with a stoichiometric CO and O_2 mixture changes dramatically as a comparison of Figure 3.33a and 3.33b demonstrates. Below 300°C, composition modulation produces substantially higher CO conversion than under steady state. Modulation provides a better distribution of CO and oxygen adatoms on the Pt surface. Gas and surface phase equilibrium is not established at the frequency used, that is, reactant species are stored on the surface temporarily in amounts that can exceed adsorption equilibria. Under steady state operation, oxidation rates are low because CO saturates the catalyst surface. Increasing temperature brings the CO and oxygen adatom densities closer to a stoichiometric balance as indicated by the sharp rise in CO conversion. This rise is, of course, the ignition or light off condition for steady state. On the other hand, the adatom distribution advantage due to periodic switches to feeds richer in oxygen is lost because of reduced adatom densities on the catalyst surface. Reduced coverage was the explanation Cho and West (1986) used in connection with the maxima in Figure 3.32. According to this reasoning, periodic composition forcing at high temperatures must approach the quasi-steady state limit because equilibrium is now established instantaneously. Figure 3.33b suggests that this is the case indeed.

Under oxidizing conditions (Figure 3.33c), the arguments advanced for SN = 1 still apply, so below 300°C, modulation provides higher CO conversion than steady state operation. At higher temperatures, conversion went to 100% for both types of reactor operation so a comparison is no longer possible. Cho and West (1986) extract further information from Figure 3.33 by calculating the excess CO oxidation for composition modulation versus stoichiometric, steady state operation. From this they conclude that the Pt surface is fully utilized below 300°C so the advantage of modulation is to increase the density of the oxygen adatoms relative to CO adatoms. Above 300°C, CO coverage is reduced so the availability of CO adatoms governs the oxidation rate. Densities of both oxygen and CO adatoms are sharply reduced above 400°C so that either oxygen or CO availability becomes rate governing.

E. SILVER CATALYST

CO oxidation over a supported silver catalyst was investigated by Shanks and Bailey (1987) as part of a study on ethylene epoxidation. Periodic composition forcing was employed by these workers as a means of measuring parameters for a model of the reaction system and as a source of data for model verification. Their catalyst was obtained from an industrial source and contained 1.0 wt.% Ag on an α-Al_2O_3 support. Silver dispersion was 13% based on O_2 chemisorption. The catalyst was pretreated first in O_2 and then in H_2 at 250°C. On introduction of the reaction mixture of 10 vol.% CO and 10 vol.% O_2 (SN = 2) used for most of the experiments, catalyst activity was variable. Shanks and Bailey noted that about 36 hours of exposure was needed to attain constant and reproducible activity. An O_2 cycling strategy was used with symmetrical and asymmetrical cycles. In the former, switching between 10 vol.% CO in He and a mixture of 10 vol.% CO and 10 vol.% O_2 in He was employed, whereas in the latter with $s \neq 0.5$, 10 vol.% CO was retained for one part of the cycle but in the second part the volume percent O_2 was adjusted so that the time average O_2 was 5.0 vol.% (SN = 1).

Modulation results are summarized in Figure 3.34. The steady state reference rate of CO oxidation was taken from Shanks and Bailey (1988). An instantaneous build up of CO_2 observed experimentally upon switching to the O_2 containing stream led Shanks and Bailey (1987) to adopt an Eley-Rideal model for this oxidation system. The slow approach to the quasi-steady state seen in Figure 3.34 as well as the time needed to stabilize the reduced catalyst suggested storage of oxygen on the surface in an unreactive state, probably as an oxide, which converts slowly to an active form when the CO partial pressure rises. A kinetic model based on the above assumptions predicted the time average oxidation rate and the CO_2 mole fractions leaving the reactor closely. The model also predicted a monotonically increasing time average oxidation rate as the cycle period decreased.

OXIDE CATALYSTS

Nickel Catalyst

CO oxidation over NiO/Al_2O_3 catalyst with composition modulation was the subject of two University of Waterloo theses (Hugo, 1985; Stanitsas, 1986). Only a small portion of the data collected in these theses has been published because of reproducibility problems. The experiments undertaken and the catalyst and its preconditioning are discussed in the literature however (Hugo et al., 1986). Two commercial catalysts were used: Girdler

CARBON MONOXIDE OXIDATION

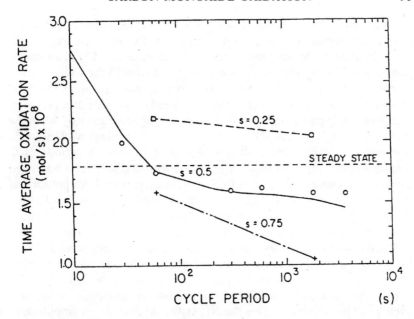

FIGURE 3.34 Time average rate of CO oxidation at 147°C over a silver catalyst supported on alumina for switching between a stream containing 10 vol.% CO and one containing 10 vol.% O_2 in addition to CO. Helium served as diluent. Cycle split based on the CO only stream. Solid line shows the model prediction; horizontal dashed line is the steady state from Shanks and Bailey (1988) (Redrawn from Shanks and Bailey (1987) and reproduced by permission of the American Institute of Chemical Engineers, © 1987, AIChE. All rights reserved)

(United Catalysts) G66 and T-310. Both catalysts employ alumina as a support. The latter contains 10–12 wt.% Ni. Experiments on the first catalyst were run at temperatures of 272°C with switching at constant amplitude around time average CO/O_2 reactant ratios from 0.05 to 0.86 (SN = 2.33 to 40). The amplitude used for most experiments was 0.24, that is, the feed ratios switched from 0.12 to 0.60 around a time mean ratio of 0.36. Symmetrical cycling ($s = 0.5$) was used throughout.

Above a time average reactant ratio of 0.6, composition modulation decreased CO conversion at periods as short as 1 minute. With a strongly oxidizing mixture $CO/O_2 = 0.36$, (SN = 5.5), the time average rate of CO oxidation increased rapidly with decreasing cycle period until at a 1 minute period the time average rate reached the steady state rate at $CO/O_2 = 0.36$. The CO_2 partial pressure in the reactor effluent was followed by IR; these measurements, however, showed that there were no overshoots in CO_2 production on switching.

Studies with the Girdler T-310 catalyst were more extensive. These were run at 150 and 250°K and used equal CO and O_2 concentrations (7.0

vol.%). Amplitude was varied but the time average reactant ratio was maintained at one so that the mixture was oxidizing. Conflicting results were obtained by different investigators. Working at 250°C and an amplitude of 2 vol.%, Hugo (1985) found a strong increase in the CO oxidation rate as the cycle period decreased. The maximum time average CO oxidation rate was about 50% higher than the steady state rate at a period of 1 minute. Cycle period influenced the time average rate strongly when the cycling amplitude increased to 4 vol.%, but the time average oxidation rate was less than the corresponding steady state rate at all periods used. Stanitsas (1986) repeated these experiments with carefully preconditioned catalyst and found just a small influence of the cycle period. Amplitude was important, however. The highest time average rates were realized at the lower amplitude, confirming the earlier observation. No improvement in CO conversion was observed under periodic composition forcing.

Stanitsas (1986) undertook a statistically designed experiment at 250°C to untangle the influences of cycle period, cycle split and amplitude. Cycling was performed around equal time average CO and O_2 concentrations (7.0 vol.%). Amplitude continued to have a significant influence on the time average CO oxidation rate for the Ni/Al_2O_3 catalyst. At an amplitude of 0.05 vol.%, modulation increased the time average CO oxidation rate by 6 to 10 % over the steady state rate, whereas at an amplitude of 2.0 vol.%, the time average rate was slightly below the steady state rate for all periods and cycle splits studied (60 to 1800 seconds, $0.35 < s < 0.65$). Cycle split and period had little influence on the time average oxidation rate.

Conflicting results appear to be caused by differences in the state of the catalyst surface. It was observed that exposing the catalyst to either oxygen or hydrogen for extended times at 400°C deactivated the surface. Thus, the catalyst was preconditioned by exposure to the 7 vol.% CO and 7 vol.% O_2 stream for about 24 h. Three volume percent water vapor in the feed stabilized the catalyst in about one hour. Step change experiments disclosed transients between 1000 and 3000 s after just a small adjustment in reactant concentration in the feed. Hugo et al. (1986) described transitions lasting up to 24 h after cycling is initiated until a stable, repetitive cycling condition is reached. It was proposed that this long transition is due to very slow interchange of O_2 (or of the Ni oxidation state) between the catalyst surface and bulk metal. So far this phenomenon has been observed only for this supported nickel catalyst.

Vanadia Catalyst

The first study of modulating CO oxidation over these oxides was undertaken by Abdul-Kareem et al. (1980a) using a potassium promoted vanadia. Oxidation over this catalyst appears to proceed via a redox mechanism so the composition variable used by Abdul-Kareem was the mole ratio of carbon monoxide to oxygen. The range from 0.12 to 1.1, corresponding to 1.8 < SN < 16.7, was investigated but most measurements were made at time average ratios of 0.36, 0.56 and 0.86. A helium diluent was used whose mole fraction in a feed mixture was about 0.8; temperature was held at 395°C and all experiments were done at one atmosphere. Steady state measurements indicated that the oxidation rate increased monotonically with the increasing CO/O_2 ratio. Curvature of the rate versus ratio curve was small so that the quasi steady state rate was only about 10% below the steady state oxidation rate at the time average mole ratio. In their study, Abdul-Kareem et al. considered the effect of cycle period, amplitude and the time average mole ratio on conversion.

A cycle period or frequency resonance of the oxidation rate was found for the three mean CO/O_2 ratios investigated as is evident in Figure 3.35. The period corresponding to the maximum oxidation rate depends on the mean CO/O_2 ratio, increasing from about 20 minutes at a ratio of 0.36 to 40 minutes at 0.86. A fourfold variation in amplitude was used, but this variable seemed to be important only for periods in the resonance region.

The shape of the curves in Figure 3.35 suggest other regions of resonance. Consequently, additional experiments were carried out at 440°C and a time average mole ratio of 0.86 (Abdul-Kareem et al., 1980b; Jain et al., 1982a). Results of these experiments appear in Figure 3.36. The increase in the oxidation rate over steady state at a period of 40 minutes is much larger than that seen in the previous figure. Further resonance peaks now are found at periods between 16 and 20 minutes and at 2 minutes. Indeed the heights of these rate maxima suggest a harmonic progression at intervals of about 20 minutes. There may be another small rate maximum at about 60 minutes. The two amplitudes investigated seem to have little influence on the position or magnitude of the rate maxima.

Jain et al. (1981) postulated oxygen storage in the catalyst as the source of this apparent harmonic behaviour. This could arise, for example, through the oxidation of vanadium V^{4+} to V^{5+}. They devised a model allowing for the promulgation of oxygen storage/depletion waves in a catalyst layer initiated by variation of the CO/O_2 ratio in the gaseous feed. Reflection of the waves at the catalyst support interface would create constructive and destructive interference of the moving waves in the catalyst layer which was thought to be the source of the harmonic phenomenon. Although this model predicted the dependence of the time average CO oxidation rate on cycle period, it failed to reproduce the harmonic behaviour so clearly shown in Figures 3.36.

Perhaps the behaviour is related to entrainment effects discussed in an earlier section. It is known that the response amplitude increases at the bounds of entrainment horns which would arise if there is a small region of autonomous oscillation in the range of CO/O_2 ratios spanned by the forcing used by Abdul-Kareem et al. (1980b). These increases probably

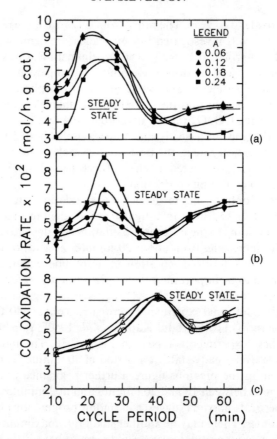

FIGURE 3.35 Variation of the time averaged oxidation rate with cycle period, cycle amplitude and mean reactant partial pressure ratio for a commercial vanadia catalyst at 395°C and atmospheric pressure: (a) $P_{CO}/P_{O_2} = 0.36$; (b) $= 0.56$; (c) $= 0.86$ (Reprinted from Abdul-Kareem et al. (1980b) with permission, © 1980, Pergamon Press PLC)

result in higher time average CO oxidation rates. Entrainment horns will arise as may be seen in Figure 3.19 as the cycle frequency or period changes. Luu (1991), as partially reported in Luu et al. (1992), searched for regions of multiplicity and autonomous oscillation with an unpromoted vanadia catalyst. They were unable to find evidence for either behaviour. Moreover, they found no evidence for multiple resonance with this catalyst. Working with a potassium promoted vanadia, similar in composition to the catalyst used by Abdul-Kareem et al. (1980a, 1980b), Luu (1993) observed an irregular, low frequency oscillation of CO conversion under steady state operation. Quasi-periodic response was seen when CO oxidation over the catalyst was forced, but the amplitude of the low frequency oscillation was

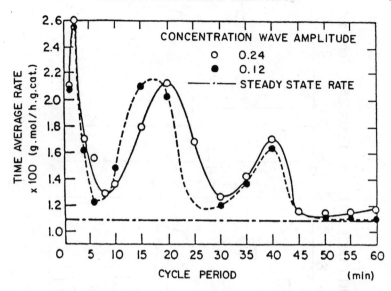

FIGURE 3.36 Variation of the time averaged rate of CO oxidation with cycling period at $P_{CO}/P_{O_2} = 0.86$ and 440°C (Reprinted from Abdul-Kareem *et al.* (1980b) with permission, © 1980, Pergamon Press PLC)

small. Unfortunately, Luu was not able to show that these oscillations arose from the catalyst system. Fluctuations in the IR detection system could have been their source. With the potassium promoted catalyst, Luu observed CO conversion varied as the forcing frequency changed. The variations were small and did not resemble the behaviour with cycle period seen in Figures 3.35 and 3.36. The source of the harmonic-like behavior in Figure 3.36 remains mysterious.

Copper Catalyst

Prokopowicz *et al.* (1988) employed a commercial CuO/γ-Al_2O_3 catalyst containing 10 wt.% CuO, and a 10 wt.% CuO on highly porous SiO_2 prepared by the investigators. The first catalyst was used in a micro packed reactor. With this catalyst, a curious "wrong way" behaviour was seen following a composition change (Prokopowicz, 1985; Prokopowicz *et al.*, 1987), Figure 3.37 shows this behaviour. "Wrong way" behaviour usually arise from thermal effects, but in this experiment the reactor was isothermal with temperature change within a cycle less than $\pm 1°C$. Prokopowicz attributed the change in rate a few minutes after a composition change in the feed to evolving reactant concentrations on the catalyst surface. An example of this is the response of O_2 in Figure 3.2a where the oxygen mole fraction in the gas decreases in the middle of the oxygen portion of the cycle.

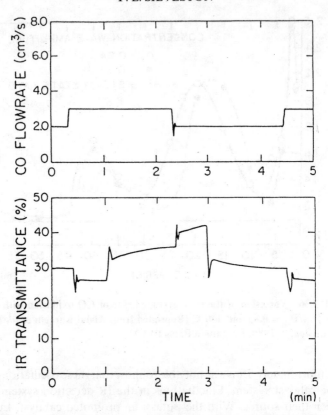

FIGURE 3.37 CO_2 transmittance variation in the reactor outlet from periodic switching of CO reactant concentration with a 10 wt.% CuO on γ Al_2O_3 catalyst at 220°C (Reprinted from Prokopowicz (1985) with the permission of the author)

The delayed, but abrupt change can also be explained by a rapid oxidation/reduction of the catalyst.

To ascertain the source, Prokopowicz *et al.* (1987, 1988) undertook a transmission FTIR study of the CuO—CO oxidation system. The second catalyst was used for this system and a special heated, flow through cell was constructed (Prokopowicz *et al.*, 1987). Unfortunately the "wrong way" behaviour was not seen for the CuO—SiO_2 wafer prepared for this cell. A CO adsorbate species was detected which changed with the CO concentration in the gas phase. However, in the presence of O_2, variations of the adsorbate intensity did not correlate with CO_2 production. Evidently, this adsorbate is not part of the CO oxidation chain. CO_2 production must occur through either an Eley-Rideal mechanism or via a surface species present in such low concentration that it was not detected in transmission IR. Prokopowicz's observations are similar to those made by Barshad et al. (1985a) for a supported Pt catalyst.

Hopcalite Catalyst

Experiments on composition modulation using this catalyst were part of an investigation of models capable of describing the dynamic behaviour of non-isothermal packed beds undertaken by Casanova et al. (1979) and Baiker et al. (1980). Hopcalite is a common mixed oxide combustion catalyst with metals Mn:Cu:Co in the mass ratios of 39:22:7, a BET surface area of 105 m^2/g and a void fraction of 0.64. The catalyst is active for CO oxidation at 25°C. For the Baiker experiments, the catalyst as 1.5 mm spherical particles was packed in a 20 mm ϕ × 300 mm pyrex vessel to a depth of 100 mm. Alumina pellets of the same size filled the remainder of the bed. The pyrex vessel was jacketed by an external cylinder held under vacuum and insulated as well as covered with aluminium foil in an attempt to realise adiabatic operation. Twelve thermocouples were located axially in the catalyst bed.

Steady state experiments were conducted by bleeding CO into an air stream flowing to the reactor. As the CO mole fraction in the feed increased to $y_{CO} = 0.027$, ignition took place resulting in an abrupt increase in bed temperature. The temperature profile for the ignited state can be seen in Figure 3.38. CO conversion approaches 100 %. Decreasing the CO mole

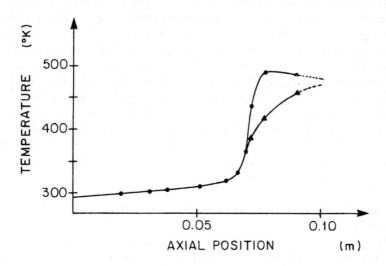

FIGURE 3.38 Axial temperature profiles for CO oxidation over a hopcalite catalyst in an adiabatic packed bed for steady state operation (A) at $y_{CO} = 0.023$ and composition modulation (B) with asymmetric switching between $y_{CO} = 0.012$ and 0.028 (Figure adapted from Baiker et al. (1980) with permission, © 1980, Verlag Chemie, GmbH)

fraction once ignition has occurred leads to switch back to the lower rate branch where conversion is only about 20 %. Hysteresis is observed. Periodic forcing was carried out asymmetrically by switching from $y_{CO} = 0.012$ to $y_{CO} = 0.028$ after 2 s at the lower mole fraction and switching back again after 4 s. The space velocity was lower for the periodic operation experiments so steady state ignition was obtained just below the time average CO mole fraction of 0.023. CO conversion was about 86 % compared to over 95 % for steady state operation at $y_{CO} = 0.023$. Composition modulation, however, smooths the temperature profile. This is shown in Figure 3.38. With steady state, the maximum temperature in the bed reaches 500°C, whereas the maximum is just 470°C with periodic forcing. Baiker's observations are important from the view point of reactor engineering as they suggest that periodic forcing may be a better operational strategy when parametric sensitivity is encountered or when there is an upper temperature limit for reactor operation. Composition modulation reduces temperature gradients in packed beds and limits hot spot development.

G. SUMMARY

Under composition modulation, CO oxidation over fixed beds of catalysts exhibits, as we have seen in this chapter, a variety of effects and phenomena that are summarized in Table 3.3. Several of these, such as 1) to 4) are general and so are found for other reactions and catalyst systems. Other effects and phenomena, such as 5) to 8) are probably general as well, but the necessary experiments to observe them have not been undertaken. With respect to 5), most modulation experiments are planned so as to avoid diffusion interference. Similarly, experiments are conducted under constant temperature so 7) has not been demonstrated for other reactions except for ammonia synthesis (Chapter 7). The remaining effects and phenomena in Table 3.3 are likely to be either catalyst or mechanism dependent or perhaps depend on both situations. Long relaxation times for forcing oxidation over nickel oxides were found when the catalyst was reduced prior to use. It was not seen when the catalyst was oxidized. Furthermore, a stationary condition is attained after just several cycles for most noble metal catalysts (Graham and Lynch (1984), Barshad and Gulari (1985a) and Zhou and Gulari (1986a)). The term excitation has been introduced in the table. It means that CO oxidation rate depends on cycle period or frequency. It does not mean rates are higher than those at a comparable steady state. We use enhancement to indicate that situation.

Surprising, perhaps, in this chapter is the broad range of cycle period or frequencies which can produce a rate resonance. These range from fractions of a second to several minutes for solids catalyzed CO oxidation. These frequencies and even cycle split are also temperature dependent as Figure 3.5 demonstrates. The wide range means attention must be given to the possibility that aperiodic response can arise if the reactor or reaction system contain parameter space where steady state multiplicity or autonomous oscillations exist and the forcing frequency is close to the system is natural frequency.

The experimental studies discussed in this review have answered the question of whether or not global enhancements in reaction rate can be achieved through composition modulation. This question has interested many investigators in the past (Nowobilski and Takoudis, 1986; Yadav and Rinker, 1989). Results for modulating CO oxidation make it clear that global enhancement is attainable in the sense that for a catalyst system at a specified temperature and pressure, rates observed under composition modulation can be greater than any rate measured at steady state whatsoever the feed composition. This may not be a general rule, however. Global increases in the CO oxidation rate were found for the Pd/Al_2O_3, $Pt-SnO_2/Al_2O_3$ and the V_2O_5/SiO_2 systems, but rates measured for the Pt/Al_2O_3 system appear to be below rates in the high rate branch in the region of multiple steady states.

Table 3.3 suggests significant differences between the noble metal and the oxide catalysts: effects or phenomena 4), 8) and 12) are seen only for the former, while responses 9) to 11) occur only for the oxide catalysts. Furthermore, rate increases between $10 \times$ and $40 \times$ have been measured for the noble metal catalysts. Increases in rate for the oxide catalyst seldom exceed $2 \times$. These contrasting observations point to differences in rate controlling steps and/or mechanism.

Explanations for the CO oxidation rate enhancement for the noble metal catalyst under composition modulation appear to involve reactant storage: abatement of CO inhibition of rate and augmentation of oxygen access to the catalyst surface. Through composition cycling, the exposure of CO to the surface is reduced so that the surface is not continuously swamped by adsorbed CO. As a result time average CO and O adatom concentrations are better balanced. The experimental evidence supporting this explanation is considerable. Mathematical models incorporating storage and competitive adsorption (Oh *et al.*, 1980; Cho, 1983; Lynch, 1983; and Graham and Lynch, 1984) and given in Tables 3.1 and 3.2 also agree well with experiment providing, thus, further evidence for our explanation.

In those systems where time average rates under periodic forcing exceed the maximum rates attainable in steady state operation, for example oxidation over Pd/Al_2O_3, reactant storage continues to be the basis for the explanation, but Gulari and his co-workers (Zhou *et al.*, 1986a) add that the spatial distribution of adatoms on the catalyst surface (as islands) is increased by composition modulation. Better spatial distribution may also

Table 3.3 Summary of Behaviour Under Composition Modulation of CO Oxidation

1) Excitation or enhancement (increase of oxidation rate above the steady state rate at the time average feed composition)
2) Rate resonance in which the oxidation is stimulated within a relatively narrow range of cycling periods or frequencies (noble metal catalysts, supported V_2O_5).
3) Global enhancement where rates exceed the highest rates attainable at steady state for a specified temperature and pressure (Pd/Al_2O_3, Pt-SnO_2/Al_2O_3, V_2O_5/SiO_2)
4) Excitation where the cycle split or duration of exposure to one reactant is as important as frequency or period (noble metal catalysts).
5) Excitation or enhancement dependence on intra particle diffusion and/or the distribution of the catalytically active material in the porous catalyst (Pt/Al_2O_3).
6) Excitation dependence on the oxidizing/reducing (stoichiometric number) character of the reactants (Pt/Al_2O_3, Pd/Al_2O_3).
7) Decrease in temperature gradients in non-isothermal packed beds (Hopcalite)
8) Suppression of the natural frequency, period doubling, quasi periodic response when changing compositions cross regions of multiplicity or autonomous oscillations (Pt-$SnO2/Al_2O_3$, $Pt(110)$, Pd/zeolite).
9) Multiple resonance of rate with changing frequency with possible harmonic excitation of rate (V_2O_5/SiO_2)
10) "Wrong way" behaviour under isothermal conditions (CuO/Al_2O_3).
11) Long relaxation times to reach a stationary, reproducible cycling condition (NiO/Al_2O3).
12) Induction periods as large as 5 to 10 seconds after a composition switch has occurred (noble metal catalysts).

arise with the Pt/Al_2O_3 catalysts and this could account for the quantitative failure of the Graham and Lynch (1984) model with this catalyst.

Oxygen storage is probably part of the explanation of the CO oxidation rate increase with the oxide catalysts, but our understanding of these catalysts is much poorer than for the noble metals. Shanks and Bailey (1987) were able to find good agreement with experimental data using a mathematical model which assumes the presence of an inactive form of oxygen on the catalyst surface or as part of the surface. Jain et al. (1981) proposed oxygen diffusion into the bulk of the catalyst to explain their observation of multiple resonances with a vanadia catalyst. Certainly, the long time needed to obtain reproducible cycles for the NiO catalyst suggest that changes in the catalyst phase are important with this system. This seems to be the limit of what we now know. A great deal of work is still needed to understand how periodic operation affects oxide catalysts.

H. FUTURE CHALLENGES

The progress achieved in assessing benefits from composition modulation and in understanding the mechanism of enhancement through studies on CO oxidation have been substantial. Catalytic oxidation of CO should continue to be a favoured system for modulation studies because of the experimental convenience the oxidation offers.

Our understanding and ability to predict performance under modulation for the noble metal catalysts is much further advanced than it is for the oxide catalysts. Thus, the primary challenge is to expand investigations of CO oxidation over different oxides. Most oxides catalyze CO oxidation. What other phenomena exist in addition to those described in this chapter? More thorough research is needed to ferret out the mechanisms of how these oxides enhance oxidation rates through forcing. Clean surface phenomena, involving high vacuums, needs study. Single crystal work using sophisticated surface science tools has contributed greatly to our understanding of the modulation mechanism over the noble metal catalysts even for experiments done at atmospheric pressures. Similar experiments should be done on oxides such as nickel and vanadia. Observations of the surface in the IR spectra would be useful as well. This has been done only once for an oxide catalyst. Finding the source of the multiple resonance observed with vanadia is a challenge. Could multiple resonances be a general phenomenon? Another challenge is the "wrong way" behaviour found when pulsing CO oxidation over copper oxide.

CO oxidation because of the ease of following response is probably the best catalytic system to use for investigating more complex modulation strategies, such as using a multipart cycle with an inert flush between reactant pulses, or forcing CO and O_2 concentrations at different frequencies, or combining composition and temperature forcing. There is as well the possibility of running composition modulation inside periodic flow reversal because of the large difference in cycle durations. Cycle periods for flow reversal are typically 1–2 hours. The latter periodic operation has the property of creating a rising and a descending temperature profile in a packed bed reactor that would be advantageous for equilibrium limited reactions. The profile may be advantageous too when reactant inhibition occurs.

CHAPTER 4

Reduction of Nitrogen Oxides

A. INTRODUCTION

The nitrogen oxides form a class of chemical compounds that are minor commercial gases, but attract interest because they are implicated in smog formation and contribute to acid rain. As pollutants, they are removed from waste gases by reduction with H_2, CO, hydrocarbons or ammonia. It is reduction with CO and H_2 that has been the subject of composition modulation investigations. The oxides include nitrous oxide (N_2O), nitric oxide (NO) and nitrogen dioxide (NO_2 or N_2O_4). Because they are reduced in a reaction chain starting from the dioxide, individual oxides and the mixture are often referred to as NO_x or NOX. The largest quantities are encountered in combustion exhaust or stack gas. There, the oxides arise from nitrogen containing hydrocarbons and the O_2–N_2 reaction which proceeds to a small extent at the high temperatures encountered in combustion.

NO_x is removed from automotive exhaust by three way catalysts (TWC). The dominant reaction is reduction with CO. As mentioned in the previous chapter, catalytic converters used in TWC systems are subjected to a quasi periodic, low amplitude disturbance at about 0.5 to 2 Hz. Many studies of NO_x reduction, consequently, have been carried out under composition modulation. For a similar reason, modulation studies have been restricted to the noble metals which are used as TWCs. Just as for CO oxidation, forcing studies are relatively simple to conduct. The nitrogen oxides exhibit strong IR bands as does the oxidized product, CO_2, so IR spectroscopy can be exploited. Mass spectroscopy is also used. The only problem is that CO and the reduction product, N_2, have identical mass numbers.

B. NO REDUCTION

Experimental

The desired overall reaction in NO_x reduction with CO is

$$2\,CO + 2\,NO \rightarrow 2\,CO_2 + N_2 \qquad (4.1)$$

where nitric oxide (NO) is assumed to represent the mixture of oxides present in stack gas or automotive exhaust. Under an oxidising condition, reduction to nitrous oxide occurs

$$CO + 2NO \rightarrow CO_2 + N_2O \qquad (4.2)$$

There is much evidence that the stoichiometric reaction of Equation (4.1) proceeds through two steps. The first of these is nitrous oxide formation. This oxide is in turn reduced by CO as

$$CO + N_2O \rightarrow CO_2 + N_2 \qquad (4.3)$$

Note that Equation (4.3) is not stoichiometrically independent.

Muraki and Fujitani (1986) investigated composition modulation of NO reduction by CO for the 5 noble metal catalysts employed by Toyota researchers for CO oxidation (Muraki et al.,1985a) and hydrocarbon oxidation (Shinjoh et al.,1989). These studies have been discussed in the two previous chapters. The stoichiometric ratio, SN, is defined for NO reduction by CO as the ratio NO (vol.%) : CO (vol.%). Therefore, if the formation of N_2O can be neglected, SN = 1 for a stoichiometric mixture. If the nitrous oxide reaction proceeds, SN = 2 for the stoichiometric mixture. Just as for CO oxidation, the stoichiometric ratio is an influencial operating parameter under composition forcing. Metal loadings on the catalysts were 0.006 wt.% and the support was α-Al_2O_3. A Pd loading of 0.24 wt.% was used for several cycling studies and this loading was also chosen for the temperature programmed desorption (TPD) experiments which supported the study. In most of the Muraki and Fujitani experiments, symmetrical switching ($s = 0.5$) between NO and CO was used with each at 0.3 vol.% when present in the reactor feed. Diluent was He.

Figure 4.1 compares the conversion of NO under modulation at 400°C and SN = 1 (stoichiometric ratio) for the 5 catalysts. Steady state operation is represented by the data points at $\tau = 0$. Clearly, composition modulation leads to an enormous increase in NO (or CO) conversion for the Pt catalyst. Pd and Iridium (Ir) catalysts are also activated by forcing, but the conversion improvement is less dramatic. Both Rh and ruthenium (Ru) are effective low temperature catalysts for NO_x reduction. Figure 4.1 shows no improvement for these catalysts. Probably the high steady state conversions makes activation difficult to detect and higher forcing frequencies, greater then the 5 Hz maximum used by Muraki and Fujitani, may be necessary to provide a conversion increase. Steady state activity at 400°C and SN = 1 are in the sequence Rh > Ru > Ir > Pd ≫ Pt. It is significant that modulation is effective for Pt. Rhodium is included in TWC formulations because of its NO_x reduction activity. The metal is rare and very expensive so there has been an effort extending over many years to replace Rh in these catalysts. Figure 4.1 shows that under modulation, the activity of Pt can be raised to close to that for Rh.

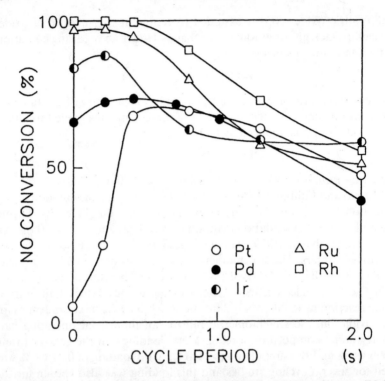

FIGURE 4.1 Comparison of NO conversion on noble metal catalysts at 400 °C, SN = 1, and SV = 30,000 h^{-1} for symmetrical forcing between 0.3 vol.% NO and 0.3 vol.% CO. (Figure reprinted from Muraki & Fujitani (1986) with permission, © 1986 American Chemical Society)

Because Pt appeared to be a possible replacement for Rh, Muraki and Fujitani examined temperature and stoichiometric number effects on CO conversion under periodic forcing for just the Pt catalyst. Their results are summarized in Figure 4.2 a–c. Under all conditions examined, there is a striking activation of the catalyst through bang-bang cycling between feeds containing just one of the reactants. This phenomenon was also observed for CO oxidation over Pt (See Chapter 3). Just as for the CO oxidation system, there is an optimal cycle period or frequency, that is, a resonance with respect to forcing is observed. Regardless of the time average stoichiometric number, increasing temperature decreases the optimal cycle period or increases the optimal frequency. These frequencies appear to be higher for oxidizing mixtures and may be about 5 Hz for both SN = 1.0 at 400°C and SN = 1.5 at 450°C. In the reducing mixture, SN = 0.25, the optimum frequency falls to 1 Hz at 400°C.

Reducing NO forms either N_2O or N_2 as mentioned above. Nitrogen formation dominates, but appreciable amounts of N_2O appear at tempera-

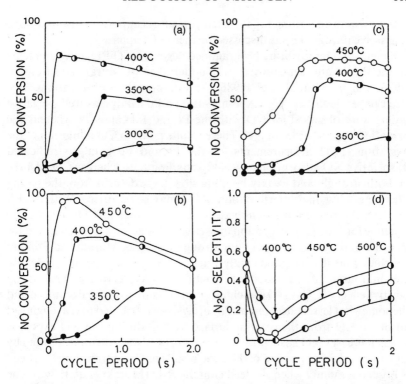

FIGURE 4.2 Behavior of a Pt-α-Al$_2$O$_3$ catalyst under symmetric composition forcing between 0.3 vol.% NO in He and 0.3 vol.% CO in He at SV = 30,000 h^{-1}: (a) conversion at SN = 1.5, (b) conversion at SN = 1.0, (c) conversion at SN = 0.25, (d) selectivity to N$_2$O at SN = 1.0, all versus cycle period and e) conversion at SN = 1.0 versus temperature. (Figure reprinted from Muraki *et al* (1986a) with permission, © 1986 American Chemical Society)

tures of 400°C and under, or when the reactor feed is oxidizing. The nitrous oxide side reaction must be considered at temperatures as low as 300°C for the Pt and Pd catalysts at steady state and it is important up to 500°C if the mixture is non-stoichiometric (Muraki *et al.*,1986a). Figure 4.2d shows that composition forcing substantially reduces N$_2$O formation. Indeed, at SN = 1 and 400°C, selectivity to nitrous oxide drops from 0.6 at steady state to less then 0.2 at a cycle period of 0.5 seconds. A comparison of Figures 4.2b and d indicates the smallest formation of N$_2$O when NO conversion is at its highest.

A cross plot of the temperature and cycle period data at SN = 1 shows that composition forcing of NO and CO sharply decreases the light-off temperature of the catalyst, arbitrarily defined as the temperature at which 50% conversion is achieved. Using a cycle period of 1.3 s, for example, reduces the light-off temperature to about 350°C from greater than 500°C in steady state operation. Light-off temperature is important in automotive emission catalysis. This is discussed further in Chapter 5.

Kinetic and temperature programmed desorption (TPD) measurements on the Pt, Pd and Rh catalysts were undertaken by Muraki and Fujitani (1986) and by Muraki *et al.* (1986a) to interpret their observations. The latter paper describes pulse experiments on a Pd catalyst as well. Negative orders were observed for CO with the Pt and Pd catalysts, whereas the oxidation was 1st order in NO. These results point to CO inhibition of the oxidation. TPD measurements for the Pt and Pd catalysts indicated dissociative NO adsorption on weakly bonding sites, whereas CO adsorbs on both strongly and weakly bonding sites. CO_2 desorbs from the strong sites suggesting disproportionation of CO may be occurring. With the Rh catalyst, the reduction reaction is about zero order in CO and slightly negative for NO. CO adsorption on weakly bonded sites is greater than for Pt or Pd and CO_2 adsorbed on the strong sites is less. Muraki and Fujitani conclude that Rh is a good catalyst under steady state operation because CO inhibition of the reduction reaction is less pervasive. The strong positive effect of CO—NO switching on conversion over the Pt catalyst and the somewhat lesser effect on the Pd catalyst are attributed to scavenging of adsorbed CO from the catalyst surface by NO during the NO exposure which increases NO adsorption. Scavenging of adsorbed atomic oxygen by CO during the CO portion of the cycle results in CO_2 formation. Nitrogen is observed mainly when the feed contains NO (Muraki *et al.*,1986a). The pulse experiments also suggested that N_2O and N_2 form on different sites. NO adsorption on a surface rich in adsorbed oxygen favors N_2O, while on a surface depleted of atomic oxygen, N_2 is the predominant product. Muraki and Fujitani explain the improved selectivity to N_2 with CO—NO switching by the exposure of NO to a surface largely free of atomic oxygen when the reactant in the feed changes from CO to NO. Composition forcing, thus, drastically decreases the time average concentration of CO on the Pt and Pd surfaces, while raising the time average NO concentration. Balancing of the reactant surface concentrations results in the higher oxidation rate and increased conversion at a constant space velocity.

Muraki *et al.*(1986a) suggest NO reduction proceeds *via*

$$NO + s \rightarrow NO \bullet s$$

$$CO + s \rightarrow CO \bullet s$$

$$NO \bullet s + s \rightarrow O \bullet s + N \bullet s$$

$$NO \bullet s + N \bullet s \rightarrow N_2O + 2s \tag{4.4}$$

$$NO \bullet s + N \bullet s \rightarrow N_2 + O \bullet s + s$$

$$CO \bullet s + O \bullet s \rightarrow CO_2 + 2s$$

They assume the first step is rate controlling and derive a rate expression which agrees reasonably with their kinetic data. The role of composition modulation is then to reduce CO inhibition of the reaction thereby permitting the first step of the above sequence to take place.

NO reduction over the rhodium constituent of TWCs was also investigated by Cho et al.(1989) employing 0.09 wt.% Rh on γ-Al_2O_3 and 0.04 wt.% Rh on CeO_2. Both catalysts were shell impregnated and each had about 50% Rh dispersion. They were used as a powder, 80 to 120 US Mesh, in a 3.2 mm stainless steel tube with the loading adjusted so equal masses of each catalyst were packed into the tube. Experiments were run at atmospheric pressure and a SV = 80,000 h^{-1} with He as the diluent. Prior to use the catalyst samples were reduced in H_2, oxidized in 4 vol.% O_2 and finally reduced at 500°C in 0.16 vol.% CO. Symmetric and asymmetric, bang-bang forcing was used switching between 0.008 vol.% ^{13}CO and 0.008 vol.% NO. A mass spectrometer was used for analysis so the ^{13}CO isotope allowed detection of both CO_2 and N_2O.

For the Rh/Al_2O_3 catalyst, the reaction proceeds rapidly above 250°C. Bursts of CO_2 production occur after the composition switch, but the burst following the introduction of CO is largest. Sharp peaks of N_2 formation occur at the same time, but the peak following CO introduction is much smaller than that after switching to NO. Increasing the temperature to 500°C causes the CO_2 production burst associated with NO introduction to vanish. Decreasing the cycle period to 20 s at 500° erases the second peaks leaving just single CO_2 and N_2 peaks in a cycle. A further decrease to 10 s gives almost complete NO reduction and CO oxidation. The explanation for these observations is essentially the same as offered by Muraki and Fujitani (1986). NO dissociates to form N_2 and O adatoms on the surface. The oxygen is scavenged by CO forming CO_2. There is no Eley-Rideal activity when NO is re-introduced, so CO_2 forms from adsorbed CO and O adatoms created by NO adsorption. N_2O appears in the NO portion of the cycle and forms through the the 4th elementary reaction in Equation (4.4) above. The elementary reactions Cho et al.(1989) propose for the reduction are essentially those given in Equation (4.4), modified only by assuming N_2O adsorbs and is in equilibrium with the gas phase. In addition, Cho et al. suggest

$$N_2O \bullet s \to O \bullet s + N_2$$
$$2N \bullet s \to N_2 + 2s$$
(4.5)

Figure 4.3 compares results obtained by symmetric bang-bang forcing at $\tau = 20 s$ with those obtained in steady state operation and shows that forcing is inferior to steady state operation for the conditions considered. It resembles Figure 3.33 for CO oxidation over Pt/Al_2O_3 (Cho and West, 1986). The explanation for both cases is that the enhancement at the

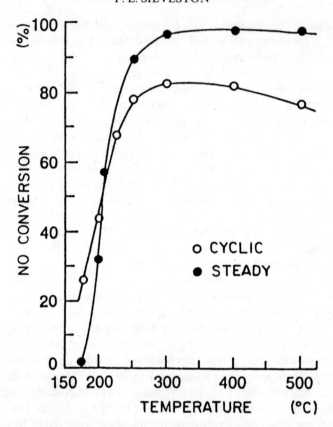

FIGURE 4.3 Comparison of NO reduction under symmetric bang-bang switching from 0.008 vol.% CO and 0.008 vol.% NO with $\tau = 20\ s$ and steady state operation at the time-average feed composition with a Rh/Al$_2$O$_3$ catalyst at 500°C (Figure reproduced from Cho *et al.* (1989) with permission, © 1989 Academic Press)

optimal cycling conditions are diminished by increasing temperature because adsorbate storage is lower and thus inhibition effects are smaller. Cho *et al.* observe that if the cycle period is dropped back to 10 s, nearly 100% NO reduction is achieved. This matches that observed at steady state so the 20 s cycle period used for the data in Figure 4.3 is sub optimal. The investigators concluded that below the lightoff temperature of ca. 200° reduction is controlled by adsorptive decomposition of NO, whereas above 200° the rate is controlled by CO scavenging of adsorbed oxygen. They suggest as well that above 300°C the overall reactions given by Equations (4.2) and (4.3) proceed so rapidly that reduction appears to go by Equation (4.1).

Asymmetric cycling permits advantage to be taken of the different rate controlling steps and this was investigated by Cho et al. for operation at SN = 1. Results are shown in Figure 4.4. Split is defined in terms of duration of the NO exposure. When a pulse has a high concentration in NO (0.008 vol.%), exposure should be short so $s < 0.5$ is desirable above 300°, but not in the range 200 to 300°, whereas for a low NO concentration, a longer exposure is needed or $s \geqslant 0.5$ between 200 and 300°. Differences are significant below 300°, but not above this temperature.

A much greater enhancement of NO reduction can be achieved by using cerium oxide as support in place of alumina. The oxide is reduced by CO so that additional capacity for NO decomposition is created. Thus, for the conditions used in the Cho experiments, NO reduction in symmetrical bang-bang cycling at 500° can be raised from 76.5% (Figure 4.3) to 93.6% by replacing Al_2O_3 by CeO_2.

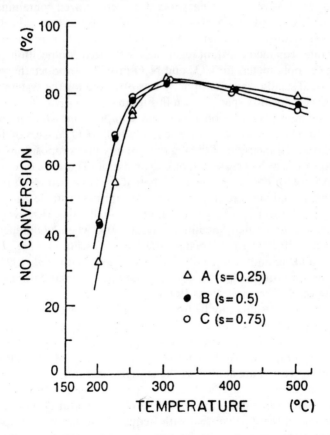

FIGURE 4.4 Effect of cycle split and temperature on NO reduction at SN = 3 (A) = 1 (B) and = 0.333 (C) for cycling conditions given in Figure 4.3 (Figure reproduced from Cho et al. (1989) with permission, © 1989 Academic Press)

Modulation of NO reduction at low temperature (485°K) over a 0.5 wt.% Pt/γ-Al$_2$O$_3$ catalyst showed a remarkable improvement over the reduction rate at a comparable steady state (Sadhankar and Lynch, 1996). Unlike the two contributions just discussed, Sadhankar and Lynch were interested more in excitation through composition modulation than in understanding the mechanism of NO reduction in this operation. Under the conditions chosen for their experiments, NO reduction on supported Pt displays multiplicity. Sadhankar and Lynch furthermore investigated the effect of phase lag when the concentrations of both reactants are forced. These researchers employed a packed bed containing 20 g of catalyst with external recycle operated such that the ratio of recycle to feed was about 110. This was sufficient to make the reactor system behave as a CSTR. Cylinders containing premixed feed compositions were used with flows and switching performed through computer controlled mass flow meter/controllers and 4-way solenoid valves. Five binary mixtures were utilized containing 2.02 or 5.94 vol.% CO and 2.04, 5.09 or 7.89 vol.% NO. N$_2$ served as the diluent for all five mixtures. A cylinder of N$_2$ provided ballast flow to ensure that the flow rate remained constant regardless of the feed composition. Separate IR spectrophometers for CO$_2$ and N$_2$O provided product analysis.

Isothermal steady state multiplicity was observed for the system in the NO—CO composition space shown in Figure 4.5. The space to the left of the hatched multiplicity region are mixture compositions which follow the high rate branch and which show essentially 100% CO conversion for the SV employed. To the right of the regions are mixture compositions which lead to low rates of NO reduction or CO oxidation. There are three steady states possible for compositions in the hatched region: two result in rates along the high and low rate branches, while the third is unstable. Sadhankar and Lynch (1996) did not observe autonomous oscillations. However, these have been seen under high vacuum at about 100°C when NO reduction is undertaken on Pt(100) single crystal surfaces. The oscillations are damped (Schwartz and Schmidt, 1987; 1988; Fink et al.,1991; Dath et al., 1992). The mechanism for both the spontaneous oscillations and their damping have been developed (Fink et al.,1990; 1991).

The reduction system may be forced using bang-bang cycling and two part cycles by either an in phase (0° phase lag) or out of phase (180° or π radian phase lag). If these strategies are applied keeping a constant time-average feed composition of 1.2 vol.% CO and 1.2 vol.% NO in the feed, the compositions visited are shown by the straight lines in Figure 4.5. Under steady state operation at the time-average feed composition, low NO reduction and CO oxidation rates will be obtained. Out of phase, bang-bang forcing, of the type we deal with frequently in this monograph, can ignite the system and should display large rate or conversion enhancement. On the other hand, in phase, bang-bang forcing appears in Figure 4.5 to be just tangent to the multiplicity region so that enhancement is unlikely.

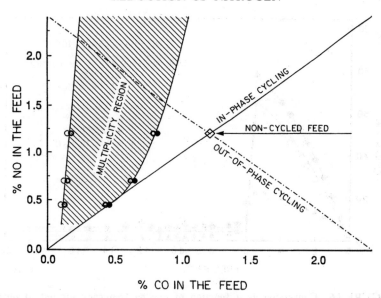

FIGURE 4.5 Boundaries of the region of isothermal multiplicity for NO reduction with CO at 212°C and 103 kPa over a 0.5 wt% Pt/Al_2O_3 catalyst and the range of composition switching used for in phase (0° lag) and out of phase (180° lag) forcing experiments (Figure taken from Sadhankar and Lynch (1996) with the author's permission)

Indeed, in phase cycling experiments over a range of frequencies failed to enhance conversion.

Out of phase bang-bang composition forcing around the mean composition shown in Figure 4.5 exhibits a pronounced conversion resonance with frequency. Figure 4.6 plots conversion of CO, total conversion of NO and NO conversion to N_2O versus frequency for the system and conditions given in Figure 4.5. CO conversion reaches a maximum of about 66% at $\omega = 0.0055$ Hz, while NO conversion attains 78% at this frequency. The CO conversion is below the 100% conversion achieved in the ignited (or high rate) branch for multiple steady states at the space time of 39 s used by the investigators, while the NO conversion about equals that measured in the high rate branch. Consequently, Sadhankar and Lynch did not find global enhancement, at least for CO oxidation. Curiously, NO conversion to N_2O at $\omega = 0.0055$ Hz is about 22% which is well below conversion in the high rate branch. Together with the observation for total NO conversion, this suggests that NO conversion to N_2 may have been globally enhanced. Perhaps this comes from excitation of the overall reaction given by Equation (4.3).

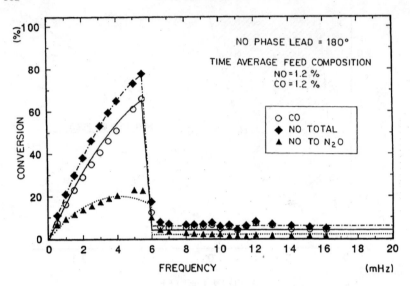

FIGURE 4.6 Conversion as a function of cycling frequency for out of phase ($\phi = 180°$) switching between feed mixtures containing 1.2 vol% CO and 1.2 vol% NO under the operating conditions in Figure 4.5 and over a 0.5 wt% Pt/Al2O3 catalyst (Figure taken from Sadhankar and Lynch (1996) with the author's permission)

Linearly increasing conversion, up to $\omega = 0.002$ Hz, falls in the region of quasi steady state operation. The increase with frequency is numerically predictable for this bang-bang modulation (Graham and Lynch, 1990; Sadhankar and Lynch, 1994; 1996). Because of mixing in the recycle reactor system used by the investigators, both reactants will be present in the environment of the catalyst for a short time after a switch between reactant mixtures so that the NO reduction proceeds. As the frequency increases or the cycle period decreases, this time during which reaction occurs becomes a growing portion of the cycle, increasing thereby the time-average conversions.

At the other end of the frequency spectrum used by Sadhankar and Lynch (1996), the relaxed steady state is observed. This appears to set in above $\omega = 0.06$. Because of mixing, composition variations in the feed are washed out by recycle (the flow ratio of the two streams is 1:10). Conversion behaviour with frequency in the rss region of operation (Figure 4.6) is interesting. Conversion appears to drop slowly with increasing frequency down to 6% for CO and 7% for NO which are the steady state values at the time-average feed composition. These limits seem to be reached at $\omega = 0.014$ Hz. Evidently, there must be a threshold effect for forcing NO reduction over the supported Pt catalyst. Thus, at $\omega = 0.006$ Hz, the system is forced out of the extinguished (or low rate) state for a short duration, but the composition change in the neighborhood of the catalyst surface is too small to ignite the reaction. Nevertheless, since the time-average conversions exceed those at steady state, the system must function for some time in a composition region unattainable in steady state operation. Only at $\omega < 0.006$ Hz is ignition consummated.

Significant amounts of N_2O appear in the qss and intermediate regions of frequency (Figure 4.6). This is not surprising in view of the results of Cho et al.(1989). Formation of N_2O via Equation (4.2) becomes favorable with respect to decomposition to N_2 below 300°C when a reductant is not present. This is the situation in the bang-bang cycling employed by Sadhankar and Lynch. Conversion to N_2O becomes negligible at high frequencies when $SN = 1$ for most of the cycle.

"Snap shots" of concentration variation with time in the qss, intermediate and rss regions of the resonance behaviour are shown in Figure 4.7. At 0.0005 Hz in (a), the double CO_2 peaks appear at the beginning of each half cycle just as observed by Cho et al.(1989) for supported Rh at 250° and 0.016 Hz. There are also double N_2 peaks, but N_2 was not measured in the Sadhankar and Lynch experiments. Just a single N_2O peak appears in (a) just after the intoduction of the CO feed mixture. Sadhankar and Lynch explain this by proposing that a surface saturated in O adatoms inhibits the fourth elementary reaction in Equation (4.4). N_2O forms only after CO exposure removes some of the adsorbed oxygen, that is, during catalyst contact with CO. These responses in qss region reproduce observations that would be made in step change experiments.

In the resonance region, (b), CO_2 double peaks remain, but the CO_2 maxima now lag the switching times by 40 to 50 s. The first peak with a ca. 50 s lag following the small NO peak simply confirms that the NO + CO reaction proceeds through the adsorbed species, presumably the O adatoms. The 40 s lag of the second peak and the observation that CO_2 proceeds through a minimum after the switch suggest that CO scavenging of adsorbed oxygen requires CO adsorption, ruling out the Eley-Rideal mechanism suggested by Cho et al. (1989) and confirming the final elementary reaction given in Equation (4.4). A small NO peak can be seen about 30 s after the switch to NO containing feed mixture. It is explained by the rising NO concentration after the switch due to mixing and the delay in creating sites for NO adsorption. The CO_2 concentration has begun to climb as the first NO peak appears. The second NO peak coincides with the switch to the CO feed mixture. The two N_2O peaks in (b) occur when the system becomes oxidizing. Their heights in Figure 4.7 are proportional to the heights of the NO peaks which lead each by several seconds. This proportionality and the lag substantiates the view that N_2O forms only through the fourth elementary step in Equation (4.4) which requires both molecular and dissociative adsorption of NO.

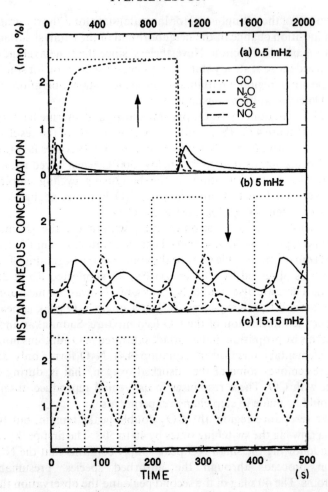

FIGURE 4.7 Transient behaviour of NO, N_2O and CO_2 species in different regions of Figure 4.6. Conditions as stated in Figure 4.6: (a) quasi steady state region, (b) resonance region, and (c) relaxed steady state region (Figure taken from Sadhankar and Lynch (1996) with the author's permission)

Finally, in the relaxed steady state region, conversions are very low corresponding to the low rate branch. The concentrations of the CO_2 and N_2O products oscillate but this is not apparant because of the scale of the figure. Variations in NO detected at the reactor outlet appear sinusoidal; they are shifted by 180° through mixing and the lower level of reaction which now occurs.

Phase lag or lead was one of the variables mentioned in Chapter 1. At the end of Section B, Chapter 3, it was shown for CO oxidation on Pt/Al_2O_3

that a phase lead less than 90° is optimal (Figure 3.18) and that the enhancement is significantly greater than that obtained with the more usual out of phase cycling. Phase lead is defined by Figure 3.17 in Chapter 3. Figure 4.8 illustrates the meaning of phase lead for the NO + CO system. When both the CO and NO concentration are periodically forced and phase lags or leads other than 0 and 180° are chosen, cycles will be composed of 4 parts as can be seen in Figure 4.10. It seems likely that

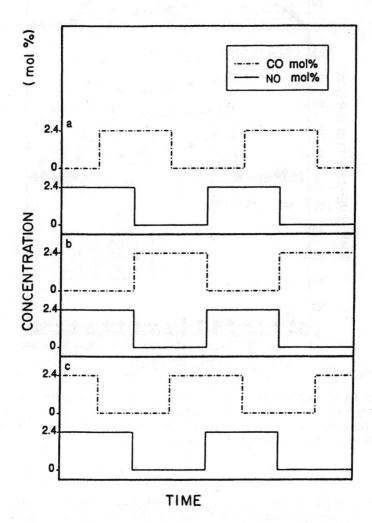

FIGURE 4.8 Interpretation of phase lead in terms of the time at switching of the NO and CO reactants in the modulated NO reduction by CO (Figure taken from Sadhankar and Lynch (1994) with permission, © 1994, Academic Press, Inc.)

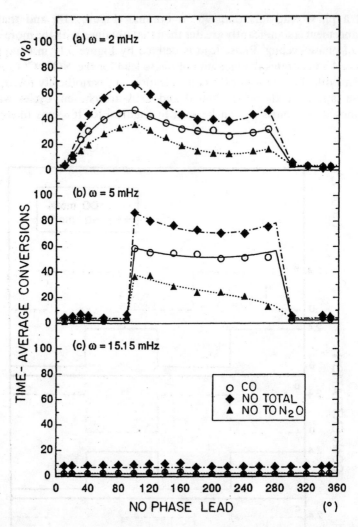

FIGURE 4.9 Effect of phase lead ($=360° -\phi$) on the time-average conversions for cycling between two feed mixtures containing 1.2 vol.% CO and 1.2 vol.% NO for catalyst and operating conditions given in Figure 4.5 (Figure taken from Sadhankar and Lynch (1996) with the author's permission)

enhancements of rates or conversions achieved with a phase lead strategy arise through multi-part cycles. Figure 4.9 shows that this is the case with NO reduction by CO over a Pt/Al_2O_3 catalyst at 485°K between 0.002 and 0.005 Hz. Figure 4.9(a) shows that with a phase lead of 90°, where the NO feed mixture is switched on 125 s before the CO feed mixture, all

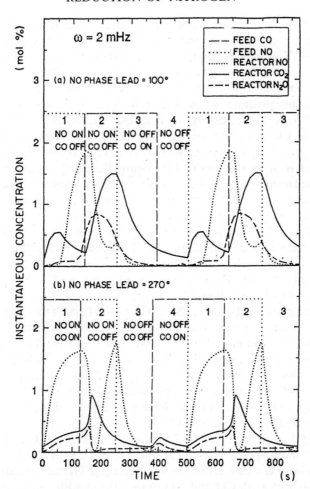

FIGURE 4.10 Transient behaviour of NO, N_2O and CO_2 species at different phase leads ($360° - \phi$) for cycling conditions in Figure 4.9: (a) NO phase lead = 100°, (b) NO phase lead = 270° (Figure taken from Sadhankar and Lynch (1996) with the author's permission)

conversions are at their maximums. CO conversion, for example, is some 52% greater than out of phase qss conversion at 0.002 Hz in Figure 4.5 which happens to be the CO conversion with a phase lead of 180°. 0.002 Hz was the boundary of the qss region in the frequency spectrum for out of phase cycling. With a 90° phase lead, the cycle has 4 parts of equal 125 s duration. These shorter exposures place forcing in the resonance region.

A phase lead of 270° also results in 4 part cycles with equal 125 s durations for each part. This achieves a CO conversion which is about the same as that obtained in out of phase cycling at 0.002 Hz in Figure 4.5. NO and N_2O conversions are slightly higher than those measured in the qss region, but both are much lower than conversions obtained with a phase lead of 90°. This implies that the sequencing of the parts is as important as their duration. Figure 4.10 compares the transient behaviour for phase leads of 100 and 270°. In the former, a CO off part cycle or quadrant proceeds the quadrant when both reactants are present. This induces double CO_2 peaks: a small one in the 1st quadrant when there is NO in the feed but no CO and a large peak at the end of the 2nd quadrant when both reactants are in the feed. Area under the NO curve is reduced and the single N_2O peak is larger than in (b) of the figure. Peak location and magnitude can be explained through the elementary reaction scheme of Equation (4.4) just as was done for the transient behaviour in Figure 4.7. For example, the CO_2 peak in the 1st quadrant results because the surface is heavily covered with CO from CO exposure in the 3rd quadrant, but the 4th quadrant in which there is only diluent in the feed allows some of the CO to desorb so that NO adsorbs rapidly and dissociatively to form O adatoms when it is introduced in the 1st part cycle or quadrant. Adsorbed CO and oxygen are required for CO_2 formation.

Above a phase lead of 300°, conversions drop back to those found in the extinguished state branch of the multiplicity region. This sharp change with phase lead is pronounced at $\omega = 0.005$ Hz where the reaction system functions in the extinguished state not only for phase leads greater than 300° but also when the lead is less than 90°. At the latter frequency, the variation of CO conversion with phase lead is quite small. The variation, however, increases for NO and N_2O conversions. With $\phi = 100°$, NO conversion seems to exceed NO conversion in the ignited state at the time-average feed composition. This is not the case for N_2O however. The explanation offered by Sadhankar and Lynch (1996) for the drop off in conversion for $\phi < 300°$ is that the duration of the 1st and 3rd part cycles or quadrants with both CO and NO in the feed or with neither in the feed become much longer than the 2nd and 4th quadrants. In the 1st quadrant the surface becomes saturated with CO which inhibits the reduction reaction and forces the system into the extinguished state. Of course with no reactants in the feed in the 3rd quadrant, conversion tends to zero. The argument applies at $\omega = 0.005$ hz, but is compounded by the shorter duration of the part cycles so that mixing becomes important and works to decrease concentration variation between the quadrants.

Mixing damps out the concentration variations between the cycle parts when $\omega = 0.01515$ and the system falls into the steady state mode which is located in the extinguished or low rate branch (Figure 4.5).

Sadhankar and Lynch (1996) report remarkably slow convergence to a stationary cyclic operation or what they call a cycle invariant state in the rss region of the frequency spectrum. This phenomenon was explored by the authors in a second contribution (Sadhankar and Lynch, 1995) using the experimental system, catalyst and operating conditions just described. Only out of phase, bang-bang cycling was used, that is, cycles had just two parts. The slow convergence is illustrated in Figure 4.11 at $\omega = 0.01$ and 0.005 Hz. In (a) of the figure, observations for N_2O reduction by CO and for CO oxidation with O_2 for the same catalyst and operating conditions are shown for comparison. In (b) both the feed rate and the switching compositions were varied. These long intervals in which the reactor or reaction system appears to be trapped at high conversion depend on the

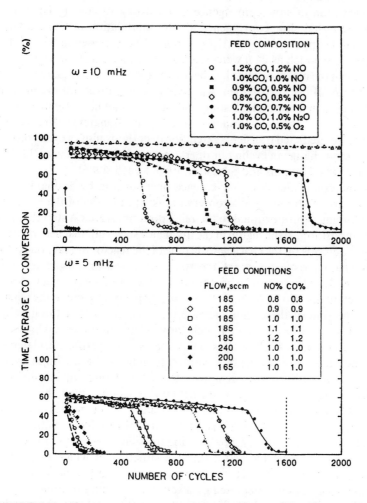

FIGURE 4.11 Slow convergence to a stationary cycling condition illustrated by CO conversion as a function of cycle number for conditions given in Figure 4.5: (a) $\omega = 0.01$ Hz, space time = 39 s, (b) $\omega = 0.05$ Hz with variable SV (Figure taken from Sadhankar and Lynch (1995) with permission, © 1995, Elsevier Science Publishers)

oxident (trapping occurs with O_2 and NO, but not with N_2O in Figure 4.11(a)), the CO concentration (Figure 4.11(a,b), and the flow rate or the space time (Figure 4.11(b). It is observed only when the catalyst is exposed to an oxidizing environment prior to use and it is not seen when NO reduction is forced at $\omega < 0.004$ Hz where the system operates in the resonance or qss regions of the frequency spectrum. Sadhankar and Lynch (1995) also report slow convergence for SN < 1, that is, in a time-average reducing environment. The data shown in Figure 4.11 for the NO + CO mixtures are all for SN = 1.

Observation of slow convergence is not a new discovery. Hugo *et al.* (1986) reported slow convergence for SN modulation of CO oxidation over a NiO supported on a γ-Al_2O_3 support. As discussed in Section F of Chapter 3, as much as 24 h were necessary to achieve a stationary cycling operation if the nickel catalyst had been exposed to a reducing environment prior to use. If the environment was oxidizing, the invariant cycling state was achieved in a few cycles lasting often less than an hour. Simulations of CO oxidation under modulation employing Langmuir-Hinshelwood models also predicted transients extending over thousands of cycles on start up or after a change in cycling or operating conditions (Jain *et al.*,1981a,b; Li *et al.*, 1984; Hudgins and Silveston, 1985; Lynch, 1986)

An analysis of the slow convergence phenomenon by Sadhankar and Lynch starts with two related observations: 1) duration of the transients under forcing is proportional to the relaxation times after a step change in feed composition, 2) the region in the frequency spectrum reached in forcing at ωs^{-1} depends on the characteristic time of the reactor or reaction system. These characteristic times are the space time and an adsorption time, $\tau_A = a_s m_{cat} L_A / Q_o (C_{CO})_o$, which is the ratio of the adsorption capacity for the key reactant divided by the mass flow rate of that component to the reactor. Using the sum of these characteristic times to reflect that the flow rate affects both, Sadhankar and Lynch form a dimensionless cycle period, $\tau/(\tau' + \tau_A)$. Figure 4.12 shows this dimensionless cycle period correlates well with the number of cycles, N_{cyc}, needed to achieve the invariant cycling state. These authors also demonstrate that a Langmuir-Hinshelwood model based on 5 of the 6 elementary reactions given by Muraki *et al.*(1986a) as Equation (4.4) and the 3 additional elementary reactions proposed by Cho *et al.* (1989) as Equation (4.5) provide a mathematical model which qualitatively predicts Figure 4.11(a). For this model, the authors assume a CSTR and a start up situation in which the catalyst surface is oxygen covered. This results in a set of 9 O.D.E.s and an algebraic relation which can be integrated by a suitable marching routine.

Sadhankar and Lynch (1995) remark that the long transients of Figure 4.11, in which the catalyst remain in at least a partially ignited state, reveal a heretofore undiscovered means of exploiting composition modulation to increase reaction rates or conversions. Their suggestion amounts to

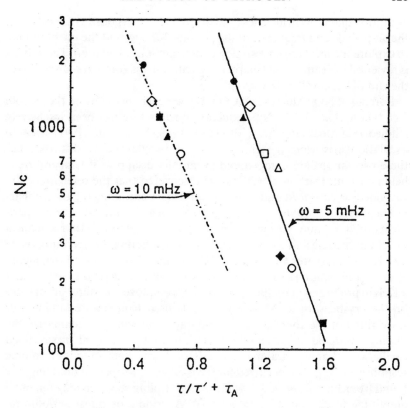

FIGURE 4.12 Number of cycles needed to achieve an invariant cycling state versus the dimensionless cycle period. Numbers of cycles are from Figure 4.11 (Figure taken from Sadhankar and Lynch (1995) with permission, © 1995, Elsevier Science Publishers)

forcing a reactor system at two frequencies. In the case of NO reduction through CO, one of these would be bang-bang cycling between NO and CO mixtures at, say, 0.01 Hz, while the second would be asymmetric exposure of the catalyst to an oxidative environment every 1000 cycles or at $\omega_2 = 0.00001$ Hz.

C. Pt SINGLE CRYSTALS UNDER HIGH VACUUM

Forcing of CO oxidation on single Pt crystal surfaces, specifically the Pt(100) and Pt(110) surfaces, has received a good deal of research attention. Both experiment and simulation have focused on forcing when autonomous oscillations occur on the unforced faces. These works discuss the excitation diagrams, which describe the frequency response as a function of the amplitude and frequency of the periodic forcing, and the use of forcing to explore the mechanism resulting in autonomous oscillation. The subject is reviewed in Section C of Chapter 3. Excitation diagrams are discussed at the end of Section B of that chapter.

Forcing of NO reduction by CO has been examined only on the Pt(100) face (Dath *et al.*, 1992). NO reduction on this face has been intensively studied (Schwartz and Schmidt, 1987; 1988; Fink *et al.*, 1991) because oscillations arise when the temperature of the single crystal is disturbed, but these oscillations are damped and eventually disappear if temperature is held constant. Dath *et al.*(1992) have discovered that the oscillations are sustained when forced and only harmonic (1:1 entrainment) and subharmonic (period doubling or tripling) response are observed. Their experiments used a 7 mm × 7 mm × 1 mm crystal prepared by electrochemical polishing, heating and cooling cycles and Ar sputtering. Temperatures were 393°K and pressures were in the UHV range of 10^{-10} bar. Low energy electron diffraction (LEED) was used to determine crystal structure and a Kelvin probe and mass spectrometry were employed to follow adsorbates on the crystal surface. Most experiments used temperature forcing, but since this not the topic of this monograph, we will just summarize the findings. Forcing at an amplitude as low as 1°K resulted in sustained oscillations. Large entrainment bands or horns were found. Harmonic response ($\omega_{response} = \omega$) occurred for $0.2 \leqslant \omega/\omega_0 \leqslant 1.8$, period doubling was found from $1.8 \leqslant \omega/\omega_0 \leqslant 2.7$, while period tripling was seen only for small amplitude forcing at $2.7 \leqslant \omega/\omega_0 \leqslant 3.4$. Aperiodic or quasi periodic response was not found. Figure 4.13(a) shows the mass spectrometer signals for CO_2 after a 5°K excursion for the system forced at a 1° amplitude and for the unforced system. Dath *et al.* report the NO + CO reaction on Pt(100) behaves as a non-linear oscillator under forcing: the maximum response amplitude is found when $\omega = \omega_0$, but the phase shift at the upper bound of the entrainment band was not 180°.

Damped oscillations could not be initiated by varying the NO or the CO partial pressures. However, a sinusoidal oscillation of the CO partrial pressure caused sustained harmonic oscillations once these were initiated by a 5° temperature disturbance. The amplitude of the response (CO_2) as a function of the forcing amplitude and frequency are shown in Figure 4.13(b). Response is harmonic. The CO_2 partial pressure oscillates with the forcing frequency. The response maxima now occurs at $\omega < \omega_0$. The phase shift ϕ varies from 0 to 180° as ω/ω_0 increase. This suggests the NO + CO reaction on the face behaves as a linear oscillator. A further set of experiments were undertaken using random variations of the CO partial pressure. This set indicated that the chemical system acts as a band path filter rejecting excitation by disturbances whose frequencies differ from ω_0. Dath *et al.* attribute the damped autonomous oscillations and the response

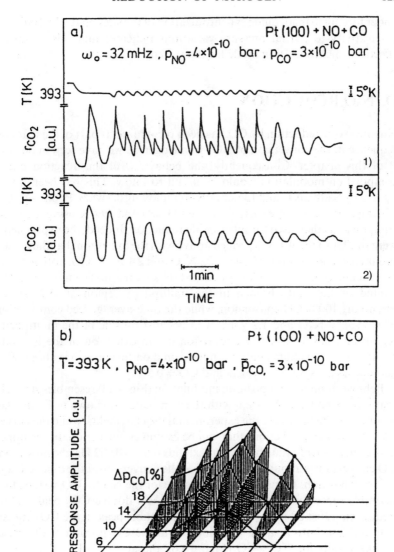

FIGURE 4.13 Response to the forcing of damped autonomous oscillations for the NO + CO reaction on the Pt(100) face at 393°K under UHV conditions: (a) mass spectrometer signal versus time for (1) sinusoidal temperature forcing at an amplitude of 1° and (2) non forced system after a 5° disturbance; (b) response amplitude for sinusoidal p_{CO_2} forcing as a function of forcing amplitude and frequuency. Temperature and partial pressures indicate in (b) apply to (a) as well (Figures adapted from Dath et al. (1992) with permission. © 1992, American Institute of Physics)

under forcing to inadequate synchronization of the sites on the Pt(100) face. Insensitivity to partial pressure variations indicates that this is not an effective synchronization mode for the system.

D. N_2O REDUCTION

Nitrous oxide reduction by CO has been recognized as part of the reaction sequence in the conversion of NO_x to nitrogen as we have seen in Section B of this chapter. Consequently, the behaviour of this reaction under composition modulation would be useful to know. This has been investigated by Sadhankar and Lynch (1994) employing 0.5 wt.% Pt on γ-Al_2O_3. Their experiments were carried out at 103 kPa and 499°K using the same equipment as they used for nitric oxide reduction. Indeed, the experiments were similar to the study described above with the exception that they used pre-mixed compositions of 2.4 vol.% N_2O and 2.4 vol.% CO, both with N_2 as diluent. The N_2O + CO system exhibits isothermal multiplicity. The ignited or high rate branch in the multiplicity region at 499° shows essentially 100% CO conversion, while the time-average feed composition of 1.2 vol.% N_2O and 1.2 vol.% CO lies just outside of the multiplicity region and has a 20% CO conversion on an extension of the low rate branch. The maximum conversion measured on this branch is about 60% and occurs at N_2O = 1.2 vol.% and CO = 0.6 vol.%.

Behaviour under composition modulation (Fig. 4.14) resembles strongly that for the NO + CO system with the same catalyst shown in Figure 4.6. Symbols in the figure are the experimental measurements, while the curves show the model predictions that will be examined further on. In the upper figure, quasi steady state behaviour extends to ω = 0.002 Hz; the maximum CO conversion is about 80%, about 4 × the conversion at the time-average feed composition. Transients within a cycle for the N_2O + CO system, too, resemble those shown in Figure 4.7. There is a double CO_2 peak at low frequencies. Just after ω_{opt} = 0.006 Hz, the double peak in the CO_2 transients collapses into a single peak located in the CO portion of the cycle. Unlike results for NO + CO (Figure 4.6), the abrupt fall to the relaxed steady state sets in at about 0.012Hz. Conversion at this point is 30% so there is evidently a slow decay to the steady state conversion as the frequency increases. Sadhankar and Lynch do not report slow convergence to a stationary cycling operation in the rss region of the frequency spectrum.

The bottom half of Figure 4.14 shows CO conversion as a function of forcing frequency for phase leads of 90 and 270°. These phase leads result in 4-part cycles in which each quadrant is of equal duration. Just as for the NO + CO system, a phase lead of 90° appears to be optimal and composition modulation at this phase lead and a cycle frequency of about 0.005 Hz gives about 92% CO conversion which is just below conversion in the high

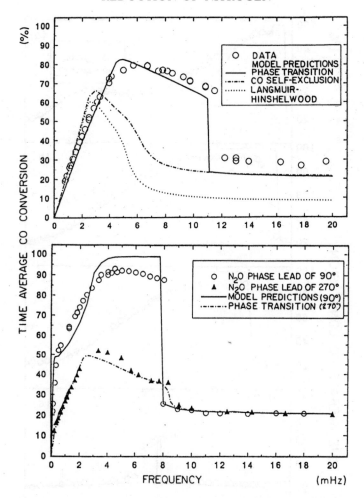

FIGURE 4.14 Dependence of CO conversion on forcing frequency for (a) symmetrical out of phase ($\phi = 180°$) cycling with two part cycles, (b) cycling with phase leads of 90 and 270° and equal duration four part cycles. N_2O reduction by CO over Pt/Al_2O_3 at 103 kPa and 499°K (Figure taken from Sadhankar and Lynch (1994) with permission, © 1994, Academic Press, Inc.)

rate branch. The system is thus essentially ignited eventhough the conversion at the time-average feed composition would be in the extinguished state. Although the optimal frequency shifts only slightly with phase lag, the qss region is now confined to frequencies below 0.0002 Hz and the drop to the rss region now occurs at about 0.008 Hz. The 4-part cycle exacerbates the effect of mixing in the recycle reactor used for the

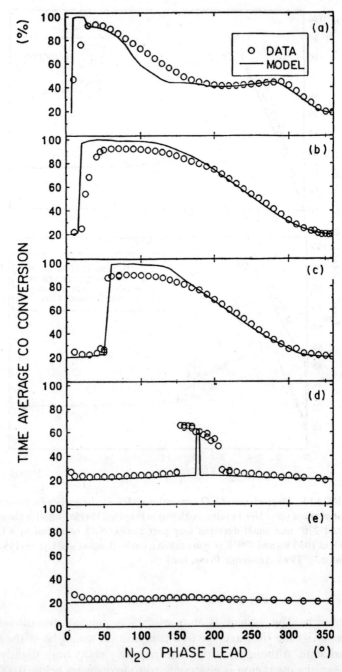

FIGURE 4.15 Dependence of CO conversion on phase lead for $\omega = 0.003$ Hz (a), $\omega = 0.004$ Hz (b) $\omega = 0.006$ Hz (c) $\omega = 0.011$ Hz (d) and $\omega = 0.016$ Hz (e). Operating conditions and catalyst as in Figure 4.14 (Figure taken from Sadhankar and Lynch (1994) with permission, © 1994, Academic Press, Inc.)

Sadhankar and Lynch experiments. Transients within a cycle are given reasonably well for the two phase leads by Figure 4.10.

The interaction of phase lead and cycling frequency is illustrated by Figure 4.15. In (a) of the figure, the system is at the boundary of the qss region for out of phase cycling. The system, however, can be further excited by decreasing the phase lead to create almost in phase cycling. Figures 4-15b,c show that in the region of ω_{opt}, phase leads from 60 to 180° give a similar performance eventhough a ϕ_{opt} can be discerned. The excitation region of phase lead becomes quite narrow at $\omega = 0.0111$ Hz and centers at about 180° where there is a 2-part cycle. This is evidentally a consequence of mixing in the recycle reactor. The important conclusion to be drawn from Figure 4.15 is that for NO_x reduction over Pt/Al_2O_3, at least, optimal reactor performance can be achieved by a combination of phase lead and cycling frequency.

Sadhankar and Lynch (1994) have developed successfully a model for their modulated $N_2O + CO$ system based on the following reactions which are assumed to be elementary:

$$CO + s \rightarrow CO \bullet s$$

$$N_2O + s \rightarrow N_2 + O \bullet s \qquad (4.6)$$

$$CO \bullet s + O \bullet s \rightarrow CO_2 + 2s$$

These steps are part of the sequence given in Equations (4.4) and (4.5) for the $NO + CO$ system. Material balances on the measured species CO, N_2O and CO_2 and on the surface species θ_{CO}, θ_O and $(\theta^s)_{CO_2}$ provide a system of 6 ODEs if the recycle reactor is treated as an isothermal CSTR (Table 4.1). The authors refer to the system of ODEs as a conventional Langmuir-Hinshewood model. Parameters were obtained from steady state experiments (Sadhankar et al.,1994). Integrating the system subject to initial conditions given in Table 4.1 gives the time variations of the gas phase concentrations. Substituting the instantaneous values in the integral

$$X = \frac{1}{\tau} \int_0^\tau C'(t) \, dt \qquad (4.7)$$

determines the time-average CO conversion. These conversions are the dotted line in Figure 4.14(a). The trends are correctly given but the data are poorly represented.

A better fit is obtained by assuming adsorbed CO prevents further CO adsorption nearby. Sadhankar and Lynch refer to this as a self exclusion effect. The resulting correction terms are shown in the CO adsorption terms in Table 4.1. In the table, N_{CO} is the exclusion constant. The authors use either 1.018 or 1.0015. On integration and substitution of C'_{CO_2} in Equation (4.7), the dashed line shown in Figure 4.14(a) results.

Table 4.1 Kinetic Model for N_2O Reduction by CO Pt/Al_2O_3 Catalyst Mounted in a Fully Back Mixed Reactor (Sadhankar and Lynch, 1994)

Gas Phase:

$$\frac{dC'_i}{dt} = (C'_i) - qC'_i - \frac{1}{Q_0}\left(\frac{S_p m}{\rho}\right)_{cat} r_i - \frac{1}{Q}\left(\frac{S_p m}{\rho}\right)_{Sup} r_i^S$$

$$i = 1, 2, 3 = CO, N_2O, CO_2$$

Catalyst Surface:

$$\frac{d\theta_4}{d't} = r_2 - r_3$$

Support Surface:

$$\frac{d\theta_3^S}{d't} = r_3^S$$

Gas Shrinkage:

$$q = 1 - y_1 r_1 - y_2 r_2 + y_1 r_3 - y_1 r_2^S$$

r_i, r_1 r_2 r_3 r_i^S depend on the kinetics assumed. Sadhankar and Lynch (1994) employed Langmuir-Hinshelwood Model with CO self exclusion:

$$CO: r_1 = S_t (k_a)_1 C'_1 (1 - \theta_1 - \theta_4)\left(\frac{1 - N_{Co}\theta_1}{1 - \theta_1}\right) - S_t(k_d)_1 \theta_1; \quad r_1^S = 0$$

$$N_2O: r_2 = S_t (k_a)_2 C'_2 (1 - \theta_1 - \theta_4); \quad r_2^S = 0$$

$$CO_2: r_3 = S_t^2 k \theta_1 \theta_2; \quad r_3^S = (k_a^S)_3 C'_3 (1 - \theta_3^S) + (k_d^S)_3 \theta_3^S$$

Initial Conditions:

$$t = 0; \theta_1 = \theta_4 = 0, \theta_3^S = 0$$

$$(C'_1)_0 = (C'_2)_0 = (C'_3)_0 = 0$$

for all values of ϕ:

$$(C'_2)_0 = 2, (n-1)\tau \leq t < \left(n - \frac{1}{2}\right)\tau$$

$$(C'_2)_0 = 0, \left(n - \frac{1}{2}\right)\tau \leq t < n\tau,$$

for $0° \leq \phi < 180°$

$$(C'_1)_0 = 0, (n+1)\tau \leq t < \left(n - 1 + \frac{\phi}{360}\right)\tau$$

$$(C'_1)_0 = 2, \left(n - 1 + \frac{\phi}{360}\right)\tau \leq t < \left(n - \frac{1}{2} + \frac{\phi}{360}\right)\tau$$

Table 4.1 (Continued)

For $180° \leq \phi < 360°$ $\quad (C'_1)_0 = 0, \left(n - \dfrac{1}{2} + \dfrac{\phi}{360}\right)\tau \leq t \, n\tau$

$(C'_1)_0 = 2, (n-1)\tau \leq t < \left(n - \dfrac{3}{2} + \dfrac{\phi}{360}\right)\tau$

$(C'_1)_0 = 0, \left(n - \dfrac{3}{2} + \dfrac{\phi}{360}\right)\tau \leq t < \left(n - 1 + \dfrac{\phi}{360}\right)\tau$

$(C'_1)_0 = 2, \left(n - 1 + \dfrac{\phi}{360}\right)\tau \leq t < n\tau$

To achieve the good fits that are represented by the solid lines in Figures 4.14 and 4.15, the adsorption rate constant for CO, $(k_a)_{CO}$, and the rate constant must be made non-linear functions of the CO coverage. Graham and Lynch (1990) used this artifice to obtain a good representation of out of phase forcing of CO oxidation over a Pt catalyst and describe it as an allowance for surface phase transformation. Single crystal studies indicate that the LEED pattern of the surface changes, often abruptly, has adsorbate increases on the surface, that is, the crystallographic structure of the surface can undergo change. Catalytic properties are structure sensitive so these can change dramatically. Under the experimental conditions used, a complicated quasi-hexagonal structure is observed on clean Pt surfaces. This restructures into a quadratic or (1×1) structure typical of Pt(100) according to Graham and Lynch. Restructuring is reversible but hysteresis exists with respect to CO coverage. Sadhankar and Lynch (1994) increase $(k_a)_{N_2O}$ and the rate constant k by $5 \times$ and $100 \times$ respectively when the surface is in the (1×1) quadratic phase. These values are used when $\theta_{CO} \geq 0.95$ and when these enhanced constants are used they continue to be used until θ_{CO} drops to 0.1. With this modification to the kinetic constants, the model summarized in Table 4.1 provides a good fit to the forcing results discussed in the earlier paragraphs. The solid lines in Figurs 4.14 and 4.15 give the predicted CO conversions. Prediction of the variations of CO and CO_2 vol.% within a cycle were also satisfactory. As might be expected, the Sadhankar and Lynch model closely reproduces the isothermal multiplicity observed by Sadhankar et al. (1994).

Cho (1994) employed an interesting cycling experiment, switching symmetrically between $N_2O + CO$ and $NO + CO$, as part of his attempt to unravel the importance of N_2O formation in NO_x reduction by CO. N_2O is observed in the $NO + CO$ reaction as we have discussed at the beginning of this section, but only in low concentrations. Much higher concentrations would be expected, however, because the $N_2O + CO$ reaction appears to be slow relative to the $NO + CO$ reaction under steady state operation.

For this experiments, Cho packed 22 mg sample of Rh/Al_2O_3 into an isothermal 3.2 mm × 10 mm bed. This reactor was modulated between feeds containing 0.04 vol.% ^{15}NO and 0.04 vol.% N_2O at $\tau = 20$ s, $SV = 84,000$ h^{-1}, $T = 275°C$ and $P = 101.3$ kPa. Both feed streams contained 0.04 vol.% ^{13}CO. Helium served as diluent. The isotopes of NO and CO permitted use of a mass spectrometer to obtain the time variations of reactants and products. Cho observed rapid consumption of N_2O and ^{13}CO and the formation of N_2 for about 4 s after the switch from the ^{15}NO mixture to the N_2O mixture. This dropped abruptly for the remaining 6 s of the half cycle. In the second part of the cycle following a switch back to the ^{15}NO mixture, the NO and CO isotopes are almost quantitatively converted to $^{15}N_2$ and $^{13}CO_2$. Small amounts of $^{15}N_2O$ were observed immediately after the switch back to the ^{15}NO mixture. The interpretation made by Cho for his results is that the N_2O reduction is strongly inhibited by CO under the conditions employed whereas NO reduction is not inhibited. Switching from NO to N_2O exposure provides vacant sites for N_2O adsorption so this can proceed before CO adsorption covers the Rh surface. Consequently, only small amounts of N_2O are observed in NO reduction by CO over Rh catalysts because any N_2O formed is rapidly reduced by CO. This reaction proceeds rapidly in the absence of inhibition.

E. DISCUSSION

Despite the presence of sequential reactions for NO reduction by CO, the modulation results for the noble metal catalysts resemble those for CO oxidation discussed in the previous chapter. Modulation strongly enhances rate and conversion for Pt at 400°C. The enhancement is much smaller for Pd and absent for Rh. Enhancement is possible for the latter catalyst, but only at temperatures near 200°C. Just as for CO oxidation, increasing temperature reduces the optimal cycle period and for Rh, at least, rate enhancement through periodic composition forcing. The importance of composition modulation for NO reduction by CO with a Pt catalyst is consistent with the negative order with respect to CO. This signals a strong CO inhibition of NO reduction just as observed for CO oxidation. Indeed, models proposed for both reactions over Pt are similar. Each models assumes competitive adsorption, a surface reaction between adsorbate species and CO_2 adsorption on the catalyst support. Both models include surface restructuring as a function of CO coverage. With this allowance, the models closely predict isothermal multiplicity and product variations with time under forcing. Consequently, composition modulation improves reactor performance for NO reduction by CO through mitigating CO inhibition by providing surface access for the less strongly adsorbed reactant, NO for the reduction reaction and O_2 for CO oxidation. Under conditions for which inhibition is absent, for example with an Rh catalyst above 300°C, composition modulation diminishes reactor performance.

There are several observations for the NO + CO system that call for further investigation. Unlike CO oxidation over Pt, Figure 4.2 indicates increasing temperature increases conversion enhancement through periodic forcing. What is the explanation? Does CO inhibition increase with rising temperature?

Sadhankar and Lynch (1994) invoke surface reconstruction for their successful model of N_2O reduction by CO over Pt/Al_2O_3. Reconstruction has been observed for spontaneous oscillations in the $CO + O_2$ system on Pt single crystals under UHV and it is imputed to explain these oscillations for the NO + CO system under similar conditions. However, Fink *et al.* (1991) and Dath *et al.* (1992) did not find that reconstruction was associated with oscillations of NO reduction by CO on the Pt(100) surface under UHV. Is the use of markedly different rate constants for different surface compositions just a modeling artifice? Single crystal studies remain to be undertaken for the N_2O + CO system.

Certainly further study of the extended transients observed by Sadhankar and Lynch (1995) for NO reduction is needed. These transients require an explanation and modeling to see for what systems they may arise. The observations suggest another approach for using composition modulation to improve reactor performance. Modulation using two substantially different frequencies would be one means to achieving the approach discovered by Sadhankar and Lynch. Is this an effective means? How general is the approach? Are there other systems which exhibit an unstable state that drops so remarkably slowly to the stable condition?

CHAPTER 5
Automotive Exhaust Catalysis

A. BACKGROUND

Catalytic converters for automobile exhaust and the air/fuel (A/F) ratio control loops that permit them to operate efficiently are a remarkable achievement. There are currently several million in service and, in contrast to industrial catalytic reactors, they function under a regime of benign neglect by their owner-operators. They merit discussion in this monograph because the A/F ratio feedback control loop subjects the catalytic converter to high frequency composition modulation. In addition, the converter often experiences irregular cycles of varying flow rate and inlet temperature.

The exploration of composition modulation of these devices has taken various directions in the decades since the first studies were made. One direction has been the study of CO oxidation as a model for all the reactions taking place. This was certainly the first direction pursued. A second direction, an extension of the first, was investigations of individual three way catalyst (TWC) reactions, then two reaction and finally three reaction systems. The third direction has employed simulated automotive exhaust, while the final direction has been to use a motor with flow and A/F ratio control. In the latter, no attempt has been made to examine the individual reactions going on. Most of the single reaction studies have been discussed in earlier chapters: hydrocarbon oxidation in Chapter 2, CO oxidation in Chapter 3 and NO reduction in Chapter 4.

Engine Behavior, Driving Cycles And Exhaust Condition

Unlike a chemical plant, an automotive vehicle operates routinely under a variety of conditions: such as cold start, acceleration, deceleration, idle, or constant speed cruising. This variety will only change with the development of a hybrid system in which the gasoline engine drives a generator with power storage so that the motor can operate under constant load. The three way catalyst and exhaust gas recirculation systems, now widely used, are capable of controlling emissions in the constant speed cruising periods. However, variations in the A/F ratio and in gas flow rate during the

---

[1]An early version of this chapter appeared in *Catalysis Today*, Volume 25 (1995)

acceleration or deceleration events lead to fluctuations in pollutant concentration, exhaust temperature and flow rate which can produce pollutant breakthrough. The A/F ratio controller is generally set to an A/F ratio = 14.7 (as weight), which is the stoichiometric ratio for most automotive fuels. CO, HC and NO conversions are sensitive to the A/F ratio (Figure 5.1) and to the gas flow rate through the converter. Flow rate increases sharply on acceleration and somewhat less with deceleration. Both the A/F ratio and space velocity force excursions in the CO emissions when the ratio is below 14.5 and excursions for NO when the ratio approaches 15. Total gas composition, with the exception of the ppm species, appears in Table 5.1. It can be seen from this table that the catalytic converter operates on just 0.5 to 2% of the exhaust gas. Typical gas conditions entering the converter are 480°C and 103 kPa. Gas flow rates are about 33,000 cm^3/s (STP) which for most converters works out to SV = 28,000 h^{-1} (Taylor, 1984). Temperature of the gas entering the catalyst bed depends on the A/F ratio, increasing as the ratio decreases. Temperatures approach 1000°C and flow rates reach over 100,000 h^{-1} when a vehicle operates at full throttle, that is, with a rich mixture. Both temperature and flow rate also increase

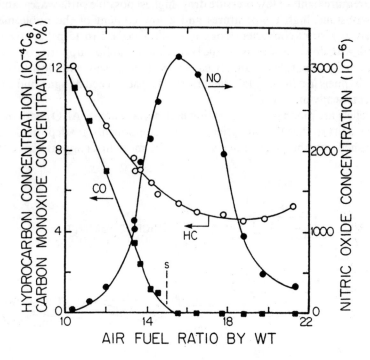

FIGURE 5.1 Variation of CO, hydrocarbon and NO concentrations in automotive exhaust with the A/F ratio (Adapted from Wei (1975) with permission, © 1975, American Chemical Society)

Table 5.1 Conversion of Target Components in a Monolith Catalytic Converter with Automotive Fuels Containing Sulphur and A/F Modulation of the Engine Exhaust.

Exhaust Temp.	Wt.% Sulphur	% Conversion					
		0.5 Hz			2 Hz		
		HC	CO	NO	HC	CO	NO
400	0.001	85	49	69	91	80	78
–	0.10	76	49	68	85	73	78
500	0.001	93	81	69	98	97	94
–	0.10	85	58	61	96	97	90

rapidly with motor rpm and vehicle speed. Cyclic discharge of the engine cylinders means that both flow and temperature oscillate at a high frequency.

Catalytic Converter Systems

The requirements of low pressure drop, highest possible surface area, ability to withstand high temperatures and a low content of the noble metal catalysts restrict converter design to shallow beds, 5 to 15 pellet layers in depth with down flow through the beds, or to monolith supports consisting of an array of parallel channels, ca. 3 to 8 channels/cm, with flow through the channels in a horizontal orientation. The packed bed design is indicated schematically in Figure 5.2.

Pellets are made by impregnating high surface area γ-Al_2O_3 with dilute metal salts so that the catalytically active components are deposited as an "eggshell" on the pellet outer surface. The monolith supports have thin

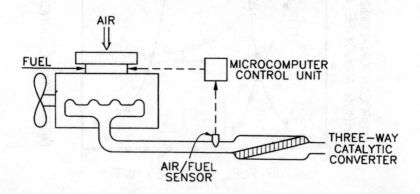

FIGURE 5.2 Schematic of a three-way catalytic converter, oxygen sensor and the A/F ratio feed back control loop (Adapted from Herz (1982) with permission, © 1982, American Chemical Society)

walls and thus require strong support materials such as cordierite or mullite. These materials have low porosity so a washcoat of γ-Al_2O_3 is applied prior to the impregnation step. Three-way-catalysts (TWC) in the active, reduced state are mixtures of noble metals: platinum (Pt), rhodium (Rh) and sometimes palladium (Pd). Many formulations add transition metal oxides such as cerium oxide to increase oxygen storage capacity. Base metals are often added to increase stability. Roles of the noble metals are well established: Pt catalyses CO and HC oxidation, but is required to oxidize paraffins, Pd is a more active CO, olefin and oxygenate catalyst than Pt, but it has poor tolerance for poisons, while Rh improves the NO reduction ability of the catalyst. Rh is also an active catalyst for the other primary reactions, including the water gas shift reaction.

The three-way catalyst functions efficiently only in a narrow range of exhaust gas composition. This range is essentially fixed by the A/F ratio. Consequently, the converter requires a controller to maintain the A/F ratio within the necessary range. A schematic of the controller appears in Figure 5.2. The sensing device measures the oxygen partial pressure in the exhaust. Its active element is a solid electrolyte, a doped zirconium oxide. The sensor signal is periodically monitored by the controller which initiates a change in the A/F ratio depending on signal magnitude. This system maintains the A/F ratio at its stoichiometric value under constant speed, cruising operation, and quite close to this value even when abrupt changes in motor rpm occur. Herz and Shinouskis (1985) present data showing that the ratio is within 0.1 of the stoichiometric value for about 70% of the time for typical throttle variation once the motor-sensor-converter system has warmed up.

Sensor hysteresis and measurement lags combined with lags due to the flow path, which change with acceleration/deceleration, result in a low amplitude oscillation of the A/F ratio even under constant motor speed. The frequency varies with the exhaust gas flow rate from 0.5 to 2 Hz.

Oscillation In Exhaust Gas Composition

The fluctuating A/F ratio caused by the feedback control loop results in time varying concentrations at the converter inlet. Variations in the CO concentrations measured in situ by Herz et al. (1983) at the converter inlet using IR spectrophotometry are shown in Figure 5.3a. The amplitude is about 0.5 for a time average concentration of 0.75 vol.%, but depends on the A/F ratio and motor operating conditions. Other data presented by Herz (1982, 1987) and Herz et al. (1983) show amplitudes as low as 0.05 vol.%.

In addition to the regular ca. 1 Hz oscillations and the irregular oscillations at 0.01 to 0.1 Hz caused by variation in motor speed and throttle control, Herz (1982) mentions a higher frequency oscillation at about 10 Hz caused by fuel maldistribution and modulation in a carburetor or fuel

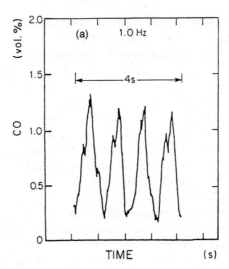

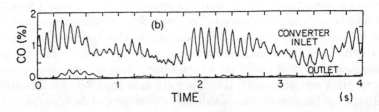

FIGURE 5.3 Time variation of CO concentrations in exhaust systems: (a) modulation of CO in the exhaust of an dynamometer mounted automotive engine by intentionally cycling the A/F ratio at 1 Hz to represent forcing by the feedback control loop, (b) time record of catalytic converter inlet and outlet CO concentrations for an exhaust from a dynamometer mounted automotive engine under A/F ratio feedback control (Figures adapted from Herz *et al.* (1983) with permission, © 1982, 1983, American Chemical Society)

injection pump. This quite regular disturbance is shown in Figure 5.3b along with a ca. 0.7 Hz oscillation resulting from the A/F ratio, feedback control loop. Variation from the cyclic discharge of the engine cylinders appears as low amplitude noise in the upper trace. The significance of this figure is that the 10 Hz variation is damped out in the outlet CO concentration; the 0.7 Hz signal is damped too, but it is still present in the outlet.

Objectives Of Composition Modulation Research On Automotive Exhaust

The automotive catalytic converter is modulated in two frequency ranges: ca. 1 Hz and ca. 10 Hz. Irregular disturbances in composition, flow rate and temperature occur at 0.01 to 0.1 Hz. These latter disturbances diminish the converter performance somewhat, but there is little in the literature about this essentially non-periodic forcing and it will not be dealt with further. Figure 5.3b suggests that the regular 10 Hz modulation does not affect the converter performance so it too will be disregarded.

The central question posed in research on composition modulation has been whether modulation enhances or depresses pollutant conversion. Secondary questions are the conditions under which enhancement is observed, the magnitude of the enhancement and, of course, the sources of improved performance. An alternative statement of the central question is how this apparently unavoidable modulation affects the choice of catalyst or catalyst design. In recent years, it is this question that has dominated the research effort. Nevertheless, our attention in this chapter focuses on the first formulation

Answers to both the central and the secondary questions have been sought in diverse ways, as mentioned above: 1) studies of CO oxidation, 2) sequential studies of individual reactions and then groups of reactions, 3) studies of simulated exhaust under carefully controlled conditions, and 4) studies on the exhaust of test engines. The shortcoming of 2) is the extrapolation to an exhaust under actual operating conditions, while for 3) and 4) the shortcoming is that these approaches often make it difficult to explain observations. For the purposes of this monograph we will follow the sequential examination of individual reactions, then groups, and finally move on to simulated and real exhausts.

B. EXPERIMENTAL SYSTEMS

The equipment and procedures described in Chapter 1 are limited to modulation frequencies well below 1 Hz. Using ordinary solenoid valves at 1 Hz produces a sinusoid rather than a square wave which diminishes amplitude (Muraki *et al.*, 1985c). General Motors (GM) researchers (for example, Schlatter *et al.*, 1983) employed special fast acting solenoid valves which allowed them to reach 4 Hz before square waves were damped into sinusoids. Feeds for both parts of a cycle were blended from as many as seven pure gases in the GM unit to represent automotive exhaust at different A/F ratios.

Much of the early modulation research utilized actual motor exhaust. Typically, these streams were supplied from an engine attached to a dynamometer so that the behavior of vehicles under diverse driving conditions could be simulated. The set-up used by GM is illustrated in Figure 5.4. Modulation of the A/F ratio was accomplished either through a feedback control loop or through a computer shown in the figure (Herz, 1982). The system employed by Toyota is similar (Matsunaga *et al.*, 1987).

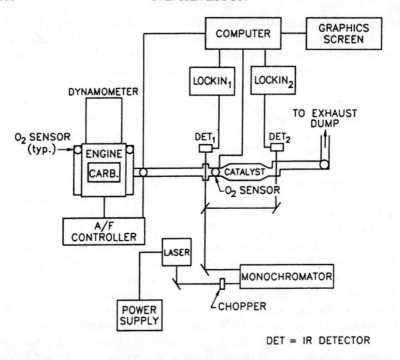

FIGURE 5.4 Schematic of a dynamometer mounted engine under either direct computer control or operated under a feedback control loop employed to provide an exhaust gas feed to a catalytic converter. The converter is equipped with sapphire or BaF_2 windows at the inlet and outlet for use in IR measurements with a tunable laser diode as the IR source. The IR system is shown schematically (Figure from Herz (1982) with permission, © 1982, American Chemical Society)

Chromatography is impractical for cycling frequencies above 0.5 Hz, so some sort of continuous on-line analysis is needed. One such method is IR spectroscopy. Figure 5.4 shows the application of a laser diode IR spectrophotometer to CO measurements. The laser diode was tuned to 2127.685 cm^{-1} and each beam in the dual beam system was chopped at 3.5 kHz to filter out background IR radiation from the gas and from equipment surfaces. IR may be used for NO, but is impractical for hydrocarbons because of the long path length required. The measurement system shown in Figure 5.4 was used for the data shown in Figure 5.3b.

Toyota researchers (Yokota *et al.*, 1985, Matsunaga *et al.*, 1987) leaked a small fraction of the reactor or motor effluent into a multichannel, quadrupole mass spectrometer in which each channel was set to a mass number of interest. This provided a continuous time record and an acceptable solution to the measurement dilemma at frequencies higher than 0.5 Hz.

C. COMPOSITION MODULATION AND SINGLE REACTIONS IN AUTOMOTIVE EXHAUST

This topic overlaps the treatment in Chapters 2 to 4 so our discussion has been divided. Experiments pertaining to reactions with two reactants, usually with a diluent, is treated in earlier chapters, eventhough the research was directed at automotive applications. On the other hand, experiments using an automotive exhaust, synthetic or actual, but concerned with just one reaction, invariably CO oxidation, will be examined now.

Schlatter and Mitchell (1980) employed synthetic exhausts corresponding to A/F ratios of 14.2 and 15.2 and a space velocity of 104,000 h^{-1} to assess the benefit of cerium oxide(CeO) in a TWC formulation. Oxygen storage in the oxide was thought to explain increased CO conversion under composition forcing. A variety of catalysts were tested at 500°C. Metal weight % in the different formulations were 0.05 wt% Pt, 5.0% Ce, and 0.005% Rh.

The experiments confirmed that Ce addition improved performance under forcing, but also showed that CO conversion was still well below conversion at steady state. Figure 5.5 illustrates the Schlatter and Mitchell results. Steady state conversion in the figure is about 95% compared to about 80% under modulation at 2 Hz with the Ce containing catalyst. Further experiments showed that water vapor must be present for Ce to affect performance. Material balances on O_2 step change experiments as

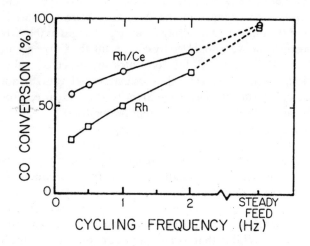

FIGURE 5.5 CO conversion as a function of forcing frequency for rhodium catalysts formulated with and without cerium (Figure adapted from Schlatter and Mitchell (1980) with permission, © 1980, American Chemical Society)

well as the cycling results indicated that the water gas shift (WGS) explained the improved performance under composition forcing when Ce and Rh were present in a TWC. Rhodium exhibits short term high activity for the shift and this activity is stabilized by Ce, according to Schlatter and Mitchell.

A series of papers by Herz and co-workers (Herz,1981; 1982; Herz et al., 1983 and Herz and Sell,1985) employed the exhaust system depicted in Figure 5.4. Various pelleted catalysts were examined in the series. Platinum content ranged from about 0.045 to 0.087 wt.%, Pd from about 0.024 to 0.032 wt.%, Rh from 0.002 to 0.006 wt.% and Ce from 1 to 2.6 wt%; γ-Al_2O_3 served as the support. Not all of the components were used in each catalyst. The noble metals were deposited in an eggshell layer 25 to 60 μm deep while Ce was uniformly distributed. Catalyst bed temperatures ranged from 450 to 500°C and space velocities were about 50,000 h^{-1}. Herz and co-workers seemed to be searching for an explanation of why CO conversion under modulation of the A/F ratio were sometimes below conversions obtained at steady state. They were also looking for ways of formulating TWCs to increase conversions. Their studies focused on the role of cerium oxide and on how composition modulation increases CO conversion when the oxide is used. Experiments generally used the computer shown in Figure 5.4 to modulate or effect step changes in the A/F ratio. Data shown in Figure 5.3b are an example of computer forcing of the ratio. Carbon storage and the oxygen capacity of several catalysts were measured by exposing the catalyst to a rich exhaust mixture, then passing O_2 through the bed and measuring the CO_2 formed to obtain the carbon content. This was followed by flushing with H_2 or with CO to determine the oxygen stored by measuring water and CO_2 carried off. Changes in the oxygen content of the catalyst across step changes in the A/F ratio were measured as well. Variations in the outlet CO resulting from step changes in the A/F ratio or modulating the ratio were analyzed by comparing the CO concentration in the outlet at any instant with the concentration expected if the catalyst acted as though at steady state for the momentary A/F ratio. In addition to these specific experiments, these researchers also measured conversions under different A/F ratios and at different frequencies.

Herz and co-workers concluded that O_2 storage occurs when Ce is present. Storage varied with A/F ratio and was large enough to account for the improved CO conversion found when the Ce containing catalysts were used under composition modulation. Furthermore, reduction and re-oxidation of CeO were rapid enough to occur in composition modulation at 1 Hz. From pulse experiments, Herz and Sells (1985) concluded that the WGS could also account for the improvement with Ce present, but they did not exclude the possibility that oxygen storage contributes to the improvement as well. Analysis of the time variation of outlet CO led the authors to observe that the catalyst performs better in the rich half cycle under A/F ratio modulation than predicted by its steady state performance. Herz and

Sell (1985) found that one of the catalysts examined, a Pt/Rh/Ce formulation, gave CO removals under cycling of 93 to 97%, which is the anticipated performance under steady state if the system shown in Figure 5.4 could be run at steady state. All catalysts showed higher conversion as the frequency increased from 0.5 to 2 Hz. Herz and Sell point out that much of the improvement is due to a reduction in the CO emission from the motor through the rapid A/F ratio modulation.

D. COMPOSITION MODULATION AND DUAL REACTIONS IN AUTOMOTIVE EXHAUST

Oxidation Reactions

The system CO—C_3—O_2 was investigated by Muraki et al. (1986b) under a modulation policy consisting of periodically and symmetrically stepping up or down either the CO or the O_2 vol.% in the reactor feed. Periods from 0.2 to 2 s were used, time-average feed composition was 1.33 vol.% CO, 0.77% O_2, and 0.0233% C_3 (a stoichiometric mixture), temperatures were

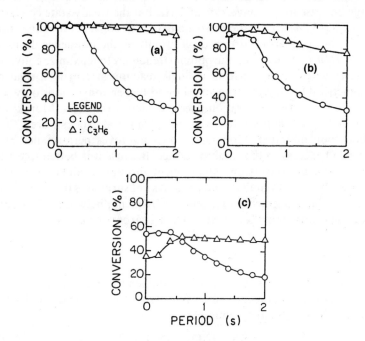

FIGURE 5.6 Effect of cycle period on conversion at 400°C with CO and O_2 modulation for CO (○) and C_3 (△): (a) Rh catalyst, (b) Pd catalyst, (c) Pt catalyst. Steady state results are at $\tau = 0$ (Figure adapted from Muraki et al. (1986b) with permission of the authors)

280 and 400°C and the space velocity was 30,000 h^{-1}(STP). The investigators used 3 catalysts: Pt, Pd and Rh supported on α-Al$_2$O$_3$ at a metal loading of 0.05 mg/cc. The low loading simulated the CO adsorption capacity of aged commercial converter catalyst.

At 400°C a relatively low SV, CO and C$_3$ conversions were high for the Rh and Pd catalysts, but just moderate for Pt under both steady state and cycling as Figure 5.6 shows. Steady state conversions are represented in the figure by the data points at $\tau = 0$. Generally, at conversions close to 100%, improved performance through composition modulation is hard to measure. Thus, in Figure 5.6b, modulation at about 2 Hz increases C$_3$ conversion slightly for the Pd catalyst used by Muraki *et al*. There is no improvement for the Rh catalyst, however, the steady state conversion is 100% for both CO and C$_3$. With the Pt catalyst, a small increase in conversion can be seen in Figure 5.6c for CO at 2 Hz, but a large increase in the C$_3$ oxidation rate is evident through cycling. Conversion rises sharply for C$_3$ at periods greater than 0.5 s. Single reaction studies (Muraki *et al.*, 1985a) showed composition forcing did not increase CO conversion over Rh and that large increases occur over a Pt catalyst, so only the Pd formulation was investigated at 280°C. Muraki *et al*. (1986b) observed a large increase in C$_3$ conversion at 2 Hz, but the steady state conversion was already 55%. Modulation also increased CO conversion by about 10% at 2 Hz, but conversion dropped off rapidly as the frequency decreased. Muraki *et al*. comment that the optimal frequency to maximize conversion decreases when changing from a single reaction system to a dual one. Asymmetrical forcing experiments showed higher conversions only when the high O$_2$ exposure duration was extended. From this, Muraki *et al*. conclude that oxidation rates over the noble metal catalysts are CO and hydrocarbon inhibited. Modulation then improves performance by increasing O$_2$ access to the catalyst surface. Because it is better access that leads to rate or conversion enhancement through modulation for both single reactions and simultaneous ones, it is not surprising that there is little difference in enhancement as a function of frequency between CO oxidation proceeding alone or in the presence of hydrocarbon oxidation or vice versa.

Reduction/Oxidation Reactions

The NO $-$ CO $-$ O$_2$ system has been examined by several investigators as a simple model set of reactions for TWCs. One of the variables used in these studies is the stoichiometric number:

$$SN = (2(O_2) + NO)/CO \qquad (5.1)$$

where $SN > 1$ signifies an oxidizing feed. The stoichiometric number was used also in Chapters 3 and 4, but was then defined only in terms of O$_2$ and CO, or NO and CO.

AUTOMOTIVE EXHAUST CATALYSIS

Hegedus et al. (1980) examined the $NO - CO - O_2$ system for a Pt/Al_2O_3 catalyst (1 wt.% Pt) in an integral reactor, simulating a catalytic converter, and using an IR transmission cell with the catalyst in wafer form. With the latter arrangement, these investigators followed the response of a $CO-Pt$ stretching band (corresponding to linearly adsorbed CO) and an isocyanate band to a periodically changing gas phase composition. Isocyanate was thought to be an intermediate in NO reduction by CO. More recent work, reviewed in the previous chapter, has established that it is not an intermediate. Nevertheless, the density of the isocyanate surface species is probably proportional to the NO or O adsorbates that are intermediates. Measurements with IR were made at about 510°C with feed shifting between a SN of 0.58 and 1.42. Figure 5.7a shows the response of the CO band in (1) for a 2 s and a 6 min cycle period and, in (2), the effect of the switching time ($= 1/2\,\tau$) on the response amplitude. In slow cycling, the surface CO species (the only one observed at 510°C) follows the square input closely, but the response becomes sinusoidal for fast cycling with a small diminution of amplitude. Indeed, Figure 5.7a shows amplitude decreasing sharply only near a frequency of 1 Hz. The isocyanate surface species, Figure 5.7b, is unable to follow the gas phase in slow cycling. The slow response of this species means that under fast cycling, the response is virtually constant. The amplitude plot shows the signal becomes constant at a switching frequency of 1 Hz. Hegedus et al. comment that Figure 5.7 reveals that the time-average surface concentrations decrease with switching time.

The integral reactor experiment carried out at a mean bed temperature of 525°C and a SV of 104,000 h^{-1}(STP) showed conversion increasing as switching time decreased. This may be seen in Figure 5.8. Steady state conversion is not shown, but the quasi steady state limiting conversion is the conversion as the 1/2 period approaches 1000 s. Hegedus et al. interpeted the IR and conversion data to indicate CO and isocyanate inhibition of the oxidation-reduction reaction. They note that surface species are responding to gas phase composition fluctuations so that the behavior and performance of catalytic converters are governed by surface processes.

Muraki et al. (1986b) used the catalysts and the reactor system discussed above on the $NO - CO - O_2$ reaction system, but made measurements only at 400°C. These measurements are presented in Figure 5.9. Although the steady state conversions are absent, it is not likely that cycling increases CO or NO conversion for the Rh and Pd catalysts. Conversions, however, are 90 and 100% for Rh and 90 and 80% for Pd at the experimental conditions used. Figure 5.9c shows a 15% increase in NO conversion and just under a 10% increase for CO with the Pt catalyst at 3–4 Hz.

The $Pd/\gamma - Al_2O_3$ catalyst (2 mm particles) used in the above experiments was singled out for detailed study by Muraki et al. (1985b). These investigators were concerned with changes in the partitioning of CO between O_2 and

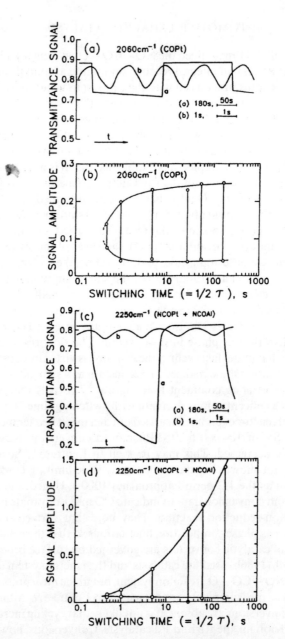

FIGURE 5.7 Transmission IR response of adsorbate species to gas phase composition modulation over a 0.1 wt% Pt/γ-Al_2O_3 catalyst at 510°C. Response of linear adsorbed CO at 2060 cm^{-1} to modulation of O_2 concentration: (a) time variation of transmittance at periods of 2 s (switching time = 1 s) and 6 minutes (switching time = 180 s), (b) amplitude of transmittance variation as a function of switching time. Response of an adsorbed isocyanate species at 2250 cm^{-1} to modulation of O_2 concentration: (c) time variation of transmittance at periods of 2 s (switching time = 1 s) and 6 minutes (switching time = 180 s), (d) amplitude of transmittance variation as a function of switching time (Figure adapted from Hegedus *et al.* (1980) with permission, © 1980, American Chemical Society)

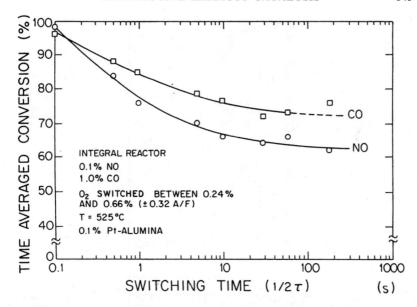

FIGURE 5.8 Time-average CO and NO conversions under periodic forcing of the O_2 concentration as a function of switching time for 0.1 wt% $Pt/\gamma\text{-}Al_2O_3$ catalyst pellets (3.2 mm) in an integral reactor (mid point temperature = 525 °C) (Figure adapted from Hegedus *et al.* (1980) with permission, © 1980, American Chemical Society)

NO reduction as well as changes in kinetics between steady state and composition modulation. Symmetrical forcing of both the CO and O_2 concentrations were used as described previously. Temperature was maintained at 400°C. Their results appear in Figure 5.10 where NO and CO conversion are plotted *vs* τ for various values of SN. Modulation of SN through the CO and O_2 concentrations has a large effect (up to 100%) on NO conversion except at or near the stoichiometric mixture. Figure 5.10a shows that the optimum cycle period for conversion depends on SN, reaching 1 s when SN = 0.5. With respect to CO (Fig. 5.10b), SN modulation increases CO conversion only under strongly reducing conditions. At higher time-average SNs, steady state operation is substantially better than composition modulation.

Partitioning of CO between O_2 and NO reduction depends on SN. Muraki *et al.* (1985b) show that CO oxidation by O_2 is favored and that this oxidation is preferred at higher CO concentrations (low SN). Figure 5.11 shows that SN modulation shifts the partitioning towards NO reduction sharply and that this shift is frequency dependent. At the highest frequency used, 5 Hz, the partitioning is essentially identical to that obeyed at steady

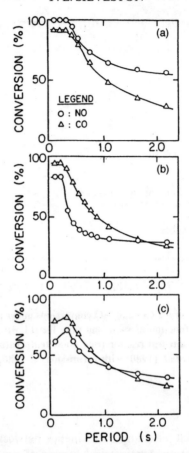

FIGURE 5.9 Effect of cycle period on conversion at 400°C with CO and O_2 modulation for NO (○) and CO (△): (a) Rh catalyst, (b) Pd catalyst, (c) Pt catalyst. Steady state is at $\tau = 0$ (Figure adapted from Muraki *et al.* (1986b) with permission of the authors)

state. This is consistent with the observations of Hegedus *et al.* (1980) that the formation of the isocyanate or any other intermediate in NO reduction by CO is slow relative to CO oxidation and neither the intermediate or the isocyanate can follow fast composition fluctuations. Figure 5.12 shows how τ influences the kinetics of the two reactions. Figure 5.12a shows that under fast cycling (5 Hz), the effect of O_2 partial pressure and the order of the reaction with respect to oxygen is the same as at steady state. Consistent with Figure 5.10, increasing τ decreases the rate of the CO-O_2 reaction. It also eliminates the zero order dependence on O_2 partial pressure. Behavior of the NO reduction rate with O_2 partial pressure appears in Figure 5.12b.

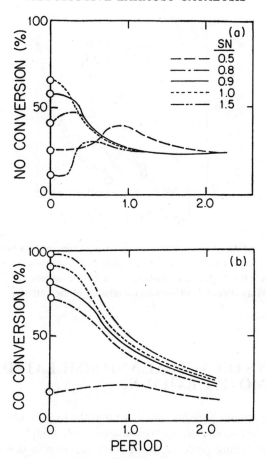

FIGURE 5.10 Effect of cycle period on NO and CO conversion over Pd/α-Al$_2$O$_3$ catalyst pellets (2 mm and 0.006 wt.% Pd) in an integral reactor at 400°C with CO and O$_2$ modulation: (a) NO conversion, (b) CO conversion. Steady state is at $\tau = 0$ (Figure adapted from Muraki *et al.* (1985b) with permission of the authors)

At steady state and at $\tau = 0.2$ s, $p_{O_2} > 0.01$ bar decreases the reduction rate dramatically. With SN modulation at 1 Hz, oxygen partial pressure decreases the NO reduction rate, but at $p_{O_2} = 0.018$ bar, the reduction rate is almost 15 × the steady state rate. Muraki *et al.* (1985b) conclude from these results that under steady state operation CO inhibits NO reduction. Modulation of the stoichiometric number, which is equivalent to A/F ratio forcing, suppresses the inhibition and thereby promotes NO reduction. It is also evident that the presence of O$_2$ in the reactor feed does not deter NO reduction significantly.

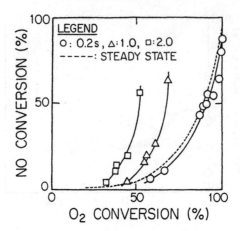

FIGURE 5.11 Effect of cycle period on the partitioning of CO between NO and O_2 reduction at SN = 1 with CO and O_2 modulation. Conditions and catalyst given in Figure 5–10. Steady state partitioning given by the dashed curve (----) (Figure adapted from Muraki *et al.* (1985b) with permission of the authors)

E. STUDIES ON ACTUAL AND SIMULATED AUTOMOTIVE EXHAUST

Some of these studies have been dealt with at the beginning of this chapter because only CO oxidation was considered even though other converter reactions were taking place. Dividing the discussion in this way allowed a comparison of the behavior of the CO oxidation reaction in the catalytic converter environment to the behavior when only CO, O_2 and a diluent were present in the reactor feed.

Studies on the modulation of actual and simulated automotive exhaust have been undertaken to develop catalysts and determine if modulation is beneficial.

Catalyst Development

The search for effective and durable three-way catalysts has been from the outset a major activity in catalytic converter R & D. Performance factors of concern are: activity at different stoichiometric numbers, light off temperature, stability with respect to poisons and temperature excursions, metal volatility and behavior under composition modulation. Because of the

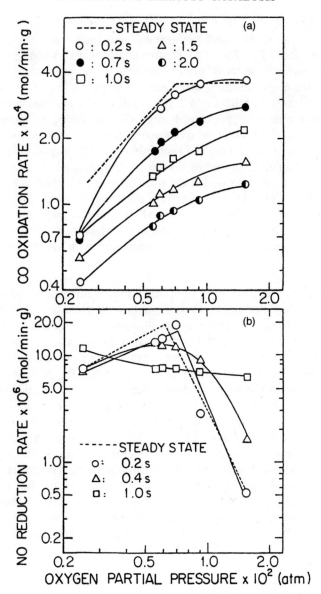

FIGURE 5.12 Effect of oxygen partial pressure on the rates of CO oxidation and NO reduction at various cycle periods with CO and O_2 modulation. Catalyst and conditions given in Figure 5.10: (a) CO oxidation, (b) NO reduction. Steady state rates given by the dashed curve (----) (Figure adapted from Muraki et al. (1985b) with permission of the authors)

latter concern, A/F ratio or SN modulation experiments are included in catalyst screening tests and performance comparisons. Hegedus *et al.* (1979) screened 7 Pt−Rh−Pd−Ce formulations with varying supports and impregnation patterns while searching for a satisfactory low Rh catalyst. One of the screening steps was to compare CO and NO conversions at different amplitudes in A/F ratio cycling. More extensive composition forcing tests were performed when Ce was added to the catalyst because it was believed that the oxide provides oxygen storage and storage was thought necessary for high CO and hydrocarbon conversion during the A/F ratio modulation caused by the feedback control loop.

Hegedus *et al.*, thus, looked at the effect of cycle frequency on hydrocarbon, CO and NO conversion for fresh and exhaust aged catalysts. For Pt/Rh and Pt/Pd/Rh catalysts with and without Ce, they observed that conversion under SN modulation was always less than the steady state conversion except for hydrocarbon conversion with the exhaust aged catalysts.

Since the 1979 study by Hegedus *et al.*, the role of cerium has been explored often and there have been other studies directed at reducing or even eliminating rhodium in TWC formulations. The use of composition modulation in these studies is nicely illustrated by a contribution from Muraki (1991). Much of the data in this paper appears also in Muraki *et al.* (1989). Muraki and co-workers observe that under modulation, Pd performs almost as well as a Rh catalyst provided Pd is promoted by lanthanum. As we have seen in Figures 5.6 and 5.9 for dual reaction systems, modulation of the A/F ratio excites NO_x reduction over Pt and Pd, but does not excite any of the primary abatement reactions with a Rh catalyst. This is shown as well in Figure 5.13. Data for the figure were collected using an engine exhaust which entered the test reactor at 340 to 360°C. The metals loading is typical of TWCs. The conversion enhancement is remarkable for the Pt catalyst. Amplitude appears to be a minor variable, but seems to affect τ_{opt}. At the optimum cycle period, NO_x reduction changes little with amplitude. Amplitude seems to be more important for Pd (Fig. 5.13b). An amplitude of 7% in terms of the A/F ratio suppresses conversion. τ_{opt} continues to depend on amplitude. Comparison of (a) and (c) shows that modulating the A/F ratio at τ_{opt} and A = 5% makes NO_x reduction over Pt about equal to that over Rh. However, A/F modulation is not that effective for Pd and NO_x conversion lags behind that obtained for the Rh catalyst.

Muraki *et al.* (1989) and Muraki (1991) show that the presence of hydrocarbons in the exhaust inhibit NO_x reduction over Pd in rich mixtures as well as lean ones. Steady state conversions for the Pd catalyst with a 2 g/L loading using a simulated exhaust at 350°C appear in Figure 5.14a. Because of different metals loading, Figures 5.13 and 5.14 are not comparable. In Figure 5.14a there is a precipitous drop in NO_x conversion at A/F > 14.6 and conversion also falls as the mixture gets richer. When

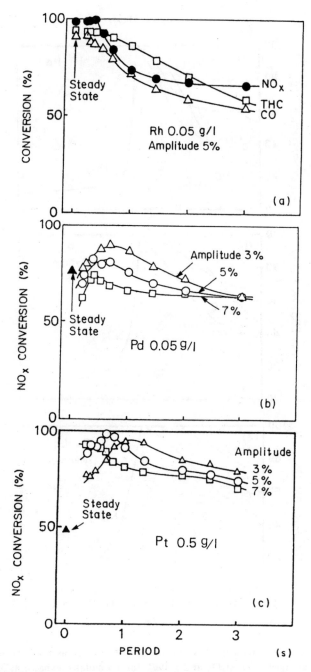

FIGURE 5.13 Conversion of NO_x and other target pollutants as a function of cycle period and amplitude for an engine exhaust entering the test reactor at 1 bar and 340–360°C: (a) Pt/α-Al$_2$O$_3$, (b) Pd/α-Al$_2$O$_3$, (c) Rh/α-Al$_2$O$_3$, but at A = 5% of A/F ratio. Conversions for hydrocarbons (THC) and CO shown as well in (a) (Figure adapted from Muraki (1991) with permission, © 1991, Society of Automotive Engineers, Inc.)

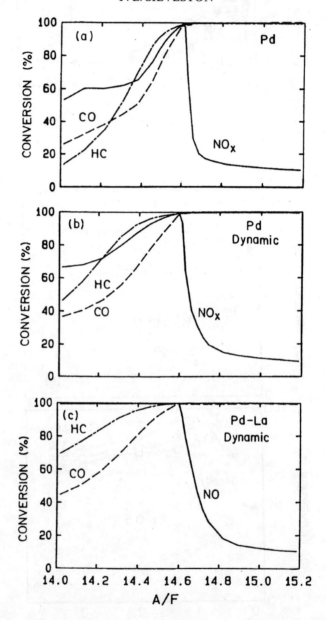

FIGURE 5.14 Conversions of hydrocarbons, CO and NO_x as a function of the A/F ratio over an aged Pd/α-Al_2O_3 at 2 g Pd/L for a simulated exhaust at 350°C: (a) steady state operation, (b) modulated at 1 Hz with A = 4% of the A/F ratio, (c) as in (b) but Pd promoted with lanthanum at a 0.1 mole/L loading (Figure adapted from Muraki (1991) with permission, © 1991, Society of Automotive Engineers, Inc.)

hydrocarbons are removed from the exhaust, conversion remains at 100% for rich mixtures and the drop after A/F = 14.6 is less severe. Figure 5.14b shows that forcing at 1 Hz mitigates hydrocarbon inhibition of NO_x reduction. CO and hydrocarbon oxidation rates also increase in rich mixtures. When the Pd catalyst is promoted with 0.1 mole/L lanthanum and subjected to A/F modulation, hydrocarbon inhibition no longer occurs; lean mixture performance improves and hydrocarbon conversion in rich mixtures increases further (Fig. 5.14c). Indeed, conversions of all the target pollutants for the catalyst and conditions in Figure 5.14c approach closely those achieved with Rh at steady state. From these results, Muraki concludes that a cheaper Pd-La catalyst could be used in place of expensive Rh containing TWCs.

We will return to catalyst development once again at the end of this chapter.

Catalyst Performance With A/F Ratio Or SN Modulation And A Simulated Exhaust

Schlatter et al. (1983) used a simulated automotive exhaust to examine how modulating the A/F ratio affects CO, HC and NO conversions at 550°C and a SV of 52,000 h^{-1} for a Pt/Rh catalyst. The results appear in Figure 5.15. The upper figure (a) shows the influence of ω on CO and NO conversion at ± 0.24 A/F ratio. The curve for 8 Hz represents steady state, whereas in the middle figure, which gives the effect of amplitude at 1 Hz, the ± 0 A/F ratio curve gives conversions at steady state. Both figures demonstrate that on the rich side of the stoichiometric A/F ratio, CO conversion increases under modulation, while on the lean side, modulation has a similar effect on NO conversion. In either case, decreasing the frequency or increasing the cycle amplitude raises the conversion. On the other hand, both figures indicate that CO and NO conversions at the stoichiometric A/F ratio are decreased by A/F ratio cycling. The decreases depend on frequency and amplitude as might be expected. HC conversions are not shown in Figure 5.15. For these oxidation reactions, cycling results in lower conversions than at steady state and does not provide improvement away from the stoichiometric A/F ratio.

Schlatter et al. also investigated the effect of catalyst aging on performance. Aging through exposure to the exhaust is a poisoning phenomenon and Figure 5.15c shows periodic forcing and steady state operations are about equally affected. Comparison with Figure 5.15a shows that modulation no longer increases conversions away from the stoichiometric A/F ratio. Schlatter et al. were not the first to observe the effects of periodic forcing on performance (see, for example, Kaneko et al. (1978)) nor were they the last.

Weibel et al. (1991) explored the role of CeO_2 in TWCs under modulation. They employed two lightly loaded catalysts: (1) Rh/γ-Al$_2$O$_3$ with 0.003

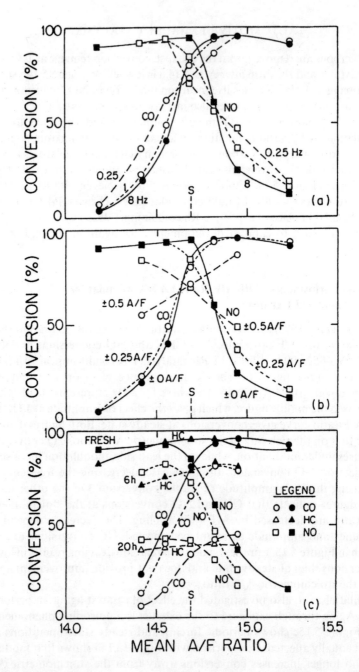

FIGURE 5.15 CO and NO conversion as a function of the time-average A/F ratio under modulation of the A/F ratio for a Pt/Rh catalyst at 550°C and a SV = 52,000 h^{-1}: (a) effect of frequency at an amplitude of $\pm$ 0.25 A/F ratio, (b) effect of amplitude at a frequency of 1 Hz, (c) fresh and exhaust aged catalyst (Figure adapted from Schlatter *et al.* (1983) with permission, © 1983, American Chemical Society)

wt.% Rh and (2) Rh-Ce with 1.02 wt.% Ce added to the Rh catalyst and ran their experiments at 450°C and SV = 100,000 h^{-1}. There was sufficient mixing in their system that the amplitude in the stream entering the reactor was negligible when modulated at 1 Hz and A = 0.25 A/F ratio. Their results in the absence and presence of water vapor in the reactor feed appear in Figure 5.16. There is a general resemblance to Figure 5.15a, but because of mixing the enhancement or supression of conversion at a specific frequency will be different from the Schlatter study. With the Rh/Al$_2$O$_3$ catalysts, modulation at 0.075 Hz with a 0.25 A/F amplitude does not increase CO conversion with respect to steady state in rich mixtures regardless of the addition of CeO$_2$ to the catalyst. Figures 15.16a, b show, however, that cerium increases conversions under modulation. When 10 vol.% H$_2$O is added to the stream, Figure 5.16c, CO conversion at $\omega = 0.075$ Hz now matches conversion under steady state operation for the Rh-Ce catalyst. This did not occur with the Rh sample. This observation and some step change experiments led Weibel *et al.* to conclude that cerium promotes the water gas shift (WGS) activity of Rh at the conditions used.

Muraki *et al.* (1985c, 1986b) extended the Toyota single and dual reaction studies, considered earlier, to explore the effect of cycling the SN or A/F ratio in a synthetic automotive exhaust. In their 1985 study, cycling was carried out by adjusting the CO + H$_2$ and O$_2$ concentrations, while in the 1986 investigation the CO and O$_2$ concentrations were changed. Their catalysts, experimental system and procedures have been described already. Muraki *et al.* (1985c) considered the effect of amplitude as well as τ on HC, CO and NO conversion at 400°C; the 1986 study only examined the effect of cycle period. The extensive measurements made by Muraki *et al.* (1985c) are reproduced in Figure 5.17. Steady state conversions are given on the vertical axes for comparison purposes. Although expressed in terms of an equivalent A/F ratio, only the CO + H$_2$ and O$_2$ concentrations in the simulated automotive exhaust were cycled. As in other papers by the Toyota researchers, forcing was undertaken around SN = 1 or A/F = 14.5. In general, composition modulation is beneficial for the Pt catalyst for NO and HC conversion between 0.5 and 3 Hz, while it is marginally beneficial for CO conversion in a narrow frequency band at or above 2 Hz. For the Pd catalyst, Figure 5.17 indicates that modulation marginally improves the NO and HC conversions between 1 and 3 Hz while the near 90 % CO conversion under steady state is equaled. With the Rh catalyst, modulation offers no improvement, however, it matches the nearly 100% steady state conversion for $\omega > 2$ Hz. Exceptions to the above generalizations are that for HC conversion over the Pt catalyst and NO conversion over the Pd catalyst, the frequency band in which forcing improves conversion becomes a function of amplitude. Increasing amplitude of forcing, in general, reduces conversions of all three reactants shown in the figure for the three

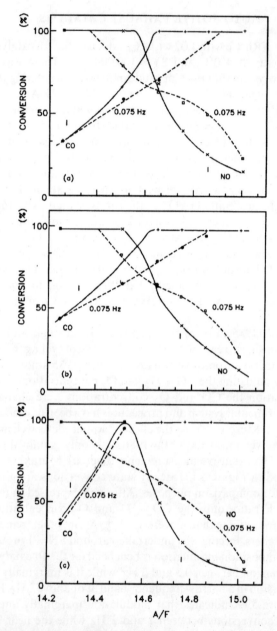

FIGURE 5.16 CO and NO conversion as a function of the time-average A/F ratio under modulation of the A/F ratio for Rh catalysts at 450°C and a SV = 100,000 h^{-1}: (a) effect of frequency at an amplitude of $\pm$ 0.25 A/F ratio for the Rh/γ-Al$_2$O$_3$ catalyst in the absence of water in the reactor feed, (b) as in (a) except that the Rh-Ce/γ-Al$_2$O$_3$ catalyst was used, (c) as in (b) except that the feed stream contained 10 vol.% H$_2$O (Figure adapted from Weibel *et al.* (1991) with permission, © 1991, Elsevier Science Publishers)

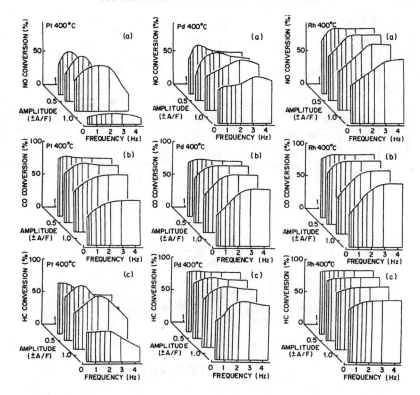

FIGURE 5.17 Effect of amplitude in terms of the A/F ratio and the cycle frequency on NO, CO and HC conversion over Rh, Pd, Pt catalysts supported on α-Al_2O_3 in an integral reactor at 400°C with CO and O_2 modulation: (1) Pt catalyst, (2) Pd catalyst, (3) Rh catalyst. Steady state conversion is indicated by the O on the amplitude axis (Figure adapted from Muraki et al. (1985c) with permission, © 1985, American Chemical Society)

catalysts with just a few exceptions. These are NO reduction over Pt where conversion diminishes only if the amplitude exceeds 0.8 in the A/F ratio, NO reduction over Pd where amplitude has just a small but varying effect on conversion, and for HC conversion where amplitude does not affect conversion. Apart from the Pt catalyst at the highest frequencies used, the amplitude effect on conversion disappears for the three reactants.

The large amount of data and their presentation in Figure 5.17 make it difficult to make comparisons. To assist interpretation, similar results from Muraki et al. (1986b) for a synthetic exhaust at 400°C are plotted in Figure 5.18. Steady state conversions are not given, but can be estimated from the data trends. Modulation of the stoichiometric number increases conversion only for the Pt catalyst. With the exception of NO reduction over Pd and C_3

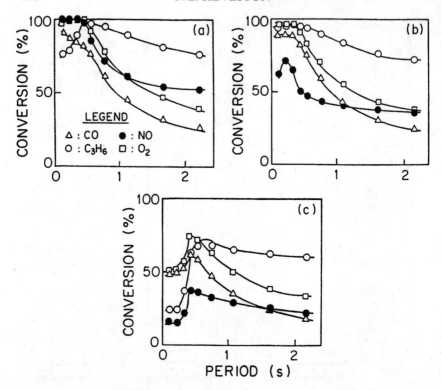

FIGURE 5.18 Effect of cycle period on NO, HC, CO and O_2 conversion over Rh, Pd, Pt catalysts supported on α-Al_2O_3 in an integral reactor at 400°C with CO and O_2 modulation: (a) Rh catalyst, (b) Pd catalyst, (c) Pt catalyst. (Figure adapted from Muraki *et al.* (1986b) with permission of the authors)

oxidation over Rh which increase, modulation suppresses conversion for the Pd and Rh catalysts. Increases for the Pt catalyst can be sizeable: for C_3 oxidation and NO reduction conversion increases about 100%. The optimal frequencies, according to Figure 5.18, are about 2 Hz. These are about the same as ω_{opt} found for the $CO-C_3-O_2$ and $NO-CO-O_2$ system. Both of these dual reaction systems were discussed in the preceding section.

Modulation experiments discussed so far have employed symmetrical cycling in which the cycle is divided into two parts of equal duration. This approximates the modulation forced on the catalytic converter by the A/F ratio control loop. Asymmetric forcing has been examined, notably by Taylor and Sinkevitch (1983). They employed a catalyst formulation: 0.68 wt.% Pt, 0.031 % Pd, 0.007 % Rh and 2.68 % Ce. The support was γ-Al_2O_3 and the catalyst was aged by extended exposure to automotive exhaust. Experiments were done on a synthetic automotive exhaust for T = 550°C, SV = 20,000 h^{-1}, ω = 1 Hz and A = 0.15 A/F ratio. Their results are shown in Figure 5.19. The variable used in earlier chapters for asymmetric operation is the cycle split. It is defined in this chapter as the ratio of the duration of catalyst exposure to the lean mixture to the cycle period. At constant temperature, asymmetric forcing at $s = 0.8$ increases CO conversion for mixtures which are rich on the time-average. For some values of SN, the increase is substantial. Periodic forcing at $s = 0.2$ improves NO conversion in mixtures which are rich on a time-average basis (Fig. 5.19).

Because steady state CO conversion drops in rich mixtures and steady state NO conversion falls in lean mixtures, periodic forcing is beneficial in that it raises conversions on either side of the stoichiometric A/F ratio. This is referred to as widening of the conversion window. Figure 5.19 confirms

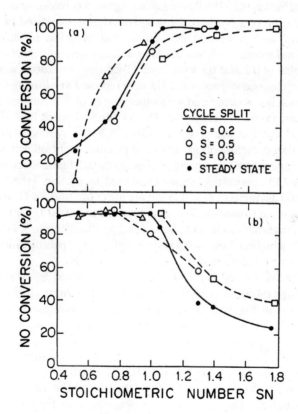

FIGURE 5.19 Effect of symmetric and asymmetric forcing of the A/F ratio at 1 Hz and an A/F ratio amplitude of ± 0.15 on conversion of (a) CO and (b) NO over a Pt/Pd/Rh/Ce/γ-Al$_2$O$_3$ catalysts in an integral reactor at 550°C. Cycle splits are $s = 0.2, 0.5, 0.8$ (Figure adapted from Taylor and Sinkevitch (1983) with permission, © 1985, American Chemical Society)

other results discussed earlier (Figs. 5.15 and 5.16) which indicate symmetrical forcing at the stoichiometric mixture suppresses conversions. Schlatter et al. (1983) found that symmetrically forcing aged catalysts did not improve CO conversion at SN < 1, or NO reduction at SN > 1. Figure 5.19, however, shows that a small improvement in NO conversion is possible at SN > 1 with symmetric forcing, but this is always less than the conversion increase when asymmetrical forcing is used.

An important property of TWCs is conversions achieved after the engine starts, but before it attains a more or less stationary temperature. This is referred to as the light-off performance. Light-off temperature is defined as the temperature at which conversion reaches 50% and it is usually measured by ramping the catalyst temperature by 2°C/min from ambient temperature until conversion reaches 100%.

Taylor and Sinkevitch (1983) investigated light-off behavior under composition modulation of a synthetic automotive exhaust, expressed in terms of the A/F ratio, and compared light-off to that observed in steady state operation. They examined symmetrical as well as asymmetrical cycling. For the latter, splits of 0.2 and 0.8 were used. Operating conditions were stated earlier. The cycling experiments at 1 Hz were repeated at a forcing frequency of 0.1 Hz which is low compared with other studies of A/F ratio modulation where conversion effects were noted. The Taylor-Sinkevitch light-off results with an aged catalyst are given in Figure 5.20 for symmetrical (a) and asymmetrical (b, c) cycling of the simulated exhaust. Light-off behavior with the time-average exhaust composition appears in each figure for comparison. Light-off for CO combustion seems unaffected by cycling. However, both symmetric and asymmetric ($s = 0.2$) forcing affect light-off for C_3 oxidation. The temperature decrease for 50% conversion is about 50°C for the former and 10°C for the latter mode. The effect of cycling is sensitive to SN or the A/F ratio for NO reduction. For $s = 0.5$ and 0.8, light-off temperature increases by ca. 20°C under modulation; at $s = 0.2$, there is no effect.

Light-off behavior for C_3, CO and NO, measured by Muraki et al. (1985c) through raising the reactor temperature at 2 C/min, are displayed as a function of ω in Figure 5.21 for each of three noble metal catalysts. The A/F ratio amplitude was ± 0.5. Modulation decreases light-off temperature for NO reduction and C_3 oxidation for the Pt and Pd catalysts, while modulation increases light-off temperature for CO oxidation for both of these catalysts and for C_3 oxidation on the Rh catalyst.

Modulation does not affect light-off on the Rh catalyst for NO reduction and CO oxidation. These results were obtained with fresh catalysts, although the metal loading was kept low so the adsorption capacities corresponded to those exhibited by used catalysts. These results agree, within reason, with the observations of Taylor and Sinkevitch, discussed above, who found cycling decreased light-off temperature for C_3 oxidation, but increased this temperature for NO reduction.

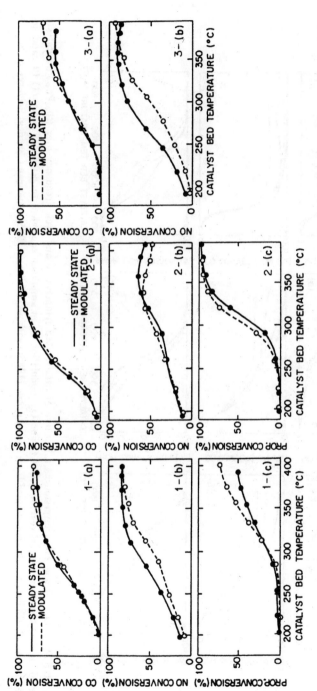

FIGURE 5.20 CO, NO and C_3 conversion as a function of catalyst bed temperature for steady state and periodic operation over a Pt/Pd/Rh/Ce/γ-Al_2O_3 catalysts in an integral reactor. A/F ratio modulation at 1 Hz and an amplitude equivalent to ± 0.15 A/F ratio: (1) symmetric forcing at $s = 0.5$, (2) asymmetric forcing at $s = 0.8$, (3) asymmetric forcing at $s = 0.2$ (Figure adapted from Taylor and Sinkevitch (1983) with permission, © 1985, American Chemical Society)

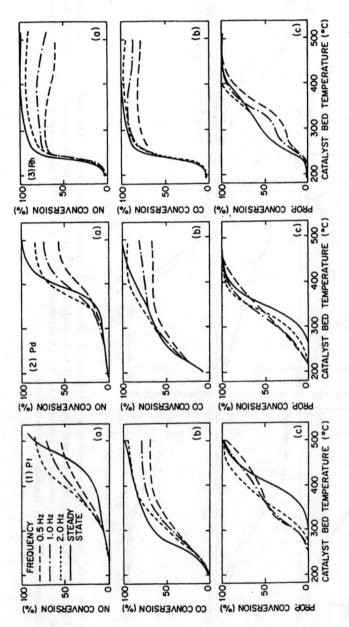

FIGURE 5.21 NO, CO and HC conversion as a function of catalyst bed temperature and cycle frequency over Rh, Pd, Pt catalysts supported on α-Al$_2$O$_3$ in an integral reactor with CO and O$_2$ modulation at an amplitude equivalent to ± 0.5 A/F ratio: (1) Pt catalyst, (2) Pd catalyst, (3) Rh catalyst (Figure adapted from Muraki et al. (1985c) with permission, © 1985, American Chemical Society)

The effect of continuous composition cycling on conversion of the target gases and of the superimposed influence of periodic exposure of the TWC catalyst to air was examined by Padeste and Baikers (1994). Periodic exposure to air arises in hybrid vehicle drive systems in which an internal combustion engine operates under constant throttle to drive an inertial energy storage device, such as a flywheel. Usually the flywheel would drive a generator so the vehicle would actually be powered electrically. Engine operation when the flywheel is fully charged in an urban environment with low speeds and idling leads to intervals of very low or zero flow into the converter and some draw in of an air charge. It is this operating condition that drew the interest of these investigators.

Padeste and Baikers first determined converter performance in continuous, but modulated operation, and then examined how the periodic exposure to air altered conversions. Two $Pt/Rh/Al_2O_3$ catalysts were tested; each was applied as a washcoat to a ceramic honeycomb monolith. One contained cerium oxide, while the other did not. Loadings on the monolith structure were 1 wt.% Pt and 0.2 wt.% Rh. When cerium was used, it amounted to 12 wt.%. Both catalysts were aged at SN = 1 and 600°C for 10 h. Experiments were performed with a simulated exhaust at inlet temperatures of 260 or 310°C and SV = 50,000 h^{-1} (STP). Modulation, around SN = 1, consisted of switching $CO + H_2$ and O_2 concentrations just as done by most investigators.

Figure 5.22 summarizes the Padeste and Baikers results for these rather low temperatures. A comparison of the Pt/Rh in (a) and Ce/Pt/Rh in (b) curves shows the addition of Ce increases conversions somewhat. Optimal frequencies of about 1 Hz are evident for the Pt/Rh formulation, but these are not seen when Ce is added to the catalyst. This is probably because conversion goes to completion for CO, C_3 and often for NO_x. Padeste and Baikers refer to the frequency at which conversion begins to decrease with decreasing frequency as a critical frequency and note that these are lower for the Ce/Pt/Rh catalyst. These critical frequencies are probably equivalent to the optimal frequency discussed in this chapter. Figure 5.22 shows that amplitude affects these critical or optimal frequencies, but the influence of temperature is unclear. The bar to the right of each plot gives steady state conversions at the time average feed composition. These decrease for the cerium containing catalyst. Padeste and Baikers explain this through improved dispersion of noble metals on the support when cerium is not present. In contrast to the results of Muraki et al. (1986b) for Rh, which is representative of the Pt/Rh catalysts, A/F ratio modulation enhances conversions of C_3 and NO at an amplitude of 0.02 SN for both catalysts. At 0.05 SN, this enhancement disappears. Explanation for this difference is probably the lower temperatures used by Padeste and Baikers. In Chapters 3 and 4, we demonstrated that rate enhancement over noble metal catalysts increases with decreasing temperature. Padeste and Baikers observe that at

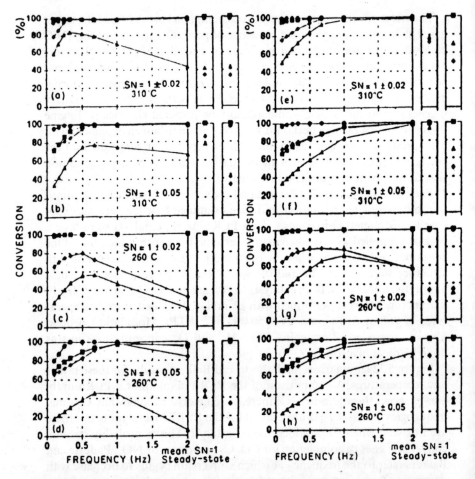

FIGURE 5.22 Conversion of target exhaust components as a function of cycle frequency at A = 0.02 and 0.05 SN for 260 and 310°C with symmetric cycling around SN = 1: (a) to (d) give results for Pt/Rh/Al$_2$O$_3$, (e) to (h) are for Ce/Pt/Rh/Al$_2$O$_3$. Bar charts to the right give conversions at steady state for the time average feed composition [Legend: ■ = CO, ● = C$_3$H$_6$, ▲ = C$_3$H$_8$, ♦ = NO$_x$] (Figure adapted from Padeste and Baikers (1994) with permission, © 1994, American Chemical Society)

or below their critical frequencies, peaks of unconverted CO appear at the end of rich half cycle, while a NO peak develops at the end of the lean half cycle. This is illustrated in Figure 5.23. Lower frequencies correspond to longer durations of a half cycle so that the appearance of peaks signals CO swamping of the metal surface towards the end of the rich half cycle and a high coverage of O adatoms at the same point in the lean half cycle (see Chapter 4).

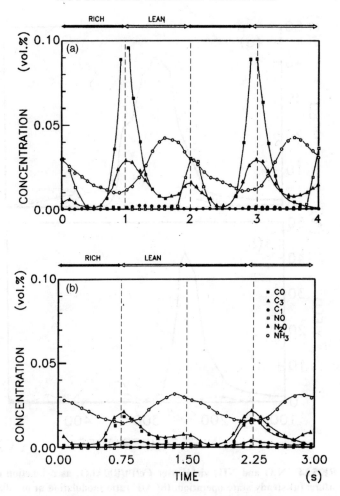

FIGURE 5.23 Variation of target exhaust components in successive cycles for a Ce/Pt/Rh/Al$_2$O$_3$ catalyst at 310°C and A = 0.05 SN: (a) ω = 0.5 Hz, (b) ω = 0.67 Hz. Solid and open arrows above the figure show the duration of the rich and lean half cycles respectively Legend: ■ = CO, ● = CH4, ▲ = C$_3$H$_8$, ◆ = NO$_x$, △ = N$_2$O, ○ = NH$_3$] (Figure adapted from Padeste and Baikers (1994) with permission, © 1994, American Chemical Society)

Formation of nitrous oxide (N$_2$O) with TWCs becomes significant at low temperatures as discussed towards the end of Chapter 4. Padeste and Baikers find that A/F ratio modulation substantially reduces N$_2$O at 0.67 Hz and A = 0.02 SN with respect to production under steady state operation. Figure 5.24 shows strong suppression of N$_2$O formation. Measurements of NH$_3$ formation were also made. The figure show yields are small and there is little difference between periodic forcing and steady state.

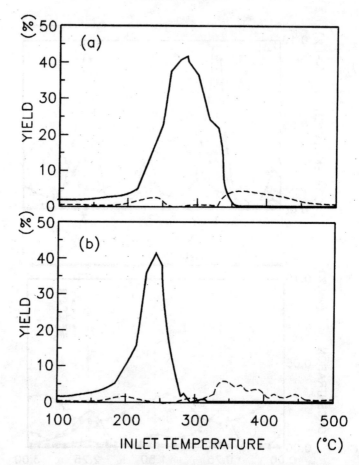

FIGURE 5.24 N_2O and NH_3 yields over $Ce/Pt/Rh/Al_2O_3$ as a function of inlet temperature: (a) steady state operation, (b) A/F ratio modulation at $\omega = 0.67$ Hz, $A = 0.02$ SN (Figure adapted from Padeste and Baikers (1994) with permission, © 1994, American Chemical Society)

Figure 5.25 compares the response to air exposure under otherwise steady state operation in (a) to (c) and in the presence of A/F ratio forcing (d) to (f). The figure shows the variation of the target components in the exhaust with time during a 3 s exposure to a simulated exhaust. In (a) and (d), the catalyst sample saw no flow for 17 s prior to the exposure, whereas in the remaining figures, a 0.2 s pulse of air followed the 17 s of zero flow. In the upper figures, the air pulse has a large effect on NO conversion. Change in the time profiles of the other exhaust components appears to be negligible. Modulation is at 1 Hz and $A = 0.05$ SN in (d) and (e) so that there are three peaks, at least in (e) and (f). Comparison of (b) and (e) shows that

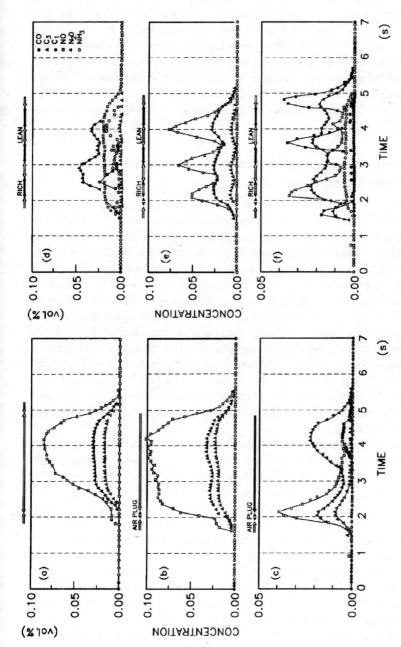

FIGURE 5.25 Variation of target exhaust components during exposure of a Ce/Pt/Rh/Al$_2$O$_3$ catalyst to a simulated engine exhaust at 310°C in the following cycles: (a) 17 s of zero flow followed by 3 s at a constant SN = 1, (b) as in (a) but the 3 s exhaust exposure preceded by 0.2 s of air exposure, (c) as in (b) but at a constant exhaust exposure at SN = 0.985, (d) as in (a) but exhaust modulated at 1 Hz, A = 0.05 SN, (e) as in (b) but exhaust modulated as in (d), (f) as in (e) but exhaust modulated asymmetrically at 1 Hz, A = 0.05 and s = 0.4 (based on lean mixture). Symbols as in Fig. 5-24 (Figure adapted from Padeste and Baikers (1994) with permission, © 1994, American Chemical Society)

modulation substantially enhances NO, C_3, and N_2O conversion in the presence of air exposure and zero exhaust flow, however, (d) and (e), just as (a) and (b) show that air exposure affects NO conversion. The bottom figures in Figure 5.25 indicate remedies to the effects of the air pulse. For steady state operation, NO conversion can be increased by using a rich mixture, while if the exhaust is modulated, asymmetric modulation with $s = 0.4$ (defined in terms of the lean mixture) further increases NO conversion, although at the expense of lower CO conversion. It is not apparent from the Padeste and Baikers contribution whether or not hybrid drive systems will require modification of the A/F ratio feedback control loop.

Catalyst Performance Under A/F Ratio Or SN Modulation With Actual Engine Exhaust

The effect of A/F ratio modulation on conversion of the usual target gases in an actual automotive exhaust are reported by Yokota *et al.* (1985). The experimental system used by these researchers and later by Matsunaga *et al.* (1987) is similar to the system shown in Figure 5.4. Modulation can be generated through the oxygen sensor and the feedback control loop, or through an external oscillator with the control loop open. The three noble metal catalysts tested by Yokota *et al.* were those used by Muraki *et al.* (1985c), except that the metal loading for the Pt catalyst was increased to 0.5 mg/mL to compensate for the lower activity of that catalyst. All catalysts were aged by an extended exposure to automotive exhaust. Converter inlet temperatures were between 340 and 360°C in the experiments. The objective of the Yokota *et al.* study was to determine if Pd or Pt based catalysts are capable of matching or exceeding the performance of a three-way catalyst containing Rh. The Rh catalyst was taken to be representative of the Rh containing TWCs.

Figure 5.26 compares conversions under modulation of the A/F ratio at 1 Hz with amplitudes from 0.3 to 1.2 in the A/F ratio with the steady state conversions as a function of the time-average A/F ratio. Comparison at the stoichiometric A/F ratio shows that modulation depresses conversions for Rh and Pd, except for NO reduction on Pd, whereas modulation elevates conversions for the Pt catalyst. These results are consistent with the other measurements discussed in the previous section. If the performance of the Rh catalyst used by Yokota *et al.* and Muraki *et al.* is representative of TWCs containing Rh, a horizontal comparison from right to left in Figure 5.26 suggests that steady state operation on Pd approaches the TWC performance. If the Pt catalyst is modulated at 1 Hz with an amplitude between 0.6 and 0.9 of the A/F ratio, it too approaches the performance measured on Rh. Nevertheless, a comparison of steady state versus modulated operation, regardless of catalyst, shows that only hydrocarbon

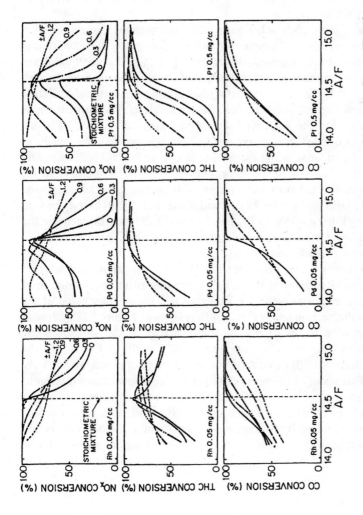

FIGURE 5.26 NO, HC and CO conversion as a function of the time-average A/F ratio and cycle amplitude as A/F ratio over Rh, Pd, Pt catalysts supported on α-Al_2O_3 in an integral reactor using an automotive exhaust with modulation of the A/F ratio at 1 Hz: (1) Rh catalyst, (2) Pd catalyst, (3) Pt catalyst. Inlet temperature to the catalytic converter was 340-360°C (Adapted from Yokota et al. (1985) with permission, © 1985 Society of Automotive Engineers, Inc.)

conversion under modulation exceeds the highest steady state conversion, and then by a small amount. With this exception and the notable enhancement of all conversions over the Pt catalyst through modulation, there is agreement with GM research which shows no conversion amelioration through modulation around the stoichiometric A/F ratio with the Rh containing TWCs.

With lean mixtures, SN > 1, Figure 5.26 demonstrates that modulation raises NO conversion well above that possible for steady state operation. Increasing amplitude increases conversions. For the Pd and Pt catalysts, conversions are enhanced away from the stoichiometric even in rich mixtures (SN < 1). With respect to hydrocarbon oxidation, modulation raises conversion in rich mixtures for all three catalysts. The improvements are substantial for the Rh and Pt catalysts, provided SN ≠ 1. Finally, modulation of the A/F ratio increases CO conversion over Pd and Pt at SN > 1 but the improvements are smaller than seen for NO reduction and hydrocarbon oxidation. Modulation effects found by Yokota *et al.* for non-stoichiometric A/F ratios are considerably greater than those reported by Schlatter *et al.* (1983) or by Taylor and Sinkevitch (1983) as may be seen by comparing Figure 5.26 with Figures 5.15 and 5.19. Measurements in the 1983 papers were made at a temperature about 100°C greater and employed, of course, a different catalyst.

At A/F ratios off the stoichiometric, Figure 5.26 demonstrates that catalysts formulated around Pd and Pt, operating either modulated or at steady state, perform as well as a Rh catalyst. This was discussed at the beginning of this section under *Catalyst Development*. Figure 5.27 addresses the question of whether or not the performance at the stoichiometric mixture can be matched using periodic forcing of the A/F ratio. In the figure, the effects of both τ and A on conversion are examined. Steady state conversions are given by the heavy dots. Figure 5.26 shows that conversions for the Rh catalyst are between 90 and 100%. The steady state conversions for Pd are close to those for Rh, except for NO reduction. This reaction, however, is stimulated by forcing so that at about 1 Hz NO conversion reaches about 95%. With respect to Pt, steady state performance is well below that exhibited by Rh. Figure 5.27 demonstrates that modulation of the A/F ratio between 0.5 and 1.5 Hz brings the performance of this catalyst up to the Rh level. Yokota *et al.* (1985) conclude on the basis of Figures 5.26 and 5.27 that a rhodium free formulation based on a Pd−Pt mixture performs about as well as the Rh containing TWCs, in agreement with our earlier comments.

Matsunaga *et al.* (1987) undertook a thorough examination of A/F ratio modulation of an engine exhaust to quantify performance improvements for a commercial 1.3 L Pt/Rh monolith catalyst (1.8 g Pt and 0.2 g Rh/L). Their results, obtained at an inlet temperature of 390°C and a SV = 40,000 h^{-1}, are shown in Figure 5.28. The lower inlet temperature and SV reflect

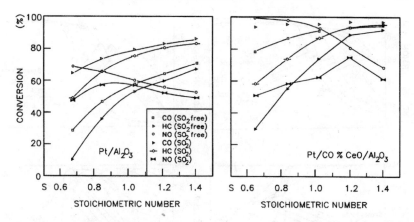

FIGURE 5.27 NO, HC and CO conversion as a function of the cycle period and cycle amplitude as A/F ratio over (a) Pd, (b) Pt catalysts supported on $\alpha\text{-Al}_2\text{O}_3$ in an integral reactor using an automotive exhaust with modulation of the A/F ratio. Inlet temperature to the catalytic converter was 340-360°C (Adapted from Yokota et al. (1985) with permission, © 1985 Society of Automotive Engineers, Inc.)

current engine size and design. They differ significantly from levels used in the 1978 to 1984 period by GM researchers. Hydrocarbon, NO_x and CO conversions as a function of the A/F ratio and amplitude are compared for fresh catalyst and a monolith aged to be equivalent to 60,000 km of driving. For the fresh catalyst, conversion enhancement increases with increasing amplitude of modulation at 1 Hz, although the effect is small for CO oxidation and an amplitude threshold seems to exist. Comparing conversions for fresh and aged catalysts shows a small loss of activity for hydrocarbon oxidation, NO reduction and CO oxidation at SN = 1 or A/F = 14.5 under both steady state and A/F ratio modulation. Unlike the results of Schlatter et al. (1983), aging reduces conversion much more under steady state operation than under cycling, except for CO oxidation. For this reaction, aging affects modulated and steady state operation about equally. At amplitudes from 0.02 to 0.04 of the A/F ratio, forcing at 1 Hz increases hydrocarbon and NO_x conversions beyond those achieved at steady state when the A/F ratio = 14.5. Improvement is even greater for NO_x with rich and lean mixtures. Modulation enhances significantly hydrocarbon conversion in rich mixtures and dramatically broadens the operating window for this TWC.

Comparison of the results for Rh in Figure 5.26 with Figure 5.28 confirms that for NO_x reduction, conversions with just Rh on a support represent the performance of TWCs containing Rh. However, Figure 5.26 suggests that although Rh alone may satisfactorily represent steady state HC and CO conversions on TWCs, results underestimate improvements through modulation for these oxidation reactions.

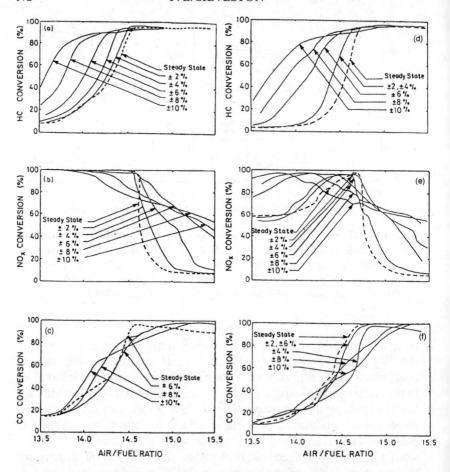

FIGURE 5.28 NO, HC and CO conversion as a function of the A/F ratio and cycle amplitude as A/F ratio at $\omega = 1$ Hz over a fresh ((a) to (c)) and an aged ((d) to (f)) commercial Pt/Rh monolith catalyst using an automotive engine exhaust. Inlet temperature to the catalytic converter was 390°C. Steady state conversions at each time average A/F ratio shown for comparison (Adapted from Matsunaga *et al.* (1987) with permission, © 1987 Society of Automotive Engineers, Inc.)

Modulation exerts a large effect on light-off. Figure 5.29 compares conversion versus temperature for fresh and aged Pt/Rh monolith catalysts. For the hydrocarbons, periodic forcing at 1 Hz with A = 0.08 A/F reduces light-off from well over 400 to under 250°C. HC conversion at any temperature increases with amplitude up to a maximum of about 0.1 A/F. Light-off for NO_x reduction decreases by just 10° under forcing at 1 Hz.

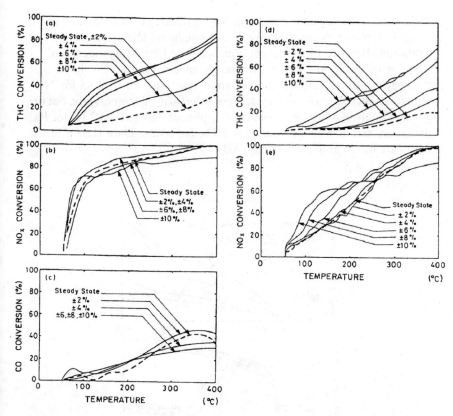

FIGURE 5.29 NO, HC and CO conversion as a function of the converter inlet temperature and cycle amplitude as A/F ratio at $\omega = 1$ Hz over a fresh ((a) to (c)) and an aged ((d) to (e)) for a commercial Pt/Rh monolith catalyst using an automotive engine exhaust modulated around A/F = 14.5. Steady state conversions at A/F = 14.5 shown for comparison (Adapted from Matsunaga *et al.* (1987) with permission, © 1987 Society of Automotive Engineers, Inc.)

Results for CO oxidation are inconclusive. Light-off in steady state operation is depressed by aging for HC oxidation and NO_x reduction, as the figure shows. It is also depressed under modulated operation, but by a lesser amount. The net effect for aged monolith catalysts is that modulation at 1 Hz with a sufficiently large amplitude in terms of the A/F ratio decreases the light off temperature by more than 100°. Neither aging or modulation seems to affect light-off for CO oxidation. Modulation influence on light-off temperatures found by Matsunaga *et al.* (1987) are much greater than those observed by Muraki *et al.* (1986b) for a simulated exhaust with a Rh catalyst for propane, but resemble what the earlier researchers found for NO_x. They are quite different from the results reported by Taylor and

Sinkevitch (1983) for NO_x, but are similar in trend at least for hydrocarbon oxidation. No investigators have found a light-off improvement for CO oxidation. Because of the large influence found by Matsunaga *et al.*, it seems worthwhile to check on frequency effects. Figure 5.30 shows that frequency is important for the oxidation reactions, but not so for NO_x reduction. Increasing frequency depresses the light-off temperature, but the positive light-off effect for modulation applies at 0.5 to about 2 Hz for HC oxidation and at all frequencies for NOx reduction. Data shown in the figure are for an amplitude of 0.06 A/F.

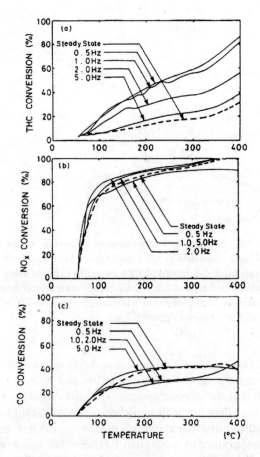

FIGURE 5.30 NO, HC and CO conversion as a function of the converter inlet temperature and cycle frequency at A = 0.06 A/F over a fresh commercial Pt/Rh monolith catalyst using an automotive engine exhaust modulated around A/F = 14.5. Steady state conversions at A/F = 14.5 shown for comparison (Adapted from Matsunaga *et al.* (1987) with permission, © 1987 Society of Automotive Engineers, Inc.)

Effect Of A/F Ratio Modulation On Sulphur Poisoning

With the large volume of data that has been collected on the effects of A/F ratio modulation on the performance of catalyst converters operating on clean simulated exhausts, attention has turned to the effect of poisons on converter performance. Sulphur, in the various forms in which it appears in exhausts or in the converter (SO_2, SO_3, sulphates or H_2S), has been the most studied poison. According to Monroe et al. (1991), sulphur poisoning is mainly a problem with rich or stoichiometric mixtures. An early study by Joy et al. (1979) found poisoning of the water gas shift (WGS) and steam reforming reactions to be severe. Joy et al. concluded that poisoning was deemed unimportant in lean mixtures because these reactions are themselves less important in TWCs and sulphur is stripped from the catalyst surface in the presence of excess O_2. A recently recognized problem is the formation of H_2S in converters when operating with rich mixtures. Prior to the comprehensive investigation of Monroe et al. (1991), A/F ratio modulation in the presence of SO_2 has been explored by Joy et al. (1979), Schlatter and Mitchell (1980) and Su and Rothschild (1986).

Two commercial catalysts, a pelleted, alumina supported catalyst containing 0.11 wt.% Pt, 0.045 wt.% Pd, 0.014 wt.% Rh and 2.6 wt.% Ce, and a monolith containing 34 $\mu g/cm^3$ Pt, 4.1 $\mu g/cm^3$ Rh, 4.7 wt.% Ce and 2.5 wt.% La, were used as fresh samples and as thermally aged samples by Monroe et al. (1991). These researchers examined SO_2 poisoning and recovery after SO_2 is taken out of the exhaust for warmed up conversions (at 500°C) and on light-off with a simulated automotive exhaust cycled around the stoichiometric mixture at 0.5 Hz. An experiment consisted of an extended exposure to clean exhaust, 180 minute exposure to an exhaust containing 5 or 20 ppm (vol.) SO_2, followed by 120 minute exposure to clean exhaust. Figure 5.31 shows the light-off behaviour for the pelleted and monolith catalysts. Light-off temperatures are severely affected by SO_2 in the exhaust for both catalysts, and in (d) to (f), it can be seen that the increase in the light-off temperature depends on the SO_2 concentration in the exhaust. Not shown but reported by the authors is that the light-off temperature drops by about 50° in the presence of ppm quantities of SO_2 when the hydrocarbon used is propane. Once SO_2 is removed from the feed, the light-off temperature decreases and the conversion vs. temperature curve shown in the figure is attained after 120 min. Recovery is virtually 100% for NO_x reduction and HC oxidation for the monolith catalyst, probably because of the presence of La in its formulation, but reaches only 50 to 60% for the pelleted catalyst. These experiments were repeated with the pelleted catalyst for a simulated lean mixture. The behavior on SO_2 addition to and removal from the exhaust were similar to the curves shown in Figure 5.31a to c, although light-off temperatures were generally higher, in some reactions as much as 100° higher. Results for warmed up pelleted

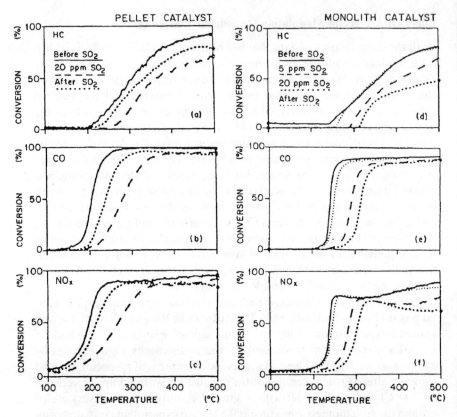

FIGURE 5.31 HC, CO and NO conversion as a function of the converter inlet temperature and the presence of SO_2 in the exhaust for thermally aged commercial Pt/Pd/Rh/Ce pelleted catalyst ((a) to (c)) and Pt/Rh/Ce/La monolith catalyst ((d) to (f)) using a simulated automotive exhaust modulated at 0.5 Hz around A/F = 14.5 (Adapted from Monroe *et al.* (1991) with permission, © 1991 Elsevier Science Publishers)

catalysts, operating at 500°C, also showed response of conversions on SO_2 addition and removal much like that seen in Figure 5.31a to c, with the exception that SO_2 poisoning of CO oxidation was not reversed when SO_2 was taken out of the exhaust. Monroe *et al.* also note that the decline in CO conversion seemed to be independent of the SO_2 concentration. Full recovery was obtained only when the catalyst was exposed to H_2; H_2S was formed. Repeating the warmed up experiments with the monolith catalyst gave responses much like those seen in Figure 5.31d to f. Almost full recovery of CO conversion was observed when SO_2 was removed from the exhaust. Recovery of HC and NO_x conversions was complete. Differences in recovery and the response to SO_2 in the exhaust were attributed by

Monroe et al. to the presence of Pd in the pelleted catalyst as well as the properties of the supports used.

Monroe et al. (1991) also undertook experiments with exhausts for an engine fed with fuels having different sulphur content. The engine was operated in open loop but under computer control so that the forcing frequency could be fixed. Amplitude was held to 0.5 A/F and the monolith catalyst used had been aged to the equivalent of 80,000 km of driving. A portion of the data obtained is presented in Table 5.1.

These data clearly show a reduction in conversion as the sulphur content of the fuel increases. Sulphur in the fuel has the largest effect on hydrocarbon oxidation, while NO is least impacted. The effect is not large for the 10 × increase in sulphur and this appears to be due to the mitigating influence of A/F modulation on sulphur poisoning. The benefit of reduced susceptibility has been recognized since the investigation of Joy et al. (1979).

As part of an investigation of the role of ceria (CeO_2) in TWCs, Lööf et al. (1991) measured the effect of A/F ratio modulation on CO, HC and NO conversion in the presence and absence of 20 ppm (vol.) SO_2 in a simulated exhaust at 450°C for a 0.8 wt.% $Pt/\gamma\text{-}Al_2O_3$ catalyst and for the same catalyst containing 30 wt.% Ce. In the absence of SO_2, the addition of ceria substantially increased conversions under modulation at 0.5 Hz, but had only a small effect in steady state operation. 20 ppm SO_2 in the exhaust decreased conversions of all target components. The decrease was largest at the lowest A/F ratio, but it was still found for lean mixtures. Lööf et al. concluded that the formation of a cerium sulphate interferes with the ability of cerium to store O_2 and thus enhances conversion in the oxidative reactions on switching between rich and lean mixtures.

F. A/F RATIO MODULATION ON EXPERIMENTAL CATALYSTS WITHOUT NOBLE METALS

The search for TWCs free of noble metals has been an important part of the development of catalytic converters for many years. Most of the testing of candidate catalysts has been carried out under steady state conditions. Now, as the promising catalysts have been identified, performance testing is being done more frequently with modulated exhausts. Catalysts exhibiting potential are copper oxides on various supports and the mixed copper-chromium oxides. With the latter class, spinels are formed which stabilize the active Cu^{2+}. Spinels also permit Cu migration to and from the catalyst surface, although it is not clear how this increases activity.

Stegenga et al. (1991) report studies of copper oxide and mixed copper-chromium oxides supported on alumina in pellet form and impregnated onto ceramic monoliths under A/F ratio forcing. Figure 5.32 compares

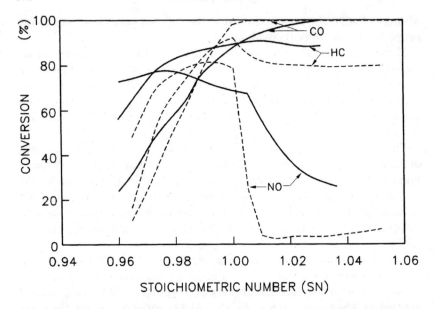

FIGURE 5.32 NO, HC and CO conversion in a simulated automotive exhaust as a function of the stoichiometric number for steady state operation at the time average exhaust composition (—) and SN modulation at $\omega = 1$ Hz (----) for a Cu/Cr monolith catalyst. Inlet temperature to the catalyst bed was 300°C (Adapted from Stegenga *et al.* (1991) with permission, © 1991 Elsevier Science Publishers)

conversions of CO, HC and NO in a simulated exhaust under modulated and steady state operation at the time average feed composition. Modulation was at 1 Hz and the tests were performed at a converter inlet temperature of 300°C. Results are given in the figure for a monolith catalyst containing 1 wt.% Cu−Cr. Steady state conversions are below those routinely measured on the noble metal TWCs, but Stegenga *et al.* comment that raising the temperature a further 50° brings the NO and CO conversions to 100%. The notable feature for the mixed oxide catalyst on comparing Figure 5.32 to Figures 5.14 to 5.16 and 5.26 is the low NO conversion in rich mixtures. The lower HC conversion in lean mixtures can be increased by higher inlet temperatures. Modulating the A/F ratio by varying CO and O_2 concentrations at 1 Hz diminishes all conversions at SN = 1 in Figure 5.32 in the same way these conversions are reduced for the noble metal TWCs under periodic forcing. On the other hand, modulation broadens the operational window in terms of the A/F ratio. NO reduction in lean mixtures is increased, while HC oxidation in rich mixtures is raised. Again, this happens with noble metal catalysts. Stegenga *et al.* conclude from their experiments that mixed oxide monolith catalysts with 1 to 10 wt.% Cu + Cr in the ratio Cu:Cr = 2: provide conversions under steady state and periodic forcing comparable to those obtained with Pt/Rh/Ce monolith catalysts. Drawback of these Cu based TWCs are poor high temperature stability and susceptibility to some poisons.

G. OVERVIEW – DOES A/F RATIO MODULATION IMPROVE THREE-WAY CATALYST PERFORMANCE ?

Section B of Chapter 3 described the large increases (10 to 40 ×) observed for the rates of CO oxidation on supported Pt or Pd catalysts when composition modulation is applied at temperatures under 200°C. As we have seen in Section C of this chapter, 2 to 3 × increases in CO oxidation rates occur on Pt/Al_2O_3 at temperatures up to 400°C under $CO-O_2$ modulation. Modulation also substantially increases C_3 oxidation rates on supported Pt (Chapter 2). In Section B of Chapter 4, 10 × enhancements of the NO reduction rate over Pt/Al_2O_3 through modulation of the CO and NO concentrations are described. On the other hand, in the last three sections of this chapter, it is quite clear that under some conditions modulation of the A/F ratio diminishes NO, HC or CO conversions on TWCs containing these same noble metals (eg. Fig. 5.14 or 5.15). Why hasn't the promise of the low temperature CO oxidation or NO reduction studies on the noble metal catalysts been realized as better performance by catalytic converters?

There appear to be several reasons. First of these is that increasing reaction temperature sharply decreases the benefit of composition modulation. This is evident by comparing the 10 to 40 × increases in rate seen for CO oxidation at low temperature over Pt or Pd with the data of Muraki et al. (1986b) in Figure 5.9 where about a 2 × conversion increase is seen. Furthermore, increasing catalyst temperature reduces the optimal cycle period at which rate or conversion is at its highest. With a Pt catalyst, τ_{opt} is 10 to 20 s at 100 to 150°C, but drops to 1–2 s in the 280 to 400°C range studied by the Toyota workers and appears to be under 1 s at 550°C. These phenomena are clearly shown in Figure 3.5. Frequency of the A/F ratio modulation caused by the feedback control loop ranges between 0.5 and 2 Hz so in the normal temperature range of converters, 350 to 600°C, the modulation imposed at the entrance to the converter is sub optimal. The explanation of these temperature effects is that most of the catalytic converter reactions are characterized by the strong adsorption of one of the reactants at temperatures below 250 to 300°C which often precipitates reactant inhibition of the oxidation or reduction reaction. For CO oxidation over Pt or Pd, oxidation is inhibited by CO. In this situation, the role of composition modulation is to provide surface access for the less strongly adsorbed reactant; in the case of CO oxidation, this reactant is O_2.

Increasing temperature diminishes the adsorption of the strongly adsorbed reactant (e.g. CO), while the adsorption of the weakly adsorbed reactant (e.g. O_2) augments. Inhibition, thus, decreases and with the decrease, the advantage provided by modulation disappears. Moreover, rates of diffusion and surface reaction grow rapidly with increasing temperature so that switching of composition must take place more quickly to offset reactant inhibition.

The model proposed by Lynch (1983) which describes the low temperature behavior on a Pt/Al_2O_3 catalyst under composition forcing, also predicts CO inhibition of the oxidation rate in steady state operation leading under some conditions to multiple steady states (see Section B in Chapter 3). This model assumes competitive adsorption of CO and O_2. It is in this situation that reactant inhibition occurs. If experimental heats of adsorption are introduced into the Lynch model, it predicts the lapse of turning point bifurcations, the disappearance of multiplicity and the end of enhancement through $CO-O_2$ modulation as temperature increases.

There are two other temperature effects which contribute to the explanation of the experimental observations. In the temperature range of converter operation, intrinsic reaction rates are high so that diffusion interference arises even with wash coat and eggshell catalyst deposition. This interference moderates the forcing amplitudes. Also, differing diffusion rates of reactants leads to partial segregation. Both of these effects lower the consequences of modulation or in some cases causes modulation to suppress rates or conversions. We discuss interference phenomenon in Section D of Chapter 3. A further contribution is that the steady state conversions are so high in catalytic converters that it is experimentally difficult to detect changes caused by periodic forcing of the A/F ratio.

Other factors may also contribute to the explanations of the lack of converter enhancement through cycling around $SN = 1$ that appears surprising in view of other experimental results. CO oxidation and to some extent NO reduction under modulation are affected by the water gas shift reaction and by oxygen storage on the surface. These contributions to the overall reaction are certainly sensitive to temperature. This sensitivity may be part of the explanation, that is, lower conversions in converters may be caused by a smaller contribution of the WGS reaction or oxygen storage. Interference between the different reactants in the exhaust can also contribute. For example, NO reduction by CO on Pd is inhibited by hydrocarbons in the exhaust.

There is another important difference between the conditions in catalytic converters and in the low temperature experiments which showed such large rate or conversion enhancements. Most of the low temperature modulation data for CO oxidation on Pt or Pd catalysts were taken in systems where SN was not unity. Studies with actual and synthetic automotive exhaust, e.g. Figures 5.15 to 5.16 and 5.26, establish that large enhancements, perhaps reaching $2\times$, continue to be found when $SN \neq 1$ or $A/F \neq 14.5$ for catalytic converters. Investigators disagree on the magnitude of this non-stoichiometric inprovement. The work of Yokota *et al.* (1985) shown in Figure 5.26, or of Matsunaga *et al.* (1987) shown in Figure 5.28 indicate large increases in conversion can be achieved at just a 0.1 to 0.2 variance in the A/F ratio from the stoichiometric, whereas Schlatter *et al.* (1983) in Figure 5.15 show smaller effects at variances of 0.1 to 0.2 from $A/F = 14.5$. An explanation for this difference is that the Toyota workers used bed temperatures 100 to 150°C lower and employed larger amplitudes. Nevertheless, involuntary modulation of the A/F ratio increases HC and NO conversions when the A/F ratio is more than 0.2 units from the stoichiometric ratio.

Does modulation of the A/F ratio improve the performance of catalytic converters? The answer is an equivocal "yes" and an equivocal "no". It is "yes" for operation at off stoichiometric A/F ratios where there is wide agreement that HC and NO_x conversions are enhanced, but CO conversions are not. However, the large enhancements are for variances greater than 0.2 in the A/F ratio, but converters operate most of the time within a variance of 0.1. It is "yes" with respect to cold starting. Studies discussed in Section E of this chapter indicate that modulation provides a substantial decrease in the light-off temperature for hydrocarbon oxidation and for NO reduction, but little change for CO oxidation. For susceptibility to SO_2 poisoning, the answer is also "yes". We show in Section E that modulation gives higher conversions in the presence of SO_2 and a better recovery of performance when SO_2 is taken out of the exhaust. On the other hand, thermal aging and SO_2 poisoning should be avoided as they always reduce conversion.

The answer to the question of improvement must be "no", however, if the A/F ratio is controlled close to its stoichiometric value for the fuel used. Figures 5.14 to 5.16, 5.26 and 5.28 all show that at $SN = 1$ or $A/F = 14.5$ all conversions under modulation are less than conversions in steady state operation. This has been observed for fresh as well as aged catalysts, for pelleted as well as monolith catalysts. Because current sensor and feedback control loops are able to maintain the A/F ratio remarkably close to the stoichiometric value over a large range of vehicle operating conditions, this is a weighty consideration.

Should the A/F ratio control be redesigned? It is likely that the modulation frequency now imposed on the converter by the control loop is sub optimal. Conversion under modulation could be increased if the frequency was between 2 and 5 Hz. Is this practical? Probably not. . . . mixing in the engine manifold and in the piping to the converter will severely damp concentration variations. Should damping be increased to smooth out the 0.5 to 2 Hz oscillations? There appears to be no incentive to do this because modulation provides benefits over some part of vehicle operating conditions.

What direction should modulation research take in this application? Study is needed on the effect of the low frequency variations due to vehicle operation that were mentioned in Section A of this chapter. These variations will not be damped in the converter and amplitudes may be significant. This question does not seem to have been addressed. Nevertheless, for the broader question of converter research, modulation is not an important consideration. The key question was and continues to be catalyst formulation.

CHAPTER 6
Hydrogenation

A. INTRODUCTION

Although study of periodic composition forcing of CO oxidation or even three-way catalysts has been much more popular with researchers, a variety of hydrogenation reactions under composition modulation have been investigated. Modulated hydrogenation reactions are interesting because with just several exceptions reactant mixtures are reducing so performance improvement will usually arise from manipulation of adsorbate or adatom concentrations and/or distribution on the catalyst surface. Surface rearrangement remains possible following a change in feed composition, but often will be overshadowed by adsorbate or adatom phenomena.

Reactions considered so far in this monograph led to single groups of products (CO_2, H_2O with possibly some CO). With hydrogenation, competing and consecutive reactions appear so selectivity must be considered. Choice of reactions or reaction systems to include in this chapter has been based on whether or not selectivity was the target for the use of composition modulation. When it was, we will delay examination of the reaction until Chapter 11. Thus, acetylene hydrogenation is examined in that chapter, while ethylene hydrogenation is discussed below. In this chapter, we will also discuss methanation of CO, methanol synthesis and NO reduction eventhough parasitic reactions leading to other products are possible. This is justified because catalysts studied under modulation have been selective enough that parasitic reactions can be ignored. Modulation was chosen as a means for increasing conversion or reactor throughput.

B. ETHYLENE HYDROGENATION

Helmrich *et al.* (1974) used this reaction to demonstrate experimentally activation of a heterogeneous catalytic reaction by composition forcing. It was one of the first systems studied. The reaction was also used by Baiker and Richarz (1976) to examine the effect of forcing other reactor operating conditions. Prairie and Bailey (1987) used composition forcing of ethylene hydrogenation to test dynamic models for the reaction.

Helmrich et al. (1974) employed a Pt/γ-Al$_2$O$_3$ catalyst but at low temperature (57 to 75°C) so that only double bond hydrogenation occurred. A 1.13 m by 6 mm (id), isothermal reactor was used with solenoid valves which switched between an ethylene-hydrogen mixture and pure hydrogen. The experimental system resembled the reactor and operating system shown in Figures 1.9 and 1.10. The tubular reactor did not operate differentially.

Figure 6.1 plots ethylene conversion versus cycle period at three different temperatures for a cycling strategy of switching between H$_2$ alone and a feed containing 15 vol.% C$_2$H$_4$ in H$_2$. All parts of the figure show a substantial improvement in conversion through composition forcing. In (a), although conversion depends on space velocity, the increase in conversion exceeds 50% and this percent increase appears to be independent of space velocity for the three velocities shown in the figure. At 75 °C in (b), the percentage increase in conversion is much smaller, but the figure suggests that conversion could have been raised further if a shorter cycle period had been used. The figure demonstrates that the time average conversion under forcing is independent of the volume % ethylene in the feed for the narrow range of concentrations used. If the forcing is unsymmetrical, as in (c), the increase in conversion can reach over 75%. In these experiments, the cycle split was varied at any period by changing the duration of exposure to pure H$_2$. The maximum conversion under hydrogen pulsing for the three experiments shown in Figure 6.1(c) varied between 75 and 83%, well above the 47% achieved under steady state. However, the notable feature of the figure is that the period needed to maximize conversion increases with the cycle split: going from several seconds at $s = 0.6$ to about 30 seconds when $s = 0.95$. This result suggests an H$_2$ exposure of ca. 1.5 to 2 seconds is needed to hydrogenate ethylene adsorbed on the Pt surface.

A curious result is that the flat portion of the conversion vs. cycle period curve, which should represent the quasi steady state behavior, indicates a conversion significantly greater than the value calculated from steady state experiments. This has been observed, occasionally, in experimental studies of other reaction systems. Flat portions of the curves in Figure 6.1(a),(b) gives conversions approximately the same as conversion observed at steady state for the time average feed composition. Steady state measurements, however, indicate ethylene inhibition above 2–4 vol.% C$_2$H$_4$; thus, if the reactor is being forced symmetrically around a mean C$_2$H$_4$ concentration of 7.5 vol.% by hydrogen pulsing, the quasi-steady state conversion should have been well below the steady state conversion. The explanation of this observation is not given by Helmrich et al. It may be that periodic operation, regardless of cycle period, activates the catalyst.

Helmrich et al. determined the steady state kinetics of the Pt catalyzed hydrogenation and demonstrated that it is reactant inhibited in the tem-

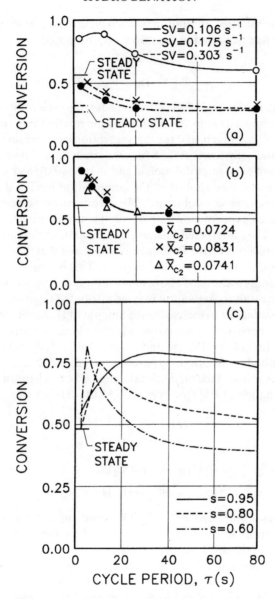

FIGURE 6.1 Ethylene conversion as a function of cycle period: (a) dependence on space velocity for $y_{ethylene} = 0.075$ and $T_{reactor} = 64$ C, (b) dependence on the time average mol fraction ethylene in the feed with $SV = 0.199\ s^{-1}$ and $T_{reactor} = 75\ °C$, (c) effect of unsymmetrical forcing (split) at $SV = 0.195\ s^{-1}$ and $T_{reactor} = 70\ °C$ and $y_{ethylene} = 0.30$ in the feed (Figures adapted from Helmrich *et al.* (1974), MS 108/74, with permission of the authors)

perature range studied. Inhibition is much greater for ethylene then for H_2 so the latter was neglected in the rate model proposed:

$$r_{ethylene} = \frac{k\, y_{ethylene}}{(K + y_{ethylene})^2} \tag{6.1}$$

where K is the reciprocal of the ethylene adsorption equilibrium constant, while k is a composite constant containing the reaction rate coefficient, the mol fraction of hydrogen, and the the adsorption equilibrium constants for hydrogen and ethylene. At 75 °C, Helmrich *et al.* determined that k = 0.2315 mol ethylene/total mols/s and that K = 0.0565 by data fitting. When this model was applied to the composition forcing data, it under predicted the conversions obtained. The investigators, however, were able to simulate their experimental data closely if k was increased, while holding K constant. They interpeted this result to mean that the adsorption and desorption processes on the Pt surface are rapid and that the slow step is the surface hydrogenation of adsorbed ethylene. This is what the unsymmetrical cycling results mentioned above imply. Consequently, hydrogen pulsing activates the catalyst. Activation may be attributed to either alteration of the catalyst surface or to increasing the amount of H_2 adsorbed. Apparently, activation persists to the quasi-steady state limit of modulation.

The objective of the Prairie and Bailey (1987) study was to examine dynamic methods of determining parameters for kinetic models. Composition forcing was one of the methods examined. The mechanism assumed for the hydrogenation of ethylene over a Pt/Al_2O_3 catalyst was:

$$H_2 + 2s \rightleftarrows 2\,H-s$$

$$C_2H_4 + 3s \rightleftarrows C_2H_2 - s + 2H - s$$

$$C_2H_4 + H - s \rightleftarrows C_2H_5 - s$$

$$C_2H_5 - s + H - s \rightleftarrows C_2H_6 + 2s$$

A model was formulated for a CSTR based on this mechanism after establishing from steady state data that the first step can be treated as irreversible and the third step is so rapid that equilibrium can be assumed. The reaction shown in the second step was treated as a reversible surface poisoning reaction. Parameters for the CSTR model were evaluated by steady state, step change and feedback-induced Hopf bifurcation experiments. A disagreement was found between parameters for H_2 adsorption evaluated by step change and Hopf bifurcation experiments. Prairie and Bailey suggest these differences indicate that there are processes with disparate time constants occurring.

Three sets of cycling experiments were performed on the 0.05 wt.% Pt, shell impregnated catalyst. All experiments were carried out at atmospheric pressure, 80 °C, with symmetrical cycling, and used He as diluent and ballast to maintain a constant space velocity. The first set cycled between an equimolar $H_2-C_2H_4$ mixture and H_2, the second set between a mixture and H_2, while the last set switched from a mixture to the He diluent. The purpose of the first set was to establish the optimal cycle period with respect to ethylene conversion. The last two sets were run at this cycle period and examined the effect of the ethylene concentration in the mixture, that is, the amplitude of the ethylene change. Both switching periodically from a mixture to H_2 and to the diluent and back again significantly raised the ethylene conversion if a suitable cycle frequency was chosen.

Figure 6.2 shows the trace of ethylene concentration in the CSTR outlet during several cycles after a reproducible cycling state has been attained. The hash markets on the abscissa indicate the switch to H_2 so results for a 30 s cycle period show desorption of C_2H_4 in the presence of H_2 requires about 5 seconds. A comparison of traces for periods of 180 s and 7200 s (2 hours) indicates that a slow deactivation, possibly poisoning, of the catalyst must be occurring. Fortunately, reducing the cycle period eliminates the deactivation. Figure 6.3(a) plots the experimentally measured time average rate of ethylene hydrogenation versus cycle period. The steady

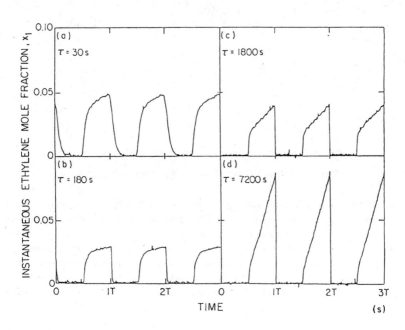

FIGURE 6.2 Ethylene mol fraction measured in the reactor outlet at a reproducible cycling state at different cycle periods for switching between a mixture of 10 vol%C_2H_4 and 20 vol%H_2 in He and 20 vol% H_2 in He over a Pt/Al_2O_3 catalyst at 80 °C and SV $= 0.25$ s^{-1} (Figure adapted from Prairie and Bailey (1987) with permission, © 1987, Pergamon Journals Ltd)

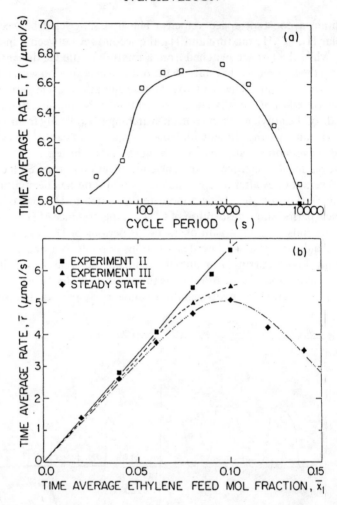

FIGURE 6.3 Composition forcing results for ethylene hydrogenation over a Pt/Al$_2$O$_3$ catalyst at 80 °C and SV = 0.25 s^{-1}: (a) time average hydrogenation rate versus period for symmetrical cycling between a 10 vol% C$_2$H$_4$ and 20 vol% H$_2$ in He and 20 vol% H$_2$, (b) time average and steady state hydrogenation rates versus time average ethylene mol fraction for two forcing policies: symmetrical cycling between a reactant mixture and H$_2$, and a reactant mixture and the He diluent (Figures adapted from Prairie and Bailey (1987) with permission, © 1987, Pergamon Journals Ltd)

state rate at the time average feed composition is 5.1 μmols/s, so that in the optimal period region there is about a 30% increase in rate. As can be seen in Figure 6.3(b), which plots the time average rate versus the time average ethylene mol fraction in the feed, the steady state rate reaches a maximum at

10 vol.% ethylene, so that the time average rate represents a global enhancement above the highest possible steady state rate under the conditions of the experiment. Figure 6.3 also shows that composition forcing increases the hydrogenation rate over the rate at steady state for all concentrations of ethylene tested. Results for the third set of experiments, switching between a reactant mixture and diluent, also shown in the figure, confirm time average hydrogenation rates are higher than the corresponding steady state rates. This means that deactivation can be reversed by removing ethylene from the CSTR feed.

Prairie and Bailey observe that their dynamic model could not reproduce the results discussed in the paragraph above and remark that a comparison with cycling data is probably the most rigorous test of model validity.

C. AROMATICS HYDROGENATION

Toluene hydrogenation was employed by Baiker and Bergougnan (1985a, b) in their ambitious study of fixed bed reactor dynamics. Their concept was that response to dynamic forcing could be used to measure rate and transport parameters of processes occurring within the reactor using just a few experiments. Periodic and ramp forcing were chosen: the former was sinusoidal variations applied to all the controlled variables at once, but without correlation of the forcing of each variable. The latter consisted of a ramp up followed immediately by a ramp down. This was applied separately to each of the variables. The reactor consisted of catalyst packed into the central 0.4 m of a 2.2 m × 0.052 m (i.d.) jacketed tube. Inert solids were packed to a depth of 0.9 m before and after the catalyst bed. Both radial and axial composition samples were collected. Also, radial and axial temperatures were measured. Objectives of Baiker and Bergougnan were to develop suitable fixed bed reactor models and devise an efficient method of data reduction for these models.

A pseudo homogeneous, two dimensional fixed bed model was used primarily, but part of the program compared predictions of this model with those of a one dimensional one. In most experiments, the toluene concentration in the feed, total pressure, bath temperature of the fluid passing through the reactor jacket, and the gas flow rate were varied sinusoidally but with different periods to give uncorrelated periodic forcing. The partial derivatives of the unsteady state models were discretized by means of orthogonal collocation to yield a set of ordinary differential equations which could then be solved by numerical integration using, for example, a variable step length Runge-Kutta-Merson procedure. Kinetic and transport parameters were obtained from radial and axial profiles measured during the periodic forcing experiments by forcing the first derivatives of the model at the collocation points to match the derivatives calculated from the data. This avoided integration of the discretized equations until all the parameters were established. The predicted overall conversion was then compared to the experimental value as a check of the estimation procedure. In this way, the computational effort is reduced enormously. Details of the procedure are beyond the scope of this chapter, but can be obtained from the Baiker and Bergougnan paper. We mention this work here because composition forcing was used in conjunction with forcing of other reactor inputs.

Results for toluene hydrogenation presented by Baiker and Bergougnan (1985b) demonstrate the efficiency of their estimation technique. The authors comment that to obtain reliable results the forcing frequencies must be chosen to reflect the relaxation dynamics of the reactor-reaction system. They observe as well that the mean axial concentration and temperature profiles are given adequately by the one dimensional fixed bed reactor model. Only the two dimensional model, however, provides the maximum temperature at any axial position.

Benzene hydrogenation was studied by Rao et al. (1992) using a single 6 mm diameter × 6 mm cylindrical pellet of a Ni/Al$_2$O$_3$ catalyst (41.7 wt.% Ni) mounted in an internal recycle reactor operating at 445 K and 2 MPa. Three sets of experiments were performed: 1) cycling of benzene flow with H$_2$ flow held constant, 2) cycling of H$_2$ flow with benzene flow held constant and 3) steady state with the benzene:hydrogen feed rates maintained at 1:4. Diluents were not used and in 1) and 2) the time average benzene:hydrogen ratio was the same as at steady state. Forcing was symmetrical and cycle periods were 20 and 60 minutes. Variation of benzene conversion and mean pellet temperature between successive cycles and the limited experimentation makes the interpetation of the investigator's results uncertain. Nevertheless, time average benzene conversions at $\tau = 20$ minutes appeared to be about 20 to 25% greater than under steady state operation, whereas at $\tau = 60$ min conversions were less than at steady state. Rao et al. attribute the increased conversion to alteration of heat storage and heat removal from the catalyst pellet through the periodic change in composition, but did not provide support for their views.

Visser et al. (1994) examined modulating the gas phase composition in a three phase reactor running the hydrogenation of phenylacetylene. Although the investigators were concerned with phenylacetylene conversion, hydrogenation is a consecutive reaction so there is a selectivity problem. We will examine the Visser et al. paper in Chapter 11.

D. METHANATION

Stuchlý and Klusáček (1990) have found that composition forcing increases the time average rate of CO hydrogenation by up to 25% for an optimal choice of the cycling frequency. The objective of their study was to determine if time average rates under forcing can exceed the maximum attainable under steady state. A nickel catalyst (29 wt.% Ni on SiO_2) was used in a thermostated microreactor. Reactor feed was modulated symmetrically between a CO in H_2 mixture and H_2. Experiments were performed at 478 and 488 K. Steady state measurements accompanied the study. Use of a quadrupole mass spectrometer permitted visualization of reactant consumption and methane formation during a cycle. Typical time traces are shown in Figure 6.4. Features of this figure are the slow rise in both reactants after a composition switch: 1–2 seconds are needed. Thereafter the slopes of the CO trace is positive during flow of the mixture over the catalyst, while the slope of H_2 is negative. When only H_2 passes over the catalyst, the H_2 partial pressure rises slowly during the half cycle and there is a significant CO tail extending 3.4 s. The behaviour of the two product partial pressures is different. The variation of methane is large and its partial pressure is highest when only H_2 passes through the reactor. Water shows little variation between the half cycles.

Stuchlý and Klusáček conclude from these observations that CH_4 is formed by hydrogenation of a carbon residue created by the dissociative

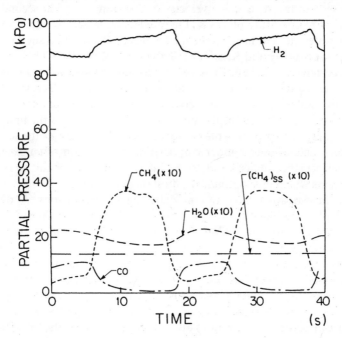

FIGURE 6.4 Variation of reactant partial pressure during symmetrical cycling between 0 and 9.6 vol% CO in H_2 over a Ni/SiO_2 catalyst at 478 K and SV = 200 min^{-1}, $\tau = 20$ seconds: Curves labelled CH_4 and H_2O are shown 10 × (Figure adapted from Stuchlý and Klusáček (1990) with permission, © 1990, VSP (Utrecht))

adsorption of CO. They suggest that hydrogenation is the rate controlling step in methanation over this Ni catalyst at the temperatures used. When CO is present in the feed, even below 10 vol.%, it floods the catalyst surface, inhibiting the adsorption of hydrogen. Adsorbed H_2 is apparently a prerequisite for hydrogenation. This appears to explain the large amount of CH_4 formed after CO is removed from the feed. Water seems to be strongly adsorbed. Their explanation of the large increase in the methanation rate under periodic flushing of the catalyst with H_2 (Fig. 6.5) is that this increases the hydrogen adatom concentration on the the surface. Carbon residues in steady state operation and after exposure to a CO containing stream significantly exceed monolayer coverage.

A resonance with respect to cycle period is evident in Figure 6.5(a). Between cycle periods of 30 to 60 seconds, the time average rate of methanation is 75% greater than the steady state rate at the corresponding time average composition of the reactor feed. Figure 6.5(b) plots the time average methanation rate normalized by the steady state rate measured at the time average composition of the cycled feed stream against molar ratios of reactants in the reactor feed. These measurements show even larger increases in the methanation rate under composition modulation than those seen in (a). At a space velocity of 200 min^{-1}, $\tau = 30$ seconds, and a catalyst temperature of 478 K, the rate has increased by 125% over the steady state rate. The improvement is smaller at 488 K and is similar to that seen in (a). Stuchlý and Klusáček (1990) mention that the $CO:H_2$ ratio for the maximum in Figure 6.5(b) is about the same as the ratio for a maximum in the steady state methanation rate. Consequently, the largest enhancement in catalyst activity under composition forcing occurs under conditions which are optimal for the steady state process. This is the first and so far the only report of this phenomenon. As will be seen in the following chapters, enhancement under composition modulation often exceed the best performance under steady state, but it is unusual for this to occur at conditions which give optimal steady state performance.

The investigators have not modeled their observations. A dynamic model of a methanation reactor, however, has been proposed by Van Doesburg and De Jong (1978) for an alumina supported nickel catalyst and tested under typical industrial operating conditions (0.6 to 2.5 vol.% CO or CO_2 in a H_2 stream, $180\,°C \leqslant T \leqslant 250\,°C$). The pseudo-homogeneous, plug flow model was able to represent both steady state and step changes in feed conditions closely. Stuchlý and Klusáček caution that a model capable of rendering steady state behaviour would not be expected to reproduce behaviour under cycling if rates are at their maxima for the the same time average feed condition and there is global enhancement as well.

Fluidized bed hydrogenation of CO over a nickel faujasite catalyst is described by Jaeger et al. (1990). Nickel faujasite is a zeolite containing

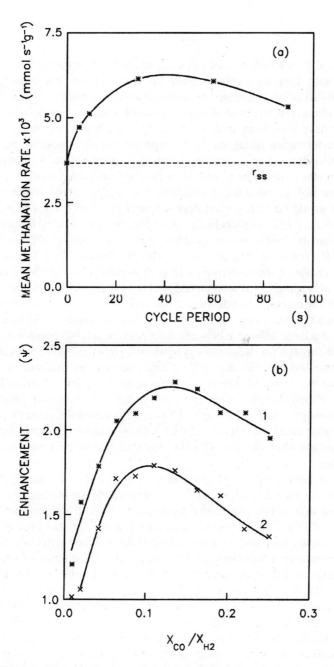

FIGURE 6.5 Methanation rates under periodic hydrogen dosing for a NiO/SiO2 catalyst: (a) variation of rate with cycle period at 478 °C, SV = 120 min^{-1}, $s = 0.5$ for cycling between H_2 and a hydrogen stream containing 7.4 vol.%CO; (b) variation of the enhancement factor with the mol ratio, CO:H_2, in the feed for 478 K (curve 1) and 488 K (curve 2). At 478 K, SV = 200 min^{-1}, $\tau = 30$ seconds, $s = 0.5$, while at 488 K, SV = 120 min^{-1}, $\tau = 20$ seconds and $s = 0.5$ (Figures adapted from Stuchlý and Klusáček (1990) with permission, © 1990, VSP (Utrecht))

strong acid sites which provoke coke formation. The catalyst deactivates rapidly with feed composition used for methanation, but coking and deactivation can be suppressed by increasing the H_2 partial pressure. While the experiments of Jaeger *et al.* are not periodic composition forcing in the sense we have been using in this chapter, they are interesting because they reveal performance which might be expected on implementing periodic composition forcing on a commercial scale. As we discussed in Chapter 1, a likely method would be to operate a two bed or at least a two zone bed with each bed or zone fed a constant, but different, feed composition. Catalyst would be transported between beds or zones. Jaeger *et al.* constructed a micro-bed of nickel faujasite powder which could be fluidized by the reactant mixture or by one of the reactants, in their case H_2. The other reactant (CO) was injected at a point within the fluidized bed. The injected reactant creates a zone relatively rich in that reactant while elsewhere the bed contains some CO, due to mixing, but the gas phase is relatively rich in the fluidizing reactant. Fluidization ensures the catalyst passes continuously through the two zones of this bed. The measurements made by Jaeger *et al.* show a large difference between fluidizing the catalyst powder with H_2 and fluidizing it with the premixed reactants. Deactivation of the catalyst was suppressed by separate feeding and the ratio of methane to CO_2 formation was low and constant, whereas the premixed operation led to a decay of methane formation and gave a high and increasing ratio with time. Although a high methane to CO_2 ratio is desirable, Jaeger's results indicate that coking has been reduced. CO_2 formation proceeds through the water gas shift reaction (WGS). This reaction is poisoned by coke formation.

Carbon dioxide can be hydrogenated to methane with virtually 100% selectivity on Ru/TiO_2. Marwood *et al.* (1994) employed modulation of the CO_2 flow rate to investigate the mechanism of this hydrogenation over a 2 wt.% Ru on TiO_2 at 383 K. A packed bed reactor was used with an external recycle loop and a means to trap H_2O. A diffuse reflectance cell in conjunction with a Fourier transform infrared spectrometer allowed the investigators to follow the time change of species adsorbed on the catalyst surface. CO_2 mole fraction in the feed varied between 0 and 0.2 and the H_2:He ratio was 4:5. The only variable considered was cycle period which ranged from 7 to 93 min; cycles were symmetrical. Adsorbed formate or carbonate and linearly bonded CO were observed and the concentration of these species varied with the forcing of the CO_2 partial pressure in the feed. Lags of signal peaks for each of these adsorbates and CO_2 and CH_4 indicate that formate or carbonate is formed on chemisorption of CO_2 and is equilibrated with that reactant. The surface formate/carbonate decomposes to the CO species which is sequentially hydrogenated to CH_4.

Sufficient data were collected to show that modulating the CO_2 concentration enhances the methane partial pressure at the reactor outlet which means rate is enhanced. This is shown in Figure 6.6 which plots the methane partial pressure normalized with the partial pressure obtained under steady state at the time average feed composition. The research of Marwood et al. will be examined further in Chapter 15.

E. METHANOL SYNTHESIS

Composition forcing of this high pressure industrial reaction has been examined by two research teams with generally complementary results. Both teams found improvement through forcing under specific forcing strategies and experimental conditions. Objectives shared by the teams were: 1) determination of the behaviour of methanol synthesis under composition forcing, and 2) identification of those cycling parameters which exert the largest effect on the time average synthesis rate. McNeil and Rinker (1994) focused on whether or not the global enhancement factor, Ψ^*,

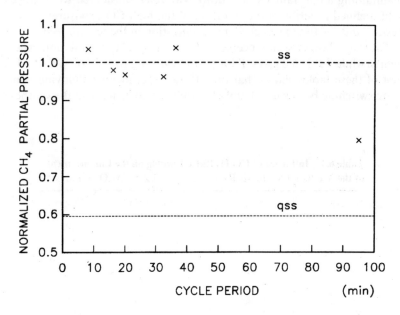

FIGURE 6.6 Methane partial pressure in the reactor off-gas normalized by the measured steady state values obtained when operating at a feed composition which is the time average of the modulated composition for CO_2 hydrogenation at 383 K, 103 kPa and a space velocity of 0.6 min^{-1} over a 2 wt.%Ru/TiO$_2$ catalyst (Figure adapted from Marwood et al. (1994) with permission, © 1994, Elsevier Science Publishers)

exceeds unity. They were also interested to see if modulation provides mechanistic insight. Different catalysts were used by each team.

Methanol synthesis has been investigated extensively utilizing flow direction switching. These investigations will be discussed in a companion monograph.

27:44:29 $CuO:ZnO:Al_2O_3$

Nappi et al. (1985) describe an exploratory study using this catalyst. The catalyst was packed in a gradientless, internal circulation (Berty) reactor and run at 2.51 MPa and 250 °C. Synthesis gas for methanol usually contains CO_2. For sometime it was believed that CO_2 acted as a promoter preventing deactivation so scrubbing of synthesis gas was operated to maintain a low level of carbon dioxide in the gas. Methanol synthesis offers a situation where there are three active components and consequently many different cycling strategies are possible. Nappi et al. tested two of these strategies: 1) modulating the ratio of CO and H_2 in the reactor feed, and 2) modulating the concentration of CO_2 from 0 to 2 vol.% while maintaining a constant $CO:H_2$ ratio. Variables considered were s, cycle split (defined in terms of the duration of the high CO exposure), τ, cycle period, and the time average CO concentration in the reactor feed.

Table 6.1 shows the enhancement, Ψ, for the $CO:H_2$ modulation experiments, while Table 6.2 gives Ψ for periodic dosing with 2 vol.% CO_2. The first of these tables shows that only those experiments employing bang-bang switching between CO and H_2 exhibited an increase in the synthesis

Table 6.1 Influence of $CO:H_2$ Ratio Forcing on the Enhancement[#] of the Methanol Synthesis Rate over a $CuO/ZnO/Al_2O_3$ Catalyst[*]

τ(min.)	s( )	$(Y_{CO})_1/(Y_{CO})_2$	Ψ( )
6	0.333	∞	1.05
9	"	"	1.05
12	"	"	1.03
30	0.325	3	0.87
60	"	"	0.86
18	0.583	4.7	0.87
30	"	"	0.87
60	"	"	0.86

[#] defined as the ratio of the time average rate of methanol formation to the steady state rate measured at the time average composition of the modulated feed
[*] at 2.51 MPa and 250 °C

Table 6.2 Effect of Periodic Pulsing of CO_2 on the Enhancement[#] of the Methanol Synthesis Rate over a $CuO/ZnO/Al_2O_3$ Catalyst[*]

τ (min.)	s ( )	Y ( )
12	0.05	1.16
”	0.10	1.27
”	0.25	1.20
”	0.50	1.05
”	0.75	1.03
30	0.50	1.04

[#] defined as in Table 6.1
[*] at 2.51 Mpa and 250 °C

rate. The increase is quite small and within the bounds of measurement error. The enhancement factor, Ψ, increases uniformly as the the cycle period is reduced so the table suggests that the cycle periods used in the $CO-H_2$ modulation experiments were too long. The authors mention that mixing in the gradientless reactor prevented the use of periods shorter than 6 minutes. Some methane was formed in the reaction, probably due to iron impurities in the Al_2O_3, but the amounts were small and interpetation was clouded by the presence of CH_4 in the CO source. Although allowance was made for CH_4 in the feed, the error this introduced made the measurement unreliable.

Periodic dosing of a constant $CO:H_2$ feed mixture with 2 vol.% CO_2 appears to be a more effective means of increasing the synthesis rate. Table 6.2 shows up to a 27% increase in the synthesis rate. Nappi *et al.* caution that the steady state synthesis rates used to calculate the enhancement in Table 6.2 are estimates. Steady state rates were measured with and without 2 vol.% CO_2 and values for calculating the enhancement factors were interpolated linearly from these data. This is equivalent to setting the quasi-steady state rate equal to the steady state rate. Table 6.2 shows that the synthesis rate is sensitive to the cycle split, s. There is an optimum at about $s = 0.1$. Split in this case is defined as the ratio of the duration of the cycle portion containing CO_2 to the cycle period.

Variation of methanol and carbon monoxide leaving the Berty reactor with time are shown in Figure 6.7 for two of the $CO-H_2$ modulation experiments. Variation of CO with time in the cycle for the bang-bang experiment in (a) of the figure with $\tau = 9$ min and $s = 0.333$ are remarkably small. Some of this is due to physical mixing in the reactor used, but the data nonetheless suggest strong adsorption of CO on the catalyst surface. Methanol variation is small as well and lags the introduction of CO

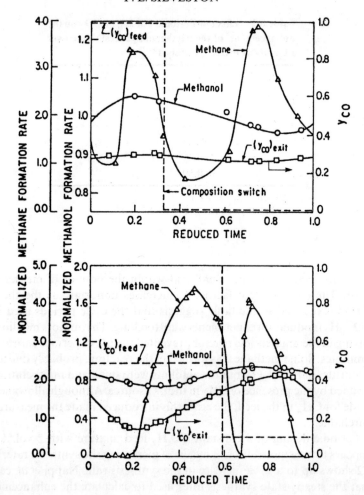

FIGURE 6.7 Variation of normalized product concentration (expressed as rates) with time within a cycle for (a) bang-bang cycling between $y_{CO} = 1$ and 0 for $\tau = 9$ min and $s = 0.333$; (b) mixture cycling between $y_{CO} = 0.486$ and 0.106 for $\tau = 30$ min and $s = 0.583$. Reactor conditions: $T = 250$ °C, $P = 2.513$ MPa. Concentrations (rates) normalized with steady state concentrations (rates) measured at the time average feed composition (Figure adapted from Nappi et al. (1985))

by 4-5 minutes indicating CO inhibition and that a surface reaction is probably rate controlling at the conditions used. Insufficient data were collected in the H_2 portion of the cycle to identify a methanol peak. A peak is evident in the portion in Figure 6.7(b) with mixture cycling. Although the modulation amplitude was much smaller in this experiment, the longer cycle period, $\tau = 30$ min, causes the CO variation to become larger and the inlet concentration is approached. The small methanol variation suggests that CH_3OH is also strongly adsorbed by the catalyst.

30:70 CuO:ZnO And 60:30:10 CuO:ZnO:Al_2O_3

During the last decade, evidence has been developed that CO_2 is directly hydrogenated over the CuO/ZnO catalysts (Chanchlani et al., 1992). This led Chanchlani et al. (1994) to re-examine composition forcing of methanol synthesis because it offers the interesting situation of two parallel reaction paths to the same product and, of course, three reactants for possible manipulation. Chanchlani et al. employed two component (CuO:ZnO = 30:70) and three component (CuO:ZnO:Al_2O_3 = 60:30:10) catalyst formulations. These oxide ratios were chosen because each exhibits the highest methanol activity in its respective two or three component class (Herman et al., 1979) and each had been previously characterized by other investigators. All experiments were undertaken at 2.86 MPa, 225 °C, and SV = 37 mL(STP)/g cat./s. An isothermal, fixed bed microreactor was used. Catalysts were prepared by the investigators and exhibited reproducible deactivation with time on stream. A steady state measurement at a constant feed condition was repeated after a set of cycling experiments to follow deactivation and provide data for correcting all periodic operation and steady state measurements back to the activity of the fresh catalyst. Composition modulation of H_2:CO_2 mixtures retarded the rate of deactivation by about 50%, compared to deactivation in steady state operation (see Chapter 1). When a H_2:CO mixture was used, modulation of the H_2:CO ratio did not affect deactivation. Deactivation was much slower when CO_2 was present in the reactor feed. This observation suggests that reduction of the catalyst causes deactivation.

Strategies tested by Chanchlani et al. (1994) were: 1) maintaining one reactant constant and periodically varying concentrations of the other two, and 2) varying just one reactant concentration while holding the other reactants concentrations constant. Experiments were performed also using just H_2 and CO, and H_2 and CO_2. All cycles consisted of just two parts, although not necessarily of equal duration. In what follows, R is the fraction of CO_2 in the carbon oxide mixture when both CO and CO_2 are present in the reactor feed. In the Chanchlani experiments, R varied from 0 to 1.0. The minimum cycle period was 2 minutes and cycle splits ranged from 0.2 to 0.8.

For the conditions used, not all the strategies tested enhanced the synthesis rate. Table 6.3 displays results of strategy testing. The strategies which provided enhancement were cycling between $H_2 - CO_2$ mixtures, $H_2 - CO$ mixtures at high CO_2 partial pressures, and between $CO - CO_2$ mixtures in the presence of H_2 but at low fractions of CO_2 in the mixtures. Most strategies were tested on the two component catalyst. It was asummed that the choice of catalyst for the test was not important. Furthermore,

Table 6.3 Results of Strategy Tests

Strategy	Cycling Variables Range	Enhancement
H_2 Constant[1] CO and CO_2 modulated	$5 \leqslant \tau \leqslant 10^*$, $0.2 \leqslant s \leqslant 0.8^{**}$, $0.051 \leqslant R \leqslant 0.787^+$	Yes, at low τ, low s and low R
$CO:H_2$ Constant[2] CO_2 and $CO:H_2$ ratio modulated	$5 \leqslant \tau \leqslant 30^*$, $0.017 \leqslant s \leqslant 0.4^{**}$, $0.182 \leqslant R \leqslant 0.705^+$	No
CO_2 Constant[1] CO and H_2 modulated	$2 \leqslant \tau \leqslant 10^*$, $s = 0.5^{***}$, $0.05 \leqslant A \leqslant 0.25^{++}$, $0.05 \leqslant y_{CO_2} \leqslant 0.15$	Yes at high y_{CO_2}
Binary – CO Absent[1] H_2 and CO_2 modulated	$5 \leqslant \tau \leqslant 15^*$, $0.2 \leqslant s \leqslant 0.8^{**}$, $0.08 \leqslant y_{CO} \leqslant 0.57$	Yes
Binary – CO_2 Absent[1] H_2 and CO modulated	$2 \leqslant \tau \leqslant 10^*$, $s = 0.5^{***}$, $0.05 \leqslant A \leqslant 0.25^{++}$	No

[1] CuO/ZnO catalyst
[2] $CuO/ZnO/Al_2O_3$ catalyst
* in minutes
** Defined in terms of CO_2
*** Defined in terms of CO
+ Fraction CO_2
++ Defined as mol fraction CO

all of the tests indicated that the enhancement increases as the cycle period is reduced so that strategies found to be unsatisfactory may be only so because composition switching was too slow.

A comparison with Table 6.1 confirms that there is little or no enhancement of the synthesis rate through modulating CO and H_2 mixture compositions in the absence of CO_2. There is no agreement between the above table and Table 6.2 when the $CO:H_2$ ratio is held constant and the CO_2 content of the feed is varied periodically. However, the experiments of Chanchlani et al. were carried out by switching between a $CO:H_2$ mixture and CO_2 at much higher amplitudes.

Chanchlani et al. (1994) carried out further experiments to measure the influence of the forcing parameters on rate and yield enhancement under modulation of the $H_2:CO_2$ ratio in the absence of CO. Two groups of experiments were performed. In the 1st, the time average $H_2:CO_2$ ratio in the feed corresponded to the ratio which maximized the methanol synthesis under steady state operation, while in the 2nd, the time average ratio was appreciably larger than the optimum steady state ratio. In both experiments, rate enhancement was observed and was as much as 30% greater than the highest synthesis rate measured under steady state operation at the same pressure, temperature and space velocity. Figure 6.8 shows the effect of cycling $H_2 - CO_2$ compositions on methanol yield for the first group of experiments and both catalysts. Yield is defined as the fraction of carbon oxide converted that is recovered as methanol. The figure indicates a

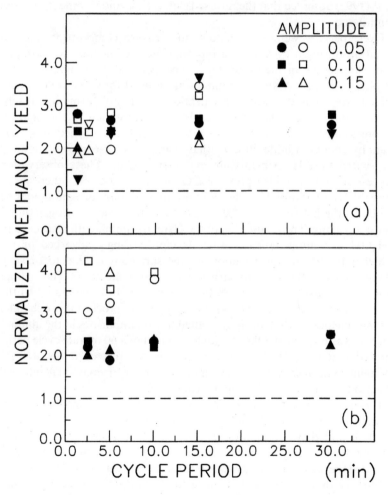

FIGURE 6.8 Normalized methanol yield as a function of cycle period and cycle amplitude for symmetrical forcing of the feed $H_2:CO_2$ ratio in the reactor feed at 2.86 MPa and 225 °C: a) forcing around the optimal steady state $H_2:CO_2$ ratio (80:20) for the CuO/ZnO catalyst, (b) as in (a) for the CuO/ZnO/Al$_2$O$_3$ catalyst (Figure taken from Chanchlani *et al.* (1994))

remarkable increase in yield through symmetrical forcing. Furthermore, in the range studied, cycle period and amplitude have just a small and uncertain influence on yield. The rate data also showed a small but consistent effect on the enhancement. Amplitude is expressed in terms of the $H_2:CO_2$ ratio. Yield results can be explained by suppression of the reverse water gas shift reaction for the two component (CuO/ZnO) catalyst. Some methanation occurs on the three component (CuO/ZnO/Al_2O_3) catalyst and this too is suppressed by feed composition modulation. Chanchlani et al. (1992) observed that the reversed shift is slow under some steady state conditions.

Cycle periods were long enough to measure composition and rate variation during a cycle after a reproducible cycling state was attained. These are shown in Figure 6.9 for a cycle periods of 10 minutes and an amplitude of 0.15. About 2.3 minutes are needed for CO_2 in the reactor effluent to reach the feed concentration. Methanol formation reaches a maximum in the CO_2-rich half cycle, well before the CO_2 in the outlet becomes constant. The peak in water formation trails the methanol maximum by about a minute. Both methanol and water rise immediately after the switch to a H_2-rich mixture and then decline. These observations suggest competitive adsorption of CO_2 with at least one other species, probably H_2O and/or H_2. The slow rise in methanol formation after the switch to the hydrogen-rich feed indicates that surface hydrogenation is rate controlling. The relative lag of the H_2O maximum with respect to the CH_3OH maximum indicates water is more strongly adsorbed or that a higher hydrogen concentration on the surface is needed to scavenge oxygen adatoms than that which is needed to form methanol. The methane time trace is just the opposite of the time behaviour of methanol so that these reactions occur on different surface sites and proceed through different intermediates. Methane is attributed to iron impurities in the alumina, but this was not tested. Observations made using a 60 minute cycle period showed that the CH_3OH concentration in the reactor outlet was still changing in the half cycle. This accounts for the surprisingly small influence of period on rate and yield (Figure 6.8).

The influence of cycle split was studied for both catalysts and for both time average $H_2:CO_2$ ratios mentioned above. Amplitudes could not be held constant and they varied with split so the observations are not entirely attributable to cycle split. For both mean $H_2:CO_2$ ratios, time average rates exceeded the maximum steady state rate for the two component catalyst, but not for the three component one. For the CuO/ZnO/Al_2O_3 catalyst, time average synthesis rates were close to or just slightly exceeded the steady state rates. With yield, results similar to those shown in Figure 6.8 were found, except that the improvement was limited for the two component catalyst with some measurements showing no yield enhancement. No influence of cycle split on yield enhancement was observed.

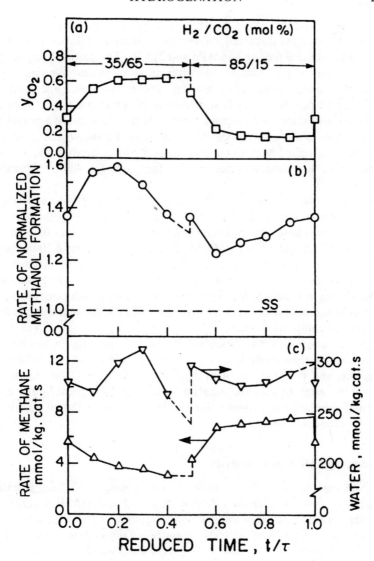

FIGURE 6.9 Variation of the CO_2 mol fraction in the reactor outlet and the rates of methanol, water and methane formation with reduced time within a cycle after a stationary cycling state was reached. Measurements for the ternary catalyst ($CuO/ZnO/Al_2O_3$) for symmetric forcing around a time average $H_2:CO_2$ ratio of 60:40 (a) mol fraction of CO_2 in the reactor outlet, (b) rate of methanol formation normalized by the steady state synthesis rate at the time average $H_2:CO_2$ ratio in the reactor feed, (c) rates of methane and water formation (Figure taken from Chanchlani et al. (1994))

Figure 6.10 sums up the Chanchlani study. The data points are for the cycling experiments discussed above dealing with modulation of $H_2:CO_2$ ratio in the reactor feed at 2.86MPa and 225 °C. The dashed line joins the extreme points and indicates the maximum methanol synthesis rates achievable under composition forcing. The heavy, solid line joins the extreme data points for steady state operation, which are not displayed in the figure, and show the maximum synthesis rate attainable at a given CO_2 mol fraction in the reactor feed. All of the periodic and steady state data have been corrected back to the activity of the fresh catalyst and are thus higher than the data reported in the literature for similar operating conditions.

Figure 6.10(a) for the CuO/ZnO catalyst shows composition forcing of $H_2:CO_2$ reactor feed is capable of doubling the methanol synthesis rate at the conditions considered by Chanchlani et al. The synthesis rate increases by up to 35% for the $CuO/ZnO/Al_2O_3$ catalyst as Figure 6.10(b) shows. Yields at low conversions are also substantially raised by composition forcing as Figure 6.8 demonstrates. Envelopes of maximum yields similar to those shown in Figure 6.10 can be drawn. Contrary to the rate results, the yield enhancement is much greater for the three component catalyst than for the two component one.

Chanchlani et al. (1994) disclaimed any intent to investigate the synthesis mechanism in their study. They remark, nevertheless, that they believe the large rate and yield enhancement they observe for $H_2:CO_2$ modulation is caused by alteration of the catalyst surface and by providing a more optimal population of carbon oxide and H atom species than is possible in steady state operation.

Commercial Methanol Catalysts

McNeil and Rinker (1994) obtained results similar to Figure 6.10(b) employing two commercial catalysts (ICI 51-2 and BASF S 2-85) and cycling between hydrogen and carbon monoxide in the presence of 2 or 3 vol.% CO_2. These two catalysts are probably three component catalysts whose compositions are close to the $CuO/ZnO/Al_2O_3$ catalyst used by Chanchlani et al. McNeil and Rinker were able to build an isothermal, plug flow micro-reactor for differential conversion which could operate with cycle periods as short as 12 seconds. Figure 6.11 shows their measurements at 2.86 MPa and 240 °C. Only the data for $\tau = 12$ and 24 seconds are given. As Figure 6.12 demonstrates, synthesis rates are the highest at these short periods. With the BASF catalyst, this bang-bang cycling gave a global enhancement of up to 25% at the conditions studied. For the ICI catalyst, the global enhancement was 15%.

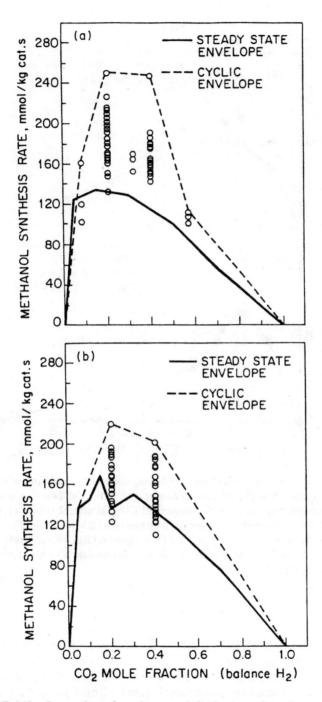

FIGURE 6.10 Comparison of maximum periodic forcing and steady state rates of methanol synthesis as a function of feed composition at 2.86 MPa and 225 °C: (a) results for the CuO/ZnO catalyst, (b) results for the CuO/ZnO/Al$_2$O$_3$ catalyst (Figure taken from Chanchlani *et al.* (1994))

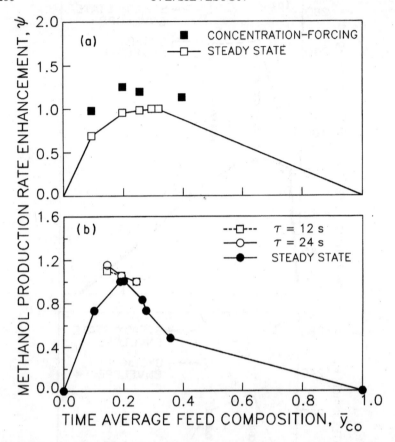

FIGURE 6.11 Comparison of periodic forcing and steady state rates of methanol synthesis as functions of feed composition at 2.86 MPa and 240 °C. Composition forcing between CO and H_2 in the presence of CO_2: (a) using 2 vol.%CO_2 over the BASF S 2-85 catalyst with $\tau = 12$ seconds (See Figure 6.12(b) for cycle split), (b) using 3 vol.% CO_2 over the ICI 51-2 catalyst (See Figure 6.12(c) for cycle split (Figure adapted from McNeil and Rinker (1994) with permission, © 1994, Gordon and Breach Science Publishers, S.A.)

Unlike the data of Chanchlani *et al.* discussed above, Figure 6.12(a) shows that the rate enhancement depended on cycle period for the BASF catalyst. Like the above discussion, however, the effect of cycle split on the enhancement is small and uncertain (Figure 6.12(b)). Cycle period, according to Figure 6.12(c), is not important in the range studied for synthesis over the ICI catalyst.

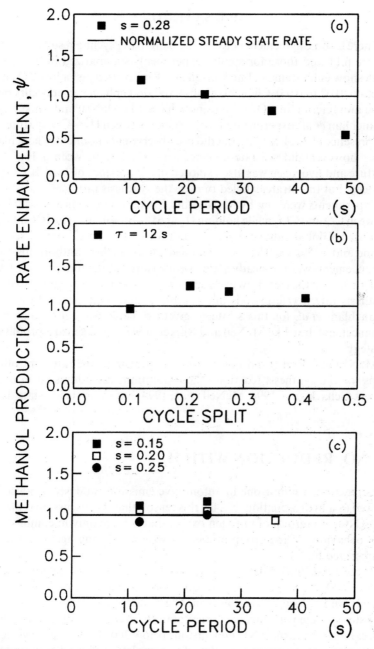

FIGURE 6.12 Influence of the cycling parameters on the global enhancement of the methanol synthesis rate for relatively fast H_2-CO modulation in the presence of CO_2 at 2.86 MPa and 240 °C: (a) effect of cycle period at $s = 0.28$, with 2 vol.% CO_2 using the BASF S2-85 catalyst, (b) as in (a) but with 3 vol.% CO_2 using the ICI 51-2 catalyst, (c) effect of cycle split with $\tau = 12$ seconds and 2 vol.% CO_2 using the BASF S 2-85 catalyst (Figure adapted from McNeil and Rinker (1994) with permission, © 1994, Gordon and Breach Science Publishers).

There is some agreement between the bang-bang cycling results shown in Figure 6.11 and those for similar experiments summarized in Table 6.3. Both show enhancement, but McNeil and Rinker used just a low CO_2 mol fraction (0.02 to 0.03). Chanchlani observed that enhancement occurs only if the mol fraction of CO_2 is high enough (> 0.1). McNeil and Rinker, like Chanchlani et al., experimented with cycling between $H_2 - CO$ mixtures in the presence of 2 vol.% CO_2, but did not observe enhancement. Chanchlani et al., however, did see rate enhancement. Cycling between a $H_2 - CO$ mixture and hydrogen was also explored in the presence of 3 vol.% CO_2 in the feed, but this strategy failed to raise the synthesis rate.

Experiments were run by McNeil and Rinker on periodically switching between CO and CO_2 while holding H_2 at 61% by volume. The ICI catalyst was used. Global enhancement was not found for the 12 and 24 second period and $0.25 \leqslant s \leqslant 0.8$ range examined, although these authors observed enhancement over the steady state rate measured at the time average CO_2 mol fraction in the feed. Low cycle splits with respect to CO_2 suppressed the synthesis rate. Cycle periods did not affect the enhancement. Results of Chanchlani et al. for this strategy, given in Table 6.3, show some rate enhancement, but, like McNeil and Rinker, it was not deemed a promising strategy.

McNeil and Rinker did not discuss mechanism or the rate controlling steps for the synthesis reaction. These questions are addressed in other papers (Schack et al.,1994; McNeil et al., 1994) but using steady state data.

F. NO_x REDUCTION WITH H_2

This reaction, a minor one in automotive emission catalysis, can also be viewed as a hydrogenation. It is for this reason that the discussion is placed here. NO_x is a group of nitrogen oxides that arise mainly in combustion. The behaviour of the group in contact with a reducing agent is usually represented by NO.

There are distinct differences between the noble metals catalyzed reduction of NO_x with H_2 and CO so observations made on the much more heavily studied former reaction cannot be transferred to reduction with H_2. The steady state performance of the catalysts changes and Pt, rather than Rh, exhibits the highest NO conversions below 300 °C. At a space velocity of 30,000 h^{-1}, conversion goes almost to completion above 400 °C and the activity of the Pt, Pd and Rh catalysts appear to be equal. Shinjoh et al. (1987) report that $NO - H_2$ switching is much less effective for the Pt and Pd catalysts than it is for rhodium.

Modulation experiments were undertaken by Shinjoh et al. at the Toyota Research Laboratories using a micro-reactor packed with noble metal catalysts in much the same way as described at the beginning of Chapter 2 for propylene oxidation. As part of the Toyota study of catalyst formulation, each of the noble metal components of the three-way catalysts was tested separately. For NO reduction by H_2, these dynamic experiments were limited to Rh.

NO conversion as a function of cycle period and reactor temperature for Rh can be seen in Figure 6.13 for symmetrical cycling between a feed containing 0.3 vol.% NO and one with 0.3 vol.% H_2. Diluent in both feed streams was He. The time average feed mixture was stoichiometric. Conversion at steady state is given on the vertical axis. Modulation exerts a large effect on NO conversion at low temperatures. At 250 °C, NO conversion doubles when composition forcing is applied. A resonance with cycle frequency is evident at this temperature. It can be seen too that the frequency which maximizes conversion increases with temperature. At 300 °C, the period which maximizes conversion has decreased to 0.2 s from about 0.5 s at 250 °C. The enhancement of conversion achieved through composition modulation decreases as the catalyst temperature increases.

Shinjoh *et al.* mention that the Rh surface appears to be oxidized by NO and this suppresses further NO adsorption. Their kinetic measurements

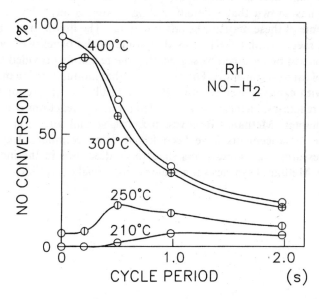

FIGURE 6.13 NO conversion as a function of cycle period at different temperatures for a Rh catalyst at a space velocity of 30,000 s^{-1} with symmetrical cycling between a feed containing 0.3 vol.% NO and one with 0.3 vol.% H_2 (Figure reprinted from Shinjoh *et al.* (1987) with permission, © 1987, Elsevier Science Publishers B.V.)

indicate a negative order with respect to NO which suggests NO inhibition of reduction. The order with respect to H_2 is large and positive which probably means that hydrogen reduces the Rh surface as well as functioning as a reactant. The role of cycling, thus, seems to be to maintain the catalyst surface in a reduced state. NO dissociation is rate controlling and this seems to proceed faster on the rhodium metal. Cycling may also equalize the mean reactant adatom concentrations on the rhodium increasing the surface reaction rate in this way.

G. SUMMARY

Probably the importance of periodic composition forcing of hydrogenation reactions will be in situations where selectivity is an issue, such as in the Fischer-Tropsch synthesis so that future work on modulating hydrogenation reactions should be directed there. This synthesis is examined in Chapter 13. We have seen in this chapter that hydrogenation reactions under composition forcing behave much like the oxidation reactions: 1) resonance with frequency is observed, 2) cycle split is an important consideration. The major difference is that the enhancements observed are much less than those described in Chapters 3 and 4. The work on methanol synthesis has shown that different cycling strategies have different outcomes, some of these may not be advantageous. The fluidized bed experiments of Jaeger *et al.* (1990) show that some of the benefits of periodic operations can be obtained by segregating the reactants provided there is a means of catalyst transport. They also show that modulation is a means of dealing with catalyst deactivation. Because results are sparse for hydrogenation reactions, further study is needed for all the reactions considered in this chapter. Methanol deserves, perhaps, special attention because significant enhancements have been found using commercial catalysts under pressures only slightly less than those described in the industrial literature. Methanol synthesis is a major commercial operation.

CHAPTER 7

Ammonia Synthesis

A. INTRODUCTION

The commercial importance of ammonia synthesis made it an early target for studies of periodic operation in both Europe and North America. At temperatures where rates of ammonia formation are promising, the reaction becomes strongly equilibrium limited so that conversions per pass seldom exceed 5 to 6% based on fresh feed. To try to bypass this limitation, Unger and Rinker (1976) attempted the synthesis in a chromatographic reactor. This effort developed into a comprehensive experimental and theoretical study of the application of periodic operation to ammonia synthesis. Recognizing that reactant inhibition reduces rates at lower temperatures where equilibrium is more favourable, other investigators (Rambeau and Amariglio, 1981a) employed rapid bang-bang switching between H_2 and N_2 to by-pass this steady state problem. The third reason for studying the synthesis under composition forcing was the encyclopedic one that ammonia synthesis is an example of hydrogenation with a single product so it deserves study to see how the reaction responds (Jain et al., 1982b).

Ammonia synthesis is a hydrogenation reaction so this chapter extends the discussion begun in Chapter 6.

B. EXPERIMENTAL METHODS

The single bed, micro reactor discussed in Chapter 1 has been used by the three research groups that have studied ammonia synthesis (See Figure 1.12). High pressure and ammonia detection require special consideration. Actuating a solenoid valve creates a brief, but strong pressure deviation if flows are dead ended or have different downstream pressure drops. Thus, all solenoid valves must switch between a reactor feed path and an exhaust path which contains a throttling valve to adjust the pressure drop. The pressure surge can be virtually eliminated if the pressure drops are exactly balanced. High pressures, as high as 10 MPa, make catalyst transport schemes for larger systems, such as pilot plants, expensive.

Three detection principles for measuring NH_3 have been pursued. Wilson and Rinker (1982) and Chiao et al. (1987) used photo-ionization, whereas Jain et al. (1982b, 1983b) employed an infrared spectrophotometer at 1030 cm^{-1}. Wet chemical methods were chosen by Unger and Rinker (1976) and Rambeau and Amariglio (1981b). The former used absorption in boric acid while Amariglio and co-workers employed absorption in sulphuric acid. Ammonia was measured by titration for acidity.

C. CHROMATOGRAPHIC STUDIES

Unger and Rinker (1976) conducted an investigation of N_2 pulses superimposed on a steady flow of H_2. Their chromatographic reactor was packed with a physical mixture of a promoted iron oxide synthesis catalyst (Girdler G-82) and an absorbent [5A, 13 X molecular sieves (Linde) or alumina]. The concept pursued by these investigators rested upon the Temkin-Pezhev model for ammonia synthesis.

$$r_{NH_3} = k_f P_{N_2} \left(\frac{P_{H_2}^2}{P_{NH_3}^2}\right)^\alpha - k_r \left(\frac{P_{NH_3}^2}{P_{H_2}^2}\right)^{1-\alpha} \tag{7.1}$$

Removing NH_3 from the reacting mixture drives the reaction forward. Maintaining a high H_2 partial pressure also forces the forward reaction, but impedes the reverse one. High pressures are no longer necessary so Unger and Rinker operated their reactor at just 928 kPa. Temperatures ranged from 455° to 479°C.

Figure 7.1 reproduces a thermal conductivity record for the pulsed operation using 13 × molecular sieves as the adsorbent. The TCD used on the column outlet senses the $N_2 - NH_3$ mixture so the peaks correspond to the N_2 pulses and NH_3 product, while the rising baseline indicates NH_3 desorbing from the 13 × sieve. N_2 is completely consumed in the first five pulses. The sieves fill with NH_3 so that by the 6th pulse, nitrogen breaks through and NH_3 desorption becomes significant. N_2 pulsing has been broken off after the 20th pulse; ammonia continues to desorb from the sieves for a further 80 minutes however.

Ammonia production was measured for the experiment shown in Figure 7.1. It corresponded to a N_2 conversion of 57.2%, whereas, if the time averaged $N_2 - H_2$ feed had been used, conversion at the equilibrium limit would have been 40.5%. Pulsed operation, therefore, results in conversions exceeding the equilibrium limit that would apply to the same flow rates of H_2 and N_2 if they had been mixed and passed over the catalyst at steady state. A continuous version of the Unger-Rinker experiment would be bi-periodic: a low frequency alternation between N_2 pulsing and flushing of the catalyst by one of the reactants. The N_2 pulsing would occur at a higher frequency during one portion of the longer cycle. Unger and Rinker (1976) observed that temperature ramping at 20°/min reduced the length of the flushing period by one half.

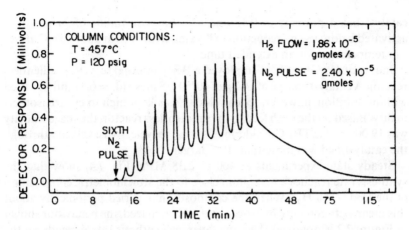

FIGURE 7.1 Total nitrogen response to N_2 pulses with an H_2 carrier gas over a mixed ammonia catalyst – 13X molecular sieve in a packed bed (Figure adapted from Unger and Rinker (1976) with permission, © 1976 American Chemical Society)

This intriguing experimental work does not seem to have been pursued further. No optimization of catalyst-adsorbent mixture, pulse frequency, pressure or temperature was attempted. Shortening the pulsing duration to, say, 5 cycles and condensing out NH_3 would yield an effluent containing only hydrogen. This recovered hydrogen can be recycled. This modification seems worthy of further study.

D. COMPOSITION FORCING

Iron Catalysts

Rinker and co-workers (Wilson and Rinker, 1982; Chiao et al.,1987) and Silveston and co-workers (Jain et al., 1982b; 1983b) experimented with a triply promoted iron catalyst, C73-1-01, obtained from United Catalyst Inc. Table 7-1 gives the composition of the catalyst in its oxidized form. The

Table 7.1 Analysis of the Oxidized C73-1-01 Catalyst

Component	Content - wt%
FeO	30–37
Fe_3O_4	65–58
Al_2O_3	2–3
K_2O	0.5–0.8
CaO	0.7–1.2

catalyst has a porosity of 0.45 and a mean pore radius of 24 nm. It is activated before use by reduction in flowing hydrogen at 25° to 50°C above run temperature for an extended time.

Jain et al. (1982b; 1983b) employed a Berty internal recycle, gradientless reactor. A catalyst charge of 14.7 g of 40/50 US mesh ($\bar{d}_p = 0.3$ mm) was used without dilution; however, space velocity was kept high to ensure isothermality. Based on the fresh feed to the gradientless reactor, the space velocity was 19,000 h^{-1} (STP). Considering internal recycle, space velocity through the catalyst bed was more than 10 × larger.

Steady state experiments at 400°C, 2.38 MPa (Fig. 7.4) show that the synthesis rate reaches a maximum close to the stoichiometric composition of the feed (75% H_2). When the composition is varied periodically about this mean between pure hydrogen and a N_2 rich feed, the behaviour shown in Figure 7.2 is obtained. The time-average synthesis rate depends on the cycle period, τ, and on the cycle split, s, the relative duration of the exposure to hydrogen. Measurements made at a split of 0.6 cover a greater than 10 × range of cycle periods. At a period of a minute, the time average NH_3 formation rate approaches the steady state synthesis rate at the stoichiometric feed composition for the P, T used. This behaviour was expected because the catalyst occupied just 10 ml of the 456 ml volume of the reactor assembly. The high internal recycle sharply reduces the composition change experienced by the catalyst. Just a small variation of composition around the mean occurs. As the cycle period approached 60 min, the time average rate moves toward the quasi-steady state limit. The gradual approach to the quasi-steady state limit is explained by the slow relaxation of the synthesis rate after an abrupt change in feed composition. Jain et al. (1983b) and Li et al. (1985a) report transients of 20 to 30 minutes after a composition change. With $\tau = 60$ minutes and $s = 0.4$, the duration of the partial periods are 24 and 36 minutes. Consequently steady state is attained only at the end of each partial period. The time average rate, therefore, must lie above the quasi-steady state limit.

Figure 7.2 clearly shows that periodic composition forcing increases the synthesis rate to levels greater than attainable through steady state operation. The maximum improvement is about 25% ($\Psi = 1.25$) for $s = 0.6$ and almost 30% for $s = 0.4$. Split is important and smaller values lead to higher rates. Large increases in the time average synthesis rate occur at time average feed H_2 fractions above and below the stoichiometric mole fraction. This may be seen in Figure 7.3. Time average rates of NH_3 formation have been normalized by the steady state rate at the time average feed composition (taken from Fig. 7.2). Only the region of the maximum time average rate is shown. For a feed containing 90% H_2 ($m = 0.9$), under time averaging, $\Psi = 1.45$. The smallest improvement is at 50% H_2 ($m = 0.5$) in the feed at $s = 0.5$. (Fig. 7.3a). This corresponds to switching between pure N_2 and pure H_2 in the reactor feed. The results show as well the

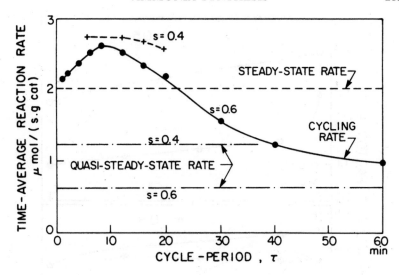

FIGURE 7.2 Effect of cycle period and split on the time average ammonia synthesis rate for a time average reactant mixture containing 75 vol. % H_2. Data for a recycle reactor at 400°C, 2.38 MPa and SV = 19,000 h^{-1} (Figure adapted from Jain et al. (1983b) with permission of the authors)

continued importance of cycle split and period, although at $m = 0.9$ the relationship of rate to cycle period weakens (Figure 7.3b).

Time average as well as steady state NH_3 synthesis rate data collected by Jain et al. (1983b) at 400°C and 2.38 MPa in their gradientless reactor are summarized in Figure 7.4 as plots of the maximum time average rates for periodic operation and maximum steady state rates vs. time average m or mole fraction H_2 in the reactor feed. Global enhancement factors are also given. Data points are not shown. The maximum rates measured show that NH_3 production rates under periodic conditions significantly exceed steady state rates at comparable conditions provided suitable values of the cycling variables are used. Furthermore, time average rates can exceed the maximum steady state synthesis rate, at the stoichiometric feed composition, by up to 30%. The single hatched region shows an estimate of the improvement possible. Data were not collected for periodic operation over this range of m.

Ammonia detection via IR allowed the measurement of NH_3 formation during a forcing cycle. Figure 7.5 shows the variation for switching between pure H_2 and pure N_2. The lack of variation at $\tau = 6$ min probably reflects mixing in the gradient-less reactor which also damped the N_2 and H_2 partial pressure variation. However, mixing cannot explain the persistence of ammonia formation up to 10 minutes after N_2 or H_2 were taken out of the feed. Increases in rate after a switch to pure H_2 were observed in other

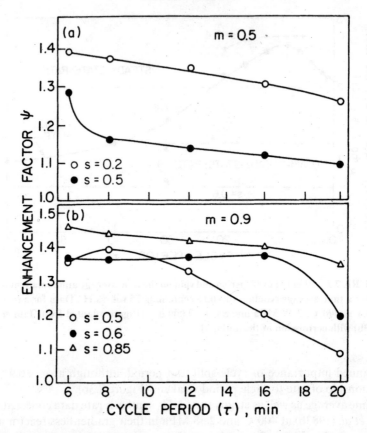

FIGURE 7.3 Effect of cycle period and split on the time average rate normalized with respect to the steady state rate at the same $y^\circ_{H_2}$. Data for a recycle reactor at 400°C, 2.38 MPa: a) $m = 0.5$, b) $m = 0.9$ (Figure adapted from Jain *et al.* (1983b) with permission of the authors)

experiments employing cycling between H_2 and mixed N_2-H_2 feeds. These observations together with Figure 7.5 suggest storage of reactants on the catalyst surface or within the bulk catalyst. This phenomenon, coupled with higher surface coverage by atomic hydrogen, was the source of the improvement under cycling according to Jain *et al.* (1983b).

In a separate paper, Jain *et al.* (1982b) report the response of NH_3 production and reactor temperature to step changes between catalyst exposed to a 50–50 vol.% mixture and pure H_2 or pure N_2. Experimental data are summarized in Figure 7.6. Catalyst temperature did not mirror the NH_3 results. It takes about 8 minutes for a H_2 source to disappear from the catalyst sample, whereas a N_2 source must be present at least 90 minutes after N_2 has been removed from the feed to the catalyst bed. These times can only be explained by dissolution of both hydrogen and nitrogen in the

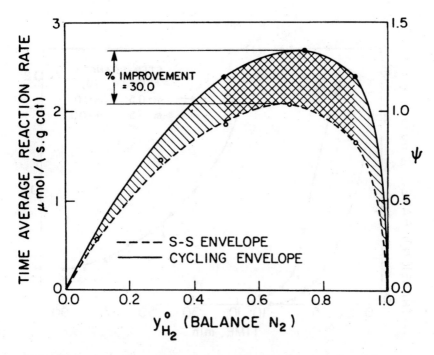

FIGURE 7.4 Comparison of steady state and time average cycling rates of NH_3 formation as a function of time average feed composition. Conditions as in Figures 7.2 and 7.3 (Figure adapted from Jain *et al.* (1983b) with permission of the authors)

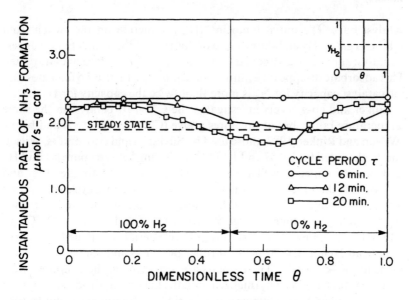

FIGURE 7.5 Instantaneous rate variations within a cycle at $m = 0.5$, $s = 0.5$ (switching between pure N_2 and pure H_2) (Figure adapted from Jain *et al.* (1983b) with permission of the authors)

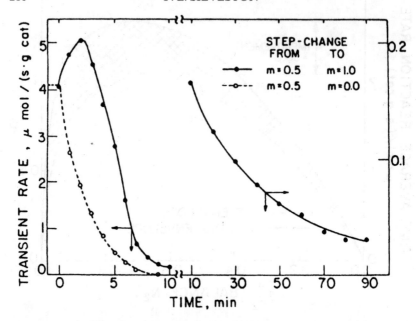

FIGURE 7.6 Response of the instantaneous rate of ammonia production to a step-change in concentration from 50% H_2 and 50% N_2 ($m = 0.5$) to pure H_2 ($m = 1.0$) and to pure N_2 ($m = 0.0$) (Figure adapted from Jain et al. (1982b) with permission of the authors)

catalyst bulk. The initial jump in NH_3 production on the switch from synthesis gas to H_2 and the absence of this rise for the switch to N_2 suggest a catalyst surface largely covered by N_2 or partially hydrogenated species. This adsorbate disappears within about 2 minutes. Figure 7.6 indicates that the catalyst capacity for N_2 is more than twice the capacity for H_2.

Rinker and co-workers investigated the C73-1-01 catalyst in the same temperature range, but at pressures above (Chiao et al., 1987) and below (Wilson and Rinker, 1982) those used by Silveston and co-workers. Wilson and Rinker used 0.13, 0.57 and 1.13 MPa. Cycling between pure N_2 and H_2 feeds was examined as well as between H_2 and a 50-50 mixture of H_2 and N_2. Most of their measurements were made at $\tau = 1$ min; cycle split varied between 0.32 and 0.75. Figure 7.7 shows Wilson and Rinker's data for H_2/N_2 cycling at a time average 50 mol% H_2. The data set at 325°C and 1.13 MPa, spanning the widest range of cycle period, are joined by a curve to indicate trends. Experiments at other pressures and temperatures were made for $40 < \tau < 60$ s. These points are also given in the figure. All time average rates have been normalized with the maximum steady state rate at the same pressure and temperature to yield a global enhancement factor.

Figure 7.7 confirms Figure 7.4 showing NH_3 formation rates which exceed significantly the highest steady state rate. A trend is evident for the

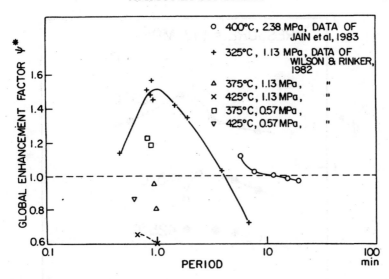

FIGURE 7.7 Normalized rates of NH_3 synthesis over a triply promoted iron catalyst with cycling between pure H_2 and pure N_2 for a time average 50 mol % H_2 in the feed (Figure taken from Silveston (1987) with permission of the author)

data collected at $\tau = 50$ s.: namely, increasing reactor temperature and pressure decreases Ψ^*. Data collected by Jain et al. (1982b) in an integral, near adiabatic, reactor for H_2/N_2 cycling are shown for comparison. There is some consistency between these independent data sets. Jain et al. (1982b) obtained $0.95 < \Psi^* < 1.10$, but operated at a higher temperature and pressure.

Wilson and Rinker found that cycle split is an important variable for bang-bang cycling, that is switching between H_2 and N_2. Figure 7.8 shows the variation in the normalized rate of synthesis, Ψ, over $0.33 < s < 0.75$, at 1.13 MPa and 325°. The bar in the figure represents 5 data points at $s = 0.55$ and $\tau = 58$ seconds. The variation persists at 375°C, but apparently disappears at 425°C. Jain et al. (1982b) also observed an influence of s on Ψ in this cycling mode but made measurements only at two cycle splits.

Cycling between pure H_2 and a 50-50 reactant mixture was also studied. This corresponds to forcing around the stoichiometric composition (75 mol % H_2). Improvement was either not observed or was just marginal for this cycling mode as Figure 7.9 shows. Data reported by Jain et al. (1982b) for the same cycling mode in an integral reactor are plotted for comparison. Jain et al. observed a small increase in the synthesis rate. Wilson and Rinker's data suggest they may have been able to observe $\Psi > 1$ at $\tau < 40$ seconds if mixing had not interfered.

Based on their pure N_2, pure H_2 data (Figure 7.10), Wilson and Rinker (1982) conclude that increasing synthesis temperature and pressure de-

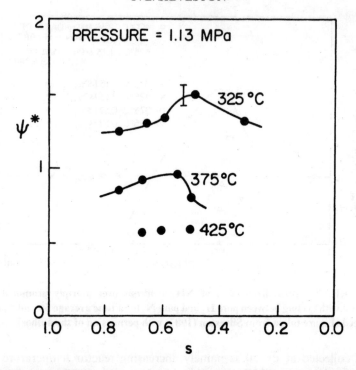

FIGURE 7.8 Influence of cycle split on the rate of NH_3 synthesis with cycling between pure H_2 and pure N_2 at $39\,s < \tau < 106\,s$ over a triple promoted iron catalyst (Figure adapted from Wilson and Rinker (1982) with permission, © 1982 Elsevier Science Ltd.)

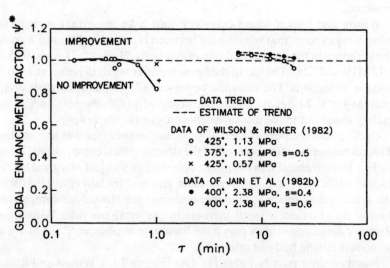

FIGURE 7.9 Variation of the NH_3 synthesis rate normalized by the maximum rate at steady state with cycle period for cycling between 50 vol.% H_2 mixture and pure H_2.

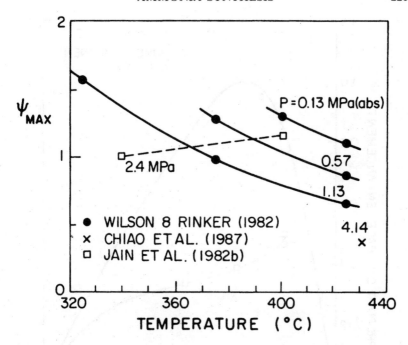

FIGURE 7.10 Influence of synthesis temperature and pressure on the maximum observed value of the enhancement factor for bang-bang cycling between H_2 and N_2 (Figure adapted from Wilson and Rinker (1982) with permission, © 1982 Elsevier Science Ltd.)

crease Ψ. A single data point from a later study (Chiao et al., 1987) at 430°C and 4.14 MPa is shown. This point seems to confirm their contention. Data from Jain (1981) have also been added to the figure. These results, from a near adiabatic integral reactor operating in the same cyclic mode as used by Wilson and Rinker, show synthesis rates increasing as temperature increases. Jain's data on the effect of total pressure is limited. Nevertheless, they show Ψ decreasing with increasing pressure. The explanation of the difference with respect to temperature is not known.

Chiao et al. (1987) extended experiments to 4.14 MPa, much higher than previous experiments, but still less than half the pressure used in commercial synthesis reactors. Measurements were made using an integral reactor for two cycling modes: 1) switching between pure H_2 and pure N_2, and 2) switching between different $H_2 - N_2$ mixtures, namely 60% and 80% H_2. Figure 7.11 shows their normalized rate of NH_3 synthesis for switching between pure reactants at $\tau = 12$ seconds. Clearly, composition modulation using this mode is unattractive. Results for $\tau = 20$ and 120 seconds exhibited even lower values of Ψ. The data point at $m = 0.5$ is used in Figure 7.10. Chiao et al. (1987) explain their results by catalyst deactivation

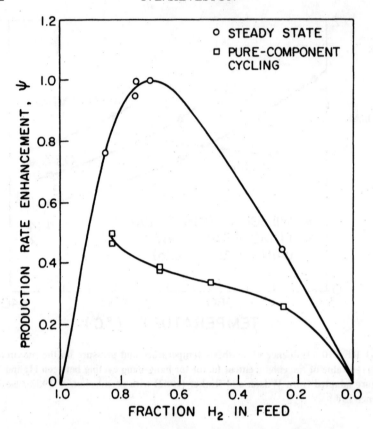

FIGURE 7.11 Comparison of normalized rates of NH_3 production at steady state and bang-bang H_2-N_2 cycling as enhancement factors at $\tau = 12$ seconds at 430°C and 4.14 MPa (Figure taken from Chiao et al. (1987) with permission, © 1987 Gordon and Breach Science Publishers)

through exposure to either pure H_2 or pure N_2. Their explanation is illustrated by Figure 7.12 which shows the normalized NH_3 concentration leaving their reactor following a switch from operating at steady state with 75% H_2 in the reactor feed to cycling between pure H_2 and pure N_2 at $s = 0.5$ and $\tau = 12$ seconds (this corresponds to the point at m = 0.5 in the previous figure). Chiao et al. speculate that this reversible deactivation arises either from electronegative nitrogen dissolving into the iron catalyst and thereby affecting the electronic structure of the active site or, possibly, via restructuring of the catalyst surface on a microscopic scale.

The second mode of periodic operation, switching between H_2-N_2 mixtures, appears to improve reactor performance slightly. This may be seen in Figure 7.13 which shows Ψ as a function of cycle period τ for forcing

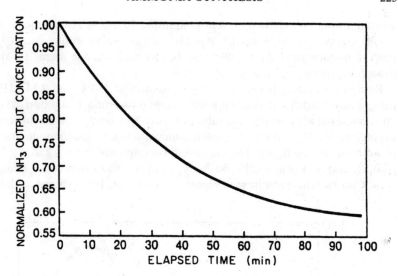

FIGURE 7.12 Variation of NH_3 concentration in the reactor outlet after a switch from steady state operation to bang-bang $H_2 - N_2$ cycling. Conditions as in Figure 7.13 (Figure taken from Chiao *et al.* (1987) with permission, © 1987 Gordon and Breach Science Publishers)

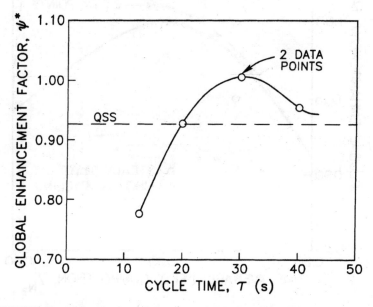

FIGURE 7.13 Enhancement of the rate of NH_3 synthesis for cycling between a 60% and 80% H_2 feed as a function of cycle period for $s = 0.5$ at 430°C and 4.14 MPa (Figure taken from Chiao *et al.* (1987) with permission, © 1987 Gordon and Breach Science Publishers)

around a time average reactor feed containing 70 mol% H_2. The peak at $\tau = 30$ seconds corresponds to $\Psi = 1.01$, a very small improvement in reactor performance. At $\tau > 40$ seconds, the performance seems to approach the quasi-steady state (QSS) limit.

Results for cycling between reactant mixtures at 430°C and 4.14 MPa using a near isothermal, near plug flow integral reactor are summarized in and compared with steady state rates in Figure 7.14 as functions of the time average mol% N_2 in the feed. A modest increase in rate due to modulation is indicated by the figure. The maximum corresponds to $s = 0.2$ at $\tau = 30$ seconds and occurs at a different % N_2 than the maximum under steady state. Contrast these results with Figure 7.4 (Jain et al., 1983b) based on data

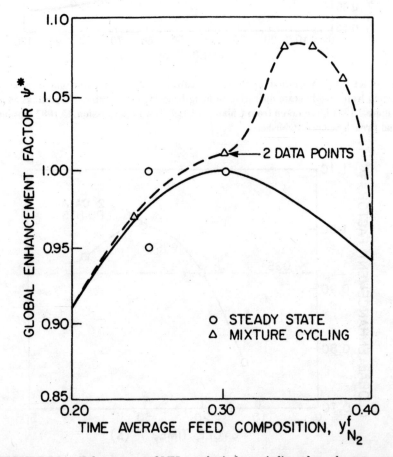

FIGURE 7.14 Enhancement of NH_3 synthesis for periodic and steady state operation as a function of time average feed composition. Curves connecting data points indicate trends only (Figure taken from Chiao et al. (1987) with permission, © 1987 Gordon and Breach Science Publishers)

collected at 400°C and 2.38 MPa which show a 30% increase in the synthesis rate through modulation and virtually no shift in the composition for the maximum synthesis rate.

Amariglio and Rambeau (1977) and Rambeau and Amariglio (1978a, 1978b, 1978c) studied ammonia synthesis on pure iron powder and the industrial KMI catalyst, an iron catalyst triply promoted with K_2O, CaO and Al_2O_3. In a discussion of their work with respect to composition modulation, Rambeau and Amariglio (1984) mention that their investigation of forcing the synthesis over an iron catalyst, presumably the KMI catalyst they have studied intensively, showed no rate improvement over the synthesis at steady state. Unfortunately, details of their experiments are not given.

Krylova (1990) summarizes work published in Russia (Krylova and Abdeligani, 1987) in which various composition modulation strategies were explored using the Russian SA-IR catalyst. These strategies included symmetrical cycling between a synthesis gas mixture and hydrogen, or between the mixture and nitrogen, and bang-bang cycling between H_2 and N_2. Pulsing was also examined: H_2 pulses were introduced into a stoichiometric synthesis gas mixture as well as N_2 pulses into the same mixture. The reverse strategy, synthesis gas pulses into a H_2 or a N_2 feed, were also studied. Krylova mentions combining concentration with temperature forcing, but how this was done was not described (Krylova, 1990). Increasing H_2 conversion/pass appeared to be the objective of her work.

Figure 7.15 shows H_2 conversion/pass reported by Krylova for 0.125 to 20 cm^3 pulses of one reactant introduced into a constant flow, 10 to 222 cm^3/min, of the other reactant. The other variable was catalyst temperature. Time between pulses ranged from 1/2 to 60 minutes and the space velocity was 1000 h^{-1}. The figure shows pulsed operation results in H_2 conversions 2 to 3 × higher than the best under steady state, although it is not indicated by Krylova how the conversions under this pulsing are defined. Temperatures at which the maximum conversions are obtained shift towards lower temperatures.

Krylova explains higher conversions in terms of surface condition. The surface is assumed to be covered by adsorbed N_2 or partially hydrogenated N. These are swept from the surface in a H_2 pulse providing H_2 access. A similar argument applies to N_2 pulsing. The shift in the temperature of the maximums in the figure arises, according to Krylova, because the order of the reaction in H_2 decreases with lower temperature. Substantial storage of N_2 in the catalyst is mentioned also as an explanation for the enhancement under pulsing and/or composition forcing. Krylova passed H_2 over the catalyst previously exposed to N_2 in a TPD experiment with temperature ramped from 250 to 500°C and measured the NH_3 evolved. The SA-IR catalyst was brought to room temperature over 12 hours. The catalyst was then reheated to 250°C and the TPD experiment repeated. Further NH_3 was evolved.

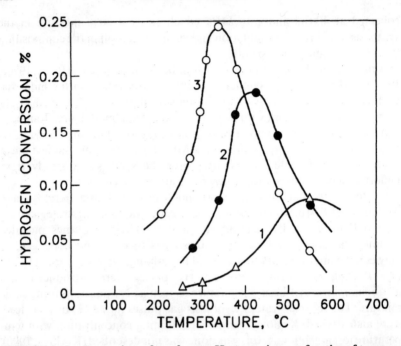

FIGURE 7.15 Comparison of steady state H_2 conversion as a function of temperature with periodic pulsing of either H_2 or N_2. Curve 1–steady state with a stoichiometric synthesis gas mixture, Curve 2–N_2 pulses onto a H_2 feed stream, Curve 3–H_2 pulses onto a N_2 feed stream. (Figure adapted from Krylova (1990) with permission of the publisher)

Symmetric composition forcing at $s = 0.5$ showed that Ψ reached 1.5 at a cycle period of 1/2 minute for cycling between a stoichiometric synthesis gas and H_2. If the cycle is between synthesis gas and N_2, Ψ reaches 2.4, but at $\tau = 40$ minutes. Few details about the Krylova work are given and there is no indication of how conversion or rate under pulsing and composition forcing is determined. Thus, her results cannot be compared directly with those of Wilson and Rinker (1982), Chiao et al. (1987) and Jain et al. (1982b, 1983b). Data for pressures from 0.1 to 5 MPa are given by Krylova. These and Figure 7.15 confirm the trends shown in Figure 7.10.

Ruthenium Catalyst

Rambeau and Amariglio (1981a) comment that the activity of an unsupported ruthenium catalyst is about 500 × greater than the activity of conventional iron catalysts at 400°C and 2000 × greater at 250°C. The high activity of ruthenium for ammonia synthesis had not been noted earlier, according to these investigators, because activity is evaluated at the stoichiometric composition for the synthesis (75 mol % H_2). This composition leads to the maximum rate for iron catalysts. For ruthenium, however, ammonia synthesis is strongly inhibited by hydrogen, much more so than over iron catalysts. Thus, at any temperature, the rate increases as the H_2 partial pressure decreases. This inhibition led Rambeau and Amariglio (1981b) to explore periodic switching between H_2 and N_2.

Hydrogen inhibition and the high intrinsic activity of ruthenium are illustrated in Figure 7.16 which plots the measured synthesis rates for different feed compositions versus reciprocal absolute temperature. Experiments were conducted with 1 gm of ruthenium powder in a shallow packed bed, formed by a 2 cm (id) disk of silica frit, at a space velocity of 0.001 m^3/g minute. Although rates are highest at the stoichiometric composition, high temperatures are needed and equilibrium keeps conversions low. Curves 1 and 1' show the effect of a 10 × change in space velocity. Curves 2 and 3 show the effect on rates of reducing the hydrogen in the syn gas mixture to 1.3 and 0.065 mol% respectively. Apart from the region where the reverse reaction is significant, reducing the mole percent H_2 from 75% to 1.3% increases the rate at any temperature by 10 to 50 ×. Rambeau and Amariglio (1981a) conclude that the synthesis is limited on the Ru catalyst by the rate of N_2 adsorption. The maximum synthesis rate, then, is the rate of N_2 adsorption on a clean Ru surface. This rate is shown as the line (Curve 4) on the upper right of Figure 7.16. Curve 5 gives the N_2 adsorption rate on a clean Fe surface of the same area as the ruthenium sample used and is a basis for Rambeau and Amariglio's claim of a high intrinsic activity of ruthenium for the NH_3 synthesis.

The importance of N_2 adsorption on the rate under $N_2 - H_2$ cycling may be seen from Figure 7.17. Measured time average rates of ammonia synthesis for a cycle are plotted versus duration of N_2 exposure within a cycle. For Curves 1 to 4, H_2 exposure has been limited so as to hydrogenate adsorbed nitrogen, but not saturate the surface. Both the maxima in the figure and the temperature dependence reflect the interaction of adsorption equilibrium and adsorption rates. At 158°C, the adsorption rate is slow so a long exposure is needed to approach saturation of the surface. Higher adsorption rates leads to high N_2 surface coverage making longer N_2 exposure ineffective. At 305°C, the maximum time average synthesis rate occurs for an N_2 exposure of just about 10 s duration, whereas at 205°C, a 60 s duration is needed and the maximum is 1/10 that measured at the higher temperature.

Curves 1 and 5 examine the effect of less than optimal H_2 exposure. If less than the amounts needed to hydrogenate adsorbed N_2 is used (Curve 5), lower rates results. If more than optimal H_2 is used, the synthesis rate does not change, but the conversion of H_2 in the cycle drops. Indeed, in Curve 5, the utilization of H_2 increases significantly although with a loss in the rate of NH_3 production. Even though the maximum production rate of

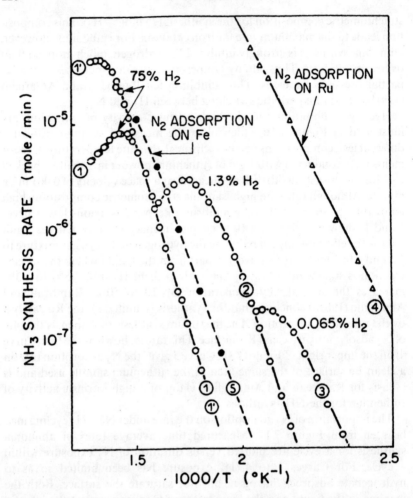

FIGURE 7.16 Effect of temperature and composition on synthesis rate over Ru powder. All curves for 1 g of catalyst, 1000 cm³/min flow rate: Curve 1—stoichiometric composition (75 vol.% H_2), but Curve 1' is at 100 cm³/min; Curve 2—H_2 is 1.3 mol % (balance He) and Curve 3—H_2 is 0.065 mol %. Curve 4 is the rate of N_2 adsorption on Ru for a clean surface and Curve 5 is for Fe powder with the same specific surface area as Ru (Figure adapted from Rambeau and Amariglio (1981a) with permission, © 1981 Academic Press)

NH_3 is not achieved, control of the H_2 exposure can lead to more efficient use of hydrogen.

Considerations summarized in Figure 7.15 are shown as synthesis rate vs. temperature in an Arrhenius plot (Fig. 7.18). In this figure, the time average rates under composition modulation are plotted as a function of

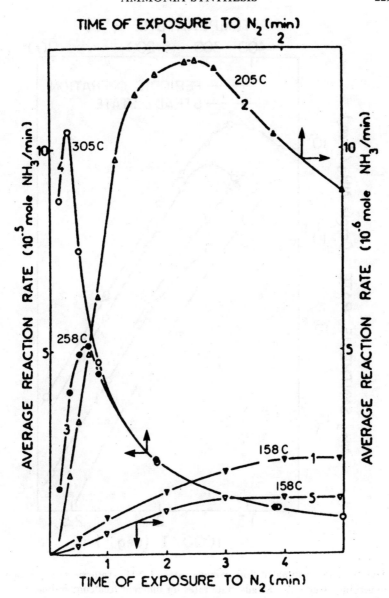

FIGURE 7.17 Estimated time average rates of NH_3 synthesis as a function of duration of the N_2 partial cycle and temperature 1 = 158°C, 2 = 205°C, 3 = 258°C, 4 = 305°C, all for 10 g catalyst and optimum H_2 exposure, 5 = 158°C and exposure to 4 cm³ H_2 (Figure adapted from Rambeau and Amariglio (1981b) with permission, © 1981 Elsevier Science Ltd.)

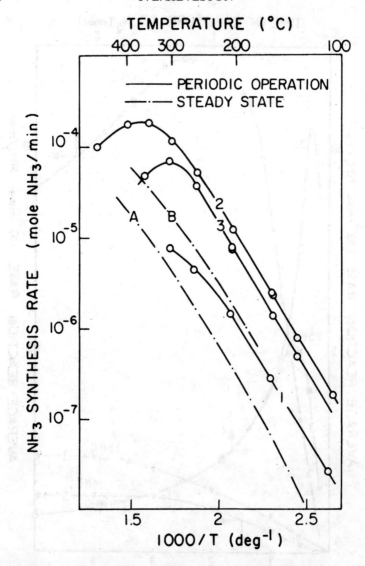

FIGURE 7.18 Estimated time average rates of NH_3 synthesis as a function of temperature: Curve A – Steady state rates for different feed compositions, flow of 1000 cm³/min with 1 g catalyst; Curve B – as for A but 10 g catalyst; Curve 1 – optimum time average cycle rate for 1 g catalyst; Curve 2 – as for Curve 1 but with 10 g catalyst; Curve 3 – as for Curve 2 but with optimum H_2 conversion (Figure adapted from Rambeau and Amariglio (1981b) with permission, © 1981 Elsevier Science Ltd.)

temperature (Curves 1,2) for optimum conversion of N_2 (maximum rate of NH_3 synthesis) and for optimum H_2 conversion (Curve 3). The maximum synthesis rate is achieved at a low utilization of H_2. H_2 utilization is 12% of the feed for optimum H_2 conversion (Curve 3). Except in the region where equilibrium becomes important, the cost of the higher H_2 utilization is a ca. 35% loss in the NH_3 production rate. Curves 1 and 2 are for catalyst samples of 1 g and 10 g respectively, and indicate the effect of space velocity, that is, conversion on both steady state and time average cycling rates. Time average rates under bang-bang switching between N_2 and H_2 are compared in the figure to the envelope of the maximum steady-state rate at different temperatures between 100 and 400°C. These curves, A for a 1 g sample and B for a 10 g sample, were taken from Figure 7.16. The compositions at each reciprocal temperature for the steady-state envelopes A and B in Figure 7.18 change. Thus, between $1000/T = 2$ and 2.5, the $H_2:N_2$ ratio is 0.00065; between 1.8 and 1.9, the ratio is 0.013; while below 1.5, it is 3.0. Similarly, cycle period and split vary with reciprocal temperature for Curves 1 to 3 in the figure. At the upper end of these curves, cycle periods are in the range of 5 to 10 seconds. It may be seen that the optimal time average rates of synthesis under forcing exceed the maximum steady state rates at any temperature by 2 to 5 × in the region of kinetic control. Even where equilibrium becomes important (for the 10 g sample data), the increase through cycling is 2 ×. These improvements must be compared to the ca. 10% improvement at 430°C and 4.14 MPa for the iron catalyst (Figure 7.14). According to Rambeau and Amariglio (1981b), enhancement of the synthesis rate can reach 1000 × if compared to the steady state rate corresponding to the stoichiometric mixture at 200°C.

Osmium Catalyst

This metal, like ruthenium, is an active ammonia synthesis catalyst. Rambeau et al. (1982b) estimate it has 10 to 100 × the activity of commercial iron catalysts in the temperature range of 200° and 400°C. Synthesis on osmium is also inhibited by H_2 but to a lesser extent than on ruthenium. Rambeau et al. (1982b) conducted experiments parallel to those just discussed using 1 g of very pure Os powder. Concentration of active sites measured 7.5×10^{-6} sites as mol NH_3/g. Prior to use, the catalyst was conditioned in a stoichiometric N_2-H_2 mixture and the temperature brought to 500°C. No loss of activity with time on stream was observed.

Steady state rate of synthesis as a function of feed composition or coverage of the catalyst surface by N_2 appears in Figure 7.19 at 301° and 398°C (Rambeau et al. 1982a). Less adsorption of hydrogen at 398°C explains the ca. 10 × higher rate at 398°C at the same N_2 coverage. The rapid increase in rates as the H_2 partial pressure decreases at both temperatures indicates H_2 inhibition, whereas the drop in rate after the maximum is

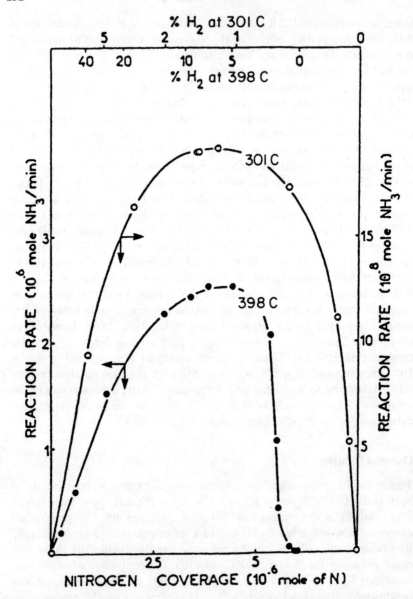

FIGURE 7.19 Steady state rates of synthesis over osmium powder as a function of H_2 in the feed or the N_2 coverage of the surface at 301°C and 398°C (Figure adapted from Rambeau et al. (1982b) with permission, © 1982 Elsevier Science Ltd.)

caused by NH_3 inhibition as well as low H_2 coverage on the surface. Optimum volume % H_2 in the synthesis feed increases from 1.5% at 301°C to 5% at 398°C and to the stoichiometric mixture at 500°C. Data are re-plotted with further experiments added in Figure 7.20. Hydrogen inhibition is again evident as lowering the H_2 partial pressure increases the NH_3 formation rate at any temperature above 230°C. The figure shows that the reverse reaction becomes important at lower temperatures as the H_2 content of the feed decreases; also lowering H_2 below 0.8 vol.% does not increase the rate.

Figure 7.20 compares the synthesis rate to the maximum possible rate, the adsorption rate of N_2 on a clean surface. Just as on Ru, N_2 chemisorption is rate controlling with osmium powder. Comparison of Figures 7.16 and 7.20 or the N_2 adsorption rate curves in the latter figure suggest synthesis rates are almost 100 × slower on Os than on Ru. Potentially, Os is a better catalyst than Fe by a factor of 10, but considerably poorer than Ru when compared by N_2 adsorption rates.

Bang-bang cycling between H_2 and N_2 increases the time average rate of synthesis significantly. Cycle optimization was undertaken by Rambeau et al. (1982b) at 202° and 301°C. These experiments were performed by exposing the catalyst that had been flushed with He for an extended period to N_2 for different durations followed by exposure to H_2. Ammonia production during the H_2 exposure was measured and is shown in Figure 7.21. The figure indicates high time average synthesis rates can be attained by a cycle consisting of a 10-15 s exposure to H_2 and a 30 to 120 s exposure to N_2 at 301°C. A similar split is optimal at 202°C, but slower adsorption and reaction rates increase the optimum period to 11 to 22 minutes. Using N_2 containing some H_2 (0.2 vol. %) does not change the optimal cycle, but does lower the rate by a half. N_2 containing 0.8 vol % H_2 was also studied by Rambeau et al., who found a further reduction in the time average rate of synthesis (see Figure 7.22). These investigators examined decreasing hydrogen usage by operating with H_2 exposure durations well below the 10-15 s optimum for N_2 conversion. They observed that only 25 cm^3 (STP)/g of H_2 exposure decreased time average NH_3 synthesis rates shown in Figure 7.21 by about 50%, but resulted in better utilization of H_2. 25 cm^3 (STP)/g corresponds to an H_2 exposure duration of 1-2 s, that is, a H_2 pulsing cycle.

Performance under periodic operation, as the time average synthesis rate, is compared in Figure 7.22 with the maximum steady state rate at atmospheric pressure using an Arrhenius plot. Curve A is the envelope encompassing all steady state rates and the composition of the synthesis gas corresponding to each point varies as discussed for Figure 7.18. Time average cycling rates also reflect different cycle periods. Forcing with just a short exposure to H_2 in a period leads to rates about 10 × higher than the optimal steady state rate in the region of kinetic control when only N_2 is used in the second part of a cycle. This is indicated by Curve 1. As the

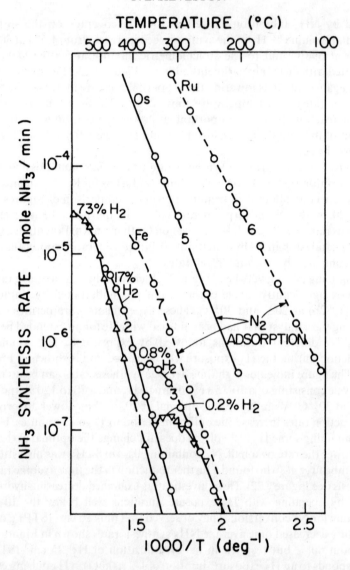

FIGURE 7.20 Effect of temperature and composition on the steady state synthesis rate over Os powder. All curves for 1 g of catalyst and a 1000 cm^3/min flow rate. Curve 1 is for a near stoichiometric mixture (27% N_2-73% H_2). Curve 2 is for a 25% N_2-17% H_2 mixture, while Curve 3 is for N_2-0.8% H_2 and Curve 4 for N_2-0.2% H_2 mixtures. Curve 5 is the rate of N_2 adsorption on a clean Os surface. Curves 6 and 7 are adsorption rates on Ru and Fe respectively (Figure adapted from Rambeau *et al.* (1982a) with permission, © 1982 Academic Press)

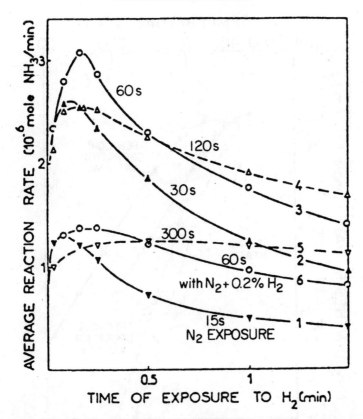

FIGURE 7.21 Time average rates of NH_3 formation over osmium powder as a function of the H_2 exposure duration at 301°C after exposure to N_2 for 15 s (Curve 1), 30 s (Curve 2), 1 min (Curve 3), 2 min (Curve 4), 5 min (Curve 5), exposure to $N_2 + 0.2\%$ H_2 for 1 min (Curve 6) (Figure adapted from Rambeau et al. (1982b) with permission, © 1982 Elsevier Science Ltd.)

reverse reaction becomes important this improvement drops to 1.5 to 2 × the optimal steady state rate. Using N_2 containing some H_2 (to 0.8 vol. %) still gives at least a 2 × improvement (Curves 2 and 3). However, in the region where equilibrium begins to dominate, Figure 7.22 shows that small amounts of H_2 in the N_2 stream do not affect the time average rates. The figure also shows that the performance under composition modulation over Ru (Curve 4 using data from Figure 7.18) is much better than over Os up to the temperature region where equilibrium begins to dominate.

Mass Transfer Interference

Concentration gradients arise in porous catalysts under both steady state and periodic operation through diffusion of reactants towards the particle

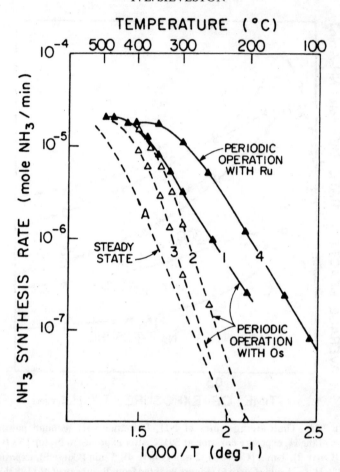

FIGURE 7.22 Time average rate of ammonia synthesis versus reciprocal temperature. Curve 1 employed cycling with pure N_2, 2 with $N_2 + 0.2$ vol % H_2 and 3 with $N_2 + 0.8$ vol % H_2. Curve A is the envelope of the maximum steady state rates at 1 atm. Curve 4 is time average rate for cycling with pure N_2 over a Ru catalyst (see Figure 7.18) (Figure adapted from Rambeau et al. (1982b) with permission, © 1982 Elsevier Science Ltd.)

centre. These gradients usually mean that volumetric reaction rates, at least time-average ones, are lower than rates that would be predicted from concentrations prevailing at the external surface of the particle. Mass transfer interference through diffusion has been considered in Chapter 3 and it was pointed out there that diffusion could alter the ratio of time average rates to rates at steady state from greater to less than unity depending on cycle period, particle size and time average composition. Effects of the interference on selectivity will be examined in Chapter 11.

To investigate diffusional interference for ammonia synthesis, Jain *et al.* (1985) undertook experiments in a gradientless reactor at 400°C and 2.38 MPa using the C73-1-01 promoted iron catalyst, but with particles of 5.0 mm diameter rather than the 0.297 mm diameter employed in their other studies. They observed that rate enhancement through composition forcing is reduced, but does not vanish. This is demonstrated in Figure 7.23 for cycling around $m = 0.75$, the composition corresponding to the maximum synthesis rate at steady state. There is a substantial reduction in Ψ at each cycle split, s, and period, t, for the 17 × increase in particle size.

The steady state rate used to define Ψ in Figure 7.23 is the rate measured for the 0.297 mm particles. Consequently Ψ in the figure is actually an "effectiveness factor" for periodic composition forcing in the absence and presence of diffusion interference. If both steady state and modulated data are normalized with the fine particle rates at steady state, the resulting 'effectiveness factors' are given versus composition in Figure 7.24a. Evidently, composition modulation, like higher particle temperatures due to heat generation within the catalyst particle, can result in $\eta > 1$. Steady state data presented in the figure indicate significant diffusional interference for synthesis employing the 5.0 mm particles.

As a practical matter of representing diffusional interference, use of the steady state rate for fine particles to define effectiveness factor is not satisfactory. The opposing effects of modulation and intra-particle diffusion

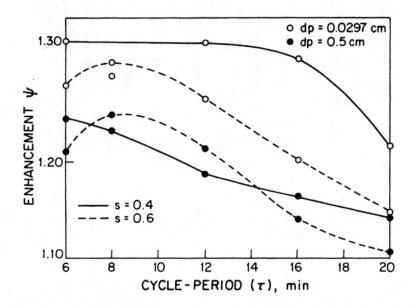

FIGURE 7.23 Effect of particle size on NH_3 synthesis rate enhancement as a function of cycle period and split (Jain 1981).

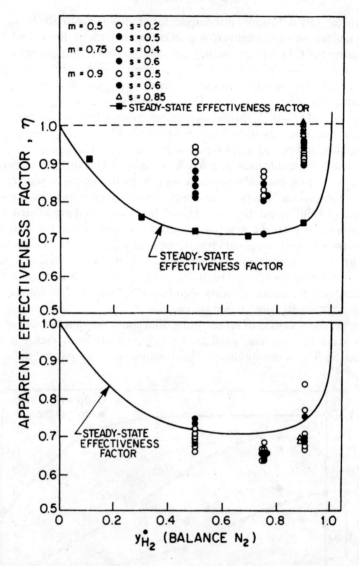

FIGURE 7.24 Effectiveness factor analysis of particle size effects under composition modulation: (a) factor defined from steady state data, (b) factor defined using data for periodic operation (Figure adapted from Jain *et al.* (1985) with permission of the authors)

are not separated. If, however, an apparent effectiveness factor, η^*, is introduced, defined as $\eta^* = \bar{r}_{cyc}/\bar{r}_{cyc'\text{fine particles}}$, the diffusion interference can be represented. Apparent effectiveness factors for the composition forcing data on the 5 mm particles are plotted in Figure 7.24b against the time average feed compositions. Effectiveness factors for the steady state data with the larger particles, from Figure 7.24a are shown for comparison. Figure 7.24b demonstrates that steady state effectiveness factors provide a satisfactory estimate of the apparent effectiveness factor under composition modulation, at least for the reaction conditions and cycling variables examined by Jain et al. (1985).

Near Adiabatic Operation

Temperature profiles in full scale synthesis reactors reflect the exothermicity of the synthesis, emergence of the reverse reaction at conversion greater than ca. 10%, and heat exchange with fresh feed. These profiles exhibit a sharp rise in temperature just after the bed entrance to a maximum followed by a slow drop in temperature with increasing axial depth in the bed. Jain et al. (1982b) utilized a near adiabatic fixed bed, non differential reactor to examine composition forcing under conditions, except for pressure, representing commercial operation. Jain et al. employed a 12.7 mm i.d. by 250 mm packed bed (0.3 mm particles of C73-1-01 catalyst - see Table 7.1), heavily insulated and equipped with six zone, temperature controlled counter heating to minimize heat loss. Entrance temperature was held at 400°C. The questions addressed by Jain's study were 1) can composition modulation raise the time average synthesis rate and how does the improvement compare with that found under isothermal, differential operation, and 2) how does modulation affect the temperature profile in the reactor?

Figure 7.25 compares the time average, space average rates under periodic forcing in Jain's near adiabatic integral reactor with the space average steady state rates of NH_3 synthesis as a function of the time average feed composition. Conversion varied widely, but the reverse reaction was significant for most of the steady state experiments. A total of 64 experiments under periodic switching of feed composition were performed to define the upper curve in the figure. Cycle period, τ, varied between 1 and 60 minutes; cycle split, s, was between 0.2 and 0.85. Experiments were carried out only in the cross hatched region and the experimental points with the largest time average rate at y_{H_2} were connected to form the upper curve. This curve was extrapolated to $y_{H_2} = 0$ and 1.0. Variation among replicate rate measurements did not exceed 4%.

Figure 7.4 is reproduced as dotted lines in Figure 7.25 for comparison purposes. Because of the ca. 15°C higher mean bed temperature, the space average rates for both steady state and modulated operation are higher despite the contribution of the reverse reaction. The absolute increase in r due to composition modulation is greater for the near adiabatic fixed bed reactor than for the differential unit. The right hand scale of Figure 7.25, however, shows that the percent increase in the synthesis rate through

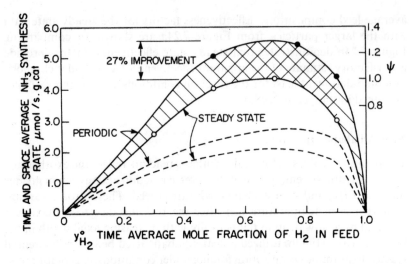

FIGURE 7.25 Comparison of time average, space average synthesis rates with space average steady state rates for 0.3 mm particles, entrance temperature = 400°C, P = 2.38 MPa -○- = steady state data, -●- = composition forcing data, --- differential reactor results from Figure 7.4.

composition switching is 27% which is slightly less than the increase found for the differential reactor. Jain's experiments do suggest that isothermal experiments in differential reactors under periodic composition forcing are adequate indicators of performance for commercial scale reactors.

Composition modulation affects substantially the temperature distribution in the fixed bed. Time average temperatures at uniformly spaced points throughout the 12.7 mm i.d. by 250 mm long reactor are shown in Figure 7.26 for composition forcing at $m = 0.5$ and $s = 0.2$ and for steady state operation at the same 50% H_2 in the feed. The increase in the maximum temperature and the higher spatial average temperatures are a consequence of the higher time average rate of this exothermic reaction. Nevertheless, the maximum bed temperature, in the space $0.2 < \ell/L < 0.4$ is only about a degree higher than the steady state maximum despite a 20 to 30% increase in the time average synthesis rate. Composition modulation, thus, flattens the temperature distribution and moves the temperature maximum deeper into the bed.

Jain (1981) has characterized flattening by a relative temperature standard deviation,

$$\sigma_{\Delta T}/\Delta \bar{T} = \left[\frac{\sum_{\ell=0}^{n}(\Delta T_\ell - \Delta \bar{T})^2}{n-1} \right]^{1/2} / \Delta \bar{T} \qquad (7.2)$$

where $\Delta T_\ell = T_\ell - T_{\ell=0}$ is the increase in temperature at axial position ℓ above the inlet temperature and $\Delta \bar{T}$ is the increase in the spatially mean bed

AMMONIA SYNTHESIS

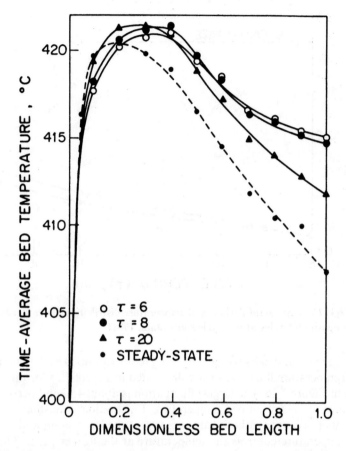

FIGURE 7.26 Time average temperature distribution in a near adiabatic packed bed as a function of cycle period (Jain, 1981)

temperature above the inlet temperature. For periodic composition forcing, all temperatures are time averaged. If there is no spatial variation in bed temperature, the ratio $\sigma_{\Delta T}/\Delta T = 0$. Figure 7.27 plots the ratio for the range of τ shown in Figure 7.26, but at other values of m and s. At $\tau = 6$ min., where the largest Ψ was observed, $\sigma_{\Delta T}/\Delta T$ is just 1/3 of the ratio at steady state. The figure shows that longer cycle periods increase the relative temperature standard deviation. Cycle split has a small effect on the deviation.

The significance of these observations is that composition modulation can avoid or mitigate hot spots in non-isothermal or near adiabatic synthesis reactors.

Explanation of the levelling of the temperature profile is the axial movement of temperature waves in the bed initiated by changing rates of NH_3 formation following a switch in composition. These temperature

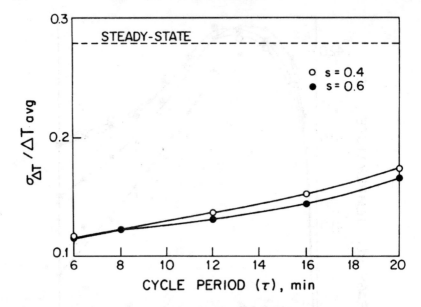

FIGURE 7.27 Standard deviation of temperature distribution under steady state and composition forcing at two cycle splits (Jain, 1981)

waves are manifested by rising and falling temperatures at each point in the bed. Temperature fluctuations are illustrated in Figure 7.28 for the $\tau = 20$ curve in Figure 7.26. The largest fluctuation is about 4°C and occurs near the point of maximum bed temperature. Fluctuations vanish at the inlet and outlet because synthesis rates are low at the latter point and because feed temperature controls the temperature at the former point. Thermal inertia dampens the temperature waves. For relatively fast cycling, at $\tau = 6$ min., fluctuations in bed temperatures disappear.

E. ANALYSIS

Several attempts to model ammonia synthesis under composition modulation have been made, but with less success than modeling SO_2 oxidation (Chapter 8). Nonetheless, these models have been applied to investigating what chemical and physical steps are responsible for rate enhancement when this is observed.

Relaxed Steady State

Wilson and Rinker (1982) contend that the relatively high cycling frequency used in their experiments forced their single pass, packed bed reactor to operate in the relaxed steady state. The relaxed steady state is a concept

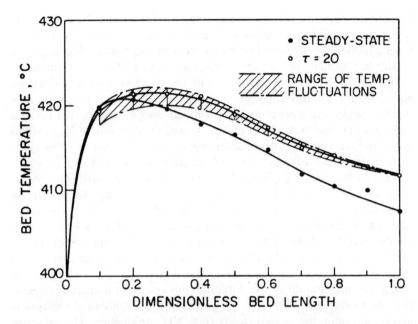

FIGURE 7.28 Range of temperature fluctuations in a near adiabatic packed bed under composition forcing at $\tau = 20$ min., $m = 0.5$, $s = 0.2$, (Figure adapted from Jain, et al. (1982b) with permission, © 1982 American Chemical Society)

brought into catalysis from control theory and is applied to reaction models to obtain the asymptotic performance as the forcing frequency becomes large (Bailey and Horn, 1968, 1971; Bailey, 1977). It is assumed that at a sufficiently high frequency, adsorbate concentrations on a catalyst surface are unable to follow gas phase composition variations and attain thereby a "relaxed steady state" that may differ from the true steady state at a gas phase composition corresponding to the time average composition of the modulated system. Bailey and Horn (1971), early investigators of periodic reactor operation, used the relax steady state (RSS) assumption to predict the performance of a model reaction system in an idealized reactor. Bailey (1977) suggests that the relaxed steady state will be encountered when the cycling period, τ, is much smaller than the characteristic relaxation time of the reactor system. There is a further discussion of RSS in Chapter 1.

Transient experiments reported by Wilson and Rinker gave a pseudo 1st order time constant of about 300 seconds. This value is comparable to estimates from Figure 7.6 taken from the data of Jain et al. (1982b). A time constant is a reasonable estimate of the characteristic time of relaxation (t_L) after an alteration in system condition. Wilson and Rinker used cycle periods of approximately $\tau = 1/5$ to $1/10$ t_L. Thus, Bailey's condition of $\tau \ll t_L$ for the relaxed steady state seems fulfilled.

If these authors are correct in assuming that they achieved relaxed steady state operation, their experimental results offer a discriminatory test of kinetic models for ammonia synthesis. Wilson and Rinker examined rate models proposed respectively by 1) Temkin, 2) Ozaki, Taylor, and Boudart, 3) Brill and 4) Tamaru. Their procedure consisted of writing the balance equations for the three species on a differential element of a plug flow reactor, neglecting storage in the gas for all species and on the catalyst surface for H_2 and NH_3. Their model allowed for variation of N_2 adatoms on the surface with time and a change in gas velocity because of shrinkage in the reaction. This leads to the set of equations shown in Table 7.2 which represents the Ozaki, Taylor and Boudart (OTB) rate model. This model assumes that dissociative adsorption is rate controlling. The system of 6 equations is solved by choosing the constant k_3 and k_4 so that hydrogenation of surface species (Eley-Rideal mechanism) and NH_3 desorption are at equilibrium. Values for other constants in the model were obtained from Ozaki, Taylor and Boudart (1960). The relaxed steady state assumption that the surface concentrations are unable to follow gas phase composition changes means that $d\theta/dt = 0$. Although this simplifies numerical integration along the axis of the reactor enormously, a numerical solution is needed to obtain the relaxed steady state NH_3 production. The subscript i distinguishes the two parts of a composition cycle. The same set of equations, disregarding the subscript i and using invariant initial conditions, must be solved to obtain the steady state production rate. The initial N_2 and H_2 concentrations must be varied to determine the maximum steady state rate which is used then to calculate Ψ^*.

Results of calculations for the OTB model appear in Figure 7.29. From this plot, Wilson and Rinker concluded that composition modulation enhances NH_3 synthesis and that to maximize Ψ with switching between pure N_2 and H_2 the cycle split, s, should be about 0.67. At this split, Ψ^* would be about 1.14. They comment that the OTB model predicts that Ψ will be independent of reactor pressure and temperature. Comparing these predictions with their experimental results, they conclude that the OTB kinetic model is not adequate for predicting performance under composition modulation.

The relaxed steady state procedure, described briefly above, was applied by Wilson and Rinker to the Brill model and with several further assumptions to the Temkin model and the Tamaru model. Predictions similar to the OTB model were obtained except that the Tamaru model allowed for a total pressure and temperature dependence of Ψ. On the grounds above, Wilson and Rinker conclude that the Tamaru model best describes NH_3 synthesis under periodic operation, but they justifiably question the applicability of steady state models to dynamic systems.

Examination of the asymptotic limit of forcing as $\tau \to 0$ has been extended by Rinker and co-workers to a simplified Eley-Rideal kinetic scheme:

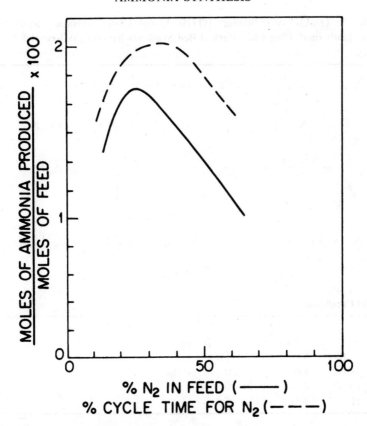

FIGURE 7.29 Comparison of NH$_3$ production predicted for relaxed steady state and steady state at the same time average feed composition using the NH$_3$ synthesis kinetics of Ozaki et al. (1960) (Figure adapted from Wilson and Rinker (1982) with permission, © 1982 Elsevier Science Ltd.)

$$A + s \rightarrow A - s$$
$$A - s + vB \rightarrow \text{products} + s \tag{7.3}$$

where s is a vacant surface site. In this situation A could represent N$_2$ and B could be H$_2$ so that the model of Equation (7.3) resembles the OTB and Brill models for NH$_3$ synthesis used by Wilson and Rinker. Assuming the reactions are elementary, two reactor systems were considered: 1) a differential reactor (Thullie et al.,1986), and 2) an isothermal plug flow reactor (Thullie et al., 1987). Provided $v > 2$, both studies confirmed that $\Psi > 1$ is possible and that maximum values of Ψ are obtained as $\tau \rightarrow 0$. Nonsymmetric cycling, that is $s \neq 0.5$, was found to give some what higher

Table 7.2 Ozaki-Taylor-Boudart (OTB) Model Under Relaxed Steady State for an Isothermal, Plug Flow Packed Bed Synthesis Reactor (Wilson and Rinker, 1982).

$$u_i \frac{d(y_{N_2})_i}{dz} = k_2(\theta)_i^2 - k_1 (y_{N_2})_i (1-\theta)^2 - (y_{N_2})_i \Delta_i / P \tag{7.4}$$

$$u_i \frac{d(y_{H_2})_i}{dz} = 1.5 k_4 (y_{NH_3})_i (1-\theta_i) - 1.5 k_3 (y_{H_2})_i^{3/2} \theta_i - (y_{H_2})_i \Delta_i / P \tag{7.5}$$

$$u_i \frac{d(y_{NH_3})_i}{dz} = k_3 (y_{H_2})_i^{3/2} \theta_i - k_4 (y_{NH_3})_i (1-\theta_i) - (y_{NH_3})_i \Delta_i / P \tag{7.6}$$

$$\frac{du}{dz} = \Delta_i / P \tag{7.7}$$

$$\Delta_i = k_2(\theta_i)^2 - k_1 (y_{N_2})_i (1-\theta_i)^2 + 0.5 [k_4 (y_{NH_3})_i (1-\theta_i) - k_3 (y_{H_2})_i^{3/2} \theta_i] \tag{7.8}$$

$$\frac{d\theta_i}{dt} = 0 = 2k_1 (y_{N_2})_i (1-\theta_i)^2 + 2k_2(\theta_i)^2 + k_4 (y_{NH_3})_i (1-\theta_i) - k_3 (y_{H_2})_i^{3/2} \theta_i \tag{7.9}$$

Initial Conditions

	$i=1$ $t \geqslant n\tau$	$i=2$ $t \geqslant (n+s)\tau$	$(n = 0, 1, 2 \ldots)$
$(y_{N_2})_i(t, 0)$	0	1.0	
$(y_{H_2})_i(t, 0)$	1.0	0	
$(y_{NH_3})_i(t, 0)$	0	0	
$\theta_1(t, z)$	$\theta_2(z)$	$\theta_1(z)$	

values of Ψ. The optimal s depended on the kinetic parameters and the forcing amplitude used.

The relaxed steady state approximation of composition forcing was extended to a non-isothermal reactor by Chiao and Rinker (1989). They considered an autothermal operation in which the reactor feed flows upward in an annulus and then downward through the catalyst bed located in the centre of the reactor. A pseudo homogeneous, plug flow model was used with kinetic expressions for the synthesis that had been tested successfully against steady state and dynamic NH_3 synthesis data. Chiao and Rinker demonstrated that their simplified model correctly predicts temperature and concentration profiles as well as multiple steady states that have been found when more sophisticated models are used.

Their simulation demonstrated that large enhancements of the synthesis rate are achievable by high frequency composition forcing. Figure 7.30 reproduces the Chiao and Rinker global enhancement factor predictions for iron catalysts in three different states of activation. These states are discussed further on in this section when the Chiao-Rinker kinetic model for NH_3 synthesis is considered in detail. Enhancement factors exceed

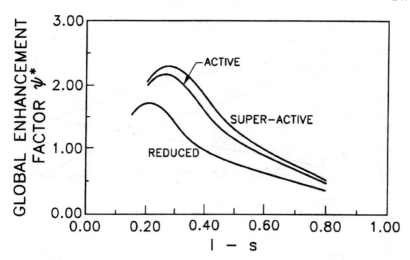

FIGURE 7.30 Global enhancement factor as a function of cycle split for a relaxed steady state simulation of composition modulation between N_2 and H_2 feeds for an autothermal reactor operating at 10.1 MPa, 9.4 s^{-1} space velocity (STP) and a feed temperature of 755°K. Curves refer to the assumed activity of the catalyst (Figure taken from Chiao and Rinker (1989) with permission, © 1989 Elsevier Science Publishers)

considerably those predicted by the Wilson-Rinker analysis of an isothermal system. The factors are plotted vs 1-s, where s, the cycle split, is defined as the fraction of cycle taken up by a H_2 rich feed. The enhancement factor is strongly influenced by split. Chiao and Rinker find that the forcing amplitude also affects Ψ. High frequency forcing, of course, affects the temperature and concentration profiles in the annulus and in the bed. This is illustrated by Figure 7.31. Figure 7.31a compares the temperature profiles for a relaxed steady state operation whose time average feed composition is the same as the composition which yields the maximum steady state ammonia concentration in the reactor off gas. Feed temperatures and feed rates are also identical. The surprising feature of this simulation is that the reactor temperature rises towards the exit rather than falls as it does in steady state operation. The concentration profile in the reactor becomes virtually linear in high frequency forcing (Fig. 7.31b). Chiao and Rinker report even higher Ψ^* can be obtained if the time average feed composition differs from the composition for the MSS NH_3 concentration.

The relaxed steady state approach to the analysis of periodic operation used by Rinker and co-workers appears to yield useful, but probably only approximate predictions. The analysis predicts $\Psi > 1$ and that $s > 0.5$ maximizes Ψ for an isothermal reactor using the NH_3 synthesis models mentioned above. Wilson and Rinker's near isothermal experiments con-

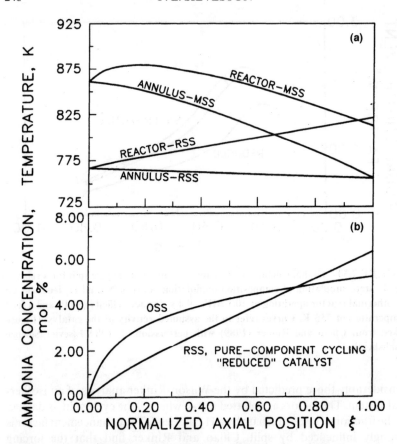

FIGURE 7.31 Comparisons of predicted steady state and relaxed steady state temperature and concentration profiles in an annular, autothermal NH_3 synthesis reactor with cold flow through the annulus. Assumed operating conditions as in Fig. 7.30, but with $s = 0.68$. Catalyst activity is "reduced": a) temperature profiles; MSS = composition corresponding to the maximum NH_3 concentration under steady state operation, RSS = relaxed steady state, b) NH_3 concentration profile in reactor (Figure adapted from Chiao and Rinker (1989) with permission, © 1989 Elsevier Science Publishers)

firm these predictions. The kinetic models used, however, result in low values of Ψ at low reaction pressures and high values at 4.14 MPa just opposite to what is observed. An interesting, but untested prediction of the analysis and the kinetic models is composition forcing should result in higher NH_3 production when the reaction is equilibrium limited in steady state operation.

Unfortunately, there are several short comings in the relaxed steady state analysis. Steady state models were employed by Rinker and co-workers and,

AMMONIA SYNTHESIS 249

as will be shown further on, these models do not represent the dynamic behaviour well. It is uncertain whether or not the relaxed steady state was achieved experimentally. The test for this state is that Ψ should be independent of τ, but all of the data presented by Wilson and Rinker show variation of Ψ with τ. Finally, the analysis used by these investigators assumed isothermal plug flow. The data reported, however, indicate temperature variations as great as 5°C and significant axial dispersion.

Chiao and Rinker (1989) in their autothermal reactor study employed a kinetic model which represented dynamic behaviour, but their pseudo homogeneous, plug flow model is probably inappropriate for the dynamic behaviour of the reactor. The reasonably close approach of annulus and reactor temperatures mean that significant radial fluxes must occur. This implies that the axial conduction fluxes will be important as well. Simulations also show particle-gas temperature differences of 5 to 10°C whereas the pseudo homogeneous model requires negligible differences. Chiao and Rinker report some bizarre temperature profiles in the high frequency limit such as a physically impossible temperature jump of several hundred degrees for the feed entering the annulus. This simulation result certainly casts doubt on the relaxed steady state solution, the reactor model, or both.

Reactant Inhibition Model

From their study of the hydrogenation of adsorbed nitrogen on Ru and the rate of N_2 adsorption (Rambeau and Amariglio, 1981a), Rambeau and Amariglio concluded that the rate limiting step for NH_3 synthesis is the dissociative adsorption of N_2. Adsorbed nitrogen is rapidly hydrogenated stepwise, probably through adsorbed atomic hydrogen, to NH-s, NH_2-s and finally NH_3-s. Only desorption of the fully hydrogenated species is possible within the usual range of NH_3 synthesis temperatures. These hydrogenation and desorption steps are so fast that they can be considered to occur under equilibrium conditions. Atomic hydrogen, according to Rambeau and Amariglio dominates the surface even at very low H_2 partial pressures. The abundance of surface hydrogen, forces N-s towards NH_3-s. Nevertheless, according to Rambeau and Amariglio (1984), N-s is the most abundant nitrogen species on the surface.

Under steady state operation, assuming competitive adsorption of H_2 and N_2, the mechanism yields

$$r \propto \frac{kp_{N_2}}{(1+(Kp_{H_2})^{1/2})^2} \tag{7.10}$$

where k is the adsorption rate constant for nitrogen and the p's partial pressure (Rambeau and Amariglio, 1984). For all but the lowest H_2 mole fractions, the adsorption constant is so large the Equation (7.10) can be represented as

250   P. L. SILVESTON

$$r \propto \frac{k p_{N_2}}{(K p_{H_2})} \qquad (7.11)$$

In the limit, consequently, for cycling between pure N_2 and H_2,

$$r = k p_{N_2} \frac{t_{N_2}}{t_{N_2} + t_{H_2}} \qquad (7.12)$$

if the duration of the H_2 partial period, t_{H_2}, is just long enough to hydrogenate all the N_2 adsorbed during the N_2 partial period, t_{H_2}, and any excess H_2 adsorbed is quickly converted to NH_3 when N_2 is readmitted to the catalyst bed. The basis of (7.12) is that if the reaction with residual H_2 is neglected and if t_{N_2} is kept short enough so that the ruthenium surface does not become saturated with nitrogen, the rate of adsorption of N_2 on to a clean surfacen surface is:

$$r_{ad} = k p_{N_2} \qquad (7.13)$$

and will approximate the adsorption rate during t_{N_2}. Thus r_{ad} in Eqn. (7.13) is an estimate of the time average synthesis rate, r, under periodic operation. No N_2 reaches the catalyst surface during the H_2 portion of a cycle.

From Equation (7.11) and (7.12), the NH_3 synthesis catalyst will be activated by composition forcing if

$$K p_{H_2}(1 - s) > 1 \qquad (7.14)$$

where s = cycle split defined in terms of the duration of the pure H_2 feed. If N_2 adsorption is the slow step, $s < 0.5$, so that the product of the H_2 partial pressure and the adsorption equilibrium constant for H_2 on the catalyst determine if periodic operation can lead to an increased synthesis rate. This approximate analysis can be extended readily to predict the magnitude of the improvement possible through cycling.

Conditions for the analysis are that the steady state reaction is severely inhibited by H_2 and that there is an equilibrium limitation to the maximum synthesis rate. Figures 7.16 and 7.20 demonstrate that these conditions are met for ruthenium and osmium powders. Figure 7.32 shows an interpretation of the analysis. A space velocity that is independent of gas composition and temperature is assumed. The dashed curve labelled '0' shows the rate of nitrogen adsorption (Eqn. 7.13) on a clean catalyst surface. The curves labelled 1' to 4', each showing a maximum, plot the average rate of NH_3 formation at the specified space velocity for four different H_2 mole fractions decreasing from 1' to 4'. The maximum rate at each mole fraction occurs at the temperature which just achieves equilibrium at the reactor outlet. At lower temperatures, to the right of the maxima, the synthesis is kinetically controlled so the rate approaches closely the limited rate of N_2 adsorption on a clean surface except for the curve 1'. For this curve, $y_{H_2} = 0.75$. Thus,

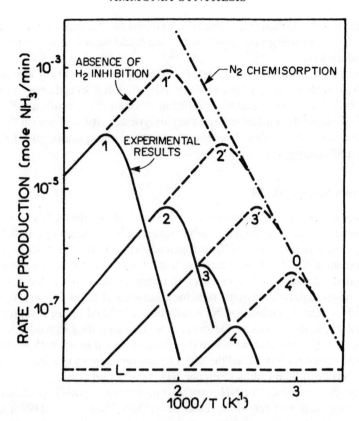

FIGURE 7.32 Schematic of the inhibition and equilibrium limitations for NH_3 synthesis over Ru (1 atm, SV = 12000 h^{-1}): — experimental results, --- kinetics without H_2 inhibition, ---- limit imposed by N_2 chemisorption rate on clean surface of Ru. Curves 1-4 = H_2 % in feed: 75%, 1%, 0.1%, 0.01%. (Figure adapted from Rambeau and Amariglio (1984) with permission, © 1984 Elsevier Science Ltd.)

the rate at this mole fraction is 1/4 of the maximum adsorption rate at P_{N_2} (Rambeau and Amariglio, 1984).

If ammonia formation is inhibited by the competitive adsorption of hydrogen, the rate follows Equation (7.11) rather than Equation (7.13). Curves 1 to 4 represent the average rates for the reactor as previously but increasing inhibition through the kP_{H_2} term shifts each rate curve to the left. Each curve exhibits a maxima which corresponds, as before, to equilibrium at the reactor outlet.

By separating the reactants competing for the same adsorption site, cycling between N_2 and H_2 eliminates inhibition, causing the system to follow the rates given by the dashed curve in Figure 7.32.

The Rambeau and Amariglio analysis allows switching between N_2 and H_2 mixtures. The presence of H_2 during the exposure of the catalyst to N_2

would shift the dashed line to the left. A criteria for the improvement of the synthesis rate through composition forcing could be developed from these considerations.

The mechanism proposed at the beginning of this section applies, it seems, to ruthenium powder and osmium powder. It does not apply to NH_3 synthesis over iron. Reactant inhibition is small if it occurs at all. The synthesis is product inhibited and the catalytic activity depends on the gas phase composition. With these complications, the simple analysis of Rambeau and Amariglio no longer applies.

Reactant Storage Model

In their search for an explanation of $\Psi > 1$ for NH_3 synthesis, Silveston and co-workers focused on the remarkable storage of N_2 in the catalyst bulk (Jain et al., 1982c). Kinetic control by nitrogen adsorption near the stoichiometric composition could reflect limited access of nitrogen to the iron surface saturated with adsorbed H_2 or reaction intermediates and ammonia. Composition modulation through switching between H_2 and N_2 would then provide a "clean" surface for N_2 adsorption and build-up an inventory of "dissolved" nitrogen below the surface. In the succeeding exposure to H_2, dissolved nitrogen diffuses back to the surface where it reacts with adsorbed H_2. The inventory is depleted thereby, but this represents a new supply of N_2 to the surface which results in higher NH_3 production rates.

A predicative model for NH_3 synthesis incorporating this hypothesis has been proposed and tested by Li et al. (1985b). Nam et al. (1990) later modified the Li proposal. The starting point is the following mechanism (Li et al., 1985a):

$$N_2 + 2s \underset{k_{d_N}}{\overset{k_{a_N}}{\rightleftharpoons}} 2N-s$$

$$H_2 + 2s \underset{k_{d_H}}{\overset{k_{a_H}}{\rightleftharpoons}} 2H-s$$

$$N-s + H-s \underset{k_{-1}}{\overset{k_1}{\rightleftharpoons}} NH-s+s \qquad (7.15)$$

$$NH-s + H-s \underset{k_{-2}}{\overset{k_2}{\rightleftharpoons}} NH_2-s+s$$

$$NH_2-s + H-s \underset{k_{-3}}{\overset{k_3}{\rightleftharpoons}} NH_3-s+s$$

$$NH_3-s \underset{k_{a_{AM}}}{\overset{k_{d_{AM}}}{\rightleftharpoons}} NH_3+s$$

N-s was assumed to be the dominant nitrogen species and the subsequent hydrogenation of N−s to NH_3−s was assumed to be so rapid that N − s and NH_3 − s could be related through an equilibrium constant. The use of s to represent an adsorption vacancy implies competitive adsorption. Adsorption on separate sites could also be considered. As a first step towards a dynamic synthesis model, a decision between these two competing representations of adsorption was needed. Li et al. (1985a) made use of heretofore unpublished step change data (Jain, 1981) for this decision. The Elovich rate model was assumed to apply for separate site adsorption. Bulk storage would contribute equally to the response for both models so it was not included. Jain's step change experiments were performed in an isothermal gradientless reactor so the models for the two cases consisted of sets of five ODE's. Each equation was formulated as a mole balance. The gas phase species were N_2, H_2, NH_3 while the surface species were adsorbed atomic nitrogen and atomic hydrogen. Model parameters were evaluated by fitting experimental data. Other data were used to test the models.

Figure 7.33 compares model predictions and experimental observations for a step change from a H_2 feed to a feed containing H_2:N_2 in the proportion 10:90. The separate site representation fails to reproduce the maxima in NH_3 mole fraction leaving the reactor, but it reproduces the measurements before and after quite well. On the other hand, competitive adsorption fails to reproduce the shape of the response. Indeed, for all tests, the dual sites or separate adsorption model predicts the response well. Reactant inhibition cannot occur if reactants adsorb on separate sites, but Rambeau and Amariglio (1978a, b) do not report reactant inhibition for their promoted iron catalyst. The Temkin-Pyzhev model, widely used for the design of synthesis reactors using iron,

$$r_{ss} = k_1 (p_{N_2}) \left(\frac{p_{H_2}^3}{p_{NH_3}} \right)^{1/2} - k_2 \left(\frac{p_{NH_3}^2}{p_{H_2}^3} \right)^{1/2} \qquad (7.16)$$

does not indicate reactant inhibition.

In a second paper, Li et al. (1985b) applied the separate adsorption model to the cycling results published by Jain et al. (1983b). This model, formed from the equations in Table 7.3 with all terms containing C_N removed, failed to predict the improvement in the time average synthesis rate shown in Figure 7.2.

Introducing storage of dissolved nitrogen in the bulk of the catalyst by adding a mole balance for dissolved N_2 and a transport term to the θ'_N balance gives the Li model shown in Table 7.3. Physically, the model assumes that atomic nitrogen present on the surface during the N_2 rich portion of the cycle diffuses into the bulk of the catalyst. A PDE should be used to describe this process; however, the mathematical complications

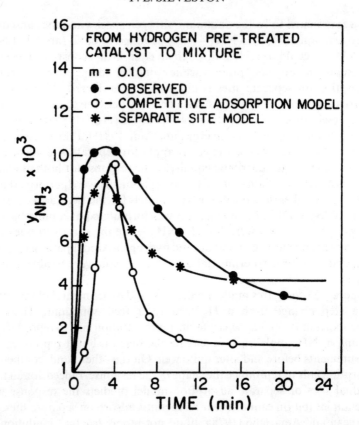

FIGURE 7.33 Comparison of the predicted response to experimental data for a step change from H_2 to a syngas mixture at 400°C and 2.38 MPa in a gradientless reactor (Figure taken from Li et al. (1985a) with permission of the authors)

introduced are large so to provide an approximation, a lumped parameter model was used (Li et al., 1985b):

$$\frac{dC_N}{dt} = \left(\frac{\rho \tau_o}{M_c}\right)\left(\frac{k_D F_t}{\delta}\right)(\alpha P_{N_2} - C_N) \qquad (7.17)$$

where C_N is a mean concentration of nitrogen in a unit volume of catalyst, α is a Henry's law form of solubility and $(k_D F_t/\delta)$ is the lumped mass transfer coefficient.

Parameters for the transport term were obtained by fitting the modified model to Jain's time average data. All other parameters were evaluated using the step change data as discussed above. With the addition of bulk storage, Li et al. (1985b) report a good fit of all of Jain's gradient-less reactor

Table 7.3 Separate Adsorption Site, Bulk Storage Model for a Gradientless Reactor Under Periodic Operation (Li et al., 1985b, Nam et al. 1990)*

$$\frac{dy_{N_2}}{d\tau} - (y_{N_2}^\circ - y_{N_2}) = 0.5\phi k_1 P y_{N_2} \exp(-b_N \theta_N) \quad (7.18)$$

$$\frac{dy_{H_2}}{d\tau} - (y_{H_2}^\circ - y_{H_2}) = -0.5\phi k_2 P y_{H_2} \exp(-b_H \theta_H) \quad (7.19)$$

$$\frac{dy_{NH_3}}{d\tau} - (y_{NH_3}^\circ - y_{NH_3}) = \phi k_3 \theta_N \theta_H \quad (7.20)$$

$$\frac{d\theta_N}{d\tau} = \xi_H [k_1 P y_{N_2} \exp(-b_H \theta_N) - k_3 \theta_N \theta_H - k_D (N_N \theta_N - C_N)] \quad (7.21)$$

$$\frac{d\theta_H}{d\tau} = \xi_H [k_2 P y_{H_2} \exp(-b_H \theta_H) - 3k_3 \theta_N \theta_H] \quad (7.22)$$

$$\frac{dC_N}{d\tau} = \tau_0 k_D (N_N \theta_N - C_N) \quad (7.23)$$

where $\tau_o = V/Q_o$, $\tau = t/\tau_0$
$\phi = \tau_o W/VC_T$, $\xi_i = \tau_0/N_i$

Initial Conditions

Dimensionless Time	$\tau \geqslant n$	$\tau \geqslant n+s$	$(n=0,1,2...)$	
$y_{N_2}^\circ$	0	$1-m$		
$y_{H_2}^\circ$	1.0	m		(7.24)
$y_{NH_3}^\circ$	0	0		
$\theta_k(\tau=0)$	0	0	$k = H, N$	
$C_N(\tau=0)$	0	0		

*Differential, isothermal operation assumed so change in gas volume neglected.

time average data. The model, however, failed to accurately reproduce the time changes in ammonia concentration during a cycle.

A later review of the Li work revealed a scaling error in converting the model shown in Table 7.3 to a form convenient for numerical integration (Nam et al., 1990). When this error was corrected, Nam observed that the model no longer predicted resonance with respect to period although it still predicted $\Psi > 1$. Chiao and Rinker (1989) recalculated the transport term (Eqn. 7.17) parameters and tested the importance of the term by including it in their models (see next section) They observed that the response to step change was altered, but the effect on Ψ was negligible.

Retaining the assumption that storage of atomic nitrogen on either an inactive portion of the catalyst surface or subsurface is essential to observe resonance and $\Psi > 1$, Nam et al. (1990) systematically examined the models shown in Figure 7.34. These are separable into surface storage models and bulk dissolution or subsurface storage ones. In the former, nitrogen is adsorbed on sites on which reaction is possible and on others where

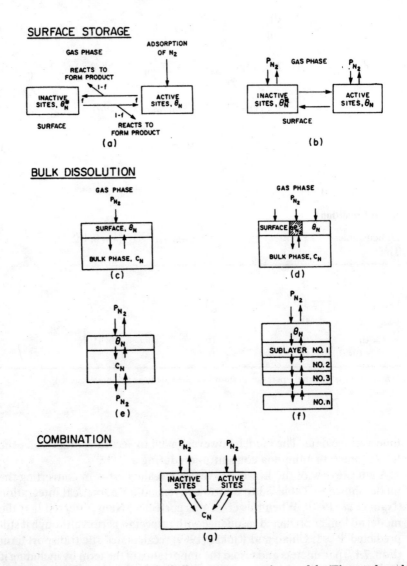

FIGURE 7.34 Schematic of the bulk storage – reaction models (Figure adapted from Nam et al. (1990) with permission, © 1990 Elsevier Science Ltd.)

reaction cannot take place. Two versions of the model differ in where adsorption/desorption and reaction to form NH_3 occur. Four versions of the subsurface storage model are given in Figure 7.34. The single site version (c) is identical to the model considered by Li et al. (1985b). An extension of this version with more than one storage region (distributed storage) appears as (f). The dual site versions (d) and (e) differ in the presence or absence of direct dissolution in the subsurface. The bottom schematic in the figure (g) shows both inactive surface and subsurface storage. In all of these phenomenological models, adsorption of the reactants H_2 and N_2 occurs on different surface sites.

All of the models considered by Nam require modification of the equations and initial conditions set out in Table 7.3. Parameters for each model were obtained in steps using the steady state, step change, and composition forcing data of Jain et al. (1982b, 1983b) and Jain (1981). Predictions were tested against the data shown in Fig. 7.2; these data were not used for fitting. Only two of the models, surface storage (a) and bulk dissolution (d) successfully predicted resonance and $\Psi > 1$. Figure 7.35a shows the behaviour of the surface storage model as a function of τ and the ratio of the inactive and active surface adsorptive capacities. The inactive capacity must be about $10 \times$ the active capacity to fit the experimental measurements in Figure 7.2.

The surface storage models replaces Equation (7.23) in Table 7.3 by

$$\frac{d\theta_N^*}{d\tau} = \xi_N^* f J \tag{7.25}$$

representing the time change of the occupancy θ_N^* of the inactive surface storage sites whose maximum capacity is N_N^*. J is the exchange flux between the active and inactive sites and is given by

$$J = k_s [\theta_N (1 - \theta_N^*) - \theta_N^* (1 - \theta_N)] \tag{7.26}$$

where f is the fraction of atoms of N which does not react to form NH_3 in the exchange. Terms $(1-f) J$, $3\xi_H (1-f) J$ must be added to Equations (7.20) and (7.22), while the term $\xi_N^* f J$ replaces the $\tau_o k_D (N_N \theta_N - C_N)$ term in Equation (7.21). These changes have only a minor effect on the two site curve in Figure 7.33.

Figure 7.35b shows the dual N_2 adsorption site, bulk dissolution model gives a good fit of the experimental data for $s = 0.6$ in Figure 7.2. The curve at $y_{N_2}^\circ = 0.75$ is consistent with the results shown in Figure 7.4 at $y_{H_2}^\circ = 0.25$. For this model, the only change in the Li model of Table 7.3 is that the driving force for diffusion of atomic nitrogen into the catalyst bulk is the partial pressure in the gas phase rather than the nitrogen concentration on the surface so that $(\alpha P_{N_2} - C_N)$ replaces $(N_N \theta_N - C_N)$ in Equations (7.21) and (7.23). Storage in the bulk and the flux to and from the catalyst surface

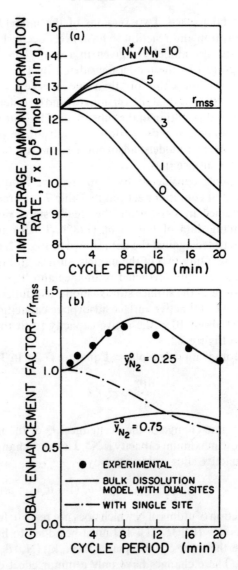

FIGURE 7.35 Prediction of models: a) surface storage model (a) in Figure 7.34 as a function of the ratio of inactive to active storage capacity, b) bulk dissolution model (d) in Figure 7.34 as function of cycle period (Figure adapted from Nam *et al.* (1990) with permission, © 1990 Elsevier Science Ltd.)

are increased by about an order of magnitude through this change. The direct result is $\Psi > 1$ for low values of τ.

An interesting observation of this modelling study is that $\Psi > 1$ requires composition forcing to activate new sites for either N_2 adsorption or the

$N-H_2$ reaction. Storage of reactant on the catalyst in places where reaction does not occur alone is not sufficient for $\Psi > 1$.

Surface Activation/Restructuring Model

Chiao *et al.* (1987) suggest the mechanism resulting in improvement of the synthesis rate through composition switching must be complicated because of the strange behaviour after a switch to periodic from steady state operation has been made. If feed composition is switched between pure H_2 and pure N_2, catalyst activity decreases as illustrated by Figure 7.12. However, if feed composition cycles between mixtures of H_2 and N_2, the response to the change to periodic operation is shown in Figure 7.36. The overshoot-undershoot seen in the figure over about an hour, or 30 minutes for overshoot alone, is far too slow to be caused by temperature excursion resulting from the switch. The overshoot, sharp deactivation and gradual reactivation were observed often according to Chiao. The overshoot without deactivation corresponds to the data at $\tau = 30$ seconds and $s = 0.2$, shown as points at $y_{H_2}^\circ = 0.62$ in Figure 7.14. The outlet NH_3 concentrations

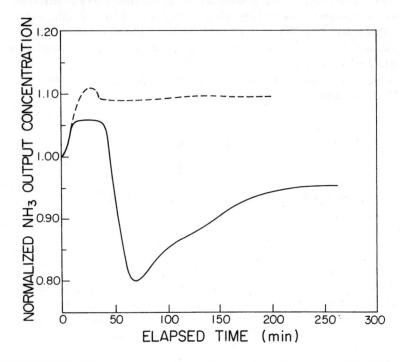

FIGURE 7.36 Time response to change from steady state to composition forcing for no change in catalyst state (—), for change to super active state (- - -) (Figure taken from Chiao and Rinker (1989) with permission, © 1989 Elsevier Science Ltd.)

have been normalized by the NH_3 concentration at steady state for 75 mol % H_2 in the feed.

To explain the experimental observations of Chiao *et al.* (1987), Rinker and co-workers pursued the mechanistic conclusions of earlier studies of the step change dynamics of NH_3 synthesis over iron (Amariglio and Rambeau, 1977; Rambeau and Amariglio, 1978 a–c). These authors reported an increase in catalyst activity in the presence of NH_3, whereas, exposure to either pure H_2 or N_2 deactivates the catalyst. Amariglio and Rambeau (1977) speculate that surface restructuring is the cause of the activity change but there is no indication whether restructuring changes the turn over frequency or the active site density. Several investigators have noted large differences in N_2 adsorption rates on the 100 and 111 faces of iron (Amariglio and Rambeau (1977) concluded that the triply promoted iron catalyst behaves in the same way as pure iron). Chiao *et al.* associate the switch in surface structure to nitride formation Although nitride formation is thermodynamically impossible in the presence of only N_2 for the condition used by Chiao *et al.* (1987), it can form on an iron surface in the presence of ammonia, according to Chiao and Rinker (1989).

On the basis of the response observations and differences in adsorption rates, Chiao and Rinker (1989) suggest that composition forcing induces a super active state, but they offer no further evidence for this state. They speculate that the mechanism resulting in the super active state involves surface restructuring.

Chiao and Rinker considered two kinetic models in their study:
M1 – Competitive Adsorption/Gas Phase Hydrogenation

$$H_2 + 2s \rightarrow 2H - s$$

$$N_2 + 2H - s \rightarrow 2NH - s \qquad (7.27)$$

$$NH - s + H_2 \rightarrow NH_3 + s$$

M2 – Nitrogen Adsorption/Gas Phase Hydrogenation

$$N_2 + 2s \rightarrow 2N - s$$

$$2N - s + 3H_2 \rightarrow 2NH_3 + s \qquad (7.28)$$

The model can be written for either a gradientless reactor to give a set of ODEs, or for an isothermal plug flow packed bed reactor to give a set of PDEs. The right hand side of both sets are the same rate terms, written assuming the steps in Equations (7.27) and (7.28) are elementary reactions. They are given in Table 7.4 for the M1 model. Reverse reactions are neglected.

The number of parameters estimated by fitting steady state and/or time average cycling data was reduced by introducing N_2 to a surface exposed previously to helium, hydrogen or a sequence of $H_2 - N_2$ switches and

Table 7.4 Chiao and Rinker Model M1 for Ammonia Synthesis over a Promoted Iron Catalyst (Chiao and Rinker, 1989)

$$\Gamma'_{N_2} = \tau_o N_i^2 \left(-k_1 y_{N_2}(1-\theta)^2 + \frac{k_{-1}}{C_T}\theta^2 \right) \quad (7.29)$$

$$\Gamma'_{H_2} = \tau_o N_i - k_1 N_i y_{N_2}(1-\theta)^2 + \frac{k_{-1}}{C_T}\theta^2 \quad (7.30)$$

$$-k_2 y_{H_2}\theta + (k_{ad})_{NH_3} y_{NH_3}(1-\theta))$$

$$\Gamma'_{NH_3} = \tau_o N_i(k_2 y_{H_2}\theta - (k_{ad})_{NH_3} y_{NH_3}(1-\theta)) \quad (7.31)$$

$$\Gamma'_\theta = \tau_o N_i(2N_i k_1 y_{N_2}(1-\theta) - \frac{2N_i k_{-1}}{C_T}\theta^2 \quad (7.32)$$

$$-k_2 y_{H_2}\theta + (k_{ad})_{NH_3} y_{NH_3}(1-\theta))$$

$$\Delta' = \tau_o N_i^2 \left(-k_1 y_{N_2}(1-\theta)^2 + \frac{k_{-1}}{C_T}\theta^2 \right) \quad (7.33)$$

where $\tau_o = V/Q_o$, i = dormant, active, superactive
For a plug flow reactor

$$\Gamma'_k = \frac{\partial y_k}{\partial \tau} + u\frac{\partial y_k}{\partial z} + y_k \Delta' \quad k = N_2, H_2, NH_3 \quad (7.34)$$

$$\Gamma'_\theta = N_i \frac{\partial \theta}{\partial \tau} \quad (7.35)$$

Initial Conditions - see Table 7.2

---

using such data for parameter estimation. During these step experiments, bulk storage of N_2 was observed for catalyst activated by NH_3 production and prolonged $H_2 - N_2$ switching, but not in the case of a dormant catalyst reduced by extended exposure to H_2 (Chiao and Rinker, 1989).

Using a Langmuir isotherm to fit the adsorption/desorption data, differences in both the adsorption and desorption rate constants were found for different catalyst states. These data were used by Chiao and Rinker in two ways: rate constants obtained with hydrogen pretreatment were used in a classical rate expression (Table 7.4), while constants obtained with helium pretreatment of the catalyst were used with an effective driving pressure (EDP) model introduced by Ozaki et al. (1960) to account for the widely observed increase in N_2 adsorption and desorption rate when carried out in the presence of NH_3. H_2 adsorption was assumed to be an equilibrium process. Chiao and Rinker set the adsorption equilibrium constant for H_2 to unity.

The remaining model parameters were evaluated from differential reactor data under the assumption that the transition from a dormant to an active catalyst or to a super active one increased the density of active sites by a factor independent of temperature. Thus, the activation energy and the rate constant of the forward reaction were measured for the dormant catalyst only. Measurements on "active" and "superactive" catalyst then were used to determine the appropriate site density factors.

Figure 7.37 plots the NH_3 mole fraction in a differential reactor outlet for different pretreatment of the iron catalyst: reduction in H_2, corresponding to the "dormant or reduced" catalyst; exposure to a $H_2 - N_2$ syngas mixture, corresponding to an "active" catalyst, and exposure to composi-

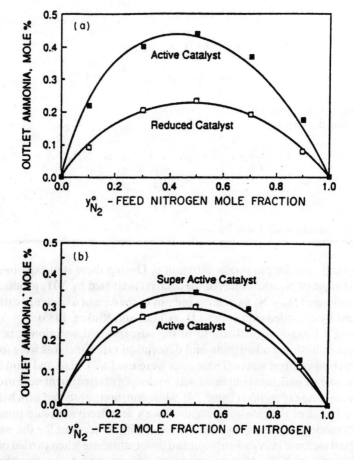

FIGURE 7.37 Variation of NH_3 concentration vs mol fraction N_2 in the feed to a differential reactor operating at 4.24 MPa. Solid lines are the fit of Model M1 (Equation 7.27) to the experimental data: a) comparison of catalyst in the "active" and "reduced" state at 703 K, b) comparison of "super active" and "active" catalyst at 673 K (Figure taken from Chiao and Rinker (1989) with permission, © 1989 Elsevier Science Ltd.)

tion forcing, corresponding to the "super active" catalyst. As Amariglio and Rambeau (1977) predicted, there is about a 100 % increase in activity or site density with the "dormant" to "active" transition at 673°C and 4.24 MPa. The increase in site density for the "active" to "superactive" transition is 18%, however.

Chiao and Rinker (1989) tested their M1 and M2 models by comparing the model predictions with data obtained from their integral reactor at steady state. Models incorporating the effective driving force (EDP) term for N_2 adsorption performed poorly and were not tested further. When cycling data was used, only the M1 model provided satisfactory predictions of performance. Indeed, the predictions are remarkable as Figure 7.38 demonstrates. The upper panel (Fig. 7.38a) compares the prediction to the experimental data in Figure 7.11, while the comparison in the middle panel (Fig. 7.38b) relates to Figure 7.15. Modelling of change with time within a cycle in the bottom panel (Fig. 7.38c) with two repetitive cycles shown is very good when allowance is made for mixing in the experimental equipment which decreases the slopes of ramp changes and reduces peak height. Neither Li et al. (1985b) or Nam et al. (1990) were able to obtain adequate prediction of the behaviour within a cycle with their models.

F. OVERVIEW AND RESERCH CHALLENGES

Despite the different interests which initiated the experimental work described in this chapter, the results obtained by each of the four teams of researchers establish that:

a) composition forcing activates ammonia synthesis catalysts
b) synthesis rates under periodic operation can exceed the maximum steady state synthesis rates at a specified temperature and total pressure.

With the triply promoted iron catalyst, the rate improvement is largest away from the syn gas composition for the maximum steady state rate. The enhancement factor is large, reaching 1.5 to 1.6 × at low temperatures and pressures (relative to industrial practice). As pressure and temperature increases, the factor drops below 1.1. The enhancement appears to be negligible for isothermal operations under the usual industrial temperature and pressure for this catalyst.

Periodic operation is impractical at any level of Ψ if conversion per pass is low and recycle of the reactants is necessary. To implement the operation would require either hold-up in the recycle, without mixing, to bring recycle and conditions at the reactor entrance in phase or it would require reactant separation prior to recycle. Conversion per pass is very low for ammonia synthesis over promoted iron catalysts so that there is no incentive to pursue periodic operation further from the aspect of larger throughput or

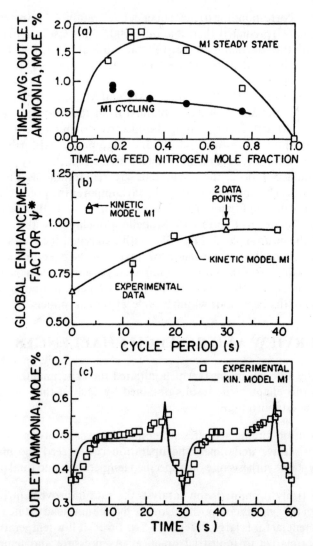

FIGURE 7.38 Comparison of M1 model prediction of the NH_3 concentration in reactor outlet or rate enhancement with differential reactor data at 430°C and 4.24 MPa: (a) steady state at the time average feed composition and bang-bang H_2 - N_2 cycling, (b) cycling between syn gas mixtures, (c) cycling between syn gas mixtures with $s = 0.8$ and $\tau = 30$ s (Figure taken from Chiao and Rinker (1989) with permission, © 1989 Elsevier Science Ltd.)

smaller reactor size. This limitation may change as ruthenium enters into commercial use as a catalyst for the synthesis.

Only the temperature levelling observations, Figures 7.26 and 7.27 provide a topic worthy of further study for the promoted iron catalysts.

AMMONIA SYNTHESIS 265

Could composition forcing raise the stability of existing synthesis reactors by damping hot spot development? This question is probably best pursued through simulation now that a dynamic model for the synthesis is available. What amplitudes are needed and what are the recycle implications?

The explanation of enhancement through composition forcing appears to depend on the catalyst. For ruthenium and osmium, synthesis rates are increased by removing H_2 from the feed for part of a cycle to provide access to the surface for N_2. N_2 adsorption is rate limiting for these catalysts. Greater access allows lower temperatures to be used which changes favorably the equilibrium conversion. This needs further study to determine effects on conversion per pass. For iron catalysts, the surface is somehow activated by composition modulation. This is clearly the outcome of the experiments of Chiao et al. (1987) and is supported by the modelling undertaken by Chiao and Rinker (1989). Modelling studies carried out by Silveston and co-workers lead to the conclusion that composition forcing enhances the synthesis rate by opening up additional N_2 adsorption or $N-H$ reaction pathways. This is essentially catalyst activation.

If, indeed, activation occurs through surface restructuring, what is the mechanism? Does the turnover number at a site change or are new sites created? Rambeau and Amariglio (1978c) and Chiao and Rinker (1989) suggest that activation occurs through the formation of new reaction sites. Direct evidence for this is lacking. How are new sites created? Is storage of N_2 or H_2 involved in restructuring?

A curious observation is that alternating exposure of the catalyst to H_2 and a synthesis gas mixture leads to $\Psi > 1$. However, this is inconsistent with the Eley-Rideal mechanism, Equation (7.28], proposed by Chiao and Rinker (1989) for the synthesis. Exposure to H_2 alone must result in H_2 adsorption. Since this policy increases Ψ, adsorbed hydrogen must play a role in the synthesis mechanism. What is this role?

A further research challenge is exploration of diffusion-reaction interaction in composition forcing of NH_3 synthesis. Cho (1983) found a strong interaction for CO oxidation as discussed in Chapter 3. Under some diffusion conditions, composition switching suppressed oxidation while under others the oxidation rate was enhanced. Jain et al. (1985) have demonstrated an interaction is present, but their experiments were limited. The availability of a model means a numerical investigation of diffusion interference is feasible, although the computational requirements will be formidable.

Does a relaxed steady state, different from the steady state, exist for ammonia synthesis over a promoted iron catalyst? Wilson and Rinker (1982) claim there is a difference for NH_3 synthesis, but this has not been demonstrated in measurements heretofore presented. Indeed, the experiment on NO reduction by CO described in Chapter 6 indicate steady state is obtained when high forcing frequencies are used. Fast cycling experiments need to be undertaken for NH_3 synthesis that are free from mixing interference. Both reactor design and choice of temperature and pressure need to be considered. If the existence of a distinct relaxed steady state can be demonstrated experimentally and if the conditions for the state are easily achieved, prediction of performance at the relaxed steady state could be as discriminatory a test of rival models as predicting the improvement under composition modulation now appears to be. Whatever the shortcomings of the Chiao-Rinker relaxed steady state analysis may be, their prediction of large enhancements for the forcing of the synthesis in an autothermal reactor should be followed up by an experimental investigation.

With the apparent commercialization of a ruthenium synthesis catalyst by the M. W. Kellog orgainzation, the remarkable enhancements observed by Rambeau and Amariglio (1981b) for bang-bang forcing call for a follow-up. These investigators estimate that ruthenium is as much as 2000 × as active as iron. Symmetrical bang-bang $H_2 - N_2$ switching leads to synthesis rates ca. 8 × greater than the maximum steady state rate (Figure 7.18). Can even a greater improvement be achieved by optimizing the cycling variables? Figure 7.16 suggests the potential improvement could be as much as 50 ×. A ruthenium based process would be certainly competitive with current NH_3 synthesis technology if a single pass and much lower temperatures and pressures could be used. Rambeau and Amariglio (1984) point out that separating the reactants removes the equilibrium constraint. This is potentially important for NH_3 synthesis. Further experimental work seems urgent. An economic study would set the goals in terms of operating conditions and rates that must be met for a feasible process based on Ru. Further experiments should employ supported, highly dispersed ruthenium to maximize the specific surface.

CHAPTER 8

Sulphur Dioxide Oxidation

A. INTRODUCTION

Three different types of periodic operation have been considered for this important industrial reaction, including composition modulation. Despite the wide interest in non-steady state methods, only three experimental composition forcing investigations are described in the literature. Nevertheless, sulphur dioxide oxidation was one of the first catalytic reactions subjected to periodic composition forcing (Unni et al., 1973). According to these researchers, their investigation was inspired by an observation that a change in SO_2 concentration or temperature, in whatever direction, induced an abrupt rise in SO_3 concentration followed by a gradual diminution, lasting as long as 6 hours (Yeramian et al. (1970). This behavior suggests at once that periodic variation of either SO_2 concentration or reaction temperature would be a means of raising time average catalyst activity. The resolve of Unni et al. to explore feed composition forcing for SO_2 oxidation was buttressed by Horn's research (Horn and Lin, 1967) which showed a significant affect of composition forcing on the selectivity in a complex reaction system. Indeed, Unni's work is an example of the chain of inspiration that occurs in science. Unni's experimental investigation was stimulated by theoretical work and chance observations made in a study with quite different objectives. The unexpected results obtained, in turn, stimulated an intensive experimental and theoretical study of the mechanism of the oxidation reaction in the USSR (Meshcheryakov et al., 1982; Kozyrev et al., 1982; Balzhinimaev et al., 1985; Mastikhin et al., 1987; Balzhinimaev et al., 1989) which eventually led to a robust predicative model capable of reproducing Unni's observations.

Further experiments (Briggs et al., 1977; 1978; Silveston and Hudgins, 1981) resulted from an attempt to apply composition forcing to the important problem of decreasing emissions from SO_2 converters. Periodic flushing of the final stage in the converter, investigated in those experiments, offers a possible solution to this problem.

B. EXPERIMENTAL SYSTEMS

Equipment for studying composition forcing has been described in Chapter 1. The installations shown are applicable to SO_2 oxidation. In their periodic flushing study, Briggs et al. (1977), however, replaced the micro-reactor shown in Figure 1.7 with a 2.54 cm o.d. integral reactor. A continuation of this work used a two reactor system with flow switching so as to simulate a possible industrial application of periodic operations (Briggs et al., 1978). The two reactor system is shown in Figure 1.14.

Measuring SO_2 conversion is difficult and several methods were used in the experiments. Unni et al. (1973) passed the SO_3 containing off gas through a train of isopropanol absorption flasks held at $-72°C$. At the completion of a run, they diluted the contents with water and removed most of the SO_2 absorbed by bubbling N_2 through the solution. The isopropanol-acid solution was then titrated for total acidity. The amount of SO_3 formed was obtained by correcting for any residual SO_2 by an iodine titration. Briggs et al. (1977, 1978) studied a system in which the $SO_3:SO_2$ ratio in the reactor feed was about 9:1 so that a more accurate measure of conversion came from an SO_2 determination. This was accomplished by absorbing SO_3 in 97% sulphuric acid and determining the SO_2 concentration chromatographically (Briggs et al. (1976a). Time variation of SO_3 in the effluent was followed using a quadruple mass spectrometer. Details of the applicaiton and the SO_3 cracking pattern are given by Briggs et al. (1976b). Use of a mass spectrometer requires scrupulous attention to the removal of water from all of the feed gases.

C. COMPOSITION FORCING AT LOW CONVERSION

Unni et al. (1973) chose SO_2 oxidation as a test reaction to explore composition modulation of catalytic reactions. A commercial catalyst containing 9.1 wt.% V_2O_5, 10.1 wt.% K_2O and 0.45 wt.% Fe_2O_3, supported on a diatomaceous earth, was employed and experiments were run at 405°C and 1 bar so that conditions were similar to those found at the inlet of a commercial SO_2 converter. Catalyst size was reduced to 30/40 U.S. mesh, however, to avoid transport interference. Variables in the Unni study were cycle period, amplitude and the time average feed composition. Both amplitude and the time average feed composition were expressed in terms of the $SO_2:O_2$ mole or volume ratio.

Unni's remarkable finding was that SO_2 oxidation is excited at cycle periods measured in hours as may be seen in Figure 8.1. The largest enhancements were obtained at a time average $SO_2:O_2$ ratio = 0.6 with an amplitude of 0.3 measured in terms of the $SO_2:O_2$ ratio. Higher and lower

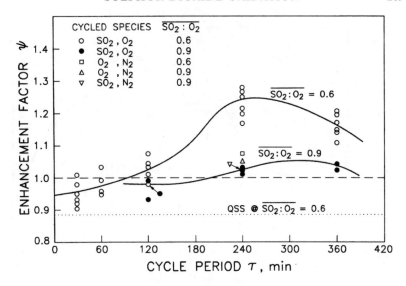

FIGURE 8.1 Rate for SO_2 oxidation over a commercial promoted vanadia catalyst as a function of forcing period, time average feed composition as $SO_2:O_2$ ratio, and forcing strategy (reactant cycled) for 1 bar and 405°C, $s = 0.5$ (Figure adapted from Unii *et al.*, 1973 with permission of the authors)

time average ratios (0.9 and 0.3) resulted in a significantly poorer performance, but this seems due to the lower amplitudes available at these means. The maximum amplitude possible was 0.1, again measured as a $SO_2:O_2$ ratio. Symmetrical cycles were used by Unni *et al.*, but because ratios rather than concentrations were used, amplitudes for SO_2 and O_2 are not the same. Regardless of the time average $SO_2:O_2$ ratio, the largest enhancements were achieved at cycle periods between 4 and 6 hours.

Strategies in which one of the reactant concentrations is held constant were examined in several runs at $\tau = 240$ min and are shown in the figure. These strategies appear to be comparable to variation of both reactants at a time average $SO_2:O_2$ ratio $= 0.9$, but variation of O_2 and N_2 holding SO_2 constant at a mean $SO_2:O_2$ ratio $= 0.6$ gave a value of Ψ less than half the value obtained when both reactants are cycled.

The influence of cycle amplitude with symmetrical forcing was examined at a time average $SO_2:O_2 = 0.6$ in the feed and $\tau = 240$ min. Figure 8.2 indicates that a threshold amplitude between 0.1 and 0.2 as a $SO_2:O_2$ ratio exists for the system. The largest amplitude shown represents switching between a feed containing just one of the two reactants and a mixture that approaches the stoichiometric ratio. The absence of one reactant appears to be detrimental just as was seen in several of the ammonia synthesis experiments discussed in the previous chapter.

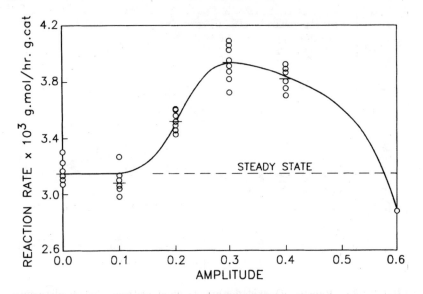

FIGURE 8.2 Dependence of rate enhancement on cycle amplitude as the $SO_2:O_2$ ratio for $\tau = 240$ min, $s = 0.5$ and a time average $SO_2:O_2$ ratio $= 0.6$ in the feed. Catalyst and other conditions as given in Figure 8.1 (Figure adapted from Unni *et al.* (1973) with permission of the authors)

The absorption method used for SO_3 made it difficult to follow the variation of the product concentration with time inside a cycle. Nevertheless, Unni *et al.* (1973) report measurements made at $\tau = 240$ min. and $A = 0.3$ and 0.6 as $SO_2:O_2$ ratios. Measurements at $A = 0.3$ appear in Figure 8.3. Data points were collected either in successive cycles or in separate experiments. The scatter is magnified by the scale used; the standard deviation at each measurement point is only 3%. For the half cycle at the high $SO_2:O_2$ ratio, the SO_3 formation rate is only about 5% greater than the steady state rate corresponding to the ratio and decays slowly towards that rate over 120 min. Figure 8.3 shows that the catalyst activity is enormously increased in the half cycle at a low $SO_2:O_2$ ratio so that the rate of formation is almost 3 × the steady state rate at the same ratio. A very slow decay in activity is evident. It is this slow decrease which accounts for the remarkable performance at cycle periods of 4 to 6 hours.

Measurements within a cycle at an amplitude of 0.6 were for switching between a feed containing only O_2 and N_2, and one with a $SO_2:O_2$ ratio $= 1.2$. At the 1.2 ratio, the catalyst operated at a rate of SO_3 formation close to the rate exhibited at steady state for this ratio. In the other half cycle, the time average rate was 0.22×10^{-3} g mols SO_3/h g.cat (compare with rates in Figure 8.3) even though there was no SO_2 in the feed to the reactor. The vanadia catalyst had a surface of area of just 1.3 m²/g and

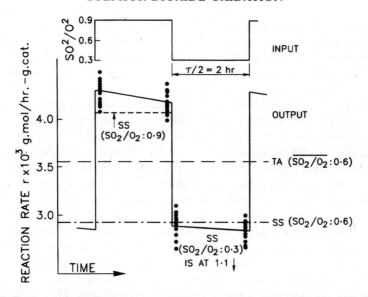

FIGURE 8.3 SO_2 oxidation rates within a cycle $A = 0.3$ expressed as the $SO_2:O_2$ ratio. Catalyst and other conditions as given in Figure 8.2 (Figure adapted from Unii et al. (1973) with permission of the authors)

would have contributed just 0.1% of the SO_3 produced in the half cycle if there had been monolayer coverage. Consequently, Unni et al. reason that SO_3 must be either absorbed in the melt at a rather high concentration or arise from the decomposition of a catalyst component.

An independent measurement (Yeramian et al., 1970) of the relaxation time after a step change in temperature or reactant concentration gave times between 4 and 8 h at 442°C. These times are about the same as the periods which result in the greatest rate enhancement. The long relaxation time suggests a diffusional process within the bulk of the catalyst must be responsible for the rate increase under composition forcing.

D. COMPOSITION FORCING AT HIGH CONVERSION

The experiments undertaken by Briggs et al. (1977) and interpreted by Silveston and Hudgins (1981) explored the application of composition forcing to multi-staged SO_2 converters. The same commercial catalyst employed by Unni et al. was chosen, but used in an integral reactor containing 30 g of 12/20 U.S. mesh particles. The initial experiments examined composition forcing at high conversion of a feed stream containing

12.4 vol.% SO_2, simulating crudely the first stage of a commercial converter. To enable comparison with Unni's work, the reactor was held in a fluidized sand bath at 405°C so that it operated almost isothermally. Cycling between mixtures of air and SO_2 at about 40% SO_2 conversion, however, provided no rate enhancement.

Because the rate of formation of SO_3 becomes inversely proportional to the SO_3 partial pressure at high SO_3 conversions, Briggs et al. (1977) considered cycling the final stage of the converter between a stream typically entering this stage and air. The role of air was to strip SO_3 from the catalyst surface and to re-oxidize the catalyst, raising thereby the reaction rate in the other half of the cycle. At the time of their work, environmental regulations on converter emissions were being tightened. Thus, their objective was to see if periodic air flushing of the final stage could raise conversion to levels which would satisfy regulations without having to resort to the expensive double contact, double absorption process. Step change experiments at 406°C involving flushing the catalyst for 18 h followed by introduction of a stream typical of that produced in the first three stages of a commercial SO_2 converter were encouraging. They showed an initial SO_2 conversion of 100% which dropped slowly over 24 h to reach the steady state conversion level.

The first composition forcing experiments using air to strip SO_3 from the catalyst were carried out in the single integral reactor described above. Feed for the cycled stage was prepared in a separate converter containing a Pt/Al_2O_3 catalyst at 470°C. This converter, simulating the first three stages and operating at steady state, converted about 90% of the SO_2 in the feed to SO_3. Figure 8.4 shows the results obtained using symmetrical cycling with $\tau = 26$ min. The feed to the steady state preconverter contained 12.4 vol.% SO_2 and was representative of acid plant feeds obtained by burning sulphur. When the vanadia catalyst bed operated at steady state, it raised the SO_2 conversion to 95.8%. The conversion under modulation, 12 min. after introducing the $SO_3:O_2$ mixture to the reactor, was 98.8%. The time average conversion in the half cycle was higher. Step change observations, mentioned above, indicate an even higher time average conversion could have been attained by shortening the half cycle, but the 12 minutes was dictated by the necessity of saturating the 97 wt.% acid absorber with SO_2 before measurements could be made. Conversion in Figure 8.4 is plotted against cycle number. The figure demonstrates that this high half cycle conversion is maintained over some 30 h of operation. Only traces of SO_2 were detected leaving the vanadia catalyst bed 2 minutes after switching back to air. Consequently, significant amounts of SO_2 are not desorbed from the catalyst in the stripping step. Therefore, the figure represents the performance under composition modulation even though just half cycle data are shown. It is notable that the higher conversion achieved by modulation implies a 53% increase in catalyst activity.

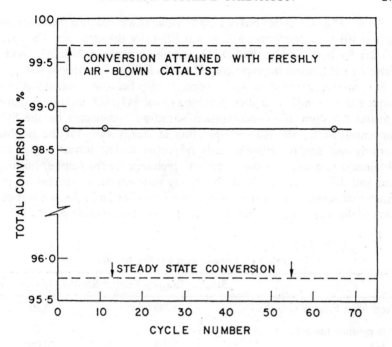

FIGURE 8.4 Total SO_2 conversion in the SO_2-rich stream leaving the second reactor as a function of cycle number and type of reactor operation for a time average feed containing 12.4. vol. % SO_2, $T = 406°C$. With composition forcing, measurement follows switch to the SO_3/SO_2 feed by 12 min. Reference lines (---) show limiting conversions for steady state operation and under periodic forcing (Figure adapted from Briggs *et al.* (1977) with permission of the authors)

The upper dashed line in Figure 8.4 shows the initial conversion on switching in the step change experiment discussed above. It gives the limiting performance for final stage periodic air flushing at 406°C with feed typical of a sulphur burning acid plant. This conversion, 99.7, exceeds slightly the equilibrium conversion of 99.4% at the bed temperature, calculated from the NBS tables of thermodynamic data.

Replacing air by either N_2 or O_2 did not affect the conversion obtained in the experiment just described. Consequently the higher conversion does not result from catalyst re-oxidation. It must be due solely to stripping SO_3 from the catalyst.

In order to explore composition modulation of the final stage of converter further, Briggs *et al.* (1978) added a second integral reactor, also holding about 30 g of the vanadia catalyst. This is the system shown in Figure 1.14. With the preconverter in place, this system was operated on a typical feed from sulphur burning with a $SO_2:O_2:N_2$ composition in vol.% of 10.8:15.2:74 and on a typical smelter effluent with a composition of 8.0:6.2:85.8. The cycled bed of vanadia catalyst was held in a fluidized sand bath at 401°C for the former feed and at 405°C for the latter one. The space velocity for both air and the $SO_3:O_2$ mixture was about 24 min^{-1} (STP). Table 8.1 summarizes the experimental results for the cycle periods tested.

SO_2 emitted from the modulated bed changes because of mixing and the rise in emission with time after switching to the $SO_3:SO_2$ mixture. Figures 8.5 and 8.6 show the concentration variation normalized by the SO_2 concentration for the reactor operating at steady state for the sulphur burning and smelter effluent feeds respectively. The lowest values are obtained 2 minutes after the composition change for the sulphur burning feed and they are about 8% of the steady state emission, whereas for the smelter effluent feed, the lowest emission occurs after 2 minutes and is about 13% of the steady state value. The figure suggests that cycle period of 4 to

Table 8.1 Composition Cycling Results

	Sulphur Burning Feed	Smelter Off Gas Feed
Feed: Mole ratio SO_2/O_2	0.709	1.30
Composition (mole frac.)		
SO_2	0.108	0.080
O_2	0.152	0.062
N_2	0.740	0.858
Bath Temperature, Final Stage	401°C	405 ± 1°C
Mass of Catalyst, Final Stage	25 g	23 g
Conversion (fractional)		
In preconverter	0.917	0.883
Final stage, steady state	0.952	0.913
Final stage, cycling (mean value)		
for $\tau_{1/2} = 10$ min	0.994	0.984
for $\tau_{1/2} = 20$ min	0.992	0.981
for $\tau_{1/2} = 24$ min	–	0.976
Equilibrium, steady state	0.995*	0.987*
$SO_{2cyc}/SO_{2s.s.}$ (mean molar ratio) from final stage		
for $\tau_{1/2} = 10$ min	0.122	0.184
for $\tau_{1/2} = 20$ min	0.164	0.229
SO_2 concentration in gas leaving scrubber (SO_3-free) at steady state	6060 ppm	7790 ppm

*Calculated

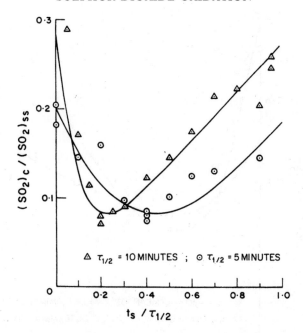

FIGURE 8.5 Time dependence of the SO_2 concentration leaving the SO_3 scrubber during symmetrical composition forcing of the final stage of a SO_2 converter with a hot air stream and hot feed from the previous stage. Measurements for $\tau = 10$ and $\tau = 20$ min. shown. Feed to the system contains 10.8 vol.% SO_2 and $T = 401°C$ (Figure adapted from Briggs et al. (1978) with permission of the authors)

5 min would be optimum for the conditions used, yielding a performance some 10% better than that shown at $\tau_{1/2} = 10$ min Table 8.1.

Using a feed similar to smelter off gas, containing 8.25 vol % SO_2, and operating the preconverter to give 86% SO_2 conversion, the lowest emission level, essentially breakthrough, is substantially delayed as the bed temperature is increased. This is illustrated by Figure 8.7. The curves for $T = 412$ and $423°C$ should be compared to the $\tau_{1/2} = 10$ min curve in Figure 8.6. The higher temperature measurements were not normalized because steady state data were not available. The explanation of the delayed minimum in the SO_2 concentration is that data in Figures 8.5 to 8.7 are for the combined streams emerging from the two parallel reactors (see Fig. 1.14), one of which is converting the $SO_2:SO_3/O_2$ mixture from the preconverter while the other is undergoing air flushing. The initially high SO_2 originates from the flushing step taking place in one of the beds. This drops with time in the half cycle and the SO_2 concentration approaches zero. The rising SO_2 concentration in the latter part of the half cycle comes from the other converter bed where SO_2 reacts to form SO_3. Adsorption of SO_2 on the air blown catalyst surface where a portion is oxidized to SO_3

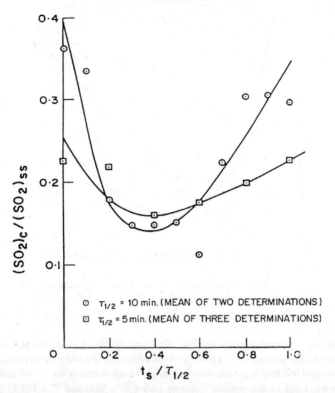

FIGURE 8.6 Time dependence of the SO_2 concentration leaving the SO_3 scrubber during symmetrical periodic flushing of the final stage of a SO_2 converter with a hot air stream. Measurements for $\tau = 10$ and $\tau = 20$ min. shown. Feed to the system contains 8.0 vol.% SO_2 and $T = 405°C$ (Figure adapted from Briggs et al. (1978) with permission of the authors)

keeps the SO_2 effluent from the active bed low. The trioxide is also adsorbed and more strongly than SO_2 so this bed becomes saturated eventually and the SO_2 concentration starts to climb. This is a breakthrough behaviour. It is the combination of the two parts of the cycle which result in the shape of the curves in the three figures.

The breakthrough behaviour just discussed is illustrated for SO_3 at 408°C in Figure 8.8 taken from Briggs et al. (1977). The curve on the left is for a catalyst that has had most of the vanadia removed by an acid wash. Breakthrough occurs in about 30 s which is somewhat longer than the transport lag indicating some SO_3 adsorption must occur on the diatomaceous support. The middle curve is for the same amount of catalyst and the same bed temperature, but the catalyst now is loaded with vanadia. More than 6 minutes are needed before breakthrough is observed. When the temperature is reduced to 309°C, below the ignition temperature for the catalyst, breakthrough is unchanged. Breakthrough begins at greater than

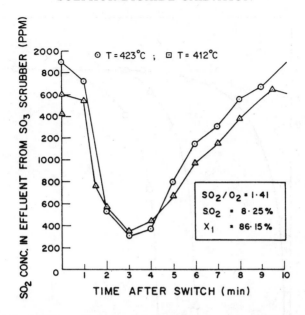

FIGURE 8.7 Influence of catalyst bed temperature on the time dependence of the SO_2 concentration leaving the SO_3 scrubber during symmetrical composition forcing of the final stage of a SO_2 converter with a hot air stream and hot effluent from the previous stage. Feed to the system contains 8.25 vol.% SO_2 and $\tau = 10$ min, $s = 0.5$ (Figure adapted from Briggs *et al.* (1978) with permission of the authors)

9 minutes but this results from a 25% increase in a amount of catalyst and changes in the adsorption equilibria with temperature. Consequently, breakthrough results from saturation of the catalyst with adsorbed SO_3.

A second group of experiments measured temperature variation in the bed and the relative concentration of SO_3 leaving the bed during the two halves of a cycle. These measurements are shown in Figure 8.9. The drop in temperature as SO_3 is stripped from the catalyst suggests desorption from a melt or glass phase or decomposition of one or more complexes in that phase. Desorption and decomposition are endothermic. When the SO_3/SO_2 reactant mixture is re-introduced (Fig. 8.9b), the breakthrough behaviour indicates SO_3 absorption. This is exothermic and it accounts for at least part of the temperature rise seen in the figure. The lack of symmetry in the feed and air flushing steps is notable because the duration of each step is the same. SO_3 desorption is incomplete after 13 minutes of an air feed to the catalyst, whereas the SO_3 breakthrough in the other cycle step takes less than 5 minutes. The temperature drop of SO_3 desorption is about 8°C, while the rise in the feed step is about 16°C. Oxidation of SO_2 when it is present in the feed step, however, explains this difference. The temperature rise after the minimum and the fall after the maximum temperatures in the

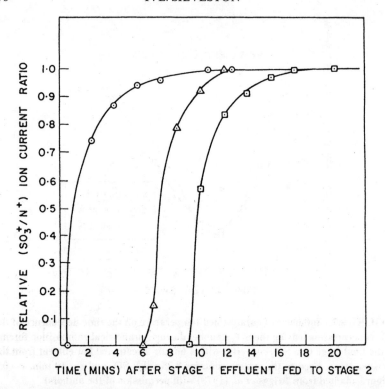

FIGURE 8.8 Dependence of SO_3 breakthrough in the catalyst bed as a function of catalyst and temperature in a step change experiment using a feed to the system containing 12.4 vol.% SO_2 with 90% conversion in the first stage after an 18 h exposure to air: ⊙ = 24 g, acid washed, 12/20 U.S. mesh catalyst particles at 406°C, Δ = 30 g 12/20 U.S. mesh catalyst particles at 408°C ☐ = 30 g 12/20 U.S. mesh catalyst particles at 309°C (Figure adapted from Briggs *et al.* (1977), © 1977, Elsevier Science Publishers)

cycle are consequences of placing the reactor in a fluidized sand bath held at 408°C. In an adiabatic reactor the temperature changes would be even larger.

Figures 8.9 suggests that an asymmetrical cycle with an air flushing step twice the length of the reactant feed step would provide even greater reduction of SO_2 emissions. The drawback of asymmetrical cycling is that more than two catalyst beds in parallel will be needed. An alternative would be to use a symmetrical cycle, but increase the air flow rate. This, however, would increase the volumetric loading of scrubbers in the acid plant. It would be possible to avoid overloading the scrubbers by using the flushing air to burn sulphur. This scheme is shown in Figure 8.10. It does not significantly affect the heat balance in the plant, but since the feed to the first bed of the converter will now contain a small amount of SO_3 there might

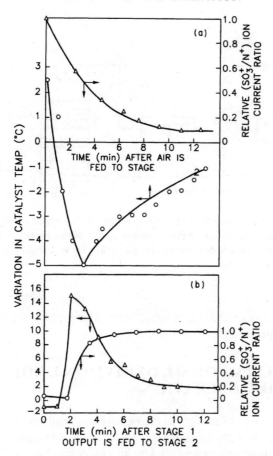

FIGURE 8.9 Time variation of catalyst bed temperature and relative SO_3 signal in the stream leaving the cycled bed for composition forcing of the final stage of a SO_2 converter with an air stream and effluent from the previous stage: a) half cycle with air feed, b) half cycle with SO_3/SO_2 feed. Feed to the system contains 12.4 vol.% SO_2, conversion in the first stage = 90% and $\tau = 26$ min., $s = 0.5$ (Figure adapted from Briggs et al. (1977) with permission, © 1977, Elsevier Science Publishers)

even be a performance benefit because SO_3 raises catalyst activity at low levels and the first stage is rarely equilibirum limited. Silveston and Hudgins (1981) used this scheme in their evaluation of the Briggs work. They concluded that the stiffest emission standards could be met at a cost below that for a double contact, double absorption plant when acid is produced by sulphur burning for either a grass roots situation or for a retro fit one. The poorer performance of periodic air flushing with a smelter off gas and the inability to recycle air probably means that feed composition

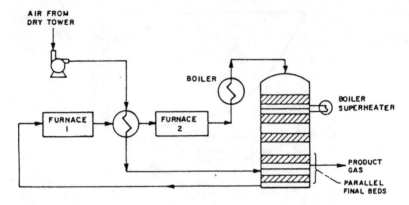

FIGURE 8.10 Flow diagram of an SO_2 converter with periodic air flushing of the final catalyst stage and recycle of the air stream to the sulphur burner (in the diagram air stripping of one final beds is shown) (Figure taken from Briggs et al. (1978) with permission of the authors)

forcing is not attractive for this application. This exploratory work so far has failed to attract industrial interest.

E. MODELLING OF SO_2 OXIDATION UNDER PERIODIC OPERATION

This reaction is one of the few studied under composition modulation for which a mechanistic model is available. Moreover, a reactor model incorporating the mechanisms and its kinetic model has been tested against experimental forcing data. This modeling success means a model is available for exploring applications of periodic air flushing. An example of such an application where the objective is to examine different forcing strategies will be discussed in what follows.

Mechanistic Model

It has been known for decades that commercial potassium oxide promoted vanadia catalysts funciton in a melt phase under reaction conditions and that the oxidation proceeds through a redox mechanism. Potassium oxide is converted to the pyrosulphate in the presence of SO_3 and this compound acts as a fluxing agent for vanadia. In recent years, progress on understanding the mechanism of SO_2 oxidation in this system has been rapid and a detailed mechanism has emerged (Balzhinimaev et al., 1989).

Potassium promoted vanadia catalysts oxidize SO_2 through a cycle in which three different types of binuclear vanadium complexes participate.

These types are 1) an oxo complex with oxygen bridging two V atoms, 2) a sulphite complex in which SO_3 forms the bridge, and 3) a peroxo complex involving O_2 in the bridge. These complexes exist as oligomers in a melt with $K_2S_2O_7$ as the fluxing agent. The redox mechanisms involves the transformation of these complexes between one another through oxidation, SO_2 absorption and the oxidation of SO_2 to SO_3. Fast side reactions involving $S_2O_7^=$ remove vanadium from the melt phase by precipitation and a slow reduciton step leads to V^{4+} which is catalytically inactive. The mechanism is summarized in Table 8.2. The bulk of the SO_3 produced comes from transfer between the V^{5+} complexes and does not involve a change in the coordination of vanadium. Dissolution of SO_3 in the melt is rapid as indicated in step (a) of the table and can be treated as an equilibirium condition. SO_3 is released from the melt by decomposition of the pyrosulphate. Oxygen enters the catalytic network through step 3 and its absorption in the melt may be rate controlling.

The kinetic model due to Ivanov and Balzhinimaev (1987) and Balzhinimaev et al. (1989) assumes that the steps in the mechanisms of Table 8.2 are elementary and that the active and inactive forms of the complexes

Table 8.2 Vanadia Complex Transformations in the Balzhinimaev Mechanism[+]

$$SO_3(g) + (SO_4^=)^m \xrightarrow{\text{rapid}} (S_2O_7^=)^m \qquad (8.1)$$

$$(V_2^{5+}O_2^=)^m + SO_2^{(m)} \underset{k_{-1}}{\overset{k_1}{\rightleftharpoons}} (V_2^{5+}O^=)^m + SO_3^m \qquad (8.2)$$

$$(V_2^{5+}O^=)^m + SO_2^m \underset{k_{-2}}{\overset{k_2}{\rightleftharpoons}} (V_2^{5+}SO_3^=)^m \qquad (8.3)$$

$$(V_2^{5+}SO_3^=)^m + O_2^m \underset{k_{-3}}{\overset{k_3}{\rightleftharpoons}} (V_2^{5+}O_2^=)^m + SO_3^m \qquad (8.4)$$

$$(V_2^{5+}SO_3^=)^m \underset{k_{-4}}{\overset{k_4}{\rightleftharpoons}} (V_2^{4+})^m + SO_3^m \qquad (8.5)$$

$$V_k + (S_2O_7^=)^m \xrightarrow[K_{in}]{\text{rapid}} V_k^{in} \text{ (inactive form)} + (S_2O_7^=)^m \qquad (8.6)$$

*All SO_3^m exists in the melt as $S_2O_7^=$; thus $C_{SO_3}^m = C_{S_2O_7=}^m$

[+] m = molten or glass-like catalyst phase (species exists in this phase) are in equilibrium. Table 8.3 presents the model. Parameters for the model are to be found in Table 8.5.

Application of the Balzhinimaev model requires assumptions about the reactor and its operation so that the necessary heat and material balances can be constructed and the initial and boundary conditions formulated. Intraparticle dynamics are usually neglected by introducing a mean effectiveness, factor, however transport between the particle and the gas phase is considered. This means that two heat balances are required. A material balance is needed for each reactive species (SO_2, O_2) and the product (SO_3), but only in the gas phase. Disappearance or formation appear as either source or sink terms, or as transport terms. Kinetic expressions for the Balzhinimaev model are given in Table 8.3.

Application To The Final Stage Of SO_2 Converter With Composition Forcing

Silveston et al. (1990) used a one dimensional plug flow model to represent the packed bed in this stage. Because the intent of their work was to model

Table 8.3 Kinetic Expression for the Balzhinimaev Model

$$\sum_k y_k + \sum_k y_k^{in} = 1 \tag{8.7}$$

$$y_k^{in} = K_{in} C_{SO_3}^m y_k \tag{8.8}$$

$$\sum y_k^{in} = \frac{K_{in} C_{SO_3}^m}{1 + K_{in} C_{SO_3}^m} \tag{8.9}$$

$$r_1 = \frac{1}{1 + K_{in} C_{SO_3}^m} \left(k_1 P x_{SO_2} y_1 - \frac{k_{-1}}{K_H^3} C_{SO_3}^m y_2 \right) \tag{8.10}$$

$$r_2 = \frac{1}{1 + K_{in} C_{SO_3}^m} (k_2 P x_{SO_2} y_2 - y_3) \tag{8.11}$$

$$r_3 = \frac{1}{1 + K_{in} C_{SO_3}^m} \left(k_3 P x_{SO_2} y_3 - \frac{k_{-3}}{K_H^3} C_{SO_3}^m y_1 \right) \tag{8.12}$$

$$r_4 = \frac{1}{1 + K_{in} C_{SO_3}^m} \left(k_4 y_3 - \frac{k_{-3}}{K_H^3} C_{SO_3}^m y_4 \right) \tag{8.13}$$

Indices on y:

$1 = (V_2^{5+} O_2^=)^m$, $\quad 2 = (V_2^{5+} O^=)^m$ $\quad 3 = (V_2^{5+} SO_3^=)^m$, $\quad 4 = (V_2^{4+})^m$

SULPHUR DIOXIDE OXIDATION

the experiments of Briggs et al. discussed earlier, they allowed for heat loss or gain in the bench scale reactor used by Briggs through wall transport terms employing an overall bed to wall coefficient. They observed from comparison of the magnitude of the terms in the balances that the gas phase storage terms could be safely neglected. Table 8.4 gives their model. Note that these researchers have incorporated reactions in the particle in their material balances rather than treating them as boundary conditions. Remaining boundary conditions and the initial conditions are given in the table. Each change of feed initiates a new solution so that two sets of initial conditions are needed: 1) $t = 0$, 2) $t = n\tau/2, n\tau, 3n\tau/2$, etc.

Parameters representing the catalyst bed were taken from Briggs et al. (1977, 1978), transport coefficients for heat and mass were calculated from well established correlations, whereas the kinetic parameters came from measurements on Russian catalysts (Ivanov and Balzhinimaev, 1987), but of about the same composition as the commercial catalyst used by Briggs. These parameters are collected in Table 8.5. The model used by Silveston et al. contains, therefore, no adjustable parameters.

Figure 8.11 gives the simulation results with the model given above for the conditions used by Briggs et al. to obtain Figure 8.9. Data points are

Table 8.4 Model of a Non adiabatic Packed Bed Reactor for SO_2 Oxidation Incorporating the Balzhinimaev Mechanism

Heat balance on the catalyst:

$$(1-\varepsilon_B)c_p^s\rho_{cat}\frac{\partial T_s}{\partial t} = \varepsilon_m(1-\varepsilon_B)H_3^{dis}\frac{\partial C_3^m}{\partial t} + \varepsilon_m(1-\varepsilon_B)S^v\sum r_j H_j + [\varepsilon_m(1-e_B)S^v H_{in}]$$

$$\left[\frac{\partial \sum y_k^\varepsilon}{\partial t}\right] + h_p S_v(T-T_s) + h_w S_w(T_w-T_s) \qquad (8.14)$$

Mass balance on gas phase:

$$0 = -\mu\rho_g c_p^g \frac{\partial T}{\partial z} + h_p S_v(T_s - T) + h_w^g S_w(T_w - T) \qquad (8.15)$$

Mass balance on SO_3 in the melt phase:

$$\varepsilon_m(1-\varepsilon_B)\frac{\partial C_3^m}{\partial t} = k_m S_m\left(\frac{K_H^3 P x_3^p}{1+K_H^3 P x_3^p/C^\circ} - C_3^m\right) + \varepsilon_m(1-\varepsilon_B)S^v$$

$$(r_1 + r_2 + r_4) - \varepsilon_m(1-\varepsilon_B)S^v\frac{\partial \sum y_k^{in}}{\partial t} \qquad (8.16)$$

Mass balance on SO_3 in the particle void space:

$$0 = -k_m S_m\left(\frac{K_H^3 P x_3^p}{1+K_H^3 P x_3^P/C^\circ} - C_3^m\right) + k_p^3 S_v(C_3^g - C_3^p) \qquad (8.17)$$

Table 8.4   (Continued)

Mass balance on SO_2 and O_2 in the particle void space:

$$0 = \varepsilon_m(1-\varepsilon_B)S^v\sum v_{ij}r_j + k_p^i S_v(C_i^g - C_i^p) \qquad i=1,2 \qquad (8.18)$$

Gas phase mass balances on reactants

$$0 = -\mu\frac{\partial C_i^g}{\partial z} + k_p^i S_v(C_i^p - C_i^g) \qquad i=1,2,3 \qquad (18.19)$$

Number balances on vanadia complexes

$$\frac{\partial y_k}{\partial t} = \sum \mu_{kj} r_j \qquad k = 1-4 \qquad (8.20)$$

and

$$\sum y_k + \sum y_k^{in} = 1 \qquad (8.21)$$

Boundary conditions

$z = 0$

$(n-1)\tau < t < n\tau/2: \qquad n = 1,2,\ldots \qquad T = T^\circ \qquad (8.22)$
$$C_i = 0 \qquad i = 1-3$$

$n\tau/2 < t < n\tau \qquad n = 1,2,\ldots \qquad T = T^\circ$
$$C_i = C_i^\circ \qquad i = 1,2 \qquad (8.23)$$
$$C_3 = 0$$

Initial conditions are

$t = 0; 0 \leq z \leq L:$
$$C_i = 0 \qquad i = 1,2,3$$
$$C_3^m = 0 \qquad (8.24)$$
$$T_s = T = T^\circ$$
$$y_k, y_1^{in}, y_k^{in} = 0 \qquad k = 2,3,4$$
$$y_1 = 1$$

$z \geq 0$

$t = n\tau/2, n\tau \quad n = 1,2,3 \quad C_i^g(t^+) = C_i^g(t^-),$
$$C_i^p(t^+) = C_i^p(t^-),$$
$$C_3^m(t^+) = C_3^m(t^-)$$
$$y_k(t^+) = y_k(t^-), \qquad (8.25)$$
$$y_k^{in}(t^+) = y_k^{in}(t^-)$$
$$T(t^+) = T(t^-),$$
$$T_s(t^+) = T_s(t^-)$$

i = species index: $1 = SO_2$, $2 = O_2$, $3 = SO_3$
j = reaction index: see Table 8.3
k = complex index: see Table 8.3

SULPHUR DIOXIDE OXIDATION

Table 8.5   Model Parameters

Specific Heats:

$$c_p^3 = 0.21\,\text{kcal/kg}^\circ\text{C} \quad c_p^3 = 1.08\,\text{kcal/kg}^\circ\text{C}$$

Heat Effects:

$$H_{in} = 26\,\text{kcal/mol} \quad H_3^{dis} = 12\,\text{kcal/mol}$$

Solubility:

$$K_H^3 = 0.0372\,\text{mol/cm}^3\,\text{atm}$$

Complex, Sulphate, Pyrosulphate Concentrations:

$$s^v = 0.0024\,\text{mol/cm}^3 \quad C^\circ = 0.02\,\text{mol/cm}^3$$

Bed and Catalyst Properties:

$$\rho_{cat} = 264.7\,\text{kg/m}^3 \quad \rho_g = 0.25\,\text{kg/m}^3$$

$$S_v = 3600\,m^{-1} \quad \varepsilon_B = \varepsilon_p = 0.4$$

$$\varepsilon_m = 0.20 \quad u = 0.1\,\text{m/s}$$

Transport Properties:

$$h_p = 0.528\,\text{kcal/m}^2\,\text{s}\,^\circ\text{C} \quad h_w = 2.2\,\text{kcal/m}^2\,\text{s}\,^\circ\text{C}$$

$$h_{wg} = 0.5\,\text{kcal/m}^2\,\text{s}\,^\circ\text{C} \quad k_p = 0.315\,\text{m/s}$$

$$k_m S_m = 6\,\text{s}^{-1}$$

shown in Figure 8.11b, but not in 11a. Mass spectrometer readings were not calibrated and only normalized data are shown in Figure 8.9a. The simulation estimates the shape of the mid bed temperature and the SO_3 vol. % variations successfully. It also reproduces the initial bed temperature lag for the first minute after the introduction of the SO_3/O_2 reactant mixure (Fig. 8.11b), as well as the absence of a lag when air is introduced to the catalyst bed displacing the reactant mixture (Fig. 8.11a). The slow adjustment of the bed temperature after the maximum and minimum temperatures is also given by the model although the rates of cooling and heating are not correct. The most serious deficiency of the model is that it over estimates the temperature rise and the drop by 15 and 8°C respectively.

Integration of Equation [8.6] for SO_2 in Table 8.4 estimates the conversion achieved. Simulation of periodic symmetrical switching between a reactant mixture and air gave an estimate of 99.4% at 12 minutes after the switch to the SO_3/O_2 reactant mixture in excellent agreement with the overall conversion of 98.8% measured by Briggs et al. (1977).

With respect to model sensitvity, it was found that bed mid point temperature was sensitive to the wall and gas to particle heat transfer coefficients. An extensive study of sensitivity, however, was not undertaken.

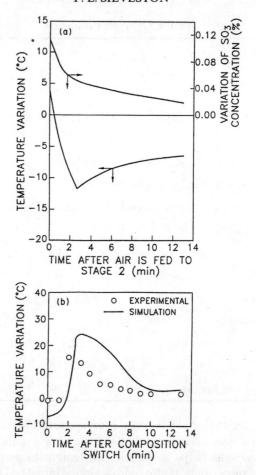

FIGURE 8.11 Simulation of the composition forcing experiment of Briggs *et al.* (1977) using the model of Table 8.4 with parameters from Table 8.5: (a) half cycle with air feed, (b) half cycle with SO_3/SO_2 feed. (Figure adapted from Silveston *et al.* (1990) with permission, © 1990 VNU, Utrecht, The Netherlands)

Silveston *et al.* conclude that their model represents the experimental behaviour well and recommend the use of the model for further exploration of the application of composition forcing to the final stage of a SO_2 converter. For example, the model should be capable of optimizing the cycle period and the cycle split as a function of mean bed temperature and the SO_3/O_2 ratio in the feed to the first stage.

Application To An Isothermal Backmixed Reactor

Strots et al. (1992) undertook a study of composition forcing employing the Balzhinimaev model given in Table 8.3 for the simplest reactor situation a back mixed reactor operating both isobarically and isothermally. These investigators also assumed equilibrium between the gas phase and the catalyst phase so heat and mass transport were neglected. With these assumptions, the equations in Table 8.4 simplify substantially to give:

$$(\varepsilon_B(1 + \varepsilon_p(1 - \varepsilon_B)) + \varepsilon_m H_i RT)(dx_i/dt) = ((x_i)_0 - x_i)/t_S$$
$$+ (\varepsilon_m C_V RT/P)(S\nu_{ik}r_k) \quad i = SO_2, O_2, SO_3 \quad (8.26)$$

$$dy_t/dt = S\mu_{jk}r_k \quad (8.27)$$

$$j = V_2^{5+}O_2^{=}, \ V_2^{5+}O_2^{-}V_2^{5+}SO_3^{=}, \ V_2^{4+}$$

where the index k signifies the reaction rates shown in Table 8.2. SO_3 is also stored in complexes present in the melt so that an additional term has to be added to the left side of Equation (8.26). This term is

$$((\varepsilon_m C_V K_L H_{SO_3} RT)/(1 + H_{SO_3} Px_{SO_3})(1/C_0) + K_L))^2)(dx_{SO_3}/dt)$$

Strots et al. used the back mixed reactor model to identify operating methods capable of improving the rate of this reaction for a given feed and set of operating conditions. They focused most of their effort on a forcing strategy using switching between an inert and the reaction mixture. The initial conditions expressing this strategy are:

$$n\tau \leqslant t \leqslant (n + 1/2)\tau: \quad (x_i)_0 = 0$$
$$n = 0, 1, 2, 3, \ldots \quad (8.28)$$
$$(n + 1/2)\tau \leqslant t \leqslant (n + 1)\tau: \quad (x_i)_0 = 2(x_i)_0$$

For their simulation of SO_2 oxidation, Strots et al. assumed $\varepsilon_B(1 + \varepsilon_p(1 - \varepsilon_B)) = 0.61$, $\varepsilon_m = 0.14$, $C_v = 0.002$, $C_o = 0.02$, $x_{SO_3} = 0$ and $x_{SO_2} = x_{O_2} = 0.1$. They examined the effect of temperature (673 to 798°K), space time ($0.01 \leqslant \tau_s \leqslant 4$ s) and cycling strategy on the rate enhancement. Resonance was observed for $\tau_s = 0.1$ s with $\Psi > 1$ for temperatures up to 770°K at a cycle period of about 1000 s. Above 770°K, $r_{qss} > r$ so that the resonance disappeared. The quasi steady state rate, r_{qss}, decreased rapidly with decreasing temperature below 770°K and this seemed to be responsible for the resonance observed.

Multiple resonance was observed at $T = 723°K$ and $\tau_s = 0.01$ s (Figure 8.12). The enhancement, Ψ, exceeds 1 at cycle periods between 0.2 and 0.3 and also at greater than 1000 s, although quasi steady state is also attained at about that period. Resonance also occurs at two different cycle periods at $\tau_s = 0.02$ s; Ψ is much smaller. As the space time increases above 0.25 s, the

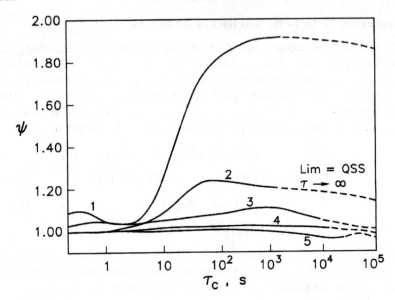

FIGURE 8.12 CSTR simulation of the effect of space time and cycle period on the enhancement factor at 723°K. Curves: (1) $\tau_s = 0.01\ s$, (2) $\tau_s = 0.02\ s$, (3) $\tau_s = 0.25\ s$, (4) $\tau_s = 1\ s$, (5) $\tau_s = 4\ s$. (Figure taken from Strots *et al.* (1992) with permission, © 1992, Elsevier Science Publishers)

rate enhancement becomes very small and the multiple resonance phenomenon seems to disappear.

The influence of cycling strategy on enhancement is given in Figure 8.13 for $T = 723°K$ and $\tau_s = 0.01\ s$. Curve (3) shows forcing by switching between an inert and a reactant feed, that is, the strategy given by the initial conditions expressed by Equation (8.15). This was the strategy used by Briggs *et al.* (1977, 1978) in the experiments discussed earlier in this chapter. Ψ is somewhat over 1.1 and compares to a curve lying between Curves (2) and (3) in the previous figure. Curves (4) and (5) in Figure 8.13 show the effect on enhancement of adding SO_3 to the reactor feed. To remain comparable to the other curves in the figure, the condition $x_{SO_2} + x_{SO_3} = 1$ and $(x_{SO_3} + x_{SO_2})/2 = 1$ is imposed. Figure 8.13 shows that for $0.09 \leqslant x_{SO_3}$ (corresponding to conversion of 90% of the SO_2 in the feed to a preconverter to SO_3) enhancement is negligible. However, enhancement is over 30% for $x_{SO_3} \geqslant 0.096$ (corresponding to 96% conversion of the SO_2 to SO_3) at a cycle period between 20 and 100 s. A multiple resonance seems to arise. This result recalls Briggs experiments on forcing the final stage of an SO_2 converter where enhancements of the order of 50% were observed at cycle periods in the range predicted by the Strots simulation.

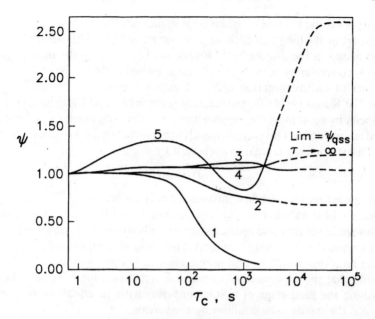

FIGURE 8.13 Influence of the cycling strategy and cycle period on the enhancement factor in a CSTR simulation at 723°K, $\tau_s = 0.1$ s. Curves: (1) SO_2 and O_2 concentration varied 180° out of phase, (2) only SO_2 concentration varied. (3) SO_2 and O_2 concentration varied in phase as given by Eqn. (8.15), (4) SO_3 present in the feed so that $x_{SO_2} + x_{SO_3} = 0.1$ and $(x_{O_2} + x_{SO_3})/2 = 0.1$ and $x_{SO_3} = 0.09$, (5) as in (4) but with $x_{SO_3} = 0.096$ (Figure taken from Strots *et al.* (1992) with permission, © 1992, Elsevier Science Publishers)

If the SO_2 and O_2 concentrations are switched 180° out of phase so that SO_2 is not in the reactor feed during one half cycle and O_2 is absent in the other half cycle, Figure 8.13 shows that Ψ is less than 1 regardless of the cycle period. Forcing just the SO_2 concentration at a constant O_2 concentration also fails to enhance the rate of SO_2 oxidation in the back mixed reactor. Even though the experiments of Unni *et al.* (1973), see above, were performed under isothermal conditions and differentially so that they could have been simulated by Strots' model, the strategy used by Unni was different from those investigated. Nevertheless, one of the experiments undertaken by Unni switched between a reactant mixture and a feed which did not contain SO_2. This experiment (at A = 0.6 in Figure 8.2) exhibited $\Psi < 1$. Strots' model predicts this observation.

Strots *et al.* (1992) studied the influence of the model kinetics on the enhancement under composition forcing and observed that Ψ significantly exceeded one when SO_3 desorbed faster than it decomposed due to the cycles involving the vanadium complexes. This suggested that one technique of raising the rate of oxidation is to increase the space velocity during the inert or flushing half cycle to strip out more SO_3. This possibility was also raised in the discussion of Briggs' study of forcing the final stage of a SO_2 converter by periodically flushing the bed in that stage with air. The results of an investigation of that technique appear in Figure 8.14 for $T = 770°K$ and $\tau_s = 4\ s$ for the reactant portion of a cycle. Raising the space velocity by shortening the flushing portion of the cycle keeps the same mean feed composition. This is conveniently represented by a dissymmetry factor, D. Dissymmetry is related to cycle split by

$$s = D/(1 + D) \tag{8.29}$$

The space time for inert flow through the catalyst bed is $(\tau_s)_{\text{inert}} = (\tau_s)_{\text{reactant}}/D$. Figure 8.14 shows that modulating the catalyst with a reactant mixture at a low space velocity and an inert stream at a high space velocity significantly increases the conversion attainable and may make it exceed the equilibrium limit that would be encountered in steady state operation. Figure 8.4 shows that the experimentally observed conversion limit for periodically flushing the final stage of a SO_2 converter with air equals or may even exceed the steady state equilibrium conversion.

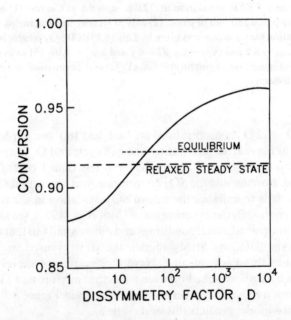

FIGURE 8.14 Influence of cycle split expressed as a dissymmetry factor and inert space velocity on the enhancement factor in a CSTR simulation at 723°K, $\tau_s = 0.1\ s$, SO_2 and O_2 concentration varied in phase as given by Eqn. (8.15) (Figure taken from Strots et al. (1992) with permission, © 1992, Elsevier Science Publishers)

Bunimovich et al. (1995) have applied the Balzhinimaev model to SO_2 oxidation over vanadia catalysts in a non-isothermal tubular reactor with periodic reversal of the flow direction. Once a stationary operation is reached, this mode of periodic operation feeds a cold SO_2 containing feed into one end of the reactor and it exits from the other end. The cold feed is heated by the catalyst bed until ignition occurs. The heat generated by the reaction is carried through the reactor heating the catalyst, particularly near the outlet which had been cooled by the previous half cycle. The front of the reactor is cooled by the entering cold feed which initiates a slowly moving cold front in the direction of flow. Before this cold front reaches the end of the reactor and extinguishes the reaction, the flow direction switches and cold feed now enters the end of the reactor. This becomes the front of the reactor in the new half cycle. Details of periodic flow reversal and its applicaiton to SO_2 oxidation are given by Boreskov et al. (1977), Boreskov and Matros (1983) and by Matros (1985, 1989).

Several commercial scale acid plants employing periodic flow reversal are operating in Eastern Europe and in Russia. They offer lower capital cost because heat recovery is performed by the catalyst bed or by a packed bed recuperator instead of a shell and tube heat exchanger. Flow reversal also provides a declining temperature profile after the maximum temperature is attained near the front of the bed. This is desirable for an exothermic, reversible reaction such as SO_2 oxidation.

A number of numerical simulations of oxidation with periodic flow reversal have been published (Boreskov and Matros, 1983; Matros, 1989; Bunimovich et al., 1990; Snyder and Subramanian, 1993). It is not our purpose to describe these here. Our interest is only in the use of the Balzhinimaev model. The earlier models just mentioned assumed that steady state kinetics apply under the slowly changing conditions of periodic flow reversal, but they over predict the time average SO_2 conversions by 1 to 3%. They also fail to predict an experimentally observed low level in the SO_3 content of the effluent after switching followed by a slow rise until the concentration exceeds slightly the SO_2 concentration in the convertor feed. After this maximum, SO_3 sinks slowly until it falls below the inlet SO_2 level. Allowing for the dynamics of the catalyst using the kinetic model given in Table 8.3, Bunimovich et al. (1995) demonstrate that the predicted conversions fall below those given by the steady state kinetic model. They also show that dissolution of SO_3 from the melt is slow and it is this that accounts for the initial low level of SO_3 in the effluent and its subsequent rise and overshoot.

These three applications of the Balzhinimaev model indicate that the model is robust and should be a promising tool for exploring the application of composition modulation to SO_2 oxidation over alkali promoted vanadia catalysts.

F. REDUCTION OF SO_3 BY CO OVER PLATINUM

Modulation of this reduction has been investigated by Olsson and Schöön (1985, 1986). Their work was motivated by the formation of SO_3 in catalytic converters from SO_2 in automotive exhausts. Sulphur containing effluent from converters results in sulphuric acid and sulphates mists when mixed with moist air and may constitute a health hazard. Pierson (1976) observed that modulation of the air/fuel ratio by the control loop used to maintain a stoichiometric mixture appears to suppress SO_3 formation relative to the steady state operation of the converter. Olsson and Schöön were interested in verifying Pierson's observation and examining how modulation might be controlled to lower SO_3 formation.

Experiments were performed in a single channel of a commerical Pt/Rh monolith catalyst mounted in a temperature-controlled furnace. The catalyst was pretreated for 40 h by alternating exposure to SO_2, O_2 in N_2 and CO in N_2 mixtures with a period of about 1 h (Olsson and Schöön, 1986). In all experiments, SO_2 partial pressures were about one thousandth of the O_2 partial pressure. Their earlier paper used Pt impregnated on the monolith support at different loadings and preteatment consisted of a short exposure to a dilute $CO + O_2$ mixture. The authors measured SO_2, CO and CO_2 in the gas leaving the monolith channel.

With Pt alone as the active catalyst, forcing consisted of switching between SO_2 and CO mixtures, each having the same O_2 concentration, at 180°C and a period of 8 min. This temperature is too low for ignition over Pt for the $CO + O_2$ mixture, but ignition did occur when the $SO_2 + O_2$ mixture was introduced. The SO_2 concentration in the off gas exhibited virtually an instantaneous peak and declined rapidly to zero. On switching to CO, another SO_2 peak was observed which took about 1 min to form and then slowly dropped towards zero SO_2 in the off gas over the next 3 minutes of the half cycle. The magnitude of this peak and the area under the curve increased proportionally with the Pt loading on the catalyst. These observations led Olsson and Schöön to propose that CO, SO_2 and SO_3 are strongly adsorbd on Pt. CO did not seem to inhibit SO_2 oxidation, rather adsorbed CO reduces the adsorbed SO_3. Assuming competitive adsorption and surface reactions between adjacent adsorbates, these researchers proposed a kinetic model based on the probability of reactant adsorbates being adjacent and demonstrated it reproduced step change measurement these researchers made. The model was not tested against modulation results.

Only the O_2 partial pressure was forced periodically when the Pt/Rh monolith was used (Olsson and Schöön, 1986). Using asymmetric forcing at periods of 180 to 300 s, a locked-in, periodic response at the forcing frequency was observed from 212 to 252°C as shown in Figure 8.15(b). The SO_2 peak appears when O_2 is taken out of the feed steam, but the SO_2

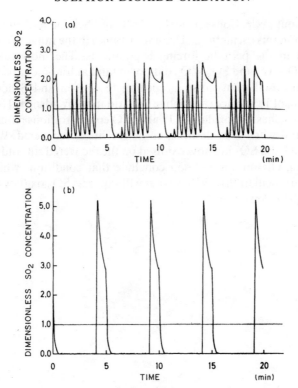

FIGURE 8.15 Periodic, asymmetric forcing of the oxygen partial pressure from 3.1 kPa to 0 with a constant flow of CO (at 1 kPa) and SO_2 (at 2.4 Pa) for $\tau = 300$ s, $s = 0.8$ (based on the relative duration of O_2 in the feed), $\tau_s = 0.16$ s over a Pt/Rh monolith catalyst: (a) response for $T = 242°C$, (b) response for $T = 212°C$ (Figure adapted from Olsssson and Schöön (1986) with permission, © 1986, American Chemical Society)

concentration drops quickly. O_2 is re-introduced after 60 s, that is at 5, 10, 15 minutes in the figure, and SO_2 is again oxidised to SO_3. The investigators attribute the peak to the reduction of the adsorbed trioxide by CO. An aperiodic response to forcing was seen, however, for $\tau = 300$ s and $T = 212°$. It is shown in Figure 8.15(a). In the portion of the cycle when O_2 is present in the feed, an irregular oscillation with a period of about 30 s appears. Modulation, apparently, initiates an autonomous oscillation of the SO_2 concentration. SO_2 formation in the absence of O_2 in the feed seems unaffected. This can be seen by comparing the areas under the SO_2 curves from 4 to 6, 9 to 11 and 14 to 16 min. in (a) and (b) of Figure 8.15. The SO_2 now produced through the oscillations corresponds to a 30% reduction in oxidation to the trioxide. CO conversion increases. Aperiodic behaviour on forcing is typical of kinetic systems exhibiting steady state bifurcations, such as a limit cycle. Consequently, Olsson and Schöön looked for steady state autonomous oscillations. These were found in the range of concentrations used in the periodic forcing experiments. The individual binary systems, $CO + O_2$ and $SO_2 + O_2$ did not exhibit autonomous oscillations. Steady state oscillations arise, according to Olsson and Schöön, from deactivation of Pt surface sites active for oxidation by adsorbed SO_3. This slows or quenches CO oxidation. Higher CO concentrations reduce adsorbed SO_3 and when SO_2 desorbs, an active site is recovered. When this happens, CO and SO_2 are now oxidised on the recovered site and the cycle starts again. Olsson and Schöön conclude that conditions which cause autonomous oscillations are those that will suppress SO_2 oxidation to SO_3 in the catalytic converter.

CHAPTER 9

Miscellaneous Reactions

This chapter completes our discussion of the use of composition modulation to activate a catalytic reaction. The examination of the literature is arranged more or less chronologically.

A. ETHYL ACETATE FROM THE ACID CATALYZED REACTION OF ETHYLENE AND ACETIC ACID

Hoechst AG investigated composition forcing of this reaction as a possible replacement for traditional processes such as esterification or the Tishtshenko process. Their experiments were carried out at 6 bar and 140 °C using sulfuric acid supported on silica (25 wt.% H_2SO_4) as a catalyst (Leupold and Renken, 1977, 1978; Leupold et al., 1979). Their reactor system differs from those illustrated in Chapter 1 and so it appears as Figure 9.1. The 17.8 mm (i.d.) by 1000 mm packed bed reactor was jacketed so as to operate isothermally. Acetic acid with 1 wt.% H_2SO_4, added to maintain catalyst activity, was brought to reaction temperature and fed to the reactor as a liquid. Under the operating conditions used, acetic acid vaporizes as it moves downward through the reactor so the reaction can be treated as a solid catalyzed, gas phase reaction. There are several ethylene side reactions (polymerization, hydrogenation and oxidation to form CO_2) as well as acid reduction to form SO_2. These reactions do not proceed to any significant extent so there is no selectivity problem. However, according to Leupold and Renken (1978), products of the side reactions are strongly adsorbed and act as poisons. The unit shown in Figure 9.1 was equipped with a solenoid valve in the HAc line to the preheater which permitted either steady state operation in the open position or composition forcing by starting and stopping the acetic acid flow. Ethylene was fed as a gas continuously to the reactor through a preheater. A diluent was not used so a small variation in space velocity accompanied the large periodic change in the acetic acid mol fraction in the feed. All forcing experiments were carried out around a time average HAc mol fraction of 0.0825. At this feed composition and at steady state, the addition reaction exhibited

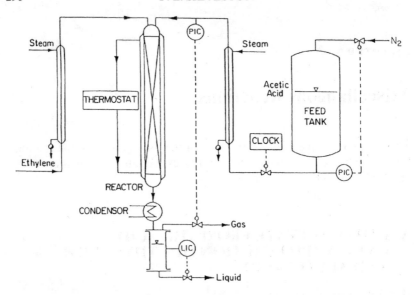

FIGURE 9.1 Schematic of the experimental apparatus used for modulated and steady state studies of the H_2SO_4/SiO_2 catalyzed formation of ethyl acetate from ethylene and acetic acid (Figure adapted from Leupold and Renken (1978) with permission, © 1978 Verlag Chemie)

a maximum in ethyl acetate production. The reaction was run in excess ethylene and HAc conversions were in the order of 90%.

In cycling runs, cycle periods, τ, varied from 4 to 30 minutes and cycle split, s, was kept between 0.167 and 0.333 where split is defined as the fraction of the cycle in which HAc passes through the reactor. Space velocity for all experiments was 3050 h^{-1} (STP). Leupold and Renken observed up to a 170% increase in ethyl acetate production through HAc flow interruption at 140 C for $s = 0.33$ (Fig. 9.2a). Cycle period seemed less important than split provided τ exceeded about 5 minutes. Figure 9.2a shows the effect of τ on the enhancement, Ψ, for cycle splits from 0.167 to 0.333. Enhancement is the time average ethyl acetate concentration in the reactor outlet divided by the steady state concentration at the time average feed composition. Because the forcing experiments were performed at a time average HAc mol fraction corresponding to the optimal steady state operation, it is the global enhancement, Ψ^*, that is given in Figure 9.2a. Although Ψ increases with s, even at $s = 0.167$, ethyl acetate production is enhanced by about 50%. At this split, there is little variation of Ψ with τ.

Measurements made under symmetrical cycling explain catalyst activation at these relatively long cycle periods. In Figure 9.2b, acetic acid (HAc) and ethyl acetate (EtAc) mol fractions in the reactor outlet are plotted versus time in a cycle after a stationary cycling state has been attained.

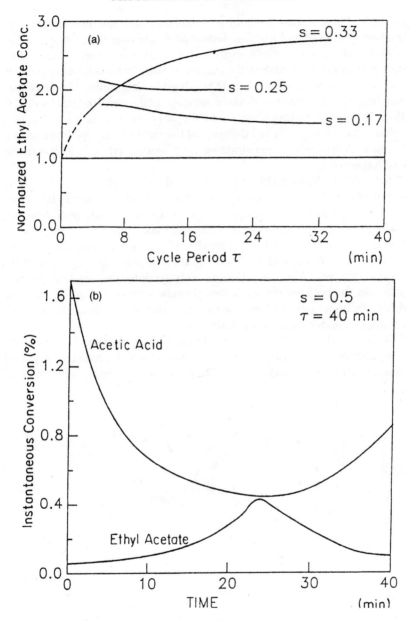

FIGURE 9.2 Ethyl acetate production under on-off cycling of the HAc feed to a fixed bed reactor packed with H_2SO_4 loaded SiO_2 catalyst operating at 6 bar and 140 °C with continuous flow of C_2H_4: (a) enhancement or normalized EtAc concentration versus cycle period for various cycle splits, (b) variation of educt and product mol fractions during a cycle for $\tau = 40$ minutes and $s = 0.5$ (Figure adapted from Leupold and Renken (1978) with permission, © 1978 Verlag Chemie)

Acetic acid flow was discontinud at $t = 0$ in the figure and resumed at $t = 20$ minutes. The HAc mol fraction decreased slowly over the 20 minutes half period and built up slowly over the same duration suggesting a slow displacement of adsorbate on the catalyst surface. EtAc behaves oppositely and reaches a maximum when HAc attains a minimum. The ester mol fraction is 7 × the maximum value achieved under steady state operation. Both the change during a cycle and the large ethyl acetate mol fraction enhancement indicate HAc inhibition of the reaction. Composition modulation improves reactor performance by providing ethylene access to the catalyst surface.

Leupold and Renken (1978) also observed that on-off cycling of the HAc feed to the reactor prevents deactivation of the catalyst. This can be seen in Figure 9.3. Deactivation measurements are for steady state operation and cycling the HAc feed at $\tau = 20$ minutes and $s = 0.25$. Ethyl acetate production data have been normalized by the steady state production after the deactivation has stopped at ca. 500 hours after start up. Startup in these experiments was from an ethylene only feed to the reactor. Activity increases for both steady state and periodic operation initially and the difference in activity between the two operations is small at the time the production under steady state reaches a maximum.

Renken and co-workers continued their investigation of the HAc addition reaction and developed a dynamic model which predicts differential reactor performance under composition forcing reasonably well (Dettmer

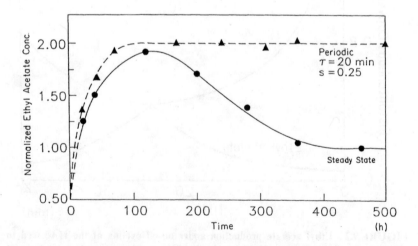

FIGURE 9.3 Comparison of normalized ethyl acetate production versus time-on-stream for steady state and on-off cycling of the HAc feed at 6 bar and 140 °C over a H_2SO_4 loaded SiO_2 catalyst with $SV = 3050 \, h^{-1}$. On-off modulation of the HAc feed used $\tau = 20$ minutes and $s = 0.25$ (Figure adapted from Leupold and Renken (1978) with permission, © 1978 Verlag Chemie)

and Renken, 1983a, 1983b; Walker & Renken, 1983, 1984; Renken et al. 1984). Based on the cyclic results summarized above, Renken concluded that ethylene absorption in the supported acid phase is rate limiting and that absorption is inhibited by HAc. The mechanism is complicated and could involve ethyl sulfate, protonated ethylene, or the ethyl ester of a sulfonic acid. One or the other of these solute species reacts with HAc dissolved in the acid phase to form the ester which then desorbs. Measurement of the sorption rate of ethylene as a function of acetic acid loaded per mass of catalyst indicated inhibition is a strong function of the HAc concentration (Walker and Renken, 1984). Renken and co-workers model inhibition as an exponential function of the HAc concentration (Renken et al. (1984)). For modelling, the ethylene intermediates in the acid phase are assumed to be in equilibrium with dissolved ethylene. A differential reactor was used to test the model so it was formulated for a continuous stirred tank reactor. Table 9.1 presents the Dettmer-Renken dynamic model. In the table, P_i is the partial pressure of a reactant or product, C_i is the concentration in the acid phase, θ is dimensionless time normalized by the space time, k_1 and k_2 are the absorption and desorption rate constants for ethylene, k_3 and k_4 are the absorption and desorption rate constants for HAc, k_5 is the reaction rate constant in the acid phase, k_6 is the inhibition constant, and $(C_A)^{sat}$ is HAc concentration corresponding to the maximum solubility of HAc in H_2SO_4.

In the model, $(C_A)^{sat}$ is the solubility limit of HAc in the supported sulfuric acid. Like the seven rate constants, it must be determined experimentally.

Table 9.1 Dettmer-Renken Model for HAc Addition to Ethylene Over a Supported H_2SO_4 Catalyst

$$\frac{dP_E}{d\theta} = (P_E)_0 - P_E + k_1 k_6 P_E \frac{[(C_A)^{sat} - C_A]}{[C_A - k_6(C_A)^{sat}]} + k_2 k_6 C_E \frac{[(C_A)^{sat} - C_A]}{[C_A - k_6(C_A)^{sat}]} \quad (9.1)$$

$$\frac{dC_E}{d\theta} = k_1 k_6 P_E \frac{[(C_A)^{sat} - C_A]}{[C_A - k_6(C_A)^{sat}]}$$

$$+ k_2 k_6 C_E \frac{[(C_A)^{sat} - C_A]}{[C_A - k_6(C_A)^{sat}]} - k_5 C_A C_E \quad (9.2)$$

$$\frac{dP_A}{d\theta} = (P_A)_0 - P_A - k_3 P_A [(C_A)^{sat} - C_A] + k_4 C_A \quad (9.3)$$

$$\frac{dC_A}{d\theta} = k_3 P_A [(C_A)^{sat} - C_A] + k_4 C_A - k_5 C_A C_E \quad (9.4)$$

$$\frac{dP_{EA}}{d\theta} = -P_{EA} + k_5 C_A C_E \quad (9.5)$$

Dettmer and Renken (1983a, b) evaluated the constants by a series of equilibrium absorption, steady state and step change response experiments. With these constants, the model predicted closely responses to step changes in C_2H_4 and HAc feed rates to an integral fixed bed reactor. Figure 9.4 compares the model predictions to measurements for a step-up in HAc from no flow and for the symmetrical step-down. The agreement is good

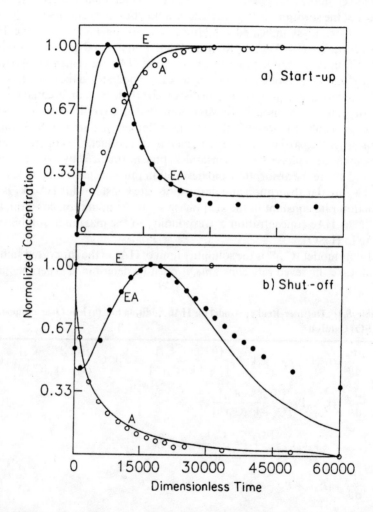

FIGURE 9.4 Comparison of predicted and measured responses to step changes in acetic acid partial pressure for the H_2SO_4 catalyzed addition of HAc to C_2H_4 to form ethyl acetate. Model is presented in Table 9.1 and measurements were performed at 100°C, SV = 1.5 mL/g cat. s, P_E = 56 mbar, and P_A was switched between 0 and 10.1 mbar (Figure adapted from Dettmer and Renken (1983b) with permission, © 1983 Verlag Chemie)

considering that no fitting of the experimental data was used to evaluate the model constants. These were obtained from different experiments.

Truffer and Renken (1986) undertook both transient and cycling experiments on the acid catalyzed HAc-C_2H_4 system using a differential fixed bed reactor and introducing acetic acid in a N_2 diluent stream saturated with the acid. By diverting the N_2 diluent stream to the reactor without first passing through the saturator, space velocity was held constant in the step change and forcing experiments. Truffer and Renken improved the model given in Table 9.1 by assuming adsorption is better described by an Elovich isotherm than by a Langmuir one. This was accomplished by replacing the inhibition terms, containing $(C_A)^{sat}$, in Equations (9.1) and (9.2) in the table by an exponential term, $\exp(-bC_A)$. Truffer and Renken re-evaluated the model constants from the same type of measurements used by Dettmer and Renken for the model given in Table 9.1. With these changes, Truffer and Renken were able to simulate the start-up of their experimental reactor under on-off modulation of the HAc partial pressure reasonably well. Their simulation is compared to experimental data in Figure 9.5. The dot–dashed curve shows the experimental measurements of the ethyl acetate partial pressure. They are about 20% below predicted values just before the step-up in HAc partial pressure. As has been mentioned in earlier chapters, simulating composition modulation is a severe challenge for a reactor model. We will discuss this further in Chapter 15. Figure 9.6 shows the

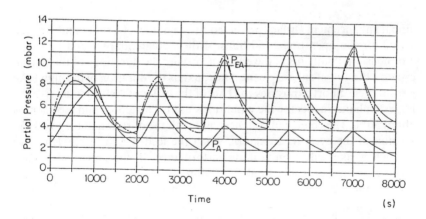

FIGURE 9.5 Comparison of predicted and measured HAc and ethyl acetate partial pressures in the start-up of a differential reactor from a C_2H_4 only feed for the H_2SO_4 catalyzed addition of HAc to C_2H_4 to form ethyl acetate. Model is presented in Table 9.1 but the Langmuir isotherm has been replaced by the Elovich isotherm (Truffer and Renken, 1986). Predicted partial pressures are shown as solid curves. Measurements were performed at 110 °C, P = 1.05 bar, P_E = 68 mbar, and P_A was switched across a mean of 10 mbar. Data are given by the dot-dash curve. (Figure adapted from Truffer and Renken (1986) with permission, © 1986 *AIChE*)

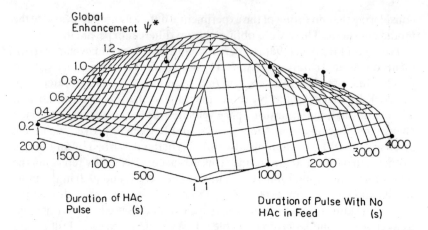

FIGURE 9.6 Comparison of predicted and measured time average global enhancements for a differential reactor running the H_2SO_4 catalyzed addition of HAc to C_2H_4 to form ethyl acetate. Model is presented in Table 9.1 except that the Elovich isotherm is used in place of the Langmuir isotherm (Truffer and Renken, 1986). Predicted values are shown as curves. Measurements were performed at conditions given in Figure 9.5. Data are given by the solid points in the figure. Vertical lines connect measured and predicted enhancements (Figure adapted from Truffer and Renken (1986) with permission, © 1986 *AIChE*)

global enhancement surface predicted by the model of Truffer and Renken (1986) for the operating conditions of their differential packed bed. The 3-dimensional plot shows the effect of varying the durations of the partial cycle on the global enhancement. The t_1 axis is the duration of the high HAc partial cycle. Short exposures to HAc, of the order of 500 seconds, give the largest enhancements for the conditions used. The global enhancement is about 30%. The change of enhancement with the duration of the low HAc portion of the cycle (t_2 axis), e.g., when the feed is rich in ethylene, has a much weaker effect on reactor performance. In the region of highest enhancement, the Truffer-Renken version of the model in Table 9.1 overpredicts the enhancement by just 10%. The largest over-prediction is about 20%.

Renken *et al.* (1984) and Truffer and Renken (1986) have used the model just discussed to predict optimal operation of the acid catalyzed HAc-C_2H_4 reaction. They find that performance under HAc modulation depends strongly on the sorption rate of HAc. At high rates, Ψ^* can be 75% greater than steady state ethyl acetate production. Enhancement also depends upon the time average HAc partial pressure. Increasing the mean partial pressure from 10 mbar (Fig. 9.6) to 740 mbar raises the predicted Ψ^* from ca. 30 to about 50%.

B. CLAUS REACTION

Composition forcing experiments on this reaction over a commercial bauxite catalyst are described by El Masry (1985). An isothermal, differential packed bed was used at 107.54 kPa and 483 K in an experimental arrangement illustrated by Figure 9.7. The $H_2S:SO_2$ ratio was modulated at cycle periods between 2 and 12 minutes; a N_2 diluent was used so the space velocity was kept constant. Other variables in the El Masry study were the cycle split, defined as the fraction of the cycle period which the feed is H_2S-rich, and the time average mol fraction of H_2S in the feed. Hudgins et al. (1988) re-analyzed El Masry's experimental data but corrections made did not change the conclusion of the El Masry paper which is that composition forcing of the Claus reaction leads to a significant enhancement of the reaction rate.

Time average and steady state H_2S consumption rates are plotted versus the time average H_2S mol fraction in the reactor feed in Figure 9.8a. Modulation data are shown in the figure and the dashed line passes through

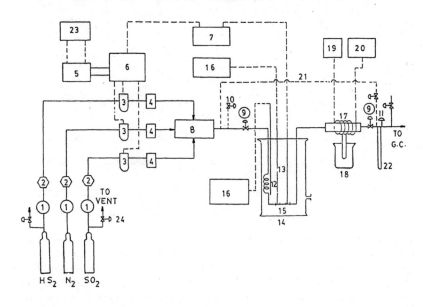

FIGURE 9.7 Schematic of the system used to study the Claus reaction: 1 – pressure regulators, 2 – macro filters, 3 – transducers, 4 – control valves, 5 – potentiometer, 6 – flow display, 7 – temperature data logger, 8 – mixing manifold, 9 – pressure gauges, 10 – safety valve, 11 – needle valve, 12 – preheater, 13 – thermocouples, 14 – fluidized sand bath, 15 – reactor, 16 – temperature controller, 17 – sulfur condenser, 18 – water condenser, 19 – variac and 20 – temperature controller, 21 – feed by pass, 22 – manometer, 23 – automatic timers (Figure taken from El Masry (1985) with permission, © 1985 Elsevier Science Publishers)

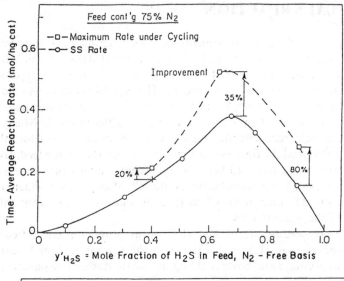

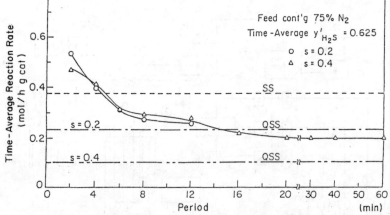

FIGURE 9.8 Composition modulation of the $H_2S:SO_2$ ratio in the Claus reaction carried out over a bauxite catalyst in an isothermal, differential reactor at 107.54 kPa and 483 K with 75% (time averaged) N_2 diluent in the feed: (a) comparison of time average and steady state H_2S conversion rates versus time average H_2S mol fraction in the reactor feed, (b) time average H_2S conversion rate versus cycle periods at a time average mol fraction H_2S of 0.625 in the feed (- - - and —-— lines are steady state and quasi-steady state rates respectively) (Figure adapted from Hudgins *et al.* (1988) with permission, © 1988 Elsevier Science Publishers)

the maximum rates measured in this mode. The solid curve is a statistical fit of El Masry's steady state data (Hudgins *et al.*, 1988). Maximum of the modulation data appears to coincide with the location of the steady state maximum. The experiments, thus, indicate about a 35% global enhancement of the H_2S conversion rate. Enhancement reaches about 80% in the H_2S rich region, but drops to 20% in the region that is SO_2 rich. Time average rates well above and well below the steady state rates indicates that the cycling variables are important. This is evident from Figure 9.8b which plots these rates as a function of cycle period for two cycle splits. The horizontal lines show the steady state and quasi-steady state rates at the time average mol fraction of H_2S in the feed. Quasi-steady state is approached at cycle periods greater than 8 minutes, while the optimal periodic performance requires cycle periods less than 2 minutes. Two minutes was the limit of El Masry's experimental equipment. The Claus reaction is believed to be poisoned by sulfur dioxide. Hudgins et al. show data for prior exposure of the catalyst to pure SO_2 which support this view. Thus, it is surprising that Figure 9.8b indicates that small cycles splits yield the largest rate enhancement. The explanation appears to be that SO_2 sorption is a slow step in the reaction sequence.

C. DEHYDROGENATION OF METHANOL

This reaction, producing formaldehyde and hydrogen, is catalyzed by sodium carbonate supported on activated carbon or on sodium aluminate. It is being studied by Renken and co-workers (Su et al., 1992, 1994; Zasa et al., 1994) as an alternative to the commercial process: partial oxidation of methanol over a $Fe-Mo-V_2O_5$ or a Ag catalyst. Step change methods have been used (Zasa et al., 1994) to explore reaction mechanism and to identify rate limiting steps in the dehydrogenation. Formaldehyde appears to form from an adsorbed methoxy species, CH_3O-s, created by the dissociative adsorption of CH_3OH. Atomic hydrogen formed in the dissociation seems strongly adsorbed so that its desorption limits the rate of formaldehyde formation.

Although there are no published accounts of modulation experiments on this dehydrogenation reaction, it would seem to be a candidate for a composition modulation process using stop flow or periodic scavenging of adsorbed H_2 with air or another oxidant.

D. DEAMINATION AND ALCOHOL DEHYDRATION REACTIONS

Experimental

At temperature above 250°C over alumina, thoria or chromia (oxide) catalysts, primary amines split out NH_3 and form the corresponding olefin. A disproportionation reaction occurs also in which a diamine is formed.

Under steady state, these reactions occur at about the same rate at 325°C. Over the same catalysts but above 150°C, analogous reactions occur with the alcohols splitting out water and forming either an olefin or an ether. Stoichiometrically, the olefin forming reactions are:

$$RCH_2CH_2NH_2 \rightarrow RCH=CH_2 + NH_3$$

$$RCH_2CH_2OH \rightarrow RCH=CH_2 + H_2O$$

Because both the disproportionation and dehydration or deamination reactions can proceed over the oxide catalysts, a selectivity problem arises. We discuss these two reactions here rather than in Chapter 11 because published forcing studies consider only activation of deamination and dehydration. The effect on selectivity, though large, has not been treated in the literature.

An unpublished Ph.D. dissertation (Hogan, 1973) reported an abrupt increase of 2 to 3 × in the rate of propylene formation from the deamination of diisopropylamine over alumina after the flow of the amine-N_2 mixture to the catalyst was replaced by N_2 alone (Koubek et al., 1980a). After the maximum, the deamination rate declined to zero over a time span of about an hour. Koubek et al. refer to this behaviour as the "stop effect" and followed up Hogan's work with a broad experimental program using C_3 to C_4 primary, secondary and tertiary amines and ethyl, propyl and isopropyl alcohol over decationized zeolites as well as alumina, chromia and thoria oxides, and binary mixtures of the oxides (Koubek et al., 1980a; 1980b). There are other mentions of the "stop effect" in the literature (Koubek et al., 1980a), but only Koubek et al. seem to have given the phenomenon thorough study.

Figure 9.9a shows successive start-stop steps for propylamine in a differential reactor packed with an alumina catalyst and illustrates the remarkable magnitude of the stop-flow effect. After removing the amine from the feed, the rate of propene formation increase 10 ×, reaching a maximum some 40 minutes later. Table 9.2 shows the magnitude of the "stop effect" for different amines and alcohols. For the former, the magnitudes are compared to the steady state rates of olefin production and an estimate of the site blockage responsible for the low steady state rates is given. The magnitude is defined as the maximum rate of olefin formation divided by the rate of formation under steady state operation. The stop effect is observed only for the primary and secondary amines. There appear to be steric effects as well. Magnitude of the "stop effect" is sensitive to temperature and decreases as temperature increases. Time needed to reach the maximum rate also decreases with temperature. The magnitude, on the other hand, is independent of the amine partial pressure. Disproportionation reactions were observed as well, but the rates decrease after the amine or alcohol is taken out of the reactor feed. The "stop effect" was observed

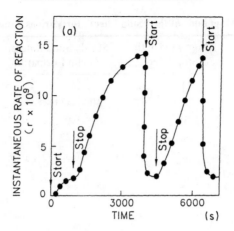

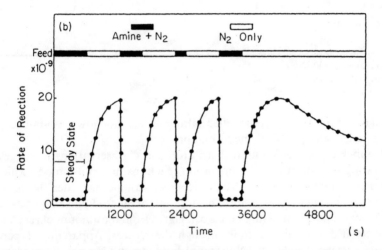

Figure 9.9 Variation of the rate of deamination for successive addition to and removal of propylamine to the N_2 feed to a differential packed bed reactor: a) with an alumina catalyst at 580.5°K and $P_{amine} = 11$ kPa in N_2 at steady state, b) with a decationized 13X zeolite catalyst at 573°K (Figures adapted from Koubek et al. (1980b) with permission, © 1980 Elsevier Science Publishers)

over a chromia catalyst for ethyl alcohol and propyl amine, for silica and zeolites only for propylamine, but it was not observed for the thoria catalyst. On an alumina-silica catalyst, the maximum rate of propylene formation reached 50× the steady state rate after stopping the propylamine feed. This was the largest "stop effect" seen. The decay of the propylene concentration in the inert flowing through the reactor to a negligible level lasted up to 6 hours. When ethanol was fed in place of the

Table 9.2 Magnitude of the "Stop Effect" for Various Amines and Alcohols

Amine/Alcohol	Magnitude of the "Stop Effect"	Steady State Rate of Olefin Formation	Estimated Site Blockage – %
Propylamine*	11.5	1.14×10^{-9} gmoles/cm^3 s	95
Butylamine	10.5	0.99	96
Isobutylamine	18.0	0.32	96
Isopropylamine	9.5	10.9	92
N-methylisopropyl amine	1.7	25.5	52
N,N-Dimethyl isopropyl amine	1.0	50.5	12
Ethyl alcohol**	6.1	–	–
Propyl alcohol	6.5	–	–
Isopropyl alcohol	1.5	–	–

*Data taken at 300°C
**Data taken at 187°C

primary amine, the maximum after halting the alcohol flow was about 5 × the steady state rate of ethene formation.

Koubek et al. (1980b) used their "stop effect" observations to suggest operating dehydration or deamination reactions in a pulse mode which consists of introducing a pulse of alcohol or amine just after the olefin formation reaches a maximum. The duration of the feed pulse depends on the reactant partial pressure and is calculated from the amount of reactant consumed during diluent flow through the reactor. Apparently, experiments were performed on cyclohexanol, but the results do not seem to have been reported. Koubek et al.(1980b) undertook successive pulsing experiments with propylamine on a decationized 13 × zeolite. Although the time-average rate enhancement was not reported, Figure 9.9b indicates it must about 10.

The "stop effect" was employed by Koubek and co-workers to investigate the deamination and dehydration mechanism. Through the use of selective poisoning of the Lewis acid and basic sites on alumina, Koubek et al. (1980b) concluded that the deamination reaction proceeds via the adsorption of the NH_2 or NH end of the molecule on an acid site (a surface Al) and the subsequent or simultaneous adsorption of the opposite end on an adjacent basic site (O^{2-} or OH^-). Rearrangement, probably involving H migration from the β carbon to give the olefin precursor, is followed by cleavage of the C–N bond at the acid site and cleavage of the C–H bond at the basic site and olefin desorption. The "stop effect" arises because the

NH$_2$ end also adsorbs on basic sites via hydrogen bonding. Adsorption is rapid but it is weak and thus reversible. Nevertheless, the basic site is blocked in this way and can no longer participate with Lewis acid site in concerted deamination. Abruptly decreasing the amine concentration by cutting off the amine feed results in desorption of the amine from the basic sites freeing these sites for the concerted deamination reaction of molecules adsorbed on adjacent Lewis acid sites. Removing amine from the feed decreases the amine adsorbed on the acid sites only slightly as these are strongly bound. The bound amine is captured by the O^{2-} or OH$^-$ site permitting deamination to proceed. Desorbing olefin frees both the acid and basic sites. Indeed, Koubek et al. (1980b) hypothesize that Lewis acid sites store the amine and the deamination rate is controlled by the availability of basic sites. These authors suggest a similar mechanism accounts for alcohol dehydration and the "stop effect" observed with that reaction. Variations of the "stop effect" for different oxides as well as for different amines on the same oxide indicate a reaction sensitive to site geometry and/or distribution.

Koubek and co-workers proposed a model for the sequence described above which adequately represents the rapid increase in rate after removing reactant from the feed and later the gradual drop in rate as only an inert flows through the reactor (Koubek et al., 1980a). A simpler form results if the acid-base site pair are treated as a site itself. Koubek and his co-workers found that the simpler form gave an equally good representation of the "stop effect" behavior.

Modeling

The two-site mechanism of Koubek et al. was formulated as a model by Nowobilski and Takoudis (1986) who used the model to argue that a steady state operating condition can be found for which rates of reaction will equal or exceed any time average rate under composition modulation. The mechanism as a set of elementary reactions is shown in Table 9.3. The model written for a differential reactor by Thullie and Renken (1990) is given in Table 9.4. These investigators examined its steady state and step change predictions. Koubek et al. (1980a) proposed a second and simpler conceptual mechanism to explain their observations. It is also given in Table 9.3. This mechanism expressed through material balances for a differential reactor by Thullie and Renken (1991) appears as Table 9.5. Thullie and Renken (1991) employed both models to optimize modulation of the deamination and alcohol dehydration reactions.

Using either the model in Table 9.4 or 9.5, the behaviour shown at the far right in Figure 9.9b can be reproduced qualitatively. This is demonstrated by the curve for $k_{-2} = 1$ in Figure 9.10. This curve assumes $k_2 = 100$, the formation of the olefin by step 3 is slow and adsorption in step 1 is more

Table 9.3 Mechanisms for the "Stop Effect"

Koubek I Version (Nowobilski and Takoudis, 1986)	Koubek II Version (Thullie and Renken, 1991)
$A + s_1 \rightleftarrows A - s_1$	$A + s \rightleftarrows A - s$
$A + s_2 \rightleftarrows A - s_2$	$A - s + A \rightleftarrows A - s - A$
$A - s_1 + s_2 \rightarrow B + C + s_1 + s_2$	$A - s \rightarrow B + C + s$

where A = amine or alcohol, s = acid-base site, B = olefin, $C = NH_3$ or H_2O

rapid than desorption. As the figure shows, the behaviour depends on the rate of desorption in step 2. If it is rapid, there is virtually an instantaneous jump in the reaction rate when amine or alcohol is taken out of the feed as can be seen in Figure 9.10.

Thullie and Renken (1991) solve the models in both tables for quasi and relaxed steady state assuming that the second step proceeds so rapidly that equilibrium is established. These authors found that the time-average deamination or dehydration rates are highest for the relaxed steady state. This is hardly a surprise as Figure 9.10 shows that for the equilibrium assumption the maximum production rate occurs at the time when reactant is withdrawn from the feed. There is an instantaneous drop in the rate when the reactant is withdrawn. Figure 9.11a plots the global enhancement for

Table 9.4 "Stop Effect" Model for Deamination or Alcohol Dehydration in a Modulated Differential Reactor (Thullie and Renken, 1990)

$$\frac{dP_B}{dt} = k_3(\theta_1)_A(1 - \theta_2) \tag{9.6}$$

$$\frac{d(\theta_1)_A}{dt} = k_1 P_A(1 - \theta_1) - k_{-1}(\theta_1)_A - k_3(\theta_1)_A(1 - \theta_2) \tag{9.7}$$

$$\frac{d(\theta_2)_A}{dt} = k_2 P_A(1 - \theta_2) - k_{-2}(\theta_2)_A \tag{9.8}$$

$$\theta_1 + (\theta_1)_A = 1 \text{ and } \theta_2 + (\theta_2)_A = 1 \tag{9.9}$$

Initial conditions: $t = 0$: $P_A = (P_A)_o,\ \theta_1 = \theta_2 = 1$

$(n-1)\tau \leqslant t \leqslant (n-1)\tau + s\tau$: $P_A = 0$ \hfill (9.10)

$(n-1)\tau + s\tau \leqslant t \leqslant n\tau$: $P_A = (P_A)_o$

Continuity conditions: $t = n\tau, n\tau + s\tau$: $(\theta_1)^- = (\theta_1)^+$, $(\theta_1)_A^- = (\theta_1)_A^+$,

$$(\theta_2)^- = (\theta_2)^+,\ (\theta_2)_A^- = (\theta_2)_A^+, \tag{9.11}$$

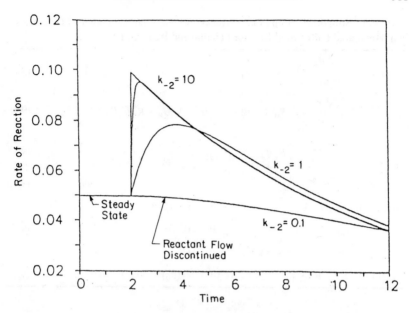

FIGURE 9.10 Predicted response of reaction rate to removal of reactant from the feed stream for the simplified "stop effect" model given in Table 9.5 with $k_1 = 1000$, $k_{-1} = 0.001$, $k_2/k_{-2} = 100$ and $k_3 = 0.1$ (Figure taken from Thullie and Renken (1991) with permission, © 1991 Pergamon Press PLC)

the bang-bang modulated operation as a function of the amine or alcohol concentration and the cycle split using the simplified model (Table 9.5). Choice of cycle split is important. For the figure it is defined as the fraction of the cycle in which only diluent flows to the reactor. The optimal split increases with reactant concentration and is close to one for most concentrations shown in the figure. Enhancement predictions for the two site model (Table 9.4) appear as a dashed line for $s_{optimal}$. These are lower than those for the model of Table 9.5. Nevertheless, they are greater than one for the range of reactant concentrations in the figure. Nowobilski and Takoudis (1986) used the same kinetic model, but did not find any global enhancement. However, these investigators did not make an equilibrium assumption for step 2 and their model was formulated for a back mixed reactor. Steady state rates are represented by the $s = 0$ curve so Figure 9.11a shows very large enhancement for $P_A/P > 0.02$ even though Ψ^* is small or less than one. There is thus agreement with Nowobilski and Takoudis in principle. Thullie and Renken mention that if equilibrium is not assumed for the second step, enhancements are smaller and frequent switching, the relaxed steady state, is not optimal.

Global enhancement at the relaxed steady state limit is sensitive to the choice of model parameters. The effects of parameter choice are explored in

Table 9.5 Simplified "Stop Effect" Model for Deamination or Alcohol Dehydration in a Modulated Differential Reactor (Thullie and Renken, 1991)

$$\frac{dP_B}{dt} = k_3 \theta_A \tag{9.12}$$

$$\frac{d\theta_A}{dt} = k_1 P_A (1-\theta) - k_{-1}\theta_A - k_3 \theta_A - k_2 P_A \theta_A + k_{-2}\theta_{AA} \tag{9.13}$$

$$\frac{d\theta_{AA}}{dt} = k_2 P_A \theta_A - k_{-2}\theta_{AA} \tag{9.14}$$

$$\theta + \theta_A + \theta_{AA} = 1 \tag{9.15}$$

Initial conditions: $\quad t=0$: $\quad P_A = (P_A)_o$, $\theta = 1$

$$(n-1)\tau \leqslant t \leqslant (n-1)\tau + s\tau: \quad P_A = 0 \tag{9.16}$$

$$(n-1)\tau + s\tau \leqslant t \leqslant n\tau: \quad P_A = (P_A)_o$$

Continuity conditions: $t = n\tau, n\tau + s\tau$: $(\theta_A)^- = (\theta_A)^+$, $(\theta_{AA})^- = (\theta_{AA})^+$ \quad (9.17)

---

the Thullie and Renken study. With an appropriate choice, enhancements exceeding 140% are possible (Fig. 9.11b).

E. OVERVIEW

Two of the reactions considered in this chapter, acid catalyzed addition of C_2H_4 to HAc and deamination, have been successfully modeled. In the ethyl acetate reaction, the model predicts substantial rate enhancement and, indeed, large enhancements were measured experimentally. The reaction, itself, appears to be an attractive alternative to existing technology and composition modulation clearly improves the acetate production rate with respect to steady state operation. It seems from the first publication (Leupold and Renken, 1977; 1978) and patents (Leupold et al., 1979) that Hoechst AG undertook their investigation for commercial purposes. Why was it not brought further along considering the advantages the modulated process offered?

Further study of the deamination and alcohol dehydration reactions would be worthwhile. Some of this is possible using the models developed so far. The first question which needs addressing is the possibility of global enhancement. Considering a fully backmixed reactor, Nowobilski and Takoudis (1986) find that there is no global enhancement through on-off modulation of the reactant, whereas Thullie and Renken (1991) find global enhancement over a wide range of reactant concentrations. That the latter investigators considered a differential reactor should not be important; it is

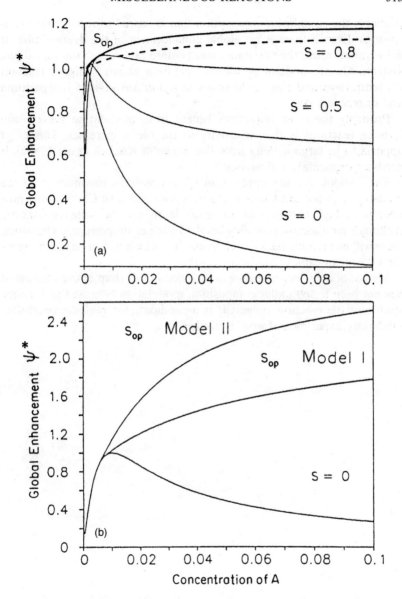

FIGURE 9.11 Dependence of the global enhancement on reactant mole fraction in the feed when reactant is present. Solid curves are for the simplified "stop effect" model (Table 9.5); dashed curves are for two site "stop effect" model (Table 9.4). Curve $s = 0$ represents steady state operation: a) rate constants are those given in Fig. 9.10; b) as in a) except $k_1 = 10$ (Figure taken from Thullie and Renken (1991) with permission, © 1991 Pergamon Press PLC)

probably the equilibrium assumption that is crucial to the different conclusions. Even if there is no global enhancement of the deamination or dehydration rates, the maximum steady state rate occurs for very dilute systems. On-off modulation allows operation at much higher reactant concentrations and this could result in significant savings in separation and/or recovery costs.

Probably the most important benefit from modulating these olefin forming reactions is that selectivity to the olefin increases. This effect appears to be large judging from the results of Koubek *et al.* (1980a). It needs experimental confirmation.

The outlook for the application of composition modulation to the industrially important Claus reaction is poor. Modern Claus plants now operate at high conversions. Emission levels are the primary concern. Although modulation probably results in lower time-average emissions, there will be periodic peaks in H_2S and SO_2 which would be much higher than the mean values. This is undesirable.

Methanol dehydrogenation was included in this chapter even though it has not been tested under composition modulation because the dynamic studies of the reaction suggest it is a candidate for periodic operation. Obviously, experimental work is needed.

CHAPTER 10

Catalytic Gas – Solid Reactions

A. INTRODUCTION

Systems considered heretofore have involved a solid phase functioning as a catalyst. We now turn to systems in which there are several solid phases: one of which is a reactant while another functions as a catalyst. Composition modulation has been applied to just one such gas-solid reaction system: the catalytic gasification of carbonaceous solids. The objective was to elevate the catalyst into its most active state, just as it has been for many of the catalytic systems considered in earlier chapters.

Composition modulation has also been applied also to the apparently non-catalytic decomposition of gypsum and anhydrite (Swift and Wheelock, 1975; Morris et al., 1987; Wheelock and Riel, 1991) and regeneration of sulfided adsorbents (Jagtap and Wheelock, 1996). The non-catalytic applications will be reviewed briefly from the standpoint of the technology explored.

B. NON-CATALYTIC GAS-SOLID REACTIONS

Decomposition of Calcium Sulfate, Gypsum and Phosphogypsum

Large quantities of low grade gypsum ($CaSO_4 \cdot 2H_2O$) are generated in the production of industrial acids, treating of acidic wastewater and in some stack gas scrubbing systems. A phosphogypsum is generated also in vast amounts as a byproduct in the production of phosphate fertilizers. Most of this waste ends up as landfill because processing to recover building grade gypsum or Portland cement, in the case of the phosphogypsum, cannot be justified economically. There has been interest for many years in reclaiming quicklime and sulfuric acid (*via* SO_2) from gypsum wastes and the growing scarcity of landfill sites has only strengthened this interest in the last decade. Recovery is possible by reductive decomposition at temperatures around 2000 °C.

The primary reaction in reductive decomposition is the formation of calcium oxide and CO_2 or H_2O or both depending on the reductant used. Sulfur is recovered as SO_2. The reduction proceeds through various mineral phases and sulfide is probably an intermediate. Indeed, it is often recovered with calcium oxide and considered as a contaminant. Reduction takes place at 1000 °C and rates become rapid above 1050 °C. According to Swift and Wheelock (1975), strongly reducing conditions and temperatures below 1100 °C favor sulfide rather than the oxide. Oxide formation is endothermic, whereas the sulfide reaction is exothermic. Sintering limits reduction temperatures to 1200 °C or below. Endothermicity and the rather narrow permissable temperature range dictate the use of a fluidized bed for reductive decomposition. Nevertheless, even when a fluidized bed reactor is employed, the calcium sulfide persists as an impurity in the lime produced and effects the economics for some applications.

Wheelock and co-workers have demonstrated that the sulfide impurity can be eliminated by using fuel injection to create oxidizing and reducing zones in the fluidized bed, or by periodically switching the air-fuel ratio in the fluidizing gas so that the entire bed becomes alternately oxidizing and then reducing. The two zone bed was studied on a bench scale (Swift and Wheelock, 1975; Morris *et al.*, 1987) and in a small pilot plant (Wheelock and Morris, 1986), while the periodically operated fluidized bed was investigated just on a bench scale (Wheelock and Riel, 1991).

Figure 10.1 presents the Wheelock two-zone concept and shows the reactions which predominate in each zone. Operating the fluidized bed in the turbulent or slugging regime provides rapid circulation between the zones, while control of the gypsum flow rate determines the SO_2 concentration in the off-gas and the gypsum conversion. Gasification of coal in the bed could furnish the reducing gas needed, or this gas could come from either synthesis gas or methane added to the fluidizing gas. A schematic of the bench scale unit is shown in Figure 10.2. The bed used by Swift and Wheelock (1975) was 12 cm i.d. × 183 cm; Morris *et al.* (1987) used a 25 cm i.d. × 210 cm bed. Desulfurized solids were removed through an overflow tube shown in the figure and as fines captured from the off-gas. Solid or gaseous fuels could be used. Because of the high temperatures employed, the beds were heavily insulated.

Swift and Wheelock (1975) demonstrated the existence of two zones in their fluidized bed by measuring the composition as a function of axial position. Their measurements are shown in Figure 10.3 for an experiment in which secondary air and the anhydrite feed were injected 100 mm above the distributor at the bottom of the fluidized bed. It is evident from the change in the CO and CO_2 mole fractions that the oxidizing zone begins at the injection point and extends upwards. The SO_2 mole fraction increases throughout the zone and has a composition-position profile different from CO_2. Swift and Wheelock conclude from these profiles that SO_2 is formed from the oxidation of CaS formed in the reducing zone at the bottom of the bed.

A large number of experiments were undertaken in the 15 years the two-zone fluidized bed reductive decomposition process has been under

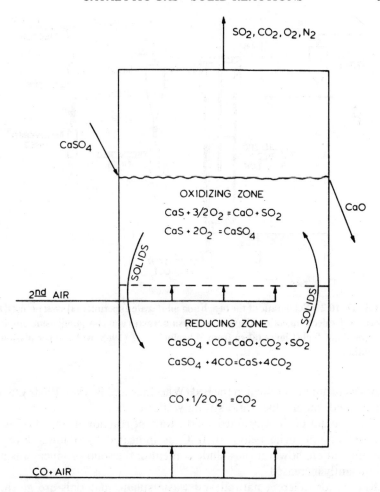

FIGURE 10.1 Conceptual schematic of a two-zone fluidized bed reactor for the reductive decomposition of gypsum and derivative materials using a carbon monoxide fuel (Figure adapted from Wheelock and Morris (1986) with permission of the authors)

development. In these, measurements were made of the composition of the overflow and entrained solids and the SO_2 content of the off gas. The percent desulfurization could be calculated from these data. Process variables studied were the $CaSO_4$ source (phosphogypsum, gypsum or anhydrite), mineral feed rate, bed temperature, air fuel ratio and the relative depths of the reducing and oxidizing zones in the fluidized bed (Swift and Wheelock, 1975). These investigators observed that the air fuel ratio had only a secondary effect on the desulfurization of the overflow solids and the sulfide content of those solids, whereas the ratio had a large effect if two

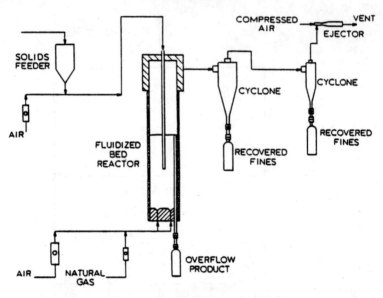

FIGURE 10.2 Schematic of the bench and pilot scale experimental systems used for the reductive decomposition of gypsum, gypsum wastes and phosphogypsum employing a methane fuel (Figure taken from Wheelock *et al.* (1988) with the permission of the authors)

zones were not maintained in the bed (Wheelock and Boylan, 1968). On the other hand, the air fuel ratio strongly affected both desulfurization and sulfide content of the entrained solids. Rate of reaction dropped off with gypsum or anhydrite below 1050 °C as indicated by a lower level of desulfurization, however the sulfide content of the solid products was not significantly increased.

Results for different natural and waste sulfate materials are given in Table 10.1. These show a high level of sulfate decomposition. Results for the two minerals used, anhydrite and gypsum are similar except for the SO_2 mole fraction, fines production and the residual sulfides in the over flow and fines solids. The higher SO_2 is explained by the additional water in the off gas from gypsum which is a dihydrate. Differences in fines formed and the sulfide distribution is the result of structure and porosity differences between the two minerals as well as decrepitation occurring during dehydration. The comparison between the gypsum mineral and waste materials is confounded by different measurements, although in general waste materials behave like the gypsum mineral. Most experiments summarized in Table 10.1 used natural gas as the source of reducing gas in the bed. One experiment used coal. Desulfurization changed little for this fuel substitution: differences in the SO_2 mole fraction and sulfide in the fines results from sulfur in the coal and the effect of coal ash.

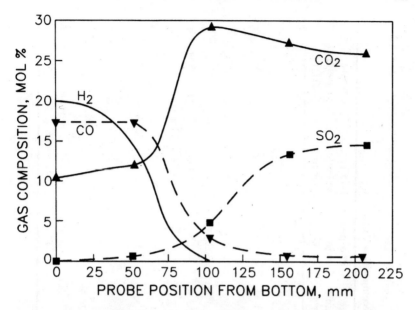

FIGURE 10.3 Variation of gas concentration as a function of height above the fluidized bed distributor. Experiment performed in a 120 mm (i.d.) bed of depth 860 mm with secondary air and anhydrite injection 100 mm above distributor. Fuel was natural gas fed with air to the distributor at the bottom of the bed (Figure adapted from Swift and Wheelock (1975) with permission, © 1975 American Chemical Society)

The first six columns in the table give results obtained in two zone fluidized beds using the system shown in Figure 10.2. The last column shows an experiment in which the fluidized bed was modulated. This was accomplished through the air fuel ratio by turning the natural gas flow on and off (Wheelock and Riel, 1991). This changed the total gas flow with time and also the bed temperature. A non-symmetric cycle was used with a cycle split of 0.6 based on the duration of the reducing portion of the cycle. Wheelock and Riel varied the cycle period 10 fold in their study and the effect of period on desulfurization and calcium recovery as lime (last row in Table 10.1) in the overflow was that longer periods seem desirable for both desulfurization and recovery, but the effect is small and cycle period seems to be a secondary variable in the range studied. Figure 10.4 shows the influence of cycle period on desulfurization of the solids overflow (Row 10 in the table) and on lime recovery. Variation of the off gas composition with time for an experiment with $s = 0.6$ and $\tau = 50$ s appears in Figure 10.5. SO_2 in the reducing partial cycle comes from reduction of gypsum by CO which yields lime, CO_2, H_2O and SO_2. In the second part of the cycle, SO_2 comes from oxidation of the sulfide to form lime. Some $CaSO_4$ is also formed.

Table 10.1 Results for Selected Reductive Decomposition Experiments

Material Used (Source of Reducing Gas)	Anhydrite (CH_4)	Gypsum (CH_4)	Waste Gypsum (CH_4)	Waste Gypsum (CH_4)**	Phospho-gypsum (CH_4)	Phospho-gypsum (coal)	Waste Gypsum (CH_4)*
Operating Conditions:							
Feed Size, mm	14/35 Mesh	14/35 Mesh	0.4 – 1.7	0.4 – 1.7	0.4 – 1.7	0.4 – 1.7	0.6 – 1.7
Temperature, °C	1205	1188	1150	1100	1150	1150	1100
Feed Rate, kg/h/m²	202	175	593	413	302	445	342
Fuel/$CaSO_4$, Mol Ratio			1.55	1.52	1.5	0.35(kg coal/kg $CaSO_4$)	1.54
Air/Fuel, Mol Ratio	9.5	10	12.9	12.6	8.3		9.1
Results:							
SO_2 in Off Gas Mol %	7.5	5.2	5.2	5.7	5.2	6.4	5.4
Overflow Rate kg/h/m²	91	72	180	136	123	165	88

Table 10.1 (Continued).

Fines Capture Rate kg/h/m²	9	28	87	61	21	47	12
Sulfide in Overflow Wt.%	0.9	0.4	Nil	Nil	Nil	0.5	Nil
Desulfurizat'n of Overflow Wt%	98.4	98.7	95.5	98.3	96.6	96.4	98.2
Sulfide in Fines Wt%	1.9	2.6	0.5	0.4	0.5	1.5	2.0
Desulfurizat'n of Fines Wt%	78	81	89.1	70.7	86.1	85.4	82.7
Calcium Oxide Yield Wt%***					91	91	86

* Measurement under composition forcing in a simple fluidized bed
** Measurement undertaken in pilot scale two-zone fluidized bed reactor
*** Based on $CaSO_4$ fed to bed

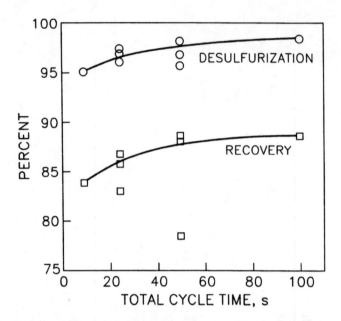

FIGURE 10.4 Influence of cycle period on waste gypsum desulphurization and lime recovery for reductive decomposition at 1100 °C, gypsum loading of 336 kg/h/m^2 and a time average air/natural gas ratio of 9:1 using a fluidized bed and asymmetric cycling of the natural gas flow (Figure adapted from Wheelock and Riel (1991) with permission, © 1991 Gordon and Breach Science Publishers).

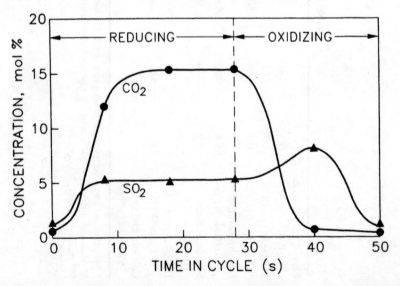

FIGURE 10.5 Variation of CO$_2$ and SO$_2$ mol % with time during an asymmetrical cycle with a period of 50 s. Conditions as stated in Figure 10.4 (Figure adapted from Wheelock and Riel (1991) with permission, © 1991 Gordon and Breach Science Publishers)

Regeneration of Calcium Oxide

Jagtap and Wheelock (1996) applied experience from Wheelock's gypsum work to the regeneration of calcium oxide, a potential sorbent in hot gas cleanup (removal of H_2S from reducing gas streams). They demonstrated that the sorbent can be regenerated at 950 to 1100°C rather than at 1400° which is required in a continuous process. The reason for 1400°, a temperature which makes calcium oxide unattractive, is that below 1400°, the calcium sulfide (CaS) formed in hot gas cleanup reacts with O_2 to form the sulfate. The sulfate is less dense and expansion of the product blocks off pores reducing O_2 access to unreacted sulfide thus throttling oxidation and sorbent regeneration. Above 1400°, calcium oxide (lime) rather than $CaSO_4$ is formed. Jagtap and Wheelock reasoned that if exposure to O_2 or air is followed by exposure to a reducing gas, some of the sulfate could be reduced to the oxide and the rest back to the sulfide, thus unblocking the pore structure. To demonstrate their concept, they carried out experiments in a thermogravimetric device using reagent grade CaS which they pelletized into 6.4 mm diameter by 1 to 2 mm discs. These discs or crushed and screened solids were placed in a TGA basket, brought to reaction temperature in a N_2 flow and then subjected to alternating flows of $O_2 - N_2$ and reducing gas-N_2 mixtures. Experiments were made with different reducing gases (CO, CH_4 and C_3H_8) and mixture compositions. Composition cycling was symmetrical and only a cycle period of 2 minutes was employed.

Essentially complete conversion of CaS to CaO was possible by periodically switching between oxidizing and reducing gas compositions even at 960°C. The rate of oxidation depended upon temperature and the gas compositions. The temperature effect is illustrated by Figure 10.6 which shows percent conversion of CaS to CaO as a function of time in the TGA using an oxidizing gas containing 20 vol.% O_2 and a reducing gas with 5 vol.% CH_4. These compositions were found to be optimal (Jagtap and Wheelock, 1996). Further experiments used repeated sulfidation and regeneration cycles. These demonstrated that the lime produced in regenera-tion did not lose its capacity for sulfur removal and times needed to reach the same level of sulfidation or regeneration did not change significantly in successive cycles. Experiments were also carried out with calcium carbonate, limestone and $CaCO_3$ mixed with coal ash.

General Comments

Separating a gas solid reaction into steps and carrying each out in an environment optimal or near optimal for that step is essentially what Wheelock and his co-workers have done for gypsum decomposition. They are able in this way to avoid sintering and still minimize impurities in their

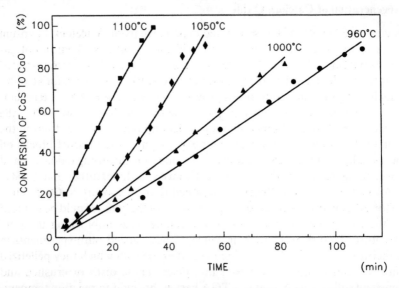

FIGURE 10.6 Recovery of lime from calcium sulfide in an atmospheric TGA with gas flow constant but switching continuously with a period of 2 minutes from 20 vol.% O_2 in diluent to a reducing gas with 5 vol.% natural gas in air + diluent (Figure adapted from Jagtrap and Wheelock (1996) with permission, © 1996 American Chemical Society)

solid product. However, it is probably not correct to conclude from their work that this can be best done through composition modulation rather than designing a reactor with zones of different gas composition. Indeed, the best solution may be to use two reactors, each operated at steady state but at a different gas composition. This offers the possibility of running each of the two reactors at different temperatures which could prove to be advantageous. It seems unlikely that recovery of lime and SO_2 from gypsum is the only gas-solid reaction that can be separated into steps. Other industrial solid conversion reactions need to be examined with the approach taken by Wheelock. The yield enhancements he has found certainly justify extending his work. These comments apply equally to Wheelock's recent application of composition modulation to overcoming pore blockage caused by density changes as the gas solid reaction proceeds. This is pioneering work well deserving of extension to other reaction systems.

Periodically changing solid compositions where catalysis is not involved are also found in chemical heat pumps utilizing low temperature waste heat. An example of such a heat pump-solid reaction couple is the methanolation and demethanolation of calcium chloride (Lai *et al.*, 1992; Lai *et al.*, 1993; Dong and Lai, 1995).

C. IRON CATALYZED GASIFICATION

Catalytic Coal Gasification

Coal gasification has been studied intensively since the 1950's as a source of synthesis gas for the Fischer-Tropsch process, methanol synthesis, and hydrogen production. Gasification is a key step in coal based integrated combined cycle systems now being developed in many countries in order to raise power generating efficiency and suppress sulfur discharge to the environment. In all commercial processes, gasification is carried out at high temperature noncatalytically, with coal ash removed as a slag. Catalytic gasification is being examined as a means of increasing carbon burnout and/or lowering the gasification temperature to decrease NO_x generation and avoid slagging of coal ash. Lower temperature could also reduce the gasifier investment. If catalytic gasification is to be competitive with current processes, a low cost catalyst is essential to avoid extensive ash processing for catalyst recovery.

Cost considerations limit catalyst choice to iron, sodium or calcium. Salts, or more specifically oxides, of these three elements have been well studied. Sodium does not look promising because its low cost form is a chloride which cannot be tolerated in a gasification environment. Calcium and sodium may act as fluxing agents to lower ash fusion temperatures. Consequently, iron seems to be the catalyst of choice for coal gasification.

Composition forcing for gasification has been investigated as a method of activating the iron catalyst. Gasification is an oxidative reaction and we have seen in Chapters 3 and 8 that modulation can enhance catalyst activity in such reactions.

Redox Mechanism For Iron Catalyzed Gasification

It is postulated that iron catalyzes carbon gasification through a redox cycle. This cycle has the following general structure (see, for example, Moulijn and Kapteijn, 1986):

$$Fe_nO_m(s) + CO_2 \longrightarrow Fe_nO_{m+1}(s) + CO \qquad (10.1)$$

$$Fe_nO_{m+1}(s) + C \longrightarrow Fe_nO_m(s) + CO \qquad (10.2)$$

where C is any carbonaceous material. Steam or O_2 can be used in place of CO_2 yielding H_2 as the gaseous product when the former is used. A redox cycle is thought to be involved in other metal-metal oxide couples that catalyze carbon oxidation.

Stable oxides of different iron valence and oxides of mixed valence exist (e.g. Fe_2O_3, Fe_3O_4, FeO). Iron can also form non-stoichiometric oxides. It is to be expected that reducibility or oxidizibility will change with the oxide so that its catalytic activity may also change. Indeed, Ohtsuka et al. (1987) have demonstrated that gasification activity depends on the oxide employed. Iron oxide present on or adjacent to carbon will be governed by the reactions given as Equations (10.1) and (10.2). Thus, the gas phase oxidant and its partial pressure, and the temperature must influence activity. Steady state operation at a given set of conditions may not engender the most active oxide. Cycling between oxidizing and reducing gas phases, however, will force iron to pass periodically through its most active oxide. The concept on which this application of composition modulation is based is that a proper choice of cycle period and split might maintain that oxide for a sufficient time to profoundly increase the gasification rate. Furthermore, cycling could periodically force the gasification reaction of Equation (10.2) to occur.

CO_2 Gasification of Low Rank Coal

Gasification under composition forcing was investigated by Suzuki et al. (1988, 1989) using a char prepared from Yallourn coal impregnated with an iron nitrate solution. Suzuki extended his study to other coal chars (Suzuki et al., 1990). Two tubular fixed bed gasification reactors placed in the same oven cavity were used as shown in Figure 10.7. Carbon dioxide was switched periodically between the two reactors. The off gas was mixed and

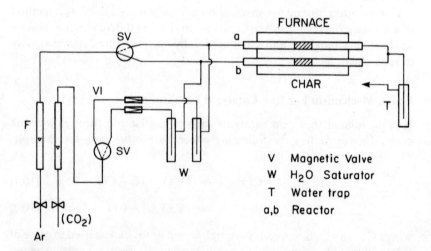

FIGURE 10.7 Schematic diagram of the coal char gasification system: a, b – quartz reactor tubes (5.5 mm i.d.), MV - mass flow controllers, V1 - stop valve, SV - three-way solenoid valve (Figure taken from Suzuki et al. (1989) with permission, © 1989 American Chemical Society)

sent to the sampling valve of a gas chromatograph. This arrangement reduced the composition variation and made sampling less arduous. Yallourn is an Australian brown coal (68.2 wt% C on a dry and ash free basis with just 1.1 wt% ash on a dry basis). Wet impregnation gave an iron content corresponding to about 0.3 mmol/g of dry coal. After drying the treated coal, char was formed by carbonizing in a sequence built of heating the coal at 20 C/min to 800°C then holding the sample at this temperature for 30 minutes. Carbonization brought the Fe level to about 0.6 mmol/g of carbon.

For the experiments, a 50 mg sample was inserted as a plug in each tube. All experiments were conducted at 800°C and the operation described in the previous paragraph meant that cycling was symmetrical. The variables explored in this pioneering work were the cycle period and the CO_2/Ar ratio in the oxidizing portion of the cycle. An experiment under steady state employing the time-average CO_2 and Ar mixture was carried out to permit a comparison of the two modes of operation.

Gasification rates were calculated from the CO recovered in the off-gas and are given in Figure 10.8 as a function of char conversion. The (a) and (b)

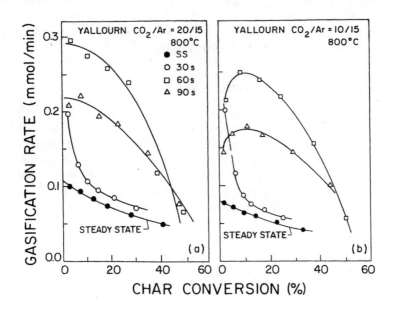

FIGURE 10.8 Rate of char gasification as a function of char conversion for the CO_2 gasification of an iron impregnated Yallourn coal char at 800°C. Cycling is symmetrical with switching between argon and an argon-CO_2 mixture: (a) CO_2/Ar = 10/15 in the mixture, (b) CO_2/Ar = 20/15 in the mixture (Figure 10.8a adapted from Suzuki *et al.* (1989) with permission, © 1989 American Chemical Society; Figure 10.8b adapted from Suzuki *et al.* (1989) with permission of the authors)

parts of the figure differ only in the CO_2/Ar ratio in the oxidizing portion of the cycle. Curves for the three cycle periods studied are shown in each figure and these indicate a 60 s cycle maximizes the gasification rate at 800°C. Up to a 40% conversion, rate for the 60 s period is about 3 × the steady state rate in Figure 10.8. Thus, the improvement achieved through this mode of operation is very large. The maxima in the curves for periods of 60 and 90 s are associated with surface area changes and indicate diffusion interference. Resemblance between the 30 s curves and the curves for steady state suggests mixing in the reactor may have damped concentration changes at this cycle period.

Suzuki et al. (1990) undertook forcing experiments with chars prepared at 800°C from iron impregnated Morwell coal and with chars from Loy Yang coals impregnated with iron, or sodium, or iron-sodium mixtures. The pulsed oxidant was CO_2, or steam, or CO_2 saturated with water vapour. At a gasification temperature of 800°C, cycle periods of 60–90 s in all of the Suzuki measurements caused about a 50% increase in gasification rate over rates found under steady state operation. Thus, the rate effect Suzuki observed for modulating the gas phase between an oxidizing and a reducing condition appears to be general. It is worth noting that gasification rates for Loy Yang char under forcing as well as steady state were 10 × the rates for Yallourn char, while the rates for Morwell char were about 50 × those for the Yallourn samples.

CO_2 Gasification with Forestburg Coal

Zhang et al. (1995) extended the Suzuki studies and improved somewhat the experimental technique. In place of the parallel reactor scheme (Figure 10.7), Zhang et al. used a single quartz tube (3 mm i.d.) together with a hold-up vessel which was located outside of the temperature-control oven and served to time-average the flow through the gasifier. When located in front, it forced the gasifier to operate at steady state, while when placed after the gasifier it smoothed concentration variations and gave the time averaged behaviour of the modulated system. Zhang et al. used a dual wave length IR to measure CO and CO_2 partial pressures instead of a gas chromatograph. When the hold-up vessel was bypassed, the IR was able to monitor the response to feed composition switches. Gas flow was measured continuously and the char sample was weighed before and after a run so that char conversion as well as CO production could be determined. A 25 to 50 mg char sample was used depending on run conditions. Samples were held firmly in place at the centre of the tube with wadded quartz wool. Char elutriation from the gasifier tube was negligible and carbon balances closed to within 5 to 10%.

Experiments were restricted to chars made from a Forestburg coal. Forestburg is a Canadian sub-bituminous coal (59.8 wt.% C on a dry basis with 5.7 wt.% ash). Most experiments employed chars prepared from an iron impregnated coal with 3 wt.% Fe based on carbon in the coal, but additional, more limited experiments were made on chars from raw coal and from demineralized coal. All chars were produced by carbonizing coal at 900°C in a micro fluidized bed under N_2 at 1 bar. In the first set of experiments conducted by Zhang et al. (1995) the gas flow rate was maintained constant and symmetrical cycles were used with switching between CO_2 and N_2. Variables investigated were cycle period and temperature. Steady state or continuous runs were made also (Zhang et al., 1994a)

Improvement in CO yields were found for the composition modulation used, but they were much less then the rate enhancements seen by Suzuki et al. (1989) and Suzuki et al. (1990) as discussed above. The amount of CO accumulated as a function of run duration is shown in Figure 10.9a for three temperatures, while in Figure 10.9b char burn-off in terms of char carbon converted is plotted vs. run duration. The cycling mode curves are for a 60 s cycle period. Data in the figure display a 15 to 20% yield or conversion improvement at 800°C over steady state for composition forcing. At higher temperatures, the increase is smaller. The reason for the temperature effect is that a 60 s cycle period is not optimal at higher temperatures. Figure 10.10 repeats the plots of Figure 10.9 but now in terms of cycle period. Run durations vary. The upper part of the figure gives CO accumulation data, while the lower plots carbon conversion. It can be seen that a 60 s period maximizes CO accumulation and conversion at 800°C, but this period is too long at 850° and 900°C.

No attempt was made by Zhang et al. (1995) to optimize iron catalyzed gasification under composition modulation. Observations reported in an accompanying paper (Zhang et al., 1994b) suggest the directions which optimization must follow. Figure 10.11 shows the IR response at the dominant line in the CO spectra to successive exposures of CO_2. In this figure, successive pulses of CO_2 increase in duration. These pulses are separated by N_2 exposures of constant duration. The sequence appears in the bottom portion of the figure. A burst of CO evolution, seen as a sharp peak, occurs 5 to 10 s after CO_2 is switched on. Height and area of the peak are determined by duration of CO_2 exposure only for durations under 30 s. For duration above 30 s, the peak height and width remain constant but after 30 s a plateau appears. The height of this CO signal plateau is greater than the height of the CO signal under steady state operation at the same degree of char conversion.

Figure 10.12 shows that peak height above the plateau. It changes little with temperature above 750°, although this is not evident because of changes in the ordinate scale. The plateau height, however, is a strong function of temperature. When a char made from raw coal is used, just a small burst or initial CO peak before the plateau is observed. There is no

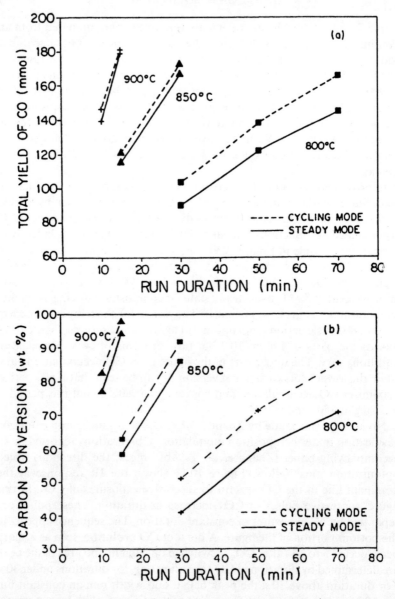

FIGURE 10.9 Comparison of composition forcing and steady state operation as a function of run duration for different temperatures. Gasification at 1 bar of Forestburg coal char loaded to an iron level of 3 wt.% Fe/g carbon in the coal. Forcing data collected at a cycle period of 60 s using symmetrical switching between CO_2 and N_2, while steady state data taken for a 50:50 mixture of these gases: (a) total CO accumulated during the run, (b) char conversion in terms of carbon in the char (Figure taken from Zhang et al. (1994a) with permission, © 1994 American Chemical Society)

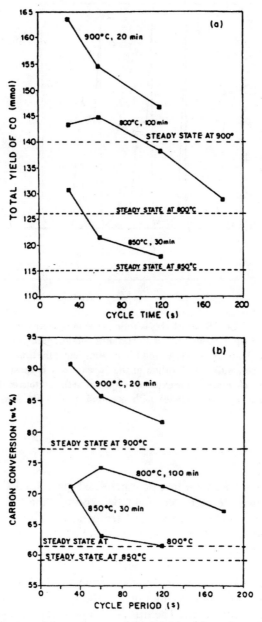

FIGURE 10.10 Comparison of composition forcing and steady state operation as a function of cycle period for different temperatures: (a) total CO accumulated during the run, (b) char conversion in terms of carbon in the char. Run durations given on the curves are the same for cycling and steady state See Figure 10.9 for experimental conditions (Figure taken from Zhang *et al.* (1994a) with permission, © 1994 American Chemical Society)

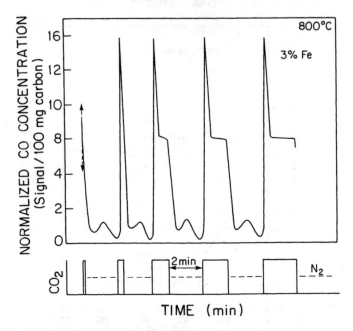

FIGURE 10.11 CO IR signal versus time in the initial exposure of an iron loaded char (3 wt.% Fe/g carbon) from a Forestburg sub-bituminous coal to alternating exposures to CO_2 and N_2 at 800°C and 1 bar. Duration of the time exposure increases from left to right, while the duration of the N_2 exposure is constant. Time scale is indicated in the schematic of the exposure sequence at the bottom of the figure (Figure taken from Zhang *et al.* (1994b) with permission, © 1994 American Chemical Society)

burst or peak when char from a demineralized coal is used. Only the step up and step down are seen. As might be expected, the CO concentration corresponding to the plateau at any temperature is largest for char from iron impregnated coal and smallest for the char from a demineralized coal. The difference is large.

On switching to N_2, a second burst of CO evolution is seen. The area under this peak is proportional to the area under the first peak. The peaks or bursts of CO evolution in Figures 10.11 and 10.12 are prima facie evidence for a redox cycle. The first peak represents the reaction of Equation (10.1). The reduction reaction (Equation (10.2)) is the second smaller peak. The existence of these bursts and their interpretation have been known for several years (see, for example, Suzuki *et al.*, 1988). The observation of a plateau by Zhang *et al.* is a new discovery and its significance is that the catalyst appears to be activated by composition switching.

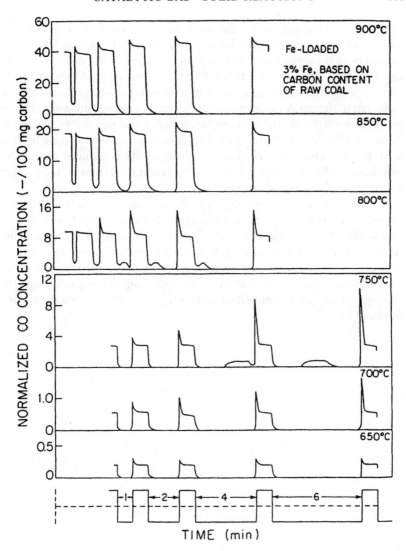

FIGURE 10.12 CO IR signal versus time in the initial exposure of an iron loaded char (3 wt.% Fe/g carbon) from a Forestburg sub-bituminous coal to alternating exposures to CO_2 and N_2 at various temperatures and 1 bar. Duration of the CO_2 exposures is constant while the duration of N_2 exposure increases from left to right. Time scale is indicated in the schematic of the exposure sequence at the bottom of the figure (Figure taken from Zhang *et al.* (1994b) with permission, © 1994 American Chemical Society)

The effect of the duration of the reducing environment, N_2 exposure, for a constant CO_2 exposure can be seen in Figure 10.12. Above 800°C, the second peak disappears and becomes an extension of the drop in the CO signal after switching to N_2. Below 800°C, the second peak is retarded by about 120 s and flattens. The figure demonstrates that oxide reduction is the slow step in the redox cycle and its rate is highly temperature dependent. If the N_2 pulse is too short, for example less than 4 minutes at 750°, the initial burst of CO production is reduced. This accounts for the small peak heights seen at 700°. Only a small fraction of the oxide is reduced during N_2 exposure so the oxidation reaction of Equation (10.1) makes just a small contribution to CO production.

The consequences of these response observations for optimizing composition forcing is that the cycle period is strongly temperature dependent and below 800–850°C, where catalytic gasification could be commercially interesting, a non-symmetrical cycle should be used with the duration of the reducing portion of the cycle exceeding the oxidizing portion. The short, unsymmetrical cycle periods mean that composition forcing would have to be implemented on a commercial scale using some version of a circulating fluidized bed.

CHAPTER 11

Multiple Reactions and Selectivity

A. INTRODUCTION

Multiple Reactions As A Target For Composition Forcing

The presence of secondary reactions inevitably causes operating problems: parasitic reactions consume feed stock, these reactions make recovery of the desired product more complicated and certainly more costly, and secondary reactions may cause catalyst fouling. These reactions sometimes have large heat effects that compromise reactor stability or require the addition of extra heat transfer surface. With multiple reactions, the objective of reactor design changes from achieving high conversion at high throughput to minimizing the extent of secondary reactions without throttling the desirable reaction. The potential of composition forcing to realize this objective was recognized in the late 1960s just as research on applications of periodic operations to reactors was getting underway.

Early studies used arbitrary model reactions and employed numerical analysis, variational or relaxed steady state techniques. Horn and Bailey (1968), for example, considered two competitive reactions of different order to demonstrate improved selectivity through composition modulation. Several years later, Renken (1972) examined a system of three consecutive-competitive reactions in a continuous stirred tank reactor and found the selectivity to an intermediate product could be improved by composition modulation. A large theoretical literature dealing with selectivity improvement through feed composition cycling has emerged since these papers. That increased selectivity is possible through composition modulation is certainly the unanimous view of this literature.

Experimental studies quickly followed theoretical investigations and the number of publications has grown so large that they cannot be dealt with in a single chapter of reasonable length. Most experimental studies have targeted important commercial reactions: (1) partial oxidation and oxidative dehydrogenation of hydrocarbons, (2) the Fischer-Tropsch synthesis, (3) polymerization, and (4) hydrogenation of hydrocarbons. Our discussion of the application of composition forcing to multiple reactions has been divided into four chapters. Each of the first three groups of reactions has been given a separate chapter. This chapter deals with the fourth and studies of other reactions that fall outside of the groups mentioned. Homogeneous, non-catalytic reactions are also examined in this chapter and are dealt with in the next section.

Selectivity And Yield

Selectivity and yield were mentioned in Chapter 1, but were not defined. Unless stated otherwise, selectivity ($S_{i/j}$) means overall selectivity calculated from conditions across a reactor. The term is an absolute quantity and refers to a product "i" and the key reactant "j". For an arbitrary reaction network, selectivity is the moles of product "i" formed per mole of key reactant "j" consumed. Thus,

$$S_{i/j} = \left| \frac{(n_i)_{\text{in}} - (n_i)_{\text{out}}}{(n_j)_{\text{in}} - (n_j)_{\text{out}}} \right| \qquad (11.1)$$

Normally, (n_i) is a species flow rate in units of mols/s; it can also be moles in the reactor if a batch reactor is considered. In this case "in" and "out" refer to start up of the reactor and completion of the run. Other flow rate units, or loading units such as flow rate per unit reactor volume may be used, but these must be consistent in the numerator and the denominator.

Yield is the moles of product "i" formed per mole of key reactant "j" fed to the reactor. It is also an absolute quantity and a product species and the key reactant must be indicated. Using the symbols given above,

$$Y_{i/j} = \left| \frac{(n_i)_{\text{in}} - (n_i)_{\text{out}}}{(n_j)_{\text{in}}} \right| \qquad (11.2)$$

Thus, the yield, $Y_{i/j}$, is the product of the selectivity, $S_{i/j}$, and the conversion, X_j, of the key reactant.

It is also possible to discuss an instantaneous selectivity, $s_{i/j}$, defined as the ratio of the rate of formation of product "i" to the rate of consumption of key reactant "j". Often a selectivity ratio, $\zeta_{i/k}$, is used in place of the instantaneous selectivity. It is the ratio of rates of formation, usually the ratio of the rate for the desired product "i" to the rate for an important unwanted product "k". Selectivity ratio comparisons lead often to large differences and so dramatize the effect of changes in operating conditions. Instantaneous selectivity and the selectivity ratio at steady state reflect local concentrations and temperatures and, of course, the reaction network and reaction kinetics. Selectivity, as an overall quantity, reflects in addition the reactor type and the conversion achieved.

Just as we have considered rate and conversion enhancement in the preceding chapters, enhancement can be defined in terms of selectivity and yield. Subscripts designate the enhancement considered. Thus,

$$\Psi_S = \frac{S_{i/j}}{(S_{i/j})_{ss}} \quad (11.3)$$

$$\Psi_Y = \frac{Y_{i/j}}{(Y_{i/j})_{ss}} \quad (11.4)$$

where $S_{i/j}$ and $Y_{i/j}$ are the time-average selectivity and yield under composition modulation and $(S_{i/j})_{ss}$, $(Y_{i/j})_{ss}$ are these measurements taken under steady state at the time-average composition of the feed used in the modulated operation.

Global enhancements of selectivity, Ψ_S^* and yield, Ψ_Y^* are defined as well by the relation just given, but the denominator of each term becomes the maximum steady state selectivity or yield in the range of feed composition covered by the modulations.

Classes of Multiple Reactions

Reaction systems can be divided into classes exhibiting like behavior with time but different levels of coupling between individual reactions forming the system. Independent reactions are not coupled. Each reaction proceeds independently of others in the system. Selectivity and yield are always unity.

Competitive or parallel reactions are those which share one or more reactants. One may be jointly consumed, such as,

$$A \begin{matrix} \nearrow^{+B} D \\ \searrow_{+C} E \end{matrix} \quad (11.5)$$

Or, two reactants can be consumed by each:

$$A+B \begin{matrix} \nearrow C+D \\ \searrow E+F \end{matrix} \quad (11.6)$$

This class of multiple reactions are coupled through reactants. Stoichiometrically, at least, many reaction systems fall into this class. A typical example would be partial oxidation where the products could be an oxygenate and water, or CO_2 and water.

Consecutive, sequential or serial reactions,

$$A \to B \to C \to D \to E \tag{11.7}$$

are rather rare. A great many reaction systems fall into the class of consecutive-competitive multiple reactions:

$$A + B \to C + D$$
$$C + B \to E + D \tag{11.8}$$
$$E + B \to F + D$$

Chlorination of hydrocarbons, forming successively the monochloro, dichloro and trichloro products, represent this class. The common reactant "B" is chlorine, Cl_2, and the common product "D" is HCl. Reactions in this class are coupled through a common reactant and through the product of the first reaction becoming the reactant of the succeeding one. Models for these systems are difficult to handle analytically and numerical methods are almost always required.

When the reactant "B" in Equation (11.8) is in large excess with respect to "A", the consecutive class is encountered, either as Equation (11.7) or one in which each step has a common product which does not react further. Hydrogenation of acetylene through ethene to ethane is an example of the consecutive reaction class.

This chapter examines multiple reactions by class, but not in the sequence just given.

B. HOMOGENEOUS REACTIONS

Although this monograph was intended to discuss just catalytic reactions, an exception will be made here because catalysis, on occasion, contributes little to the system response to forcing so that a homogeneously catalyzed reaction behaves just like a non-catalytic reaction. Our discussion deals with the consecutive-competitive class of reactions.

Concept

In the paper mentioned in the previous section, Renken (1972) found through a numerical simulation that selectivity to an intermediate product "C" in the homogeneous, non-catalytic reaction network given by Equation (11.8) can be increased, when a continuous stirred tank reactor is used, by in phase concentration cycling of reactants "A" and "B" if it is assumed that the stoichiometric equations represent elementary reactions, that is, all of the reactions are irreversible and 1st order in each reactant. Others, who have looked at consecutive-competitive reaction networks theoretically have reached the same conclusion (Farhadpour and Gibilaro, 1975; Lee and Bailey, 1980). The concept behind in phase modulation is simple: as the 1st reaction proceeds building up the concentration of "C", the rate of the 2nd reaction forming "E" accelerates. By cutting off the feed of "A" and "B", the 2nd reaction can be throttled before the concentration of "C" and the rate of formation of "E" reach maxima. Of course, this strategy decreases the conversion of the key reactant "A". Renken shows, however, that the yield of "C", $Y_{C/A}$, actually increases if a suitable choice of cycle period and split are used, although the increase is small.

The two experimental studies of homogeneous reactions were conceived, it appears, to test the theoretical predictions recognizing that kinetics are imperfectly known, mixing can give rise to inhomogeneity, and mechanically stirred tanks are not perfectly isothermal or fully back mixed. Each of the studies used a different reaction system and imposed different restrictions on comparing modulated and steady state operation.

Ethyl Adipate Saponification

This reaction system was chosen because it proceeds under mild conditions and its kinetics are reasonably well known. Product concentrations are readily measured so, despite a problem with hydrolysis in water, the system is well suited for model use. Ethyl adipate is the ethyl diester of adipic acid which has two acid functional groups. It saponifies in the presence of a strong base to the adipate salt:

$$Et_2A + NaOH \longrightarrow NaEtA + EtOH$$
$$NaEtA + NaOH \longrightarrow Na_2A + EtOH$$
(11.9)

In the above, NaOH and EtOH are sodium hydroxide and ethanol, Et is an ethyl group and A is the adipate $\begin{array}{c} CH_2-CH_2-COO^- \\ | \\ CH_2-CH_2-COO^- \end{array}$. Experiments on this reaction system were undertaken by Lee et al. (1980). Their objective was to measure global enhancement of the selectivity to the intermediate product, sodium monoethyl adipate (NaEtA), under composition forcing. Global enhancement was deemed to occur if selectivity under forcing at a specified temperature, total pressure and space velocity exceeded the maximum steady state selectivity at any concentrations of ethyl adipate and NaOH used for cycling. The space velocity condition differs from the comparison condition used by Renken and discussed in the next section.

The experimental system employed by Lee et al. (1980) is shown schematically in Figure 11.1. The design, namely size and position of the three-bladed impeller and sizes of the equally spaced vertical baffles and the external jacket, were chosen to model geometrically a standard, industrial

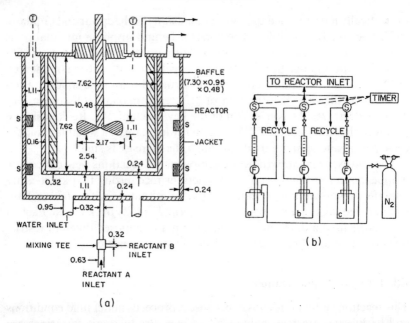

FIGURE 11.1 Schematic of (a) the experimental jacketed, mechanically stirred autoclave and (b) the reactant feed arrangement used by Lee et al. (1980) for diethyl adipate saponification with sodium hydroxide. In (a), T indicates a thermocouple and S a spacer, whereas in (b) F indicates a filter disk and S a solenoid valve. The contents of the storage vessels a, b, and c are respectively a sodium hydroxide solution, a solution of the diethyl ester and the solvent, a 50:50 wt.% mixture of isopropanol and water (Figure taken from Lee et al. (1980) with permission of the authors)

stirred autoclave according to the authors. Reactants were premixed just prior to entering the reactor vessel. Samples from the autoclave were analyzed by gas chromatograph after quenching with potassium phthalic acid. However, the mono ethyl ester was measured only after extraction with benzene. Figure 11.1b shows the feed system. Reactant and solvent storage bottles were pressurized to avoid using pumps. Solenoid valves either directed the reactant or NaOH solution to the mixer or returned them to the storage bottles. An alcohol-water solution of the diester was used. Even at room temperature, some hydrolysis occurred. This was measured and allowed for in calculating selectivity.

Although the experiments were designed to investigate cycle period, split and the ester/saponification agent ratio, the paper by Lee et al. (1980) deals only with the effect of the reactant ratio on yield enhancement. Cycle split, defined in terms of the duration of reactant flow, was set at 0.1; while the cycle period was held at 22.5 min. for all experiments. These were conducted at 298 K using a 50:50 wt.% mixture of isopropanol and water as solvent for the ester and NaOH. Use of a mixture in place of water increased the diester solubility. For the experiments, the time average diester concentration in the feed, $(C_A)_o$, was held constant at 0.023 mol/m^3. The time average NaOH concentration, $(C_B)_o$, changed, however, from 0.025 to 0.075 mol/m^3 to provide the reactant ratio variation. Amplitude was at a maximum as the flow was switched from the reactant-solvent mixture to solvent only. For comparison purposes, steady state measurements were made with the same concentrations used for the modulation study.

The experimental time average yields of NaEtA, the mono ester, under in phase modulation are plotted versus the diester/saponification agent ratio in Figure 11.2. Data lie clearly above the steady state yields, despite scatter of the modulation results which are greater than those for steady state operation. Maxima are apparent for both periodic and steady state data so that global enhancement of yield was achieved. Lee et al. (1980) modelled their experiments assuming an isothermal CSTR, irreversible reactions and that all rates were 1st order in each reactant. Separate, steady state experiments were performed to evaluate the rate constants and activation energies for each of the saponification reactions. The curves shown in Figure 11.2 are the simulated yields. With the exception of the modulation data point at $(C_B)_o/(C_A)_o = 1.6$, the agreement between model and experiment is very good. Using the simulated yields, the global yield enhancement

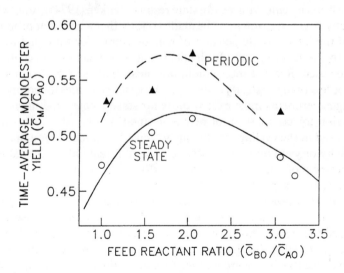

FIGURE 11.2 Comparison of time average sodium monoethyl adipate yields with steady state yields at various time average or steady state diethyl adipate/sodium hydroxide concentration ratios at 298 K and $\tau = 22.5$ minutes for the diester saponification using a 50:50 wt.% mixture of isopropanol and water as solvent (Figure taken from Lee et al. (1980) with permission of the authors)

observed by Lee *et al.* was about 10%. Because the modulation employed by these researchers decreases conversion substantially at a split of 0.1, a yield improvement means that the selectivity has been enormously increased by this modulation strategy. This analysis will be developed further in the next experiments considered.

Formation Of Diethanolamine

For his experimental investigation, Renken (1974) used the formation of diethanolamine (DEA) and triethanolamine (TEA) from the reaction of ethylene oxide (Et_2O) and monoethanolamine (MEA) in an aqueous solution. The reaction network is given by Equation [11.8] where "A" = MEA, "B" = Et_2O, "C" = DEA and "E" = TEA. There is no second product "D" in the system and the product "E" does not react further. The two reactions proceed at moderate rates around 50 °C. Parasitic reactions of Et_2O are not significant at this temperature so MEA and Et_2O are a good choice as a model system. Renken employed in phase modulation of reactant concentrations to investigate the effect of cycle period and split on selectivity to DEA and on DEA yield. As discussed under **Concept**, this modulation strategy depresses conversion, but augments selectivity so the effect on yield is uncertain.

This early paper is also concerned with the proper comparison of composition forcing with steady state results. Renken (1974) proposed that a selectivity comparison must be made (1) for either a constant production rate of the intermediate per unit of reactor volume or for equal reactant flows to the reactor, and (2) identical ranges of reactant concentrations must be used. Because these conditions mean that the time average $(C_A)_o$ must be less than the steady state $(C_A)_o$, the space velocities for composition forcing experiments must exceed those for steady state experiments. This restriction for comparison of results means that only the concentration range used in the cycling measurements must be searched for a maximum selectivity or yield under steady state to see if forcing has achieved global enhancement of one or both of these performance parameters.

In the diester saponification study discussed in the preceding section, Lee *et al.* (1980) utilized an equal space velocity comparison condition. The maximum concentration of one reactant at least with in phase cycling exceeded the steady state concentrations studied. Lee et al. used reactant ratio as a variable. In principle, they should have extended the range of diester concentrations searched for a reactant ratio which maximizes steady state yield or selectivity. This was not done. However, an extended search is unnecessary for the diester saponification system used. Lee *et al.* (1980) showed that the two saponification reactions were 1st order in each reactant. With such kinetics, yield and selectivity are independent of concentrations. They depend on just the rate constants and the reactant ratios. For other kinetics, the search for optimal steady state conditions must encompass the highest and lowest concentrations used in forcing.

Experiments on the MEA-Et$_2$O system were carried out in a thermostatted, magnetically mixed, round bottom glass flask. The feed system resembled the arrangement shown in Figure 11.1b, except that dosing pumps were used in place of gas pressurization. Reactant storage vessels for Et$_2$O, MEA and water were held at 3 °C and the pumps were cooled to avoid parasitic reactions. Electrical conductivity was monitored in the reactor outlet to identify steady state or when cyclically reproducible conditions were achieved. Product composition was measured chromatographically using an FID. When the system was forced, successive cycles were sampled to build up the concentration variation within a cycle. These data were used to determine time average conversion, selectivity and yield. Start up was from a water filled reactor. For all experiments, reactor temperature was held constant at 50 °C.

Symmetrical and asymmetrical in phase cycling of MEA and Et$_2$O concentration between 0 and 0.47 mol/L, and 0.50 mol/L, respectively, were undertaken with cycle periods from 20 to 60 minutes. Space velocities were 0.067 and 0.033 min^{-1} for the symmetrical cycling and steady state measurements. These varied from 0.042 to 0.071 min^{-1} for the asymmetrical cycling experiments, while the comparable steady state runs used a space velocity of 0.025 min^{-1}.

Measured time average yield, $Y_{DEA/MEA}$, and time average selectivity to diethanolamine (DEA), $S_{DEA/MEA}$, are plotted versus the cycle period, normalized by the space time, in Figure 11.3a for symmetrical forcing. Comparable steady state values are given on the ordinates. For symmetrical cycling, yield decreases in the range of variables studied by 9.1% at the smallest period used or by 2.6% at the highest used. The figure shows that the selectivity, $S_{DEA/MEA}$, changes in the opposite direction, but it is roughly 27% higher than the steady state selectivity for the periods tested. Conversion of MEA is sharply depressed, of course. Selectivity exhibits a small maximum at $\tau/\bar{t}$ of about 1.7.

The time average yield of DEA can be increased slightly above the yield at steady state by letting the cycle split exceed 0.5. Split (s) is defined in terms of the duration of reactant flow to the reactor. Results for asymmetrical forcing are plotted in Figure 11.3b against the cycle split for a normalized cycle period of 3.0 ($\tau = 42$ min.). Space velocity is not constant in this plot. Nevertheless, the experimental time average yield exceeds the steady state yield by 1.5% at s = 0.6. This is a global yield enhancement in the range of variables investigated. A split of 1.0 represents steady state, while at the other extreme of s, the reactor contains just water and the rate and yield go to zero. $S_{DEA/MEA}$ shows a large variation with s, unlike the situation with cycle period under symmetrical forcing. Selectivity is unity at s = 0, but drops to the steady state value as the split mounts towards unity.

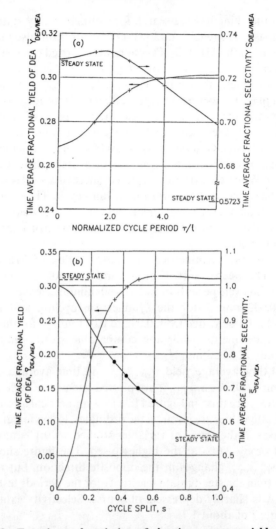

FIGURE 11.3 Experimental variation of the time average yield of DEA and selectivity to DEA versus in (a) the normalized cycle period for in phase symmetrical cycling of MEA between 0 and 0.47 mol/L and of Et_2O ("B") between 0 and 0.5 mol/L at 323 K in a CSTR, and (b) cycle split for in phase cycling of MEA between 0 and 0.50 mol/L and of Et_2O between 0 and 0.5 mol/L at a reduced cycle period of 3.0. Steady state yields and selectivity at the maximum concentrations are shown on the ordinates. Curves are from numerical simulations (Figure adapted from Renken (1974) with permission of the author).

Figure 11.4 shows the experimentally observed variation of the normalized MEA, Et_2O and DEA concentration during a cycle for the optimal yield condition shown in Figure 11.3b. Note that the DEA concentration

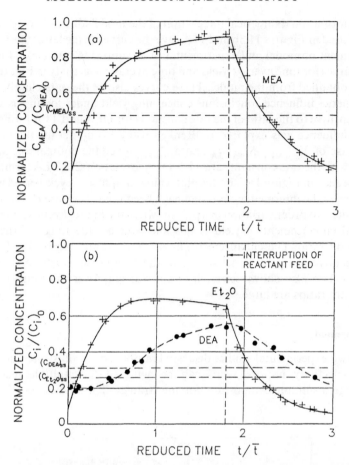

FIGURE 11.4 Experimental variation of MEA, Et$_2$O and DEA concentrations with reduced time (t/τ) within a cycle for $s = 0.6$, $\tau/\bar{t} = 3.0$ in the experiments summarized in Figure 11.3b. Comparable steady state concentrations of DEA and the reactants shown as horizontal lines. Curves are from numerical simulations (Figure adapted from Renken (1974) with permission of the author)

does not reach a maximum until the reactant flow is replaced by water. The maxima for MEA and Et$_2$O are out of phase. These phase shifts, a consequence of the choice of τ and s, augment conversion of MEA without substantially reducing selectivity to DEA and thus provide the increased $Y_{\text{DEA/MEA}}$.

Renken (1974) used the same set of assumptions as Lee et al. (1980) to model this consecutive-competitive reaction system. Separate measurements at 323 K were conducted to obtain the rate constants for DEA and

TEA formation. Numerical integration of the model for the forcing conditions used in Figures 11.3b and 11.4 gave the curves in the latter figure. The agreement of model and experimental observations is very close. Indeed, the curves for time average yield and time average selectivity in Figure 11.3 were obtained from the model. These curves also fit the data closely.

Kinetics influence conclusions concerning yield under forcing as yields are sensitive to the ratio of the rate constants for DEA and TEA formation. This influence was explored by Renken using his model for the system. Increase of $Y_{DEA/MEA}$, $S_{DEA/MEA}$ and S_{DEA/Et_2O} over the comparable steady state with the rate constant ratio (1 = DEA formation, 2 = TEA formation) can be seen in Figure 11.5. In this plot, values of split and cycle period which maximize the intermediate product have been used. They are not constant over the two order variation in k_1/k_2 shown in the figure. Results show that global enhancement of yield is possible over a wide range of the rate constant ratio. The enhancement, however, does not exceed 5% for the consecutive-competitive network and the comparison restrictions used. Figure 11.5 also shows selectivity improvements over the range of rate constant ratios are large.

Assessment

The two experimental studies discussed in this section illustrate that yield and selectivity improvements can be achieved for multiple reactions through a suitable composition forcing strategy. The studies, too, demon-

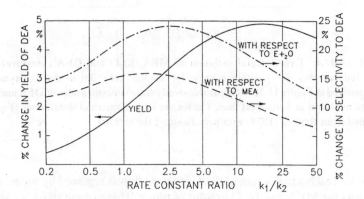

FIGURE 11.5 Calculated effect of the rate constant ratio for DEA and TEA formation steps on the change in DEA yield and selectivity under cycling for optimal values of reduced cycle period and cycle split (with respect to intermediate yield) for in phase, on-off reactant cycling in an isothermal CSTR with consecutive-competitive 2nd order reactions (Figure adapted from Renken (1974) with permission, © 1974 Verlag Chemie GmbH)

strate the magnitudes of yield and selectivity enhancements possible for the simple consecutive-competitive reaction network they considered. Despite different kinetics and restrictions on the forcing-steady state comparison, there is general agreement between the two studies.

There are other forcing strategies, such as forcing with a phase shift, that could be usefully explored using the homogeneous reaction systems just discussed. Certainly further inquiry would extend our still limited knowledge of how forcing and forcing strategy effects performance. Good agreement of model and experiment suggests that simulation would be satisfactory for such investigations. If studies are extended to more complicated networks with, say, parasitic reactions of the reactants and/or reversibility, stiffness problems might make experiments faster and cheaper than simulation. New reaction systems, of course, would be required.

C. CATALYTIC CONSECUTIVE AND COMPETITIVE REACTIONS

We begin our examination of multiple catalytic reactions with this class. Many heterogeneous catalytic reaction systems, including some of major industrial significance, are consecutive-competitive reactions. Examples are most hydrogenation reactions, many partial oxidation reactions, and perhaps all polymerization reactions. Unlike the homogeneous reactions we have just considered, it is the stoichiometric equations which fall into the class. Mechanisms occurring on the catalyst surface and transport steps certainly change the structure so that observations for homogeneous reactions of this class cannot be extended to the heterogeneous catalytic reactions with any confidence.

As a tie-in to the homogeneous reactions just discussed, we begin our considerations with ethanolamine production via a heterogeneous catalytic route.

Catalytic Reactions Of Ammonia And Ethylene Oxide

Ethanolamines are produced commercially by an acid catalyzed, liquid phase reaction between ethylene oxide (Et_2O) and ammonia. The stoichiometric relations are given by Equation [11.8] where "A" = NH_3, "B" = Et_2O, "D" is ethanol and "C" – "F" are respectively monoethanolamine (MEA), diethanolamine (DEA) and triethanolamine (TEA). In a published but not widely distributed Ph.D. thesis, Vamling (1987) explored ways to increase Et_2O conversion and selectivity to MEA using a reactor configuration and operating conditions that were close to those used commercially. Composition forcing was tested using on off switching of the Et_2O flow. Vamling's choice of this forcing mode rested upon its successful use on butadiene hydrogenation (Al-Taie and Kerschenbaum, 1978).

Unusual in the Vamling work was the choice of an integral, near adiabatic, tubular packed bed reactor. The intent was to use a reactor that could be simply scaled if the study proved successful. The philosophy of attempting to model a commercial design is apparent in the experimental system that is shown in Figure 11.6a. A piston pump feeds the liquid reactants to the reactor and a separate coil preheater is used. The system contains a continuously operated stripper to remove unconverted NH_3 from the product stream and controls are included. This mimics the industrial configuration. The reactor, shown in Figure 11.6b, was 24.3 mm (i.d.) × 1000 mm and was fabricated from 316 stainless steel. Thermocouples were inserted axially along the reactor as indicated in the figure. Glass bead packing at the inlet and outlet left 830 mm for the pelleted catalyst charge. Ethanolamine products were measured chromatographically using a fused silica capillary column with a nonpolar, methylsilicon as the stationary phase.

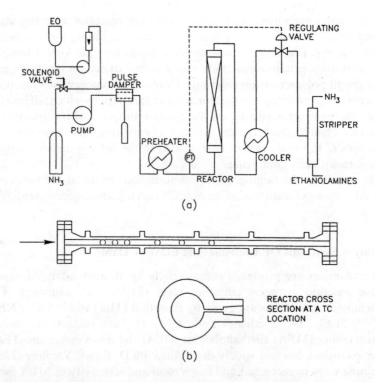

FIGURE 11.6 Schematic of the experimental system (a) and the near adiabatic tubular reactor (b) used by Vamling (1987) for the Et_2O-NH_3 reaction system (Figure taken from Vamling (1987) with permission of the author)

MULTIPLE REACTIONS AND SELECTIVITY   349

The primary catalyst was a strongly acidic, macro-porous, ion exchange resin, Amberlite 200, used as 20–50 U.S. mesh beads; a 13 X molecular sieve, in the Na form was also used. Grains of the sieve material had a mean diameter of 0.83 mm and were washed in an NH_4NO_3 solution prior to use.

Forcing experiments employed switching between NH_3 and a $NH_3 - Et_2O$ mixture at about 13 MPa in near adiabatic operation. Inlet temperatures were held close to 360 K, Et_2O conversion was about 99.9 % and the adiabatic temperature rise brought the reactor exit temperature to about 400 K. Experimental variables were cycle period, cycle split and inlet Et_2O weight fraction. Steady state experiments were undertaken for comparison purposes. Reactor modelling was also tackled.

No difference in MEA production per mol of Et_2O consumed was observed between composition forcing and steady state operation. This is evident in Figure 11.7 which plots dimensionless MEA concentration versus dimensionless consumption of Et_2O for the Amberlite 200 catalyst. Concentration are rendered dimensionless using the inlet Et_2O concentration. At the largest period used, 125 to 150 s, the Et_2O conversion decreased. A smaller number of experiments with the zeolite 13-X catalyst also failed to show a production difference.

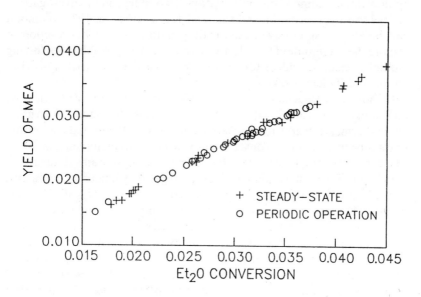

FIGURE 11.7 Relative production of monethanolamine, as a dimensionless outlet MEA concentration, as a function of the relative consumption of ethylene oxide, expressed as a dimensionless concentration change across the near adiabatic, integral reactor, under composition modulation and steady state operation. Data collected for an Amberlite 200 ion exchange resin catalyst at 13 MPa, inlet temperature of ca. 360 K (Figure taken from Vamling (1987) with permission of the author)

The lack of improvement surprised Vamling (1987) and explanations were sought. The most likely one is that the reaction system was not sufficiently disturbed for the MEA production to change. The time average Et_2O/NH_3 ratio ranged from 0.02 to 0.04. The on-off mode means that for symmetrical cycling, the ratio went from 0 to 0.04 or 0.08. High Et_2O conversion, 99.9%, caused even this amplitude to vanish by the reactor outlet. Furthermore, use of a liquid phase reaction reduced transport rates into the porous catalysts and lowered the amplitude significantly within the catalyst particles. Cycle periods were varied between 12 and 150 s. At $\tau = 20$ s, no temperature variation was observed in the bed over a cycle, while at $\tau = 50$ s, the variation was just 3 °C. These observations suggest there is a lower cycle frequency limit with solid catalyzed, liquid phase reactions. Liquid phase, catalytic reactions have been neglected in studies of composition modulation. It is unfortunate that Vamling did not proceed further with his study.

Butadiene Hydrogenation

Selective hydrogenation, an industrially important process, is a potential application for composition modulation. Two independent investigations of selective hydrogenation (Al-Taie and Kerschenbaum, 1978; Bilimoria and Bailey, 1978), however, used hydrogenation as a model reaction to explore the magnitude of the selectivity improvement possible when forcing multiple reactions and to test predictions of a theoretical analysis (Bilimoria and Bailey, 1978).

Butadiene hydrogenation over a commercial nickel catalyst (23 wt. % Ni on kieselguhr) was investigated by Al-Taie and Kerschenbaum (1978) using an isothermal, integral reactor fabricated from a 10 mm (i.d.) × 250 mm brass tube and held in a constant temperature bath maintained at 70 °C. The tube was charged with 18–20 U.S. mesh catalyst grains diluted with kieselguhr of the same grain size. The reactor was symmetrical modulated by adding either H_2 or C_4H_6 to a steady flow of a $H_2 - C_4H_6 - N_2$ mixture. This arrangement allowed different amplitudes and time average feed compositions. A mixer-integrator was used so only time average results are reported by the authors. A gas chromatograph using dimethyl sulphone on a chromosorb P column measured concentrations. Steady state runs provided conditions for the maximum steady state yield. These runs, as well as separate step change response measurements (Al-Taie, 1977), furnished parameters for a rather simplified model Al-Taie and Kerschenbaum used to simulate their experiments.

Objectives of the Al-Taie and Kerschenbaum (1978) study were to determine selectivity and yield improvement possible through periodically forcing an integral reactor and to find the cycling conditions which give the best performance. Modulated and steady state data were compared at the same C_4H_6 feed rate and a constant space velocity. Because the maximum selectivity for the first product in a consecutive-competitive reaction system is at zero conversion, selectivities to butene were compared for forcing conditions which gave yields identical to the maximum steady state yield. Variables examined by the researchers were cycle period, cycle amplitude, space time, and time average feed composition.

Steady state experiments at 343 K and 1.07 bar established that butene yield reached a maximum at a H_2/C_4H_6 ratio in the feed between 1 and 2 depending on the space velocity (SV). Conversion at the maximum was about 50% but independent of SV. Modulation runs were carried out consequently for a time average feed H_2/C_4H_6 ratio = 1.7 and conversions were about 50% for the shortest cycle periods used. Measurements were made at $1.85 \leqslant SV \leqslant 3.70\,s^{-1}$.

Cycle period and amplitude exert a large influence on butadiene conversion as shown in Figure 11.8a. Fast cycling at $\tau = 2\,s$ or less afford conversions which equal or exceed slightly the maximum conversions attained under steady state operation. Increasing the period, however, suppresses conversion. This is most severe at amplitudes (A) corresponding to switching between the single reactants and diluent. Amplitude is based on the variation of the C_4H_6 concentration between the half periods: $A = 1$ represents switching between H_2 in a N_2 diluent and C_4H_6 in this same diluent, that is, between feeds containing diluent and just one reactant. A substantial increase in selectivity was obtained, about 20% based on the optimum steady state operating conditions. This may be seen in Figure 11.8b. Apart from modulation at the amplitude extreme, neither amplitude or cycle period influences selectivity substantially. A curious result is that operating at $\tau = 2\,s$ and $A = 1$ increases conversion by about 8% over steady state, but provides the smallest improvement in selectivity. Despite the expected negative effect of forcing on conversion, Figure 11.8c shows that the selectivity enhancement leads to a yield enhancement at the shorter cycle periods. This reaches 20% at $\tau = 2\,s$, the highest cycling frequency used. Increasing cycle period decreases yield. Amplitude is an important variable. Using a small amplitude, corresponding to switching between $C_4H_6 - H_2 - N_2$ mixtures, results in improved butene yields at τ up to 30 s.

Al-Taie and Kerschenbaum (1978) attribute improvement under periodic forcing to the dynamics of butadiene adsorption, but do not discuss their explanation in any detail. Instead, they formulate a plug flow model for their essentially isothermal reactor assuming competitive adsorption of C_4H_6, C_4H_8 and C_4H_{10}, negligible H_2 adsorption and that hydrogenation occurs via a Rideal mechanism. Mass transfer interference is assumed to be absent. Adsorption rate and equilibrium constants were obtained from non-competitive adsorption measurements, while reaction constants for the surface reaction terms involving H_2 partial pressure and the C_4

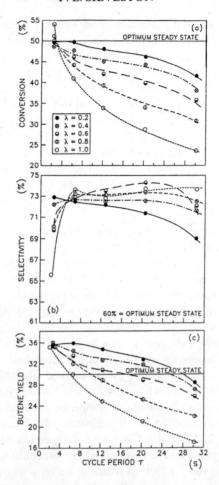

FIGURE 11.8 Variation of reactor performance under symmetrical forcing as a function of cycle period and amplitude for the consecutive hydrogenation of butadiene to butene and butane at 343 K and 1.07 bar over a Ni/kieselguhr catalyst in a near isothermal packed bed reactor: (a) conversion, (b) selectivity and (c) yield of butene (Figure adapted from Al-Taie and Kerschenbaum (1978) with permission, © 1978 American Chemical Society)

adsorbate concentrations were evaluated from steady state runs made by the authors. The Al-Taie - Kerschenbaum model predicted qualitatively the experimental observations. It showed a significant yield enhancement at low cycle periods and increasing selectivity with longer periods. The model failed, however, to reproduce quantitatively either the steady state or modulation results.

Acetylene Hydrogenation

The experimental study of this reaction by Bilimoria and Bailey (1978) using nickel catalysts was concerned with whether or not global yield and selectivity enhancement is possible. A further objective appears to have been evaluating the level of modelling that is needed to adequately predict performance under composition modulation. A simulation of acetylene hydrogenation employed steady state data, but assumed the same reactor configuration and operating conditions used by the experimenters (Lee and Bailey, 1980). The question raised, thus, is whether or not parameters based on steady state measurements simulate a reactor operating under forcing.

Experiments were carried out using the equipment and procedures discussed in Chapter 1 with the exception that the acetylene flow rate was held constant and only the hydrogen flow rate was manipulated. The flow of a diluent, N_2, was manipulated concurrently so as to maintain space time constant during a cycle. A stainless steel, spinning basket reactor was used. At the total flow rate of 5 cm^3/s employed, the mean residence time was 66 s. Primary analysis of the reactor products used a gas chromatograph with a Poropak N column and a FID. Flow leaving the reactor could be sampled or measurements could be made following a holdup vessel which served as a flow integrator. An online IR was also used to observe product concentrations within a cycle.

Two commercial catalysts, Harshaw Ni-0707T and Girdler G52, were chosen for study. The former was used by Komiyama and Inoue (1968) and furnished the kinetics adopted by Lee and Bailey (1980). The latter catalyst had a heavier nickel loading and was much more active. Both catalysts were crushed and sieved to a narrow size range, 0.84–1.2 mm, and soaked in a 5 wt. % solution of sodium thiosulfate prior to use. Once in the reactor, the catalysts were activated in flowing H_2 at 477 K. All experiments were performed at 108 KPa and 439 K. Steady state measurements were undertaken by varying the H_2 flow rate between 0.25 and 1.625 cm^3/s while holding the C_2H_2 flow rate at 0.5 cm^3/s to give a 0.1 to 3.3 variation in the H_2/C_2H_2 ratio. Variables in the forcing experiments were cycle period and split. Split was defined in terms of H_2 presence in the reactor feed and ranged from 0.133 to 0.867.

Figure 11.9 shows that the product yields increase under composition forcing for both catalysts over yields attainable at steady state. Figure 11.9a plots the mol fractions of all the hydrocarbons leaving the back mixed reactor, whereas Figure 11.9b shows only the ethane mol fractions to simplify the plot. The ethene mol fractions exhibit the same shape as the ethane curves, but are higher. A comparison of the ethane results indicates the Girdler catalyst has an activity some 10 x that of the Harshaw catalyst. The Girdler catalyst, however, underwent fouling and deactivation during the experiments. This was handled by Bilimoria and Bailey by running

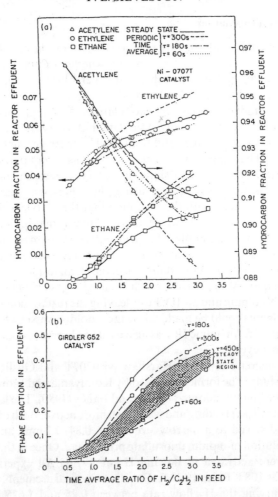

FIGURE 11.9 In (a) dependence of the hydrocarbon mol fraction in the reactor off gas on the time average hydrogen/acetylene ratio in the feed for steady state and modulated operation at 439 K, 108 KPa and a space velocity of 0.015 s^{-1} for the Harshaw Ni-0707T catalyst. Dependence of the C_2H_4 mol fraction on the H_2/C_3H_6 is given in (b) for the Girdler G52 catalyst and the conditions in (a). Modulation of the H_2 mol fraction was used and the effect of cycle period is shown (Figure adapted from Bilimoria and Bailey (1978) with permission, © 1978 American Chemical Society)

forcing experiments between steady state ones. The latter are represented by the shaded region in Figure 11.9b. Because of fouling, only a qualitative interpretation of the cycle period effect can be made. Short periods, in this case 60 s, have just a small influence on yield. Yield depends primarily on cycle split, s. It is s that is varying along the abscissa for these measurements. The largest H_2/C_2H_2 ratio in the figures corresponds to the highest s used.

In the case of the Harshaw catalyst, increasing the period, τ, increases the yields of both C_2H_4 and C_2H_6; however, for the Girdler catalyst the optimum τ for yield enhancement appears to be about 180 s. Longer periods depress the ethane yield. There is probably an optimum τ for the Harshaw catalyst, but it was not uncovered by the experiments. Increasing s increases $Y_{ethene/ethyne}$ and $Y_{ethane/ethyne}$, but not the C_2H_4 yield enhancement, Ψ_Y, over steady state at $\tau = 180$ s for the Harshaw catalyst or Ψ_Y for ethane at $\tau = 60$ s for the Girdler catalyst.

In contrast to butadiene hydrogenation where a different cycling strategy was used, forcing the hydrogen mol fraction did not increase selectivity to the intermediate product, ethene. For the Harshaw catalyst, steady state selectivity to C_2H_4, $S_{C_2H_4/C_2H_6}$ is 91 % at $H_2/C_2H_2 = 1$, but under forcing $S_{C_2H_4/C_2H_6}$ drops to 87 % at $\tau = 60$ s and to 82 % at $t = 300$ s. At $H_2/C_2H_2 = 3$, the C_2H_4 selectivity is 72 % at 60 s, but falls to 71 % at $\tau = 180$ s and to 64 % at $\tau = 300$ s. With acetylene hydrogenation over nickel, modulating the H_2 mol fraction raises the acetylene conversion and selectivity to the final product, ethane, in the reaction chain. Out of phase forcing of both butadiene and hydrogen mol fractions, however, increased $S_{C_2H_4C_2H_6}$, but raised Y_{ethene} only at the highest frequencies used. Conversion was suppressed by modulation except at high frequency and there the improvement was small.

Simulation of acetylene hydrogenation in a continuous stirred tank reactor by Lee and Bailey (1980) used as rate expressions:

$$r_1 = \frac{k_1 C_{C_2} C_{H_2}}{1 + K_A C_{C_2}} \tag{11.10}$$

for ethene from acetylene, while for ethene hydrogenation to ethane,

$$r_2 = \frac{k_1 C_{C_2} (C_{H_2})^n}{1 + K_A C_{C_2}} \tag{11.11}$$

These expressions were derived by Komiyama and Inoue (1968) from steady state measurements. The convex nature of Equation [11.11] with respect to H_2 indicates that H_2 forcing should improve $S_{C_2H_4/C_2H_6}$. Indeed, when these terms were used in material balances for C_2H_2 and C_2H_6, neglecting adsorbate storage on the nickel surface, the behaviour given in Figure 11.10 was obtained for $\tau = 60$ s and a space time of 50 s. For the simulation a catalyst density of 1200 kg/m³, reactor void fraction of 0.91, rate constants evaluated at 433 K and 1 bar were assumed. The Lee-Bailey simulation of H_2 forcing predicts a 46 % improvement in $Y_{ethene/ethyne}$ at $H_2/C_2H_2 = 0.5$. A 30 % decrease in $Y_{ethane/ethyne}$ was predicted as well so that the simulation showed enhanced selectivity to ethene. Comparison of Figures 11.9a and 11.10 shows, however, that the predictions are badly in error in magnitude as well as in the influences of τ and s.

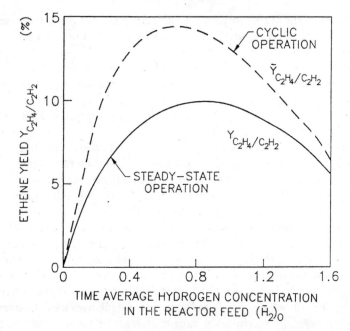

FIGURE 11.10 Predicted variation of the time average yield of ethene with the time average H_2/C_2H_2 ratio in the feed for steady state and modulated operation ($\tau = 21$ s) at 433 K, 101.3 KPa and a space velocity of 0.02 s^{-1} in a continuous stirred tank reactor using steady state data from Komiyama and Inoue (1968) (Figure adapted from Lee and Bailey (1980) with permission, © 1980 American Chemical Society)

Failure of the model results from the neglect of C_2H_2 and C_2H_4 adsorption dynamics on the catalyst surface. Adsorption dynamics furnishes an explanation of Bilimoria and Bailey's data as well. Their importance is illustrated by Bilimoria and Bailey's concentration measurements during a cycle. These are shown in Figure 11.11 for the Harshaw catalyst when $\tau = 300$ s. At $s = 0.867$ (top of the figure), a wrong way behaviour appears. Ethene and ethane mol fractions drop rather than increase after the H_2 partial pressure is stepped up at $t/\tau = 0$ or 1. The C_2H_2 mol fraction rises rather than drops. In the bottom of Figure 11.11, at $s = 0.133$, wrong way behaviour again shows up. This behaviour is explained by strong C_2H_2 adsorption relative to H_2 and the reaction products and by an adsorption process that is slow relative to the other reaction steps. The importance of C_2H_2 adsorption is suggested, of course, by the denominator terms in Equations [11.10] and [11.11] and is consistent with the experimental effect of the H_2/C_2H_2 ratio on product concentrations for steady state operation shown in Figure 11.9a and for the large swing in the C_2H_2 mol fraction in the middle portion of Figure 11.11 at $s = 0.467$. Clearly, adsorbate storage must be included if a model is to accurately represent composi-

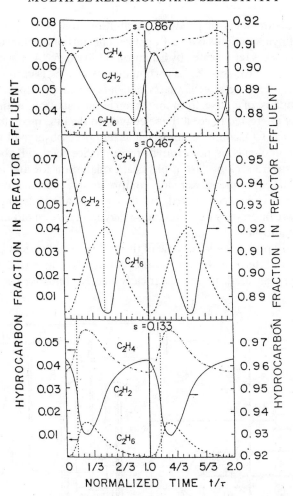

FIGURE 11.11 Variation of hydrocarbon mol fractions in the reactor off gas with time within a cycle under modulation of the H_2 concentration in the feed to a back mixed reactor employing a Harshaw Ni-0707T catalyst and operating at 439 K, 108 KPa and a space velocity of 0.015 s^{-1}. Modulation used $\tau = 300$ s and $s = 0.867$ in (a), 0.467 in (b) and 0.133 in (c) (Figure adapted from Bilimoria and Bailey (1978) with permission, © 1978 American Chemical Society)

tion forcing. The successful model proposed by Al-Taie and Kerschenbaum for their hydrogenation work allowed for adsorbate storage.

Total Oxidation

In perhaps the earliest experimental study of solid catalyzed, multiple reactions under composition forcing, Wandrey and Renken (1973, 1974) reported measurements of product distribution and conversion for the oxidation of hydrocarbons over a Pt gauze. The authors assumed that the reaction network is represented by Equation (11.8) where "A" is a hydrocarbon species, "B" is oxygen, "C" is carbon monoxide, "E" is carbon dioxide and "D" is water. The stoichiometry depends on the hydrocarbon consumed and differs, of course, from that given in the equation. Wandrey and Renken were interested to see if the selectivity to the intermediate product in the sequence, CO, could be improved as had been observed by Renken (1972) for a homogeneous, noncatalytic reaction (See Section B above). The investigators were seeking also a physical explanation for any change in selectivity. The experiments were less comprehensive than the hydrogenation ones we have just considered, but they represent a different reaction system and are interesting as well because they were carried out under near adiabatic conditions.

Figure 11.12 presents a schematic of the experimental system employed by Wandrey and Renken. Reactants were preheated or vaporized, if a liquid hydrocarbon was used, in the electrically heated fluidized bed of glass

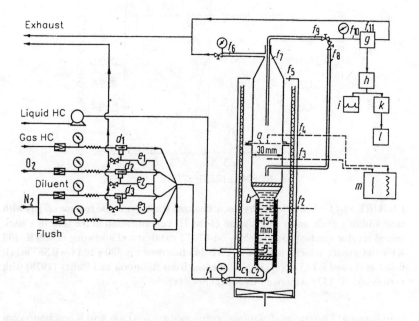

FIGURE 11.12 Schematic of the experimental equipment for the Wandrey-Renken study of the oxidation of hydrocarbons over a Pt gauze at 103 KPa. In the diagram a – Pt gauze, b – fluidized bed preheater/vaporizer, c – electrical heating, d – solenoid valves, e – flow meters, f – thermocouples, g – sampling tubes, h – gas chromatograph, i, k, l, m – recording equipment (Figure adapted from Wandrey and Renken (1973) with permission, © 1973 Verlag Chemie GmbH)

ballotini. The reactor consisted of a horizontal gauze sheet of 0.06 mm Pt wire, woven to have 1024 openings/cm^2, mounted near the bottom of a 30 mm i.d. glass tube. Although not shown in the figure, both reactor and preheater were contained in an electrically heated chamber to reduce heat loss. The temperature in this chamber was maintained at 40–65 °C below the temperature of the fluidized bed. The feed system shown in the figure allowed different forcing strategies to be used: 1) in phase cycling of the reactants, 2) holding the concentration of the hydrocarbon constant and switching the O_2 flow on and off, or 3) maintaining the O_2 concentration constant and switching the hydrocarbon flow on and off. Cycling of the reactants 180° out of phase was also possible, but such experiments were not reported by the investigators. The diluent was either N_2 or He and its flow was adjusted so the space velocity was constant throughout a cycle. Thermocouples above and below the gauze measured reaction extent through the temperature change. A gas chromatograph with TC detectors and Porapak S and 5 A molecular sieve columns in series provided feed and product analysis.

Three hydrocarbons were used in the experiments: propane (Wandrey and Renken, 1974) and butane and cyclohexene (Wandrey and Renken, 1973). Symmetrical forcing was used and the variable studied was cycle period, τ. Different preheater temperatures were used and usually set to insure ignition. Reactant concentrations and space velocities were changed in the individual experiments, but these were not treated as variables. They appear to have been chosen to facilitate the measurements.

Experiments employing propane utilized in phase cycling, that is the reactor feed alternated between the reactant mixture and a He diluent. This was the strategy used by Lee et al. (1980) and Renken (1972) for their studies of homogeneous, noncatalytic reactions. A search on cycle period established that $\tau = 5.4\,s$ gave the largest increase in selectivity to CO. A CO yield of 0.25 % was measured under steady state operation at the time average feed composition and the same space time whereas in phase cycling raised this yield to 6.45 %. For the 5.4 s period, the yields of the three oxidized products are plotted in Figure 11.13 against time normalized by the cycle period. The feed to the Pt gauze switched to a mixture of 2.8 mol/L C_3H_8 and 10 mol/L O_2 in He at $t/\tau = 0$ and 1 and to He alone at $t/\tau = 0.5$ and 1.5. Minima and maxima for the different species occur at different intervals after the feed composition switch so the plot indicates that the transport lag in the feed and sampling lines cannot be more than 0.2 to 0.4 s. The primary CO peak appears well before the maximum temperature on the gauze and also leads the CO_2 peak. The H_2O peak after the switch to He precedes the CO_2 peak. This can be explained by the initial formation of CO followed by its oxidation to CO_2. The secondary CO maximum coincides with the temperature maximum after the O_2 flow has been halted and suggests burnoff of residual carbon on the Pt surface with adsorbed oxygen or CO

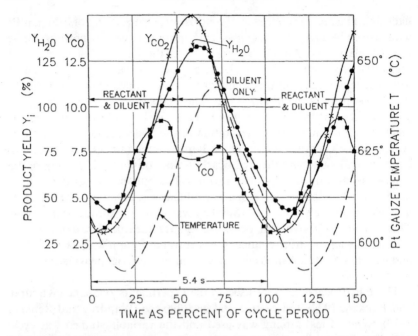

FIGURE 11.13 Variation in product yield and Pt gauze temperature with time within a cycle of 5.4 s for the oxidation of propane over Pt gauze in a near adiabatic reactor. Preheater temperature = 530 °C, total pressure = 103 kPa and the time average C_3H_8 mol fraction in the feed = 0.028. Cycles consist of alternating reactant and diluent feed streams. Times within the cycle is shown as fraction of τ (Figure adapted from Wandrey and Renken (1974) with permission, © 1974 Verlag Chemie GmbH).

desorption from the surface. The decline in CO_2 production extends for about 2.7 s during the time when no reactants are passing over the catalyst so that there must be residual matter and adsorbed O_2 on Pt to account for the CO and CO_2 formed. The surprisingly long temperature lags in the figure, ca. 1 s, may be due to the endothermic cracking of C_3H_8 on the gauze surface after the reactant mixture is reintroduced or to continuing oxidation after the mixture flow has terminated, although thermocouple heat transport lags undoubtedly contribute to displacement of the temperature maxima and minima. Evidently, the reaction network is more complicated than the sequence given by Equation [11.8].

A much larger increase in CO yield and selectivity was achieved using a strategy of cycling just the oxygen flow while holding the hydrocarbon flow constant (Wandrey and Renken, 1973). These experiments employed a feed containing 2.5 mol % cyclohexene; O_2 was switched between 0 and 17.5 mol %. Space time based on moles of Pt in the catalyst/mole of C_6H_{10} flowing was 1.16 s and the preheater temperature was 415 °C. For the shortest τ used, 16 s, mixing in the fluidized bed and in the reactor damped the variation in the O_2 mol % in the feed so that the system behaved almost as though it was at steady state. The effect of raising the cycle period is shown in Figure 11.14. Cyclohexene conversion increased from 9.5 to 45.3 % for change in τ from 16 to 120 s, while the selectivity to CO went from about 8.5 % to 55 %. However, if the selectivity change is viewed from an enhancement standpoint, in phase switching between diluent and reactants appears to be about as effective a strategy as cycling O_2 flow while holding hydrocarbon flow constant. In phase cycling strategy enhanced the CO yield by almost 2500 %, whereas the O_2 cycling strategy enhanced yield by 2400 %.

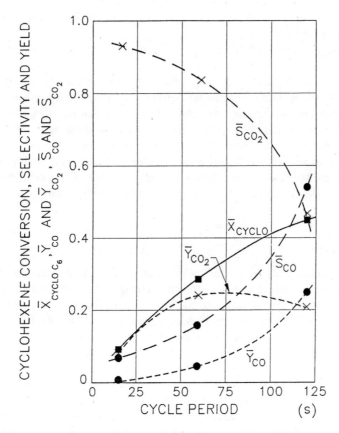

FIGURE 11.14 Influence of cycle period on time average cyclohexene conversion, carbon oxide yields and selectivities for oxidation of cyclohexene over a Pt gauze with a preheater temperature = 415 °C, pressure = 103 KPa, and time average C_6H_{10} and O_2 mol fractions = 0.025 and 0.088 respectively, SV = 0.86 s^{-1}. O_2 mol fraction in the feed modulated symmetrically between 0 and 0.175 (Figure adapted from Wandrey and Renken (1973) with permission, © 1973 Verlag Chemie GmbH)

Two explanations were offered by Wandrey and Renken for the yield and selectivity changes observed. The first was that O_2 modulation caused oxidation to proceed at a time average temperature that was difficult to attain in steady state operation. The argument draws on the existence of multiplicity in oxidation: ignited and smoldering states can exist for the same feed conditions. Composition modulation induces a time average gauze temperature in between a high temperature associated with the ignited state and a low temperature observed in the smoldering state. Presumably, as the temperature of the catalyst cycles back and forth across this mean temperature, surface conditions favor the formation of CO. Wandrey and Renken (1973) demonstrated stabilization of a time average temperature experimentally using butane oxidation where the butane flow rate was cycled while the oxygen flow rate remained constant.

Wandrey and Renken's second explanation relied on the strong adsorption of hydrocarbons and CO on Pt. This provides a reservoir of carbon and CO on the surface. With high carbon or CO coverage, O_2 adsorption is limited so the surface is oxygen deficient encouraging CO formation. CO is retained on the surface and desorbs as the surface temperature raises, but this takes place after the O_2 flow has stopped so that further oxidation to CO_2 does not proceed. Measurements of changing product concentrations within a cycle for $\tau = 120\ s$ show mol fractions of the carbon oxides in the gas leaving the reactor reach maxima some $30\ s$ after O_2 is reintroduced. This long delay can be explained only through strong adsorption.

Assessment

Experiments discussed in this section demonstrate that selectivity to an intermediate product in solid catalyzed, consecutive-competitive reaction systems can be profoundly altered through periodic composition forcing provided the forcing amplitude is adequate and the frequency is low enough so that mixing doesn't significantly reduce amplitudes. Selectivity enhancement arises from non-linear reaction kinetics on surfaces and differences in adsorption/desorption equilibria and rates for reactants and products. The dual source of improvements make it risky to adopt a cycling strategy before measurements are made or before a reliable model exists. This section has shown that three cycling strategies, in phase cycling of two reactants, single reactant cycling and out of phase cycling of two reactants, can alter selectivity to the intermediate and change its yield.

Solid catalyzed, consecutive-competitive reaction systems are revisited in Chapters 12 (Partial Oxidation), 13 (Fischer-Tropsch synthesis) and 14 (Polymerization).

D. COMPETITIVE REACTIONS

There are several commercially important reaction systems in which different products are formed from the same set of reactants. Perhaps the most prominent of these systems is methanol synthesis over the early catalysts in which methane was a secondary product. Although modern copper-zinc mixed oxide catalysts suppress completely the methane reaction, the synthesis of higher alcohols along with methanol represents stoichiometrically, at least, competitive pathways. Unfortunately, the production of higher alcohols from synthesis gas under composition modulation has not been explored. Another competitive reaction system of commercial significance is the catalytic dehydration of alcohols to form olefins where the formation of ethers is an important side reaction. This system has been examined under composition forcing.

A second type of competitive reaction system involves just one of the reactants or a primary product. Typical of such systems are gum formation or thermal cracking which consume reactant and often lead to catalyst fouling. Ethylene, one of the reactants in the Pd catalyzed formation of vinyl acetate, polymerizes under reaction conditions. Oxygen, also a reactant, apparently oxidizes some of the vinyl acetate product. We refer to these side reactions involving just small amounts of reactant or product as parasitic reactions. Feed stream impurities can also be the source of parasitic reactions. There is some oblique discussion of the effect of periodic operations on parasitic reactions in the literature, but for the most part this possible application has not been explored. Problems with parasitic reactions are usually solved by modifying or changing catalyst, or by altering operating conditions.

Methanol Synthesis

Studies of this synthesis by Nappi et al. (1985) and later by Chanchlani et al. (1994) examine enhancing the selectivity to methanol by composition modulation. In the former, Nappi et al. employed a catalyst formulated on recipes published by Herman et al. (1979) in an internal recycle reactor that they treated as a fully back mixed reactor. Two cycling modes were used: 1) periodic switching of the H_2/CO ratio in the reactor feed, and 2) periodic dosing of the feed with CO_2. Details of the Nappi experiments are given in Chapter 8. A selectivity change was observed under forcing which was at first attributed to modulation. However, it was later discovered that the CO source contained low, but still measurable concentrations of CH_4. These concentrations were the same magnitude as those measured so that a selectivity effect was uncertain.

Chanchlani et al. (1994) studied the synthesis under composition forcing using mixtures of CO and CO_2 and H_2 over two low pressure Cu/Zn oxide catalysts. These investigators looked for a selectivity effect even though methane production with the catalysts used was very small. Details of the Chanchlani experiments are also discussed in Chapter 8. It was not possible to remove the ppm levels of CH_4 from CO, but these levels were measured and used to correct the CH_4 production calculations. A large change in selectivity to methane, but a minuscule improvement in the methanol selectivity under forcing was observed after corrections were made. These results indicate composition modulation affects selectivity in this system. Once again, the effect is uncertain because variations of CH_4 concentration in the reactor off gas were entirely in phase with the imposed CO variations and methane measurements were made at the lower limit of the FID used.

With mixed carbon oxides in the reactor feed, Chanchlani et al. (1992) found an unusual competitive reaction network even if methane is not formed. Methanol can be formed from either CO or CO_2 over Cu/Zn oxide catalysts, albeit at different rates. Furthermore, H_2O is formed in synthesizing methanol from CO_2 so the water gas shift reaction can proceed. It is these alternate paths to CH_3OH which form the network. The presence of the shift reaction, however, changes the network to a more complicated competitive-consecutive system. Chanchlani et al. (1994) report large changes in the synthesis rate when the reaction system is forced by periodically switching the CO/CO_2 ratio in the feed. Some of the observed enhancement may arise from either the influence of forcing on the rates of the parallel CH_3OH reactions or on the rate of the shift reaction. The investigators did not speculate on this possibility and their experimental system did not allow measurement of the methanol formation rates by different reaction pathways.

There are sufficient hints of a cycling effect on methane formation in the two experimental studies to justify investigating the application of composition forcing to the synthesis of higher alcohols. Perhaps modulation offers a means of achieving a more desirable distribution of the alcohol products.

Ethanol Dehydration

This reaction proceeds over alumina at temperatures between 200 and 300 °C and atmospheric pressure. Two products are possible: ethene ($C_2^=$) and diethyl ether (DEE). The stoichiometric relations are

$$CH_3CH_2OH \rightarrow CH_2=CH_2 + H_2O \qquad (11.12)$$

$$2\ CH_3CH_2OH \rightarrow CH_3CH_2OCH_2CH_3 + H_2O \qquad (11.13)$$

Renken et al. (1974) studied this system at 240 °C in a reactor packed with a γ-alumina catalyst. The system was forced by periodically replacing ethanol (EtOH) in the feed to the reactor by diluent. Steady state measurement were also made so that the improvement under forcing could be determined. Comparisons were made at EtOH feed concentrations and flow rates that were the same as the time average concentrations and flow rates used in the cycling experiments. Variables in the experiments were cycle period, τ, and split, s. Split is defined as the fraction of τ in which EtOH is part of the feed stream. In discussing their work, Renken et al. (1974) mention that diethyl ether may decompose to $C_2^=$ and water or disproportionate to the olefin and the alcohol. Furthermore, the reactions given by Equations (11.12) and (11.13) are reversible. However, these investigators assumed the two side reactions and the reverse reactions do not proceed sufficiently under the conditions used. They were neglected in the system model.

Periodic interruption of the ethanol supply to the catalyst bed enhances $Y_{C_2^=/EtOH}$ as Figure 11.15a demonstrates. The maximum increase in $Y_{C_2^=/EtOH}$ is about 25 % for this symmetrical cycling and was obtained at $\tau/\bar{t} = 20$ where $\bar{t}$ is the space time. Below $\tau/\bar{t} = 10$, the yield enhancement decreases towards zero, probably because mixing in the bed diminishes the amplitude of the EtOH concentration variation. Figure 11.15a suggests that as the dimensionless period approaches the quasi-steady state limit the increase in $Y_{C_2^=/EtOH}$ goes to zero.

Increasing the relative duration of the flow interruption portion of the cycle can raise selectivity to $C_2^=$ and its yield even further. Results for the cycle split experiments of Renken et al. appear in Figure 11.15b. Steady state corresponds to $s = 1.0$. A split of 0.25, the lowest used at $\tau/\bar{t} = 73$ raised $Y_{C_2^=/EtOH}$ to 70 % which is about 33 % greater than the yield under comparable steady state operation. A much shorter dimensionless period of 7.3 still gives a small increase in $Y_{C_2^=/EtOH}$, suggesting that the absolute duration of flow interruption is important.

Although the role of flow interruption duration was not explored by Renken et al. (1974), it was recognized by them as the explanation for the large increase in selectivity they observed. Flow interruption generally reduces reactant conversion so that if yield increases, there must be a much larger increase in selectivity. Water and to a lesser extent ethanol are relatively strongly adsorbed on the alumina surface and these species inhibit the formation of $C_2^=$ more than they do the formation of DEE. This may occur through the blockage of sites for dehydration of DEE to $C_2^=$ and/or for DEE disproportionation. Evidence that $C_2^=$ selectivity enhancement arises through interference with adsorption/desorption processes comes from measurement of product mol fractions leaving the fixed bed as well as from modeling. The mol fraction measurements are given in Figure 11.16 for τ and s which gave the largest yield improvement in Figure 11.15b. The $C_2^=$ mol fraction falls immediately when EtOH is re-introduced into the reactor feed, while the DEE mol fraction displays a delay before rising rapidly. The source of this delay is not known. Surface equilibrium appears to be reached before the end of the EtOH portion of the cycle. When the

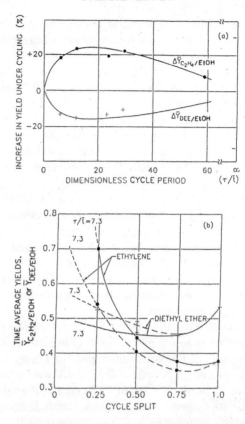

FIGURE 11.15 Effect of a symmetrical on-off modulation of the ethanol feed rate to a packed bed reactor on time-average ethene and diethyl ether yields: a) change of yield relative to the comparable steady state yields with dimensionless cycle period, b) effect of cycle split on $C_2^=$ and DEE yields at two values of $\tau/\bar{t}$. Ethanol dehydration was over a γ-alumina catalyst at 513 K and near atmospheric pressure (Figure adapted from Renken *et al.* (1974) with permission, © 1974 Verlag Chemie GmbH)

EtOH flow is shut off, DEE and $C_2^=$ mol fractions fall and rise immediately at rates which appear to be in a ratio of 2:1. Thus, $C_2^=$ forms primarily in the part of the cycle when EtOH is not fed to the reactor, whereas diethyl ether forms mainly when EtOH is in the reactor feed.

Modeling of the system by Renken *et al.* used measured adsorption equilibrium constants which were in the ratio of $(K_{ad})_W : (K_{ad})_{EtOH} : (K_{ad})_{DEE} : (K_{ad})_{C2=} = 54 : 15 : 4 : 0$. Adsorption rate constants were assumed to be in the ratios of $(k_{ad})_{DEE} : (k_{ad})_{EtOH} : (k_{ad})_W = 6 : 3 : 1$. Rate constants for the reactions given in Equations (11.12) and (11.13) were drawn from steady state

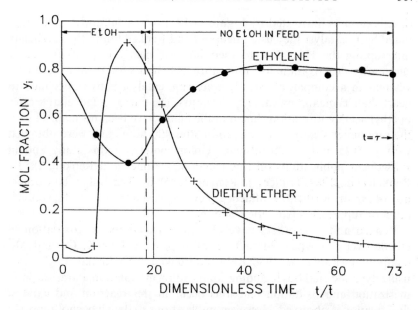

FIGURE 11.16 Variation of the ethene and diethyl ether mol fractions in the reactor off-gas with dimensionless time within a cycle under on-off modulation of the ethanol flow to the reactor. These measurements are for the experiment at $s = 0.25$ and $\tau/\bar{t} = 73$ in Figure 11.15b (Figure adapted from Renken *et al.* (1974) with permission, © 1974 Verlag Chemie GmbH)

measurements. For the reactor model, a cascade of 10 identical CSTRs was assumed. This model, which allowed for adsorption/desorption dynamics, qualitatively predicted the trends shown in Figure 11.15, but overpredicted the increase in $Y_{C_2^=/EtOH}$ and underpredicted by an order of magnitude the optimal τ. Nevertheless, prediction of the general trend of yield versus τ supports the view of Renken and co-workers that improvement of selectivity under composition modulation is due to interference in the adsorption/desorption processes on the alumina surface.

Parasitic Reaction Systems

This class of competitive reaction systems appears to have been examined just twice. Leupold and Renken (1978) mention in their study of composition forcing of the formation of ethyl acetate (EtAc) from ethene and acetic acid (HAc) over a supported sulfuric acid catalyst that parasitic reactions, such as $C_2^=$ polymerization and oxidation, lead to a slow deactivation of the catalyst. The system used for their study and the experiments performed were discussed in Chapter 9. Forcing was accomplished by periodically interrupting the HAc feed. This mode of operation prevented catalyst deactivation as illustrated by Figure 11.17. For the sulfuric acid impregnated SiO_2 catalyst they used, Leupold and Renken found that cyclically interrupting the HAc flow to their integral, nearly isothermal reactor eliminated deactivation for the entire 500 hours of their experiment, whereas in a comparable steady state run, catalyst activity declined to nearly half its maximum value in 400 hours. The time average relative rates of ethyl acetate formation shown in the figure have been normalized with the rate under steady state operation after 400 hours and were obtained with $s = 0.75$ and $\tau = 20$ minutes. Unfortunately, in this study and in subsequent publications on this reaction (Dettmer and Renken, 1983b; Renken et al., 1984; Truffer and Renken, 1986), Renken and his co workers do not explore further the influence of cycling variables on catalyst stabilization or suppressing parasitic reactions.

Renken and co-workers have also investigated the acetoxidation of ethene over a 2 wt.% Pd/SiO_2 catalyst promoted by K, Cd and Mn compounds (Rabl et al., 1986; Rabl and Renken, 1986; Renken et al., 1989) using dynamic methods including composition modulation. Parasitic polymerization and oxidation reactions occur in this reaction and catalyst deactivation is observed. However, publications to date have not reported forcing results on deactivation.

The positive effect of cycling on reducing or eliminating catalyst deactivation shown in Figure 11.17 is not an isolated observation. Chanchlani

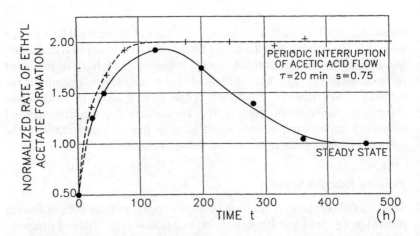

FIGURE 11.17 Rates of ethyl acetate formation under steady state and modulation of the acetic acid flow rate. Rates have been normalized using the steady state rate after 400 h. Modulation results are for $\tau = 20$ min. and $s = 0.75$. Both modes of operation employed a 25 wt.% H_2SO_4 on SiO_2 catalyst at 413 K and 6 bar, and $SV = 4880$ h^{-1} (Figure adapted from Leupold and Renken (1978) with permission, © 1978 Verlag Chemie GmbH)

et al. (1994) in their study of forcing of the methanol synthesis reaction over Cu-Zn oxides report a 2 fold decrease in the rate of deactivation. Deactivation mechanism, the effect of the cycling variables, and any trade-off that may occur between the rates of the primary reactions and the parasitic ones all need attention. Perhaps parasitic reactions are not the cause of deactivation. Because of the large amount of information already collected, either ethyl acetate formation from $C_2^=$ and HAc or the acetoxidation reaction to form vinyl acetate would appear to be good candidates for a study of the influence of composition modulation on deactivation and, presumably, suppression of parasitic reactions.

E. METHANE HOMOLOGATION

Belgued et al. (1991, 1992a,b), Amariglio et al. (1994, 1995b) and Mielczarski et al. (1993) report that higher hydrocarbons can be produced over supported Pt, Ru and Co catalysts in a cyclic process at temperatures between 300 and 600 K. A higher temperature cyclic process, at 723 K, was investigated by van Santen and co-workers (Koerts and van Santen, 1991, 1992, and Koerts et al. 1992). Similar research has been described also by Solymosi et al. (1992). A review paper by Amariglio et al. (1995a) summarizes the studies undertaken so far.

The stoichiometry indicates a competitive reaction system, the likely surface mechanism suggests, however, that the network is basically a consecutive one. Thus, our final example of periodic operations applied to multiple reactions has been singled out for separate discussion. Another reason to discuss the homologation system separately is the unusual use of a periodic operation to overcome a thermodynamic barrier by separating conversion into two quite different reactions.

The existence of a thermodynamic barrier is illustrated by Figure 11.18 taken from Koerts et al. (1992). The free energy of reaction for the reaction $2 CH_4 \rightarrow C_2H_6 + H_2$ is $+ 70.5$ kJ/mol at 500 K so if the reaction was carried out at steady state the conversion to ethane would be miniscule. The free energy gap can be overcome by carrying out the reaction in two steps using a catalyst as shown in the figure for cobalt. In the first of these steps, methane is decomposed at a low pressure on the Co surface: $2 CH_4(g) + s$ (Co surface sites)$\rightarrow 2 C - s + 4 H_2(g)$ where $C - s$ is a carbide (Co_3C) or a carbon residue. In the second step the residue is hydrogenated: $2 C - s + 3 H_2 \rightarrow C_2H_6 + 2 s$. The thermodynamic barrier appears as the distance from the $\Delta G° = 0$ line to the intersection of the $\Delta G°$ vs. T lines for the separate reaction steps.

Methane Homologation At High Temperature

Van Santen and co-workers (Koerts and Van Santen, 1991; 1992; Koerts et al., 1992) observed that the $\Delta G°$ barrier shown in Figure 11.18 can be

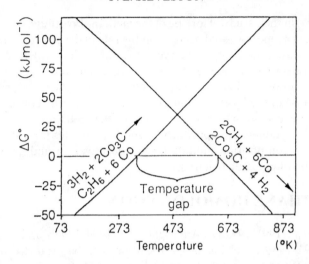

FIGURE 11.18 Variation of the Gibbs Free Energy with temperature for the decomposition of methane on cobalt and the hydrogenation of cobalt carbide to ethane (Figure adapted from Koerts *et al.* (1992) with permission, © 1992 Academic Press)

interpreted as a temperature barrier. The figure shows that ethane could be produced from methane by cracking to carbon over the metal catalyst or forming the metal carbide above 600K then cooling the carbided catalyst to about 300K and hydrogenating the carbon or the carbide to form C_2H_6. Van Santen's experimental work considered Group VIII transition metals supported on SiO_2. Typically, their catalyst sample was exposed to a 3 min pulse of 0.5 vol.% CH_4 in He at 450°C. The chamber and catalyst sample were then quickly cooled (in less than 100 s) to 150°C or less and pure H_2 at one bar was introduced to hydrogenate the residue on the metal surface. The cycle would be completed by reheating the catalyst back to 450°C. Rapid cooling was found to be necessary to avoid transformation of the carbide or carbon residue into an unreactive form, possibly the Cγ form that is widely discussed in the literature (McCarty and Wise, 1979; Winslow and Bell, 1984; 1985). Of the metals considered by Koerts and van Santen (1991, 1992) and by Koerts *et al.* (1992), ruthenium and cobalt seemed to be the most promising. Only part of a single cycle was investigated by van Santen and his co-workers. Variables studied were surface coverage, temperature of the hydrogenation portion of the cycle and exposure durations. Cycling studies as such were not performed.

Highest yield of C_{2+} products was 13% and was achieved with a 5 wt.% Ru/SiO_2 catalyst at a fractional CO surface coverage of 0.18 and a hydrogenation temperature of 95°C. The carbon laid down on the Ru surface was measured by exposing the catalyst to H_2 under temperature ramping (TPSR or temperature programmed surface reaction) and it was found that

21 % of the carbon deposited was converted to C_{2+} paraffinic products. At 450 °C, decomposition of methane proceeds readily and conversion in the adsorption part of the cycle is high. Hydrogenation is the slow step. Van Santen and co-workers investigated the influence of operating conditions on this step. Figure 11.19 shows in part (a) the variation of the hydrocarbon concentrations with the duration of H_2 exposure for a 10 wt.% Co on SiO_2 catalyst. Methane and a small amount of ethane appear instantaneously when H_2 is introduced suggesting that their surface precursors are present in the surface carbon. CH_4 concentrations are about 10 × as high as ethane vol.%s. About 20 s elapses before higher HC's appear. These peak after a 45 s exposure to H_2 and by 150 s only CH_4 and a tiny amount of C_2 are formed. Increasing concentrations up to 45 s suggest displacement of products from the surface allowing increased chemisorption of H_2 and thus continuing high hydrogenation rates. Rates drop off for C_{2+} products after 45 s exposure as reactive carbon on the surface is consumed. The decrease for C_1 is much slower reflecting either a sequence of steps to form higher molecular weight material that is hydrocracked as adsorbed H_2 floods the surface or a slow conversion of unreactive carbon to a reactive form. The rapid appearance of products in the first seconds of H_2 exposure was attributed by Koerts et al. to hydrogenation of a reactive, probably carbidic carbon known in the literature as C_α (see e.g. McCarty and Wise, 1979; Winslow and Bell, 1984).

Data in Figure 11.19a indicate most of the deposited carbon is hydrogenated back to CH_4. This is made more evident in part (b) of the figure which shows the selectivity to the C_1 to C_4 products as a function of temperature in the hydrogenation step of the cycle. For the supported Co catalyst, only about 15 % of the carbon deposited is converted to C_{2+} products. Temperature has a large influence on the product distribution and on the amount hydrogenated (Fig. 11.19c). A wide range of temperature, 60 to 120 °C, can be used for the Co catalyst. But for the 5 wt.% Ru/SiO_2 catalyst, hydrogenation temperatures between 65 and 95 °C give the best C_{2+} yields. Although the amount of carbon hydrogenated increases with temperature, the percentage of CH_4 in the off gas increases as well. Temperatures above 150 °C result in just CH_4. Indeed, above 200 °C, formation of C_{2+} hydrocarbons is thermodynamically impossible according to Koerts et al.

Not all of the deposited carbon can be recovered through hydrogenation. For example, at 95°, only about 50 % of the surface carbon deposited at 450 °C on 5 wt.% Ru/SiO_2 is hydrogenated. This means that a carbon burn-off step would have to be inserted in the homologation cycle.

Wang et al. (1993) have kept up interest in high temperature homologation by demonstrating that methane can be forced to benzene at temperatures above 700°C using a ZSM-5 zeolite that incorporates either molybdenum or Zn. Commercial zeolites were employed for the Wang

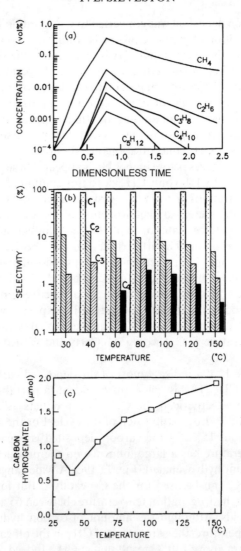

FIGURE 11.19 Hydrogenation of carbidic carbon residues from the dissociative adsorption of CH_4 on a 10 wt.% Co/SiO_2 catalyst at 450°: (a) concentration as volume % of C_1 to C_5 products leaving the micro fixed bed catalytic reactor as a function of dimensionless time in the hydrogenation part of a multipart cycle for methane homologation, (b) Selectivity to the C_1 to C_4 hydrocarbon and (c) the amount of surface carbon residues converted at various temperatures with a multi-part cycle for methane homologation (Figure adopted from Koerts *et al.* (1992) with permission, © 1992 Academic Press)

experiments. These were converted to the ammonium forms by washing with 1 N NH_4NO_3 at 95° and then through ion exchange to the Mo or Zn zeolites. Metal loadings were 2 wt.% Homologation experiments were carried out in a packed bed, continuous flow reactor. The MoZSM-5 was the most active giving 7.2 % CH_4 conversion at 700°, 2 bar and a 1440 mL/g h space velocity with benzene, H_2 and small amounts of C_2H_4 as products. Selectivity to benzene was virtually 100 %. Only the CH_4 and aromatization step was studied. Presumably periodic hydrogenation or possibly oxidation of the zeolite would be necessary to restore catalyst activity.

Embedding molybdenum in a molecular sieve does not seem essential for aromatization. Solymosi et al. (1995) achieved benzene selectivities up to 56 % using a 2 wt.% MoO_3/SiO_2 catalyst in a packed bed flow reactor. Just as in the experiments of Koerts et al. (1992), hydrogen was the main product recovered from the reactor. Benzene was detected at 675°, but significant amounts were formed only at 700°C. In addition to benzene, ethane and ethene were found in important amounts. Selectivity to the C_2's was high initially, but dropped with time so that on completion of the run, selectivity to each C_2 was half the selectivity to benzene. Molybdenum must be in the 4 + state for benzene to form. The catalyst is reduced during the initial exposure to CH_4 yielding CO, CO_2 and H_2O as the only products. Methane conversion decreases with time on stream due to carbon deposition. Solymosi et al. (1995) report that a graphitic carbon forms in the decomposition reaction at 700° and that this carbon is hydrogenated slowly to methane at temperatures between 100 and 700°. Methane conversion reached about 5.4 %. Solymosi and co-workers (Erdöhelyi et al., 1993; Solymosi et al., 1992; 1994; 1995; Solymosi and Cserényi, 1994; 1995) have studied the decomposition step of a methane homologation cycle over various supported and transition metal catalysts. Most of this work was done at about 250° and will be examined in the following section.

Methane Homologation At Low Temperature

Formation of small amounts of ethane or ethene on adsorption of methane on Group VIII transition metals has been frequently mentioned in the catalysis literature. This and the ease of the CH_4-D_2 exchange appears to have lead Amariglio and co-workers to consider a relatively low temperature two-step cyclic process (Belgued et al., 1991). In this cycle, methane is dissociatively adsorbed on a transition metal catalyst splitting off H_2 at a low partial pressure. Carbon-carbon bond formation takes place between adsorbed CH_x fragments so when the catalyst is flushed with hydrogen higher hydrocarbons are recovered. In the hydrogenation step H_2 is supplied at a partial pressure of at least one bar so that the free energy barrier for CH_4 homologation is overcome by the work supplied to compress H_2 from the low recovery pressure in the adsorption step to the one bar or greater pressure in the flushing step.

The experiments of Belgued et al. (1991; 1992a,b; Amariglio et al., 1995a,b) were performed in a quartz U tube flow reactor containing from

100 to 600 mg of catalyst powder held on a fritted silica disk. The valving system allowed for the flow of one of four gases through the reactor (CH_4, H_2, He or Ar). A thermostatically controlled furnace provided preheat and kept the reactor at the reaction temperature selected. GC analysis of the off-gas was employed with a temperature programmed Porapak column and a FID for hydrocarbons, and a molecular sieve column and a TCD for H_2. Catalysts chosen for the experiments were Euro-Pt 1 a standardized material with 6.3 wt.% Pt on SiO_2, a 4.7 wt.% Ru on SiO_2 prepared by impregnating $RuCl_3$ onto a Spherosil support, and a commercial Co catalyst with 2 wt.% Co on kieselguhr. All samples were reduced in flowing H_2 before use. Prior to an experiment, carbonaceous residues and adsorbed H_2 were removed from all surfaces by passing H_2 over the sample at 400 C and then switching to Ar for experimentally determined time intervals. Variables explored by Amariglio and his co-workers were temperature (temperatures of the CH_4 adsorption and the H_2 flushing steps could be controlled independently), space velocities (again the velocities in each step were independent), cycle period, and cycle split. One experiment was undertaken to observe if the catalysts deactivate and separate measurements of the amount of methane adsorbed were made.

With the Pt catalyst, methane adsorption and dissociation began at 150 °C. Both H_2 and C_2H_6 were detected in the off gas from the adsorption step at a molar ratio of about 10:1 although this ratio varied with the duration of exposure to CH_4. Increasing temperature increased strongly the rate of methane dissociation. C_2H_6 formation rate at 250°C passed through a maximum after about 2 minutes of exposure and then fell steadily, whereas H_2 formation decreased immediately as CH_4 began to pass over the catalyst. The mol fractions of C_2H_6 and H_2 in the off gas when the ethane maximum occurred correspond to attainment of about 40 % of the 2 $CH_4 \rightleftharpoons C_2H_6 + H_2$ equilibrium according to Belgued et al. (1991). Hydrocarbon production was found to be very sensitive to the methane space velocity, approaching zero as the space velocity decreased (see Fig. 11.21a). Belgued et al. (1992b) suggest this confirms that the adsorption step is equilibrium controlled on the Pt surface. Ethane was not observed in the adsorption step off gas for the Ru and Co catalysts.

Hydrogenation in the second part of the cycle occurs at 1 bar and is so rapid that Belgued et al. (1992a) were only able to measure the cumulative amounts of hydrocarbons formed. When the temperature is constant over a cycle, Figure 11.20 summarizes the performance of the three catalysts tested. The figure shows an optimal temperature window for each catalyst: 225–275 °C for Pt, 140–180 °C for Ru and 260 to 320 °C for Co. The Schultz-Flory-Anderson distribution of the molar amounts of products in terms of carbon number is not followed for the C_2 to C_5 species. This distribution is widely observed for the Fischer-Tropsch synthesis, which proceeds readily on Ru and Co. Koerts et al. (1992) also find the distribu-

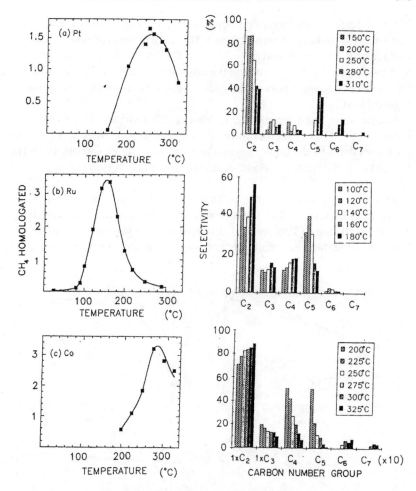

FIGURE 11.20 Amount of methane converted and selectivities to C_{2+} products as a function of temperature for a two part cyclic operation in a micro packed bed reactor: (a) Pt – 100 mg of 6.3 wt.% Pt on SiO_2 (EUROPT-1) and CH_4 adsorption for 1 min. at 1 bar and 400 mL/min.; (b) Ru – 100 mg of 4.7 wt.% Ru on SiO_2 and CH_4 adsorption for 5 min. at 1 bar and 375 mL/min.; (c) Co – 600 mg of 2 wt.% Co on kieselguhr and CH_4 adsorption for 4 min. at 1 bar and 250 mL/min. With all catalysts hydrogenation was for 2 min. at 1 bar and 50 mL/min. (Figure adopted from Amariglio *et al.* (1995a) with permission, © 1995 Elsevier Science B.V.)

tion is followed for higher temperature homologation. All catalysts show a low selectivity to C_3 while the Pt and Ru materials exhibit a high C_5 selectivity. Selectivity in the figure is based on CH_4 homologated, that is the total C_{2+} hydrocarbons formed. On the basis of the methane adsorbed in the first part of the cycle, the yields of C_{2+} hydrocarbons, Y_{C_{2+}/C_1}, were 19.3 % with Pt at 250 °C, 36.9 % for Ru at 160 °C and 7.5 % for Co at

275 °C. Cycle periods and split differed for each catalyst. For Pt, $\tau = 3$ min. and $s = 0.333$, defined in terms of the CH_4 adsorption portion of the cycle, while for Ru, $\tau = 7$ min. and $s = 0.714$, and for Co, $\tau = 6$ min. and $s = 0.667$. Space velocity differed as well for the adsorption part, but it was constant for the hydrogenation part of the cycle as can be seen from the caption to Figure 11.20.

Temperature has a large influence on the selectivity as can be seen from Figure 11.20. Higher temperatures decrease selectivity to C_2, S_{C_2/C_1}, but raise selectivity to the higher carbon numbers for the Pt sample. With the Ru and Co catalysts, the effect of temperature is reversed for S_{C_2/C_1} and S_{C_5/C_1}.

Belgued et al. (1992a) observe that the CH_4 space velocity and the duration of exposure effect the methane homologated and the selectivity. This is illustrated in Figure 11.21. Increasing both generally increases the CH_4 homologation, although the increase in homologation becomes quite

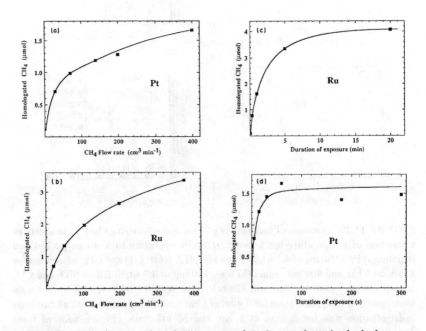

FIGURE 11.21 Amount of methane converted to C_{2+} products in the hydrogenation portion of a two part cyclic operation in a micro packed bed reactor as a function of (a) and (b) CH_4 flow rate, and (c) and (d) time of CH_4 exposure. Temperature in the adsorption portion is constant at 160 °C in (c) and (d), but is 250°C in (a) and 160°C in (b). Data in (a) and (d) are for 100 mg of 6.3 wt.% Pt on SiO_2 (EUROPT-1), while in b) and (c) they are for 100 mg of 4.7 wt.% Ru on SiO_2; Flow and exposure conditions for the two catalysts are those given in Figure 11.20 (Figure adapted from Amariglio et al. (1995a) with permission, © 1995 Elsevier Science B.V.)

small for the Pt catalyst after a 1.5 min exposure. Higher CH_4 flow rates gave higher C_{2+} production and shifted the product distribution toward higher carbon numbers. Increasing the CH_4 exposure duration increases S_{C_2+/C_1} as well as the CH_4 homologated. For the Ru catalyst, S_{C_4+/C_1} jumps from 1.7 to 39.6 % when the exposure duration increases from 5 to 20 min. (Amariglio et al., 1995a). The H_2 flow rate also effects selectivity: selectivity to the higher hydrocarbons increases as the flow rate increases.

Using a lower temperature in the hydrogenation portion of the cycle has a remarkable influence on both homologation and selectivity for the Ru catalyst (Belgued et al., 1992b). This is evident in Figure 11.22. Dropping the temperature to 100 °C from 160 °C used in the adsorption part of the cycle, raises the CH_4 homologation by almost 50 % and raises S_{C_5/C_1} to about 55 % among the C_{2+} hydrocarbons formed. At 25 °C, the CH_4 converted matched that at the isothermal operation. With the Pt catalyst, CH_4 homologated decreased with decreasing temperature of the hydrogenation portion of the cycle. Selectivity, however, shifted to C_5 just as with the Ru catalyst.

Experiments with 2 sequences of 10 successive cycles were reported by Belgued et al. (1992a) to test for catalyst deactivation. No apparent deactivation between the two sequences was observed for the Pt sample at 250 °C and the Ru material at 160 C when symmetrical cycles with periods of 3 minutes were used. The Co catalyst at 300 °C using a 6 minute period, however, experienced a 25 % loss in homologation activity. The investigators report cycle periods as short as 10 s can be used, but that there is about a 10 fold drop in the moles of methane converted per cycle.

Amariglio et al. (1995b) explored inserting a helium flush between the CH_4 and H_2 exposure steps for the Eurocat Pt catalyst they employed. Some CH_4 was swept from the Pt surface and this reduced the amount of C_{2+} material formed in the subsequent H_2 exposure step. This suggests either adsorbed CH_4 remains on the surface after the methane flow is discontinued, or that there is a dynamic equilibrium among the H deficient hydrocarbons on the Pt surface. Replacing He by CO at 1 bar and omitting the hydrogenation step produced about 55 % of the homologation achieved in two part CH_4 (30 s) – H_2 (15 min) cycle for the same CH_4 exposure duration. The product distribution, however, changed dramatically as may be seen in Figure 11.23. Hydrocarbons above C_{3+} are primarily olefins and carbon numbers as high as C_8 were detected. Both alkanes and alkenes were branched and small amounts of cyclopentene were found. No aromatics, alkynes or oxygenates were seen. The amount of CO or the duration of the exposure did not significantly effect the homologation or the product distribution. Diluting the CO with He reduced homologation. Following the CO dosage with an extended exposure to H_2 increased CH_4 homologation to 76 % of that obtained in

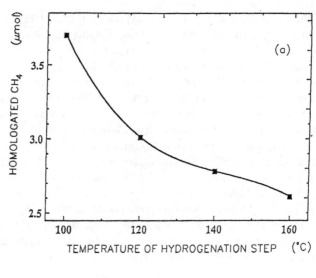

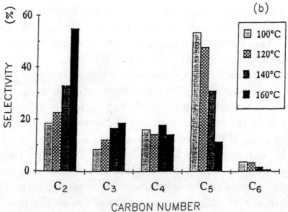

FIGURE 11.22 Amount of methane converted and selectivities to C_{2+} products as a function of the hydrogenation temperature in a two part cycle using a micro packed bed reactor with 100 mg of 4.7 wt.% Ru on SiO_2 and CH_4 adsorption for 3 min. at 1 bar and 375 mL/min. Hydrogenation was for 2 min. at 1 bar and 50 mL/min. (Figure adopted from Amariglio et al. (1995a) with permission, © 1995 Elsevier Science B.V.)

the two part cycle. Only alkanes were recovered during the hydrogenation step.

Belgued et al. (1992a) speculate that the relatively low temperature adsorption generates adsorbed CH_3, CH_2, CH and H species on the

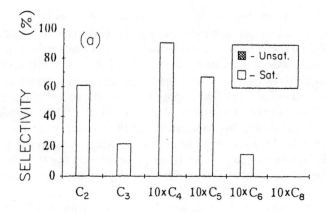

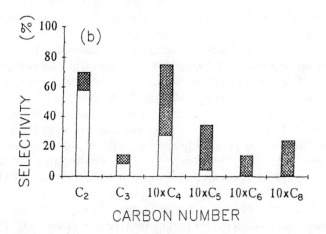

FIGURE 11.23 Comparison of selectivities to C_{2+} products for two part cycles: a) $CH_4 - H_2$ cycle with a 30 s CH_4 exposure followed by a 15 min. H_2 exposure at 1 bar and 50 mL/min, b) $CH_4 - CO$ cycle with the identical CH_4 exposure, but followed by dosing with 1.4 cm^2 of CO at 1 bar. Measurements used a micro packed bed reactor with 100 mg of 6.3 wt.% Pt on SiO_2 (EUROPT-1) (Figure adopted from Amariglio et al. (1995b) with permission, © 1995 J.C. Baltzer AG)

surface. The H species combine and desorb as H_2. Neighboring partially hydrogenated carbons primarily dimerize to form adsorbed C_2H_6, CH_2-CH_2, and $CH-CH$ species. The latter react in turn to form the higher carbon number precursors which desorb on hydrogenation. The mild temperatures limit the dehydrogenation to carbon in the adsorption step. The surface population is thought to be similar to that obtained from the adsorption of C_2 and C_3 olefins on the same metal surfaces. The failure of the product distribution to follow the Schultz-Flory-Anderson distribution is attributed to dimerization as the primary process for $C-C$ bond formation and to hydrogenolysis during the hydrogenation step. Even though Ru is a better hydrogenolysis catalyst then Pt, adsorption of CH_4 is faster so that lower temperatures can be used with Ru (160 vs 250 °C). It is the reduced hydrogenolysis activity at the lower temperature which accounts for high selectivity to C_5 and the opposite selectivity trend with temperature that was found for the Ru catalyst. Amariglio et al. (1995b) explain their results with CO dosing by displacement of the olefinic precursors on the surface by CO. Desorption of alkanes and alkenes requires a dynamic equilibria and the exchange of an H surface species on the Pt surface. The appearance of C_8's and olefins confirms the importance of hydrogenolysis in determining the product distribution in the two-step cycle.

How attractive is the Amariglio cyclic process for methane upgrading? There has been a great deal of industrial interest in this subject for several decades. In the next chapter, we consider oxidative coupling of methane, while in Chapter 13 we will discuss the Fischer-Tropsch synthesis, both alternative routes for methane upgrading. The key consideration is methane conversion per pass or the yields of C_{2+} hydrocarbons. These are very low in the experiments reported by Belgued et al. (1991, 1992a,b), approaching under the best conditions used with the Ru catalyst just 1 %. Amariglio et al. (1994, 1995a) argue that these can be substantially improved by reactor design. High CH_4 flow rates are necessary to strip H_2 from the catalyst surface thereby permitting chain growth among the hydrocarbon residues. Amariglio et al. (1994) demonstrate that yields of about 40 % have been achieved by using a batch reactor with external circulation of CH_4 and adding 5 g of 5 wt.% Pd on Al_2O_3 as a H_2 trapping material in a separate contactor following the bed of 1 g of Eurocat Pt supported on a silica aerogel. Methane recirculates for 16 min until the dissociation has virtually ended. Flow through the system is then switched to H_2 and the hydrocarbon residues on the catalyst and on the trap (if any) are hydrogenated. With higher yields, the Amariglio cyclic process is indeed interesting with the Ru catalyst. Two fluidized beds could be used in a continuous analogue of the batch system with catalyst transport between the beds. The first bed would operate at a higher temperature and dissociatively adsorb methane. The off gas would flow to a membrane separator to strip out H_2 so that the unconverted methane could be recycled. The relatively pure H_2 would be compressed and fed to the second fluidized bed, operated at a lower temperature and a higher pressure, to produce hydrocarbons. These could contain as much as 50 % methane and would probably have to be separated by cryogenic distillation. The recovered methane would be recycled. It would appear that make up hydrogen would not be required. The only process inputs would be methane and power.

Choice of catalyst also seems important if yields are to be improved. Solymosi et al. (1992) investigated methane decomposition on a group of noble metal catalysts supported at 5 wt.% on a SiO_2 aerogel using an isothermal packed bed. At 250°C, Rhenium proved to be more active than platinum in the decomposition step. Hydrogen formation was about an order of magnitude greater than that obtained with Pt, while the initial rate of formation and the total C_2H_6 formed was the highest with Pt. Rh/SiO_2 gave the second highest initial formation rate and total amount of C_2H_6 formed. Pd/SiO_2 followed the Rhenium catalyst, but unlike Pt and Rh, it continued to form C_2H_6 at a steady rate after 20 min on stream. Contrary to the observations of Belgued et al. (1992a), H_2 production on all the catalysts with the exception of Pt continued at a low, but steady rate after about 5 minutes on stream. In the case of Rh/SiO_2, this rate was about 1/4 of the peak rate that occurred on the initial exposure to the 12.5 vol.% CH_4 in N_2 feed. This observation suggests carbon mobility on the metal surface and possibly spillover on to the support. Although the time resolution afforded by the GC used by Solymosi et al. (1992) was poor, C_2H_6 formation did not begin on introduction of the 12.5 vol.% CH_4 feed. With exception of Ru/SiO_2 and Ir/SiO_2, ethane production peaked sharply some 5 to 10 minutes after CH_4 flow commenced. This led Solymosi et al. to propose that methane is stepwise dehydrogenated. Only the CH_3 fragments are loosely enough bonded to dimerize and desorb.

After flushing with Ar, the flow was switched to H_2 at 1 bar and 250 °C. In the Solymosi experiments, only Pt produced C_2H_6 in this step. Other catalysts produced copious amounts of CH_4 which endured for several hours. A TPR experiment using H_2 disclosed that the metals, with the exception of Ir, had various amounts of active carbon, referred to as C_α, that readily hydrogenated below 250 °C. Most deposits on the catalysts (except Rh) needed temperatures above 300 °C and important amounts required temperature greater than 600 °C in these TPR measurements in order to be removed from the catalyst surface. These observation suggested the existence of refractory C_β and C_γ carbons (see, e.g., McCarty and Wise, 1979; Winslow and Bell, 1984; 1985).

Solymosi et al. (1994) using the 5 wt.% Pd catalyst demonstrated that the support oxide is important. Using Ti/SiO_2 in place of SiO_2 increased the rate of C_2H_6 formation about 5 × at 250 and led to the production of C_2s and C_3s in the subsequent hydrogenation step. The improvement was believed to be largely due to better dispersion of Pd on the support. The Ir catalyst was investigated by Solymosi and Cserényi (1994) who found that MgO and Al_2O_3 supports gave the greatest C_{2+} production in a flow reactor at 250° during both the methane decomposition and hydrogenation steps. Small amounts of C_6 hydrocarbons were detected with the Al_2O_3 support. Increasing the temperature to 500°C increased the CH_4 conversion by 15 to 30 × depending on the Ir support, however, S_{C_{2+}/C_1} decreased and on hydrogenation at 250° only CH_4 was produced. The carbonaceous deposited became primarily C_γ. Ir dispersion on the support was 60 % for

Al$_2$O$_3$ and just 12 % for MgO indicating that the nature of the support is important too.

Indications in the research literature that copper promotes the surface coupling of CH$_3$ species and suppresses dehydrogenation caused Solymosi and Cserényi (1995) to explore CH$_4$ homologation on Cu–Rh catalysts supported on a silica aerogel. Cu/Rh ratios of 0.5 to 1 gave S_{C_2+/C_1} up to 5 %, based on CH$_4$ decomposed, after 20 minutes on stream at 250 °C whereas a catalysis containing the same amount of Rh metal alone gave 1 %. S_{C_2+/C_1} increased monotonously with the duration of CH$_4$ exposure at 250°, but at 300 and 350 °C, selectivity reached 3 to 4 % after about 5 minutes and then remained constant. Cu/Rh between 0.5 and 1 dramatically increased S_{C_2+/C_1} in the hydrogenation step by about 3 to 10 × depending on the duration of CH$_4$ exposure. Cu/SiO$_2$ cannot decompose CH$_4$ between 250 and 350 °C so Solymosi and Cserényi suggest the CH$_3$ and CH$_2$ species migrates to Cu sites where dimerization proceeds. Significant amounts of ethene were formed in both the decomposition and hydrogenation steps.

In general, the experiments of Solymosi and co-workers support the results and mechanistic implications reported by Belgued et al. (1991, 1992b). Yields of higher hydrocarbons are quite low in flow reactors. Decomposition of adsorbed CH$_4$ proceeds stepwise with concurrent dimerization of the partially dehydrogenated CH$_4$. Hydrogenation of the C$_\alpha$ species can produce C$_{2+}$, whereas much of the carbon produced even at 250 °C is refractory and hydrogenates only slowly to CH$_4$.

Related to methane homologation by catalysts and conditions, Erdöhelyi et al. (1992) explored a two step process using CH$_4$ and CO$_2$. Co-feed of these gases produces synthesis gas between 250 and 500°C. Erdöhelyi et al. observed that the presence of CO$_2$ in the CH$_4$ feed to the catalyst bed suppressed formation of the higher hydrocarbons and reforming to CO and H$_2$ results. Clearly, a cyclic process is not advantageous.

Assessment Of The Cyclic Methane Homologation Processes

From an engineering standpoint, the higher temperature van Santen process has a problem with more than a 300 °C temperature difference between the adsorption and hydrogenation parts of the cycle. Further difficulties are the accumulation of graphitic carbon which will eventually deactivate the catalyst and the low S_{C_2+/C_1}. On the other hand, the yield of these products can reach 13 % in a flow reactor which is an order of magnitude greater than the yields measured for the moderate temperature Amariglio version. Clearly more exploration of both cyclic processes is needed. The Amariglio version is more attractive from an engineering standpoint because of the lower temperatures used. The research target for this process must be increasing the yield and this must start by achieving greater CH_4 adsorption in the CH_4 exposure portion of the cycle and increased selectivity to the C_{2+} products in the course of methane decomposition. This should probably proceed through a search for better catalysts as the work of Solymosi and co-workers demonstrate that the choice of catalyst and support has a large effect on product distribution.

F. OLIGOMERIZATION OF ETHENE

Their success with methane homologation induced Amariglio and his co-workers to explore application of their cyclic process to ethane and ethene. Only their experiments employing C_2H_4 have been published (Lefort et al., 1994). It is difficult to determine which class of multiple reactions ethene oligomerization falls into, but the stoichiometry, displaying products from C_1 to C_6, indicate a competitive reaction system.

The reactor system and procedure utilized by Belgued et al. (1992a) was used, but the Pt catalyst was Europt-2 containing 6.3 wt.% Pt on silica. For a cycle consisting of a 1 minute exposure to a stream of 1 vol.% C_2H_4 in Ar at $SV = 1$ cm^3/mg-min followed by a 30 s exposure to H_2 at half the previous space velocity, Lefort et al. found about 42 % of the C_2H_4 adsorbed at 100°C was oligomerized to C_{3+} products, while at 150° the percent oligomerized increases to 64 %. Co-feeding C_2H_4 and H_2 at 100 to 150° and the average composition of the separate feeds results only in C_2H_6. Figure 11.24 describes the products formed in terms of the selectivity to each product. The selectivity to butane, hexane and cyclohexane and octane indicate the primary reaction is oligomerization. Hydrogenation of C_2H_4 occurs as well and the presence of CH_4, propane and pentane means some hydrogenolysis also takes place. About 18 % of the C_2H_4 adsorbed is not removed in the H_2 portion of the cycle. This means that either this portion of the cycle must be extended or carried out at a higher temperature.

The presence of small amounts of C_2H_6 in the off gas leaving the flow reactor during the C_2H_4 portion of the cycle as well as TPD measurements which detected only CH_4, C_2H_4 and C_2H_6 after exposure of the Pt catalyst to the argon-ethene mixture lead Lefort et al. to conclude that an ethyne (C_2H_2), strongly adsorbed on the Pt surface, and H_{ad} form in the adsorption of C_2H_4. The adsorbed hydrogen is available to hydrogenate some of the adsorbed C_2H_4 to C_2H_6 which was detected in the off gas. Precursors of the higher hydrocarbons result from the oligomerization of this adsorbed ethyne species.

The experiments of Lefort et al. (1994) are exploratory in nature. The selectivities to higher hydrocarbons are substantial so that further examination of this composition modulation process would seem worthwhile at least so far as to obtain sufficient results to make a comparison with conventional ethene dimerization technology.

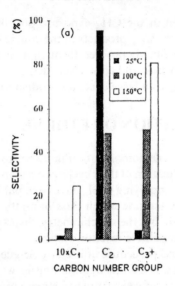
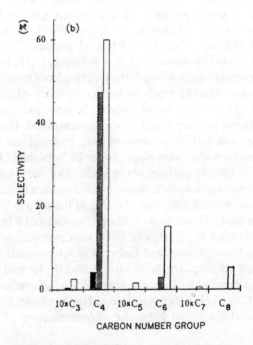

FIGURE 11.24 Selectivity to different hydrocarbon products for a two step, cyclic process over a 6.3 wt.% Pt/SiO$_2$ catalyst with 12.5 vol.% C$_2$H$_4$ in Ar for 60 s followed by 100 % H$_2$ for 30 s: (a) breakdown of the products among C$_1$ to C$_{3+}$, (b) breakdown of the C$_{2+}$ products (Figure adopted from Lefort *et al.* (1994) with permission, ©1994 J.C. Baltzer AG)

CHAPTER 12

Hydrocarbon Dehydrogenation and Partial Oxidation

A. INTRODUCTION

With the exception of processes with rapid catalyst fouling, such as catalytic cracking, where regular regeneration is required, hydrocarbon partial oxidation and oxidative dehydrogenation are the only process systems to date that have seen full scale application of composition modulation, or for which development has advanced to the pilot or demonstration plant stage. There are several reasons for industrial interest in this group of applications. Thermal cracking, the conventional process for producing olefins from paraffins, deposits carbon on cracking surfaces which wastes carbon and requires periodic decoking. Selectivity also suffers from the loss of carbon to methane and ethane. Thermal cracking is endothermic and occurs at moderately high temperatures so that a cracking furnace must be used. The process is energy intensive. Oxidative dehydrogenation, on the other hand, is exothermic offering energy recovery and thus a lower net energy burden. Unless the addition of oxygen is carefully controlled, however, some total oxidation occurs, reducing selectivity. In a conventional continuous process, thus, there may be little or no selectivity advantage for oxidative dehydrogenation. Composition modulation is a method of oxygen control to achieve higher selectivity and explains presumably industrial interest in this method of reactor operation.

Several of the commercially produced hydrocarbon oxygenates are profitable synthesis feeds or monomers. Most of these commercial oxygenates, such as the anhydrides and aldehydes, can be synthesized by multi-step routes, but partial oxidation has been found to be cheaper so that bulk oxygenates generally are made by the latter route. Selectivity is an important concern in these processes. The problem arises because total oxidation often occurs in parallel with the desired partial oxidation route. In some cases an intermediate or even the desired product is more susceptible to oxidation than the reactant. Thus, there are side reactions in almost all partial oxidation processes which lower selectivity. In some systems, total oxidation is attributed to the state of oxygen on the surface: adsorbed O_2, possibly as a peroxide, has been identified as the source of CO_2 in both butane and benzene partial oxidation. Composition modulation may offer a means of controlling total oxidation by segregating the hydrocarbon feedstock from oxygen. The catalyst provides, in this application, oxygen transport. The catalyst is oxidized by exposure to air in one part of a cycle and is then reduced by the hydrocarbon in the other portion. In pilot or full scale implementation, the alternate exposures of the catalyst is more conveniently done by using two vessels, one for oxidation and the other for reduction, and circulating the catalyst between the vessels as discussed in Chapter 1. It is no surprise then that the industrial applications of composition modulation described in this chapter employ circulating bed systems.

Partial oxidation under composition modulation has been studied for many hydrocarbons. All of these studies plus two application to paraffin oxidative dehydrogenation will be considered in this chapter. The discussion will be organized more or less in terms of the carbon number of the reactant. Thus, the chapter begins with the oxidative coupling of methane, proceeds to epoxidation reactions of ethylene and propylene, the partial oxidation and ammoxidation of propylene, continues with oxidative dehydrogenation of butane and then ends with the partial oxidation and ammoxidation of aromatics.

B. OXIDATIVE COUPLING OF METHANE

Background

During the last decade, the conversion of methane to higher value products was intensively explored by many research teams around the world. The work was motivated by the demand for ethylene (e.g., Keller and Bhasin, 1982) as well as for means of economically bringing remote natural gas to markets (e.g., Jones *et al.*, 1987b). Two routes, oxidative coupling to ethane and/or ethylene and partial oxidation to methanol or formaldehyde, have received the most attention. A third route, cracking on metal surfaces and forced desorption, was considered in the previous chapter. Both of the routes to be considered now have a severe selectivity problem: products formed are more easily oxidized than methane under the reaction conditions used. For oxidative coupling, the catalyst's primary function is to break the C−H bond to create a methyl radical. Most of the C_2 formation occurs in successive gas phase reactions. Selectivity drops rapidly as conversion rises above 10 % resulting in what has been called the yield barrier (Amenomiya *et al.*, 1990). Yields of C_2 products appear to be limited to less than 25 % which is believed to lie below the target for economic feasibility. Control of oxygen addition should be important. Several 1979 and 1980 U.S. patents issued to Mitchell and Waghorne (see Sofranko *et al.*, 1987) recognized an opportunity to use composition modulation. Keller and Bhasin(1982) employed on-off methane and air modulation in experiments described in an early publication on oxidative coupling emanating from the R & D Department of the Union Carbide Corporation. Similar experiments were reported by ARCO Chemical Company in 1984 (Sofranko et al., 1987). A considerable portion of the literature on this application of composition modulation is in the form of patents.

Experimental Considerations

Composition modulation experiments in oxidative coupling have been carried out through flow control or opening and closing of valves in the manner described in Chapter 1. The high temperatures needed to activate the C – H bond require the replacement of the temperature control bath in the systems discussed earlier by a furnace capable of maintaining temperatures between 600 and 1000°C. The set up used by Mortazavi et al. (1992, 1996) is representative of the equipment used for composition modulation studies of oxidative coupling. Mortazavi's system is shown schematically in Figure 12.1a. The reactor used by Mortazavi appears in b) of the figure. The feed inlet at the top of the reactor would be outside the furnace. Quartz chips shown above the catalyst bed provide preheating of the feed. The

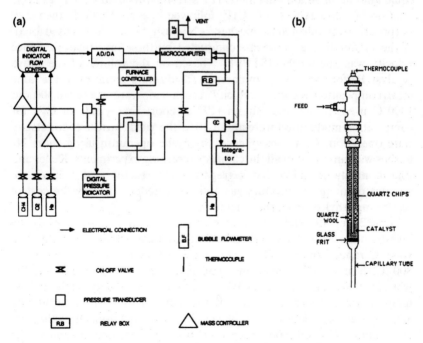

FIGURE 12.1 Schematic of a) experimental arrangement and b) quartz tube reactor used by Mortazavi et al. (1992, 1996) for methane coupling studies (Figures taken from Mortazavi (1990) with permission of the author).

reactor off gases leave through a capillary tube which extends below the furnace. Use of a capillary reduces the post reactor void volume and the residence time of the product gases in the high temperature region. Some experimenters employ an inert packing in place of the capillary tube after the catalyst bed. Somewhat poorer heat transfer characteristics of furnaces in comparison with constant temperature baths used at lower reaction temperatures means that temperature fluctuations of about 2°C can occur.

Union Carbide And ARCO Studies

A primary objective of the milestone Union Carbide study (Keller and Bhasin, 1982) was to determine if switching between methane and air would be advantageous. The second objective was to screen a wide variety of catalysts for activity and selectivity to the C_2 products. Screening experiments were performed in a reactor made of 1/4 inch o.d. stainless steel tubing containing about 2 mL of 8×20 mesh catalyst particles with about 4 mL of inert packing before and after the active bed.

Investigation of composition modulation was made using a PbO/α-Al_2O_3 catalyst in a 1/2 inch o.d. tube containing 47 mL of charge and an equal amount of an alumina inert. The investigators used a complex cycle built up of a 20 s exposure to CH_4, followed by a 5 s N_2 flush, then a 20 s exposure to air followed by another 5 s N_2 flush. This cycle was used for all of the modulation experiments conducted by these investigators. Space velocity was about 1900 (STP) h^{-1} based on the volume of the active catalyst. Performance was compared to steady state operation at the same mean composition and space velocity at each temperature between 800 and 1000°C used. Keller and Bhasin (1982) report that yields of C_2's for composition modulation were about 10 × the yields measured for steady state operation. As a consequence, the cyclic operation using short N_2 flushes was employed in all the catalysts screening experiments. Keller and Bhasin mentioned also that segregating the reactor feed reduces the amount of air in the product gas and could reduce thereby the cost of separating ethylene from the reactor output.

Some 25 metal-metal oxide on α-alumina systems were screened by the investigators. More than half exhibited coupling activity in the 500 to 1000 °C range explored. Only Mg and Sr showed activity at 600°C; at 800 °C, the most active metal–metal oxide systems were Mn, Cd, Tl and Pb, while at 1000 °C, Li, Tl, Mg, Sn, Pb, Sb showed good activity. At the highest temperature, the inert packing and the alumina support also exhibited activity, but just about half of the activity of the metal-metal oxide pairs given above. Carbon oxide formation was also measured in some of the experiments. Keller and Bhasin observed that the oxygen removed during the CH_4 and N_2 pulses in the cyclic experiments exceeded monolayer coverage for most of the catalysts by an order of magnitude and concluded that the catalysts function via a redox mechanism. An examination of the Gibbs free energy change for the reaction:

$$CH_4 + MO \rightarrow 1/2\ C_2H_4 + H_2O + M \tag{12.1}$$

indicated the redox mechanism was feasible. The systems showing the largest free energy decrease were not, however, those that were the most active for the coupling reaction.

Work undertaken by ARCO is reported in two research publications (Sofranko et al., 1987 and Jones et al., 1987a) and a summary paper (Jones et al., 1987b). Like the Union Carbide program just examined, this effort used complex cycles of CH_4 and air exposures of the catalyst separated by N_2 flushes. However, the duration of CH_4 cycle step was varied and the effect of its duration on conversion and selectivity was investigated. N_2 flush duration was also examined, but little is reported so it is probably not a significant variable. Cycle periods are not given. With the exception of some process development described in an ARCO patent (Jones et al., 1985), the research effort was aimed primarily at catalyst choice and optimization. Research on the pathways to the various products is mentioned also in the ARCO papers.

Measurements were made in a micro reactor consisting of a 14 mm (i.d.) quartz tube packed with a 10 mL sample of the test catalyst at a 14 × 30 U. S. mesh size. Analyses were made by sampling the off gas to give instantaneous values of selectivity and conversion as well as by sampling a gas bag which collected the products of several cycles. This gave the time average result. Temperatures were between 700 and 800°C and the CH_4 space velocity was 600 to 800 h^{-1}.

Screening tests identified manganese oxide supported on silica as an attractive catalyst. This oxide is not volatile at 800°C and exhibited good selectivity to C_{2+} at CH_4 conversion of the order of 10%. The best selectivity was obtained by promoting the catalyst with sodium oxide or pyrophosphate (Sofranko et al., 1987; Jones et al., 1987a). Selectivity was further enhanced by adding methyl chloride to the gas feed (Jones et al., 1987b) A mineral phase, Braunite (Mn_7SiO_{12}) identified by XRD, was thought to be the active component.

The ARCO investigators did not find a large selectivity advantage for the complex composition modulation they used. Figure 12.2 plots the selectivity vs. conversion data for their best silica supported, alkali promoted manganese catalysts under composition modulation and under steady state (or co-feed) operation. Switching the feed between CH_4 and air offers about a 10 % selectivity advantage below 10 % conversion; this drops to less than 3.5 % by 30 % conversion. Steady state data for other oxides, measured by other investigators, are shown in the figure. These results corroborate the high selectivities observed with some oxides at steady state, but contradict the finding of Keller and Bhasin discussed above.

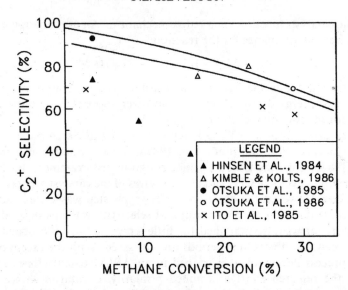

FIGURE 12.2 Comparison of selectivity for composition modulation and steady state operation at constant CH_4 conversion for the ARCO 15% Mn, 5% $Na_4P_2O_7/SiO_2$ catalyst at 800°C and a CH_4 space velocity of 860 h^{-1}. Data points of other investigators are for steady state operation and may be at other temperatures and space velocities (Figure adapted from Jones et al. (1987b) with permission, © 1987 American Chemical Society)

Despite the high selectivities observed, the C_2 yield which is the selectivity-conversion product is under 25 %. The difficulty of exceeding a 25 % C_2 yield in steady state operation has been referred to as a yield barrier for methane coupling (Amenomiya et al., 1990). A target of recent research on the application of composition modulation has been to see whether or not this barrier can be overcome.

Significant amounts of higher hydrocarbons were found under composition modulation. Selectivities to C_3 reached 7 % for the manganese catalysts. Selectivities to C_4 and benzene were about 4 %. The experimental data are given in Figure 12.3. Products corresponding to the 3 and 4 carbon numbers were olefinic, while at 6 and 7, the products were benzene and toluene. Sofranko et al. (1987) comment that product distribution by carbon number were similar at different conversions and for different catalysts. Their data for lead oxide and tin oxide are given in the figure. Points for empty tube runs are also given and are similar to the catalytic results. This observation led Jones et al. to conclude that the C_2's and higher hydrocarbons must be formed by a common mechanism, probably involving gas phase species. Conversion differed by three orders of magnitude for the data in the figure and is only 0.05 % for the empty tube measurements.

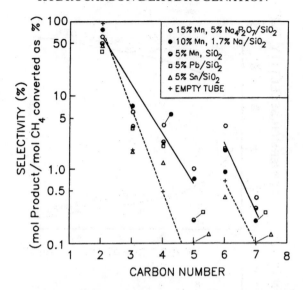

FIGURE 12.3 Product selectivity as a function of carbon number for several ARCO manganese, PbO and SnO_2 catalysts at 800°C and various CH_4 conversions. Results for the "homogeneous" reaction given for comparison. Solid lines are trend of the data for the 15% Mn, 5% $Na_4P_2O_7/SiO_2$ catalyst, while the dashed lines are trends for the homogeneous, empty tube data

Thus, the primary role of the catalyst is to generate methyl radicals according to these investigators. The lines in the figure reflect the trends of the data for the empty tube experiments and for runs made with the 15 wt.% Mn catalyst. The slopes differ significantly for the C_2 to C_5 products which suggests that the catalyst does influence product distribution.

Figure 12.4 explores the effects of the duration of the catalyst exposure to CH_4 and the contact time on selectivity and CH_4 conversion. The duration effect is similar to the influence of the O_2:CH_4 ratio in steady state operation and expresses the influence of catalyst reduction. Increasing reduction or removal of oxygen from the catalyst raises the selectivity to C_{2+} products from 65 to 95 %, but methane conversion drops from 40 to less than 10 %. Consequently, over oxidation of the catalyst should be avoided to maintain high selectivity. The contact time influence is shown in Figure 12.4b. At contact times under 0.1 s, when conversion is very small, the main product is ethane. Increasing contact time leads to the formation of ethylene, higher hydrocarbons, coke and total oxidation products. Detection of CO_2 in the off gas during exposure of the catalyst to air suggests coke like deposition on the catalyst surface. According to Jones *et al.* (1987a), carbon oxides form from oxidation of the olefins produced in the coupling reaction and/or from oxidation of condensation residues, but not

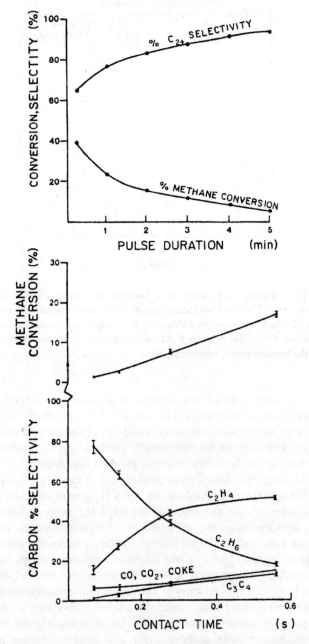

FIGURE 12.4 Effects on the C_{2+} selectivity and on CH_4 conversion of a) methane pulse duration at 850°C and 860 h^{-1}, and b) contact time at 825°C for the ARCO 15% Mn, 5% $Na_4P_2O_7/SiO_2$ catalyst (Figures taken from Jones et al. (1987a) with permission, © 1987 Academic Press)

from direct oxidation of CH_4 or methyl radicals. They reason that the primary coupling product during composition modulation is ethane which is then dehydrogenated. Higher hydrocarbons originate from methyl radical–olefin reactions and olefin dimerization. A role of sodium, the authors argue, is to reduce surface acidity thereby suppressing olefin oxidation and coke formation.

University Of Waterloo Studies

In contrast to ARCO, the objective in the Waterloo work was to improve performance under composition modulation and ascertain if a 25 % yield could be achieved. As mentioned earlier, this yield was cited by Amenomiya et al. (1990) as the boundary for the economic feasibility of methane coupling. Forcing strategy as well as the influence of cycle period, split, composition of the cycling mixtures on the C_2 selectivity and methane conversion were explored (Mortazavi et al., 1992, 1996). The apparatus used is shown schematically in Figure 12.1. Two catalysts were employed: a Li doped magnesium oxide studied intensively by Lunford and his co-workers (Ito et al., 1985) and a mixed cerium oxide – Li doped magnesium oxide investigated by Bi et al. (1988) and by Bartsch et al. (1989). Indeed, the composition of the latter catalyst was identical to the catalyst described by Bartsch et al. The purpose of selecting two catalysts was to see how the type of catalyst, irreducible vs. reducible oxide, affected the enhancement achieved through modulation.

Three forcing strategies in the sense used in Chapter 1 were explored. The first employed switching between methane and $CH_4:O_2$ mixtures (the mixtures ranged from 2:1 to 4:1; the stoichiometric mixture for ethylene formation is 2:1). This was intended to represent a catalytic system operating around its reduced state (in steady state studies high $CH_4:O_2$ ratios gave the highest C_2 selectivities), while the second strategy considered cycling for a catalyst held at its oxidizing state by switching between the same mixtures and oxygen (highest steady state CH_4 conversions are found at low $CH_4:O_2$ ratios). This strategy also explored a proposal by Bartsch et al., (1989) that the presence in the gas phase of O_2 enhances the formation of CH_3 radicals through homogeneous reactions thereby increasing rates of methane conversion. Oxygen in this strategy is always present in the feed. It can be viewed as a methane cycling strategy, while the former is an oxygen cycling one. The bang-bang strategy chosen by the Union Carbide and ARCO researchers was the third one investigated. It employed switching between CH_4:He and O_2:He mixtures.

Steady state experiments with a $CH_4:O_2 = 2:1$ reactant ratio indicated the maximum selectivity to C_2 products for the LiMgO catalysts occurred at about 750°C (Mortazavi et al., 1992) so all forcing experiments were performed at this temperature. In both steady state and modulation experiments, total pressure was 1 bar, total flow rate was 200 mL(STP)/min corresponding roughly to a space velocity of 1 L/g.cat. min. Generally, the He diluent made up 75 to 90 vol. % of the reactor feed on a time-average basis. All experiments began with the fully oxidized catalyst; the catalyst was exposed to a $CH_4:O_2 = 2:1$ reactant mixture at 750 °C for 70 h prior to use to stabilize activity. Individual forcing experiments were separated by returning to the standard 2:1 mixture and re-establishing steady state. These repeated steady state measurement checked deactivation. Only very slow deactivation was observed. About an hour under modulation was needed to attain a stationary, reproducible cycle. Cycle periods from 3 s to 30 minutes and splits from 0.3 to 0.9 were tested. Most experiments were done at $\tau = 3$ to 180 s and $s = 0.5$.

Exploratory experiments using methane cycling and oxygen cycling with a $CH_4:O_2 = 3:1$ mixture indicated enhanced formation of C_2 products at cycle periods below 5 min and attainment of quasi steady state at periods exceeding 10 minutes. Global enhancement of C_2 products was observed. Cycle split appeared to be important. Between 2 and 3 moles of ethane are formed for each mole of ethene and the effects of cycle period and split on ethane and ethene formation are similar. Further details are given by Mortazavi *et al.* (1992).

More thorough study in the forcing frequency range where the largest enhancements were observed confirmed most of the above observations. Results for oxygen cycling are given in Figure 12.5 for $s = 0.5$ and each catalyst. The figure explores the forcing frequency (presented as τ) and amplitude effects. It shows that for the cerium promoted LiMgO catalyst CH_4 conversion and C_2 formation enhancement is confined to cycle periods under 10 s. For the LiMgO catalyst, only conversion is enhanced. Unlike the exploratory experiments, the quasi steady state seems to be reached for cycle periods exceeding 60 s. Amplitude appears to have little effect on selectivity to C_2 products and on CH_4 conversion. Figures 12.5c and f show an effect on the % ethene in the C_2 products. The amplitude and cycle split variables were studied at $\tau = 30$ s and the amplitude observations were confirmed (Mortazavi *et al.*, 1996). Cycle split, defined as the ratio of mixture flow duration to cycle period, influenced CH_4 conversion but not selectivity to C_2 products. Experiments at shorter cycle periods were not possible because the duration of a portion of the cycle at high or low split would be less than a second. Mixing in the system and flow irregularity caused by valve positioning limited durations of cycle parts to about a second.

Like oxygen cycling shown in Figure 12.5 and bang-bang cycling shown in Figure 12.8, the effect of cycle period, split and amplitude on conversion, C_2 selectivity and the ethene/ethane ratio for methane cycling on the two catalysts were about the same. Thus, only results for the Cerium promoted catalyst are given in Figure 12.6. In methane cycling, C_2s are observed in the

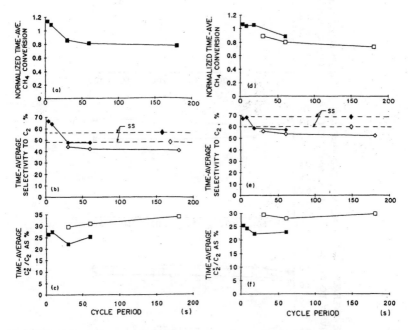

FIGURE 12.5 Influence of cycle period and amplitude on performance under oxygen cycling for the Ce/Li/MgO catalyst in (a) to (c) and for the Li/MgO catalyst in (d) to (f) at a cycle split $s = 0.5$, 750°C, 103 kPa and SV $= 1$ L/gcat min. (a) and (d) are CH_4 conversion normalized by steady state conversion at the time-average feed composition, (b) and (e) are selectivity to C_2, while (c) and (f) show the volume percent ethene in the C_2 product. Open points in (a) to (c) are for cycling between 10 vol.% CH_4 in He and a mixture with $CH_4:O_2 = 13.7:13.2$ as vol.%, while for the solid points $CH_4:O_2 = 22.7:5.4$. In (d) to (f) the open points are for cycling between 10 vol.% CH_4 in He and a mixture with $CH_4:O_2 = 14.6:11.2$ as vol.%, while for the solid points $CH_4:O_2 = 23.8:5.2$. The dashed lines, marked ss, represent the steady state selectivity to C_2 (Figure taken from Mortazavi et al. (1996) with permission of the authors)

reactor off gas only during exposure of the catalyst to the CH_4-O_2 mixture. In this situation, there are two standards for evaluating the benefit of composition modulation: 1) comparison of time-average performance under forcing to performance at a steady state corresponding to the time-average feed composition (CSS1), or 2) comparison to the performance in the half cycle that the CH_4-O_2 mixture flows to the reactor (CSS2). The former standard overstates the benefit for selectivity because the selectivity to C_2s is poor at low $CH_4:O_2$ ratios, but gives a truer picture for CH_4 conversion because conversion is high for such ratios. Figure 12.6 employs both standards. In Figure 12.6a there is only a small improvement in conversion through methane cycling at $\tau < 5$ s (the small improvement for $CH_4:O_2 = 2:1$ seen in the figure is unlikely and reflects experimental

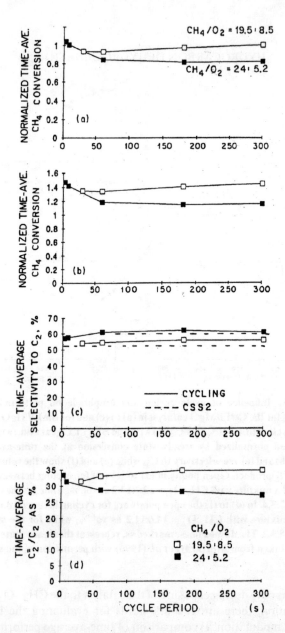

FIGURE 12.6 Influence of cycle period and amplitude on performance under methane cycling for the Ce/Li/MgO catalyst at a cycle split s = 0.5, 750°C, 103 kPa and SV = 1 L/gcat min.: (a) CH_4 conversion normalized by steady state conversion at the time-average feed composition, (b) CH_4 conversion normalized by steady state conversion at the composition in the mixture half cycle, (c) selectivity to the C_2 products, and (d) the volume percent ethene in the C_2 product. Open squares are for cycling between 4 vol.% O_2 in He and a mixture with $CH_4:O_2$ = 19.5:8.5 as vol.%, while for the solid squares $CH_4:O_2$ = 24:5.2. The dashed line, marked CSS2, represents the steady state selectivity to C_2 for a composition corresponding to that in the mixture half cycle (Figure taken from Mortazavi et al. (1996) with permission of the authors)

variance) if steady state conversion at the time-average feed composition is used. For reasons just mentioned, the enhancement becomes substantial if steady state conversion at the mixture composition is taken as the standard as is done in Figure 12.6b. C_2 selectivity as a function of cycle period is given in Figure 12.6c. For the standard, there appears to be a very small selectivity enhancement that is largely independent of cycle period. There is also a small enhancement of volume percent ethene in the C_2 product, but this cannot be determined from Figure 12.6d which just shows the amplitude effect on the % ethene. As observed in steady state operation, the % $C_2^=$ increases as the $CH_4:O_2$ ratio decreases.

Amplitude effects on selectivity and CH_4 conversion under symmetrical methane cycling are shown in Figure 12.7. In the figure, amplitude is expressed as the $CH_4:O_2$ ratio as switching occurs between the mixture and an O_2 stream. The C_2^+ curve is for ethane. For methane cycling, the amplitude effect is large for selectivity to ethene and for CH_4 conversion. With the exception of selectivity to ethane, enhancement drops off as the amplitude used in forcing increases.

Results for bang-bang cycling, shown in Figures 12.8 and 12.9, confirm the moderate enhancement seen in the ARCO data rather than the high enhancements found by Keller and Bhasin (1982). This high amplitude modulation does not increase performance for the LiMgO catalyst except for a small improvement in the C_2 selectivity at $\tau = 3$ s. There is, however, a significant increase for the cerium promoted catalyst at cycle periods less than 15 s. At $\tau = 3$ s, CH_4 conversion increases by about 16 %, the C_2 selectivity increases by 15 % and the percent $C_2^=$ in the C_2s increases by 29 % over the steady state. Cycle split has a large effect on selectivity, but just a small influence on CH_4 conversion as Figure 12.9 illustrates. Split is defined as the duration of methane exposure to cycle period. Thus, short exposure to methane, corresponding to a low split, increases the C_2 selectivity. However, the parallel lines in the figure mean that split does not affect the $C_2^=/C_2$ ratio. If these results can be extrapolated to $\tau = 3$ to 15 s, where modulation improves performance, they indicate a methane coupling design should use a very short exposure to methane, probably 1 s or less. This dictates the use of a fast fluidized bed or riser reactor. The O_2 exposure is probably not critical. The C_2 selectivities measured were well below those reported by Jones et al. (1987a).

Mortazavi et al. (1996) also examined flushing with a 3-part cycle: 10 vol.% CH_4 in He, 10 vol.% O_2 in He followed by He alone. Equal durations were used with $\tau = 30$ and 60 s. Cycle period, long for the results given in Figure 12.8, had little effect on performance. Flushing had a large influence on C_2 selectivity raising this important performance criterion to about 53 %, well above selectivities at $\tau = 30$ and 60 s seen in Figure 12.8 and even higher than the selectivity at $\tau = 3$ s. The selectivity to $C_2^=$ reached 13 % compared to 9 % without the diluent flush. Flushing had no effect on CH_4

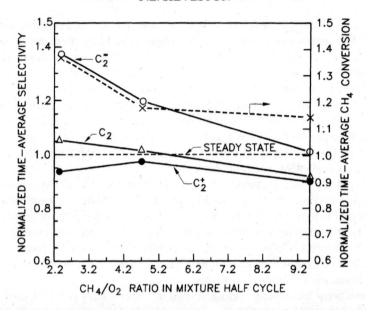

FIGURE 12.7 Normalized CH_4 conversion and product selectivity for methane cycling at $\tau = 60$ s as a function of the $CH_4:O_2$ ratio in the mixture half cycle. Catalyst and operating conditions as given in Fig. 12-6. Data were normalized with steady state results for a feed corresponding to the mixture composition in the mixture half cycle (Figure taken from Mortazavi et al. (1996) with permission of the authors)

conversion. Thus, stripping of the catalyst after O_2 exposure should be part of a commercial methane coupling process just as ARCO proposes.

The University of Waterloo investigators also explored a three part cycle with overlapping of the CH_4 and O_2 exposures so that the catalyst sees a $CH_4 - O_2$ mixture for part of the cycle. Experiments were performed with a half cycle of a feed containing 10 vol.% CH_4 in He, a third of a cycle with a feed containing 10 vol. % O_2 in He and a sixth of a cycle with a mixed feed containing 5 vol. % CH_4 and 5 vol. % O_2. Cycle periods were 30 and 60 s. Cycle period had a small effect. Shorter periods gave higher conversions and C_2 selectivities. In a 30 s cycle, overlapping raised the C_2 selectivity to 56 %, the highest measured in the Mortazavi experiments, while the $C_2^=$ selectivity went to 15 %, also the highest observed.

Can composition modulation reach the 25 % C_2 yield necessary to make methane coupling economically attractive? The Mortazavi experiments indicate that this is not possible. Highest C_2 yield were obtained with methane cycling (Figure 12.6) and were 10% at $\tau = 3$ s and $s = 0.5$. Symmetrical bang-bang cycling gave 9.5 %. Optimizing the cycle might raise the yield to 11 or 12 %, while introducing flushing or overlapping

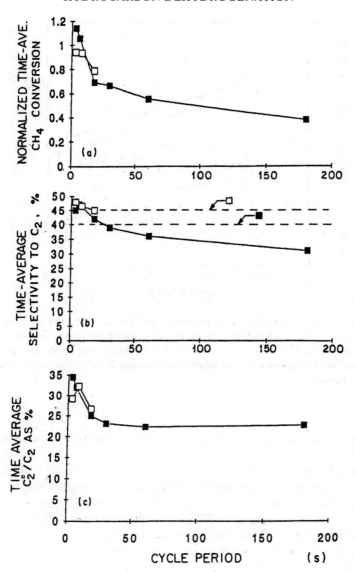

FIGURE 12.8 Influence of cycle period on performance under symmetrical bang-bang cycling between 12.9 vol.% CH_4 in He and 9.4 vol.% O_2 in He for the Ce/Li/MgO catalyst (solid squares) and for the Li/MgO catalyst (open squares) at 750°C, 103 kPa and SV = 1 L/gcat min.: (a) CH_4 conversion normalized by steady state conversion at the time-average feed composition, (b) selectivity to C_2, while (c) volume percent ethene in the C_2 product. The dashed lines represent the steady state selectivity to C_2 at the time-average feed composition (Figure taken from Mortazavi et al. (1996) with permission of the authors)

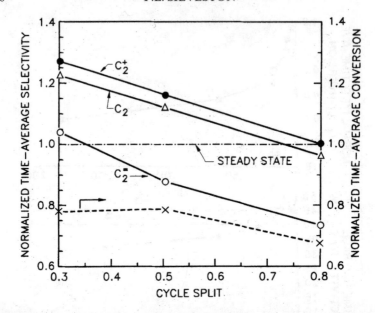

FIGURE 12.9 Normalized CH_4 conversion and product selectivity for bang-bang cycling at $\tau = 30\,s$ as a function of the cycle split. Catalyst and operating conditions as given in Figure 12.8. Data were normalized with steady state results for the time-average feed composition (Figure taken from Mortazavi *et al.* (1996) with permission of the authors)

could increase the yield to perhaps 13 or 14 %. This is well below the target yield.

Bang-bang and methane cycling are the most attractive strategies: oxygen cycling is not competitive. This observation implies that the best performance for methane coupling can be obtained by using a short exposure of the catalyst to the CH_4 containing stream. If a choice between bang-bang and methane cycling must be made, the preference is for bang-bang because diluent flushing or overlapping become possible with this mode of modulation.

Although the forcing experiments undertaken by Mortazavi do not provide any novel insight into the mechanism of methane coupling or its kinetics, they do permit judgement of mechanisms proposed by others. The cycling experiments suggest that the catalyst is involved in ethene formation. The bang-bang experiments and the contribution of flushing imply that ethane dehydrogenation must occur. They also suggest that O_2 adsorbed on the LiMgO surface causes total oxidation. Mortazavi(1993) speculates from his methane cycling results that CO_2 inhibition occurs.

Overview And Challenges

The existence of the yield barrier is attributed to the preferential oxidation of C_2 in the presence of methane by both adsorbed and lattice oxygen in the catalyst. Segregating oxygen and methane as is done by cycling increases selectivity to the desired products, as Mortazavi et al. show, but this does not eliminate product loss. Consequently, the question is whether or not other modes of operation might be possible which would reduce costs so that a lower C_2 yield would suffice to make coupling feasible economically. Mortazavi (1993) proposes the simultaneous use of flow reversal and composition modulation. Cycle periods for flow reversal are in the range of hours, while the results reported above used periods in the range of seconds. Thus, it would be possible to superimpose composition modulation on flow reversal. The reason to explore this combination is that periodic flow reversal offers efficient, low cost heat recovery. Preheating the feed to the 700–800°C reaction temperature is a major cost in methane coupling. A recent paper (Budman et al., 1996) demonstrates control of highly exothermic reactions in periodic flow reversal through heat withdrawal in the catalytic bed. Removing the heat of reaction at the highest temperature in the system could significantly reduce methane coupling cost.

The work just discussed provides enough dynamic and cyclic data to construct a reliable model for the kinetics of methane coupling using either cerium promoted or unpromoted LiMgO. Good models are available for periodic flow reversal so that it would seem desirable to investigate the combinaton of periodic flow reversal and composition modulation for methane coupling by simulation. Models could also be used to optimize bang-bang or methane cycling and to explore the effect of using air and natural gas as feeds to the coupling reaction and the use of N_2 as the diluent as well as the effect of conversion on the C_2 yield. Eventually further experimentation might be worthwhile.

Partial Oxidation To Methanol

The second route for methane conversion to higher value products is the partial oxidation to methanol and/or formaldehyde. It is surprising that the application of composition modulation to these reactions does not seem to have been tried in view of similarities between CH_4 partial oxidation and oxidative coupling. In both groups of reactions, the products are more susceptible to oxidation than methane and steady state (co-feed) operation show rapidly dropping selectivity as the CH_4 conversion increases. Pitchai and Klier (1986) do not mention composition forcing in their comprehensive review of the literature.

C. ETHYLENE AND PROPYLENE OXIDES FROM OLEFINS

Ethylene oxide, an important intermediate in the chemical industry, has been produced from the partial oxidation of ethylene over a supported silver catalyst since the 1960's. Total combustion of ethylene and ethylene oxide are significant side reactions whose heats of reaction are considerably greater than the heat of reaction for epoxidation. Activation energies are also high so the reaction is difficult to control. To insure control, The feed contains high levels of inerts and conversion of ethylene per pass is held to about 15 %. The reaction captured the attention of pioneering investigators of periodic operation and was one of the first commercially important reaction to be studied where the objective was improved selectivity rather than catalyst activity (Renken et al., 1976; Gau, 1982).

Renken's Work

In the first study, Renken et al. (1976) used a fixed bed reactor, 30 mm (i.d.) × 380 mm, containing about 349 g of catalyst and immersed in a oil bath for temperature control. Objectives were the effect of composition modulation on the selectivity-conversion behavior and on the temperature profile in the integral reactor. For the latter purpose, the reactor was fitted with 12 thermocouples. Modulation was carried out by holding the O_2 mol percent in the feed constant at 6 % and switching the ethene flow so that the feed contained either 0 and 12 mol %. The flow of N_2 was adjusted so as to maintain the space velocity constant. Symmetric cycles ($s = 0.5$) were employed in all experiments. Variables were cycle period and conversion. Space velocity and the oil bath temperature were varied to control the latter variable. A promoter in the reactant feed was not used.

At the same time average feed composition, space velocity and bath temperature, composition modulation of ethene results in lower conversions for this partial oxidation reaction, but the yield of ethylene oxide increases. This is the result of significantly higher selectivity in the modulation mode. Figure 12.10a shows higher selectivity over a wide range of conversion at a cycle period of 2 s. At longer periods, the selectivity improvement is restricted to conversions below 20 %. Period has a large effect on selectivity and Figure 12.10b shows a resonance behavior. For conversions in the 5 to 10 % range and a 200 °C bath temperature, the optimal cycle period is between 20 and 40 s. These measurements are among the few that are extensive enough to show resonance with respect to selectivity, although this phenomenon would be anticipated in periodic forcing of systems consisting of networks of competing and consecutive reactions. Resonance in catalyst activity as measured by rates of single reactions is mentioned frequently in Chapter 2 to 9.

Higher selectivity at the same feed conditions reduce the heat generated because the side reactions are much more exothermic than the epoxidation reaction. This has a marked effect on the temperature profile in the catalyst bed as illustrated by Figure 12.11 which compares the temperature profiles

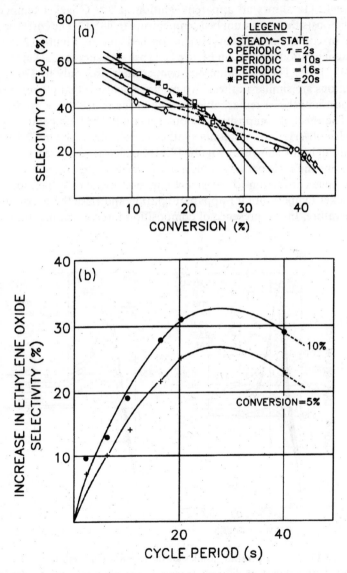

FIGURE 12.10 Selectivity to ethylene oxide in a packed bed containing a supported Ag catalyst: (a) as a function of ethene conversion and cycle period, (b) as a function of cycle period at constant ethene conversion (Figure adapted from Renken *et al.* (1976) with permission, © 1976 VCH Verlagsgesesellschaft)

for steady state and periodic operation at different bath temperatures, that is, different conversion levels. Both steady state and periodic operation exhibit parametric sensitivity. As the reactor jacket temperature increases, there is a jump in bed maximum temperature and ethylene conversion. Figure 12.11a shows an apparent ignition at 208°C jacket temperature which boosts the maximum bed temperature to 385°C. The corresponding jump in conversion appears as the dashed line in Figure 12.10a. The apparent ignition is muted in Figure 12.11b and occurs at a jacket temperature of 217°C. The maximum bed temperature rises only to 270°C. Inlet conditions are similar in these two measurements so that ethene modulation accomplishes a remarkable leveling of the temperature profile in the bed. The effect of composition modulation on the bed temperature profile was also observed for ammonia synthesis and is discussed in Chapter 7.

Although not shown in Figure 12.11b, which is restricted to $\tau = 20$ s, increasing the cycle period suppresses sensitivity of the temperature profile and ethene conversion to the jacket temperature. Shorter periods reduce the jacket temperature for apparent ignition and raise the maximum bed temperature. In the presence of composition forcing, temperature at any

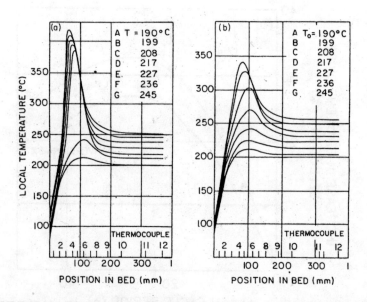

FIGURE 12.11 Temperature profiles for the epoxidation of ethene in a fixed bed of supported Ag catalyst at different reactor jacket temperatures: (a) steady state operation, (b) composition forcing at $\tau = 20$ s. Curves are labelled by the jacket temperature, while the numbers on the abscissa show the thermocouple locations (Figure taken from Renken et al. (1976) with permission, © 1976 VCH Verlagsgesesellschaft)

point in the catalyst bed changes with time. Up to a cycle periods of 40 s, these temporal variations do not exceed 3° C and the shape of the temperature profile does not change, that is, the location of the point of maximum temperature is fixed. Increasing the period to 120 s causes the maximum temperature to oscillate axially.

Renken *et al.* explain their observations through inhibition by ethene, ethylene oxide and decomposition products on the surface which reduce the adsorption of molecular oxygen. They suggest the epoxidation reaction occurs between adsorbed ethene and adsorbed molecular oxygen. The role of forcing then is to remove adsorbed species and to increase the concentration of adsorbed O_2. Figure 12.12, which plots the mol % of C_2H_4 and C_2H_4O leaving the reactor as a function of dimensionless time within a cycle, supports this explanation. Ethene takes about 5 s to equilibrate on the Ag surface after a switch from 0 to 6 mol % in the feed and about the same time to disappear when the ethene flow is cut off. The C_2H_4O plateau continues for up to a second after the C_2H_4 cut off indicating adsorbed C_2H_4 or an intermediate present in abundance. It takes more than 8 s for C_2H_4O to disappear from the off gas indicating either that it is strongly adsorbed or is still being formed on the surface.

Further experiments by Renken and his co-workers to elucidate the epoxidation mechanism and develop a model for the application of compo-

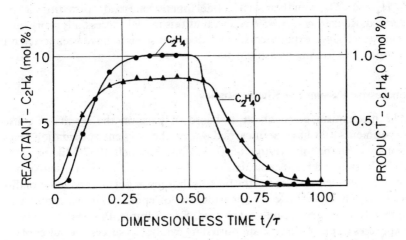

FIGURE 12.12 Ethene and ethylene oxide mol percents in the reactor off gas vs. dimensionless time within a cycle in the composition forcing of ethene epoxidation at $\tau = 20$ s over a fixed bed of supported Ag catalyst with a reactor jacket temperature of 217°C. Step up of ethene from 0 to 6 mol % in the feed occurs at $t = 0$. Data has not been corrected for transport lag (Figure adapted from Renken *et al.* (1976) with permission, © 1976 VCH Verlagsgesellschaft)

sition forcing to this reaction were undertaken in a recycle reactor with an external recycle loop and used silver supported on α-Al_2O_3 (Renken et al., 1989). Step change experiments at temperatures between 200 and 300°C using a feed containing 70 mol % C_2H_4 indicated rapid initial changes on a time scale of minutes, followed by slow changes which occurred on a scale of hours. From these observations and a large number of kinetic studies available in the literature (Kilty and Sachtler, 1974), the following kinetic model was proposed:

$$4\, Ag + O_2 = 2\, Ag_2O$$
$$2\, Ag_2O + O_2 = 2\, Ag_2O - O$$
$$Ag_2O + O_2 = Ag_2O - O_2$$
$$C_2H_4 + Ag_2O - O \rightarrow C_2H_4O + Ag_2O$$
$$C_2H_4 + 3\, Ag_2O - O_2 \rightarrow 2\, CO_2 + 2\, H_2O + 3\, Ag_2O$$
$$CO_2 + Ag_2O - O_2 = Ag_2O - \dot{O}_2 - CO_2$$

The arrows in the above scheme indicate slow steps. Basing material balances on the above scheme and assuming competitive adsorption of C_2H_4 and CO_2, a mathematical model for ethylene oxide formation can be derived. Renken et al. (1989) report that this model successfully represents their step change experiments, but it does not seem to have been tested on composition forcing measurements.

Successive Pulsing For Kinetic Studies

The mechanism given above is based in part on the results of pulse experiments that have been carried out on the ethylene epoxidation reaction by a number of investigators (Flank and Beachell, 1967; Mikami et al., 1970; Verma and Kaliaguine, 1973 and Prauser et al., 1976). A discussion of these experiments, indeed of the pulse technique, is beyond the scope of this monograph. Nevertheless, their wide use for epoxidation warrants mention. Often, successive pulses are used. For example, Prauser et al. (1976) expose a sample of alumina supported Ag catalyst to successive pulses of O_2 and monitor the transfer of O_2 to achieve an oxygen saturated surface or one that is but partially saturated. The saturated surface may then be "aged" in a flowing inert to desorb physically bound oxygen or to allow oxidation of the surface to proceed prior to exposure to C_2H_4 pulses. Analysis of the off gas measures the conversion of ethene and selectivity to the oxide or to total combustion. Use of successive pulses indicate how the saturation of the surface or the degree of reduction affect conversion and/or selectivity. In some experiments (Mikami et al., 1970), the C_2H_4 pulse is followed by "aging" and a return to an O_2 pulse. With a switch in the O_2 isotope, reactions between adsorbates can sometimes be distinguished from a gas species-adsorbate reaction.

Investigators, such as Mikami et al. and Prauser et al., conclude from the pulse experiments that C_2H_4 must adsorb on the oxide or onto an oxygen adsorbate for either epoxidation or total oxidation to proceed. The latter researchers contend that molecularly adsorbed O_2 results in epoxidation whereas dissociated species lead to total oxidation. Both groups attribute selectivity improvements in the presence of CO_2 to competitive adsorption of ethene and carbon dioxide on Ag_2O sites.

In the context of Reactor Engineering, the pulse technique is interesting because it partially simulates composition modulation. Successive pulses of one reactant followed by a pulse of the second reactant suggest the effect of cycle split in forcing. The "aging" experiments mentioned above indicate what might be attained in complex cycles using an inert sweep (See Chapter 1). There appears to be no advantage, however, in using pulse experiments to explore potential applications of periodic operations. The forcing experiments are often faster and easier to undertake. Pulse measurements do not disclose the effect of frequency, often an important variable in composition modulation.

Work Of Gau At The Institut De Petroleochimie Et De Synthese Organique Industrielle

Park et al. (1983) used complex 4 and 8 part cycles to investigate the epoxidation and total oxidation mechanisms. Modulation, extending over many hours of operation, replaced in this study the more common pulse or successive pulse technique. The 4-part cycle consisted of exposure to a reactant mixture in the C_2H_4:O_2 ratio 38:12 or to C_2H_4 alone, followed by an inert flush (N_2), then O_2 and a second N_2 flush. The duration of the inert flushes were usually equal, but the durations of the ethene containing pulse and the pulse with just O_2 were varied widely. The 8-part cycle retained the N_2 flushes after exposure to each reactant, but added a pulse with H_2 exposure and another O_2 pulse. One of the O_2 pulses was at 30 to 50°C higher than the temperature in the remainder of the cycle. Occasionally, the other O_2 pulse used a lower temperature. The off gas from the fixed bed of catalyst was sampled by an on-line gas chromatograph to give the change of product and reactant concentrations with time in each part of the cycle. The purpose of these complex cycles, particularly the 8-part one, was to investigate the role of reaction residues on C_2H_4 conversion and the C_2H_4O selectivity. The investigation by Park et al. supplemented a modelling study by Ghazali et al.(1983) that was performed at steady state. In both investigations, some experiments were carried out with a dichloroethane promoter in the gas feed to the fixed bed reactor. A commercial catalyst, supported on low surface area α-Al_2O_3, and two catalysts prepared by impregnating a high surface area SiO_2 were used.

Residues were identified that could be removed as CO_2 and H_2O in the O_2 exposure portion of the cycle or through hydrogenation. Park et al. suggest that these were strongly adsorbed oxygenates, possibly acetic acid or a surface precursor of this acid. At 180°C, with or without dichloroethane, the production of C_2H_4O and CO_2 fell with time for exposures of C_2H_4 or C_2H_4:O_2 mixtures exceeding 30 s indicating oxidation was drawing O_2 from the catalyst. In both the 4-and 8-part cycles, duration of the O_2 exposure or the O_2 partial pressure affected C_2H_4 conversion, but did not have a large affect on selectivity to C_2H_4O. In the presence of the promoter, the duration of the N_2 flush prior to exposure to C_2H_4 affected conversion but not selectivity. Park et al.(1983) explain their results by postulating 3 oxygen adsorbate species on the catalyst surface, one of these species occurs only in the presence of the dichloroethane additive. Atomic oxygen was held to be responsible for total oxidation. This species could attack the epoxide or the ethene precursor. Dichloroethane dissociates on the Ag_2O surface and inhibits the dissociation of O_2. The residues observed with the active catalysts as well as the low surface area catalyst were believed to also suppress dissociation. Catalyst activity was also low.

In several of the experiments with dichloroethane promoter using the active catalysts (some 10 to 30 × the activity of the commercial catalyst), C_2H_4O selectivities between 92 and 98 % were measured in the 4-part cycles. These are much higher than the ca. 50 % selectivity found in full scale installations. These observations apparenty led the Institut de Petroleochimie et de Synthese Organique Industrielle to patent an ethylene oxide process based on high activity catalysts, a dichloroethane promoter and a circulating fluidized bed to affect the cyclic operation used in the Park experiments.

A multitubular circulating fluidized bed for ethene or propene epoxidation using a high surface area supported Ag catalyst has been proposed and is described by Park and Gau (1986). The concept is discussed in Chapter 1 and illustrated there by a schematic of the system. An attempt was made to simulate the multitubular circulating fluidized bed with a two tube design, but operating problems were encountered as Park and Gau report. Instead, the system was simulated using a fixed bed adjusted so the contact time was the same as in the circulating fluidized bed. For this, a 4 mm (i.d.) × 2.1 m tube was employed. The cyclic operation of the circulating fluidized bed was achieved by mixing the Ag catalyst with a silica gel adsorbent which trapped the ethylene oxide as it formed. The cycle then alternated between a 50:50 mixture of C_2H_4 and O_2 and a N_2 flush.

Experiments were run at atmospheric pressure and 170°C; variables were the durations of the mixture and the N_2 exposures, and the N_2 flow rate during the flush. A steady state experiment was made for comparison purposes.

Provided the total flow of N_2 during the flush was long enough to recover the adsorbed C_2H_4O, simulation of the cyclic operation as described increased the selectivity to C_2H_4O from 64.4 % under steady state to 70.4 %; conversion of C_2H_4 increased by 27 % as well for a cycle period of 50 s. Operation of the mixed catalyst adsorbent bed under reactant mixture-inert cycling is shown in Figure 12.13 (Park and Gau, 1986). The 60 s exposure to the $C_2H_4 - O_2$ mixture is too long for the weight of adsorbent used so loss of the epoxide from the bed is large at the end of that portion of the cycle. Desorption of C_2H_4O from the silica gel is slow at the N_2 flow rate used so a 180 s flush was employed.

The Institut de Petroleochimie et de Synthese Organique Industrielle also studied propene epoxidation. Balzhinimaev et al.(1984) followed up on Kobayashi's dynamic experiments which demonstrated that propylene oxide yields could be increased by exposing the AgO/SiO_2 catalysts to O_2. O_2 apparently removes strongly adsorbed species that block the active

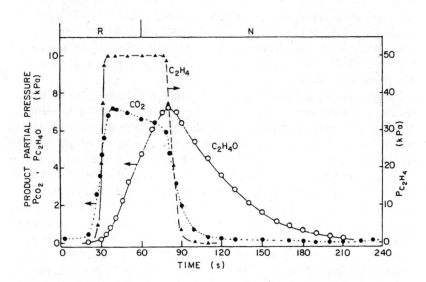

FIGURE 12.13 Variation of product and reactant partial pressures in the off gas from a mixed bed of Ag catalyst and silica gel adsorbent under switching between a reactant mixture of 50 mol % C_2H_4, 50 mol % O_2 and 100 % N_2 at atmospheric pressure and 170°C. Data have not been corrected for a transport lag of about 25 s (Figure taken from Park and Gau (1986) with permission, © 1986 Pergamon Press)

surface sites (Kobayashi, 1980). They reasoned that a cyclic operation could be attractive for propene epoxidation and undertook a study of cycling strategies, but did not attempt to optimize those which gave large improvements. Two catalysts were used: both contained 25 wt. % Ag, but one was promoted with 1.8 wt. % Ba, while other used 0.5 and 0.025 wt. % of Cd and Se respectively. The latter was less active, but showed higher propylene oxide selectivity and was the catalyst most used in the Balzhinimaev experiments. The cycling strategies explored were:

1) $C_3H_6:O_2/N_2/O_2/N_2$
2) As in 1) but with ppm quantities of dichloroethane in the $C_3H_6:O_2$ mixture
3) As in 2) but with the O_2 pulse at a 20°C higher temperature
4) $C_3H_6:O_2/N_2/H_2/N_2$
5) $C_3H_6:O_2/N_2/O_2/N_2/H_2/N_2$ with ppm quantities of dichloroethane in the $C_3H_6:O_2$ mixture and with the O_2 pulse at a 20°C higher temperature

The purpose of the H_2 pulse was to remove surface carbonates just as in Gau's ethene epoxidation studies. Hydrogen flushing also removes adsorbed oxygen from the surface and may reduce the oxide. Raising the temperature in the O_2 pulse increased the removal of residue from the surface. The barium promoted catalyst was used at 413°C, while the other catalyst was held at 435°C in the 4-part cycles and at 393°C in the 8-part ones. Variables in the experiments were the duration of the pulses, in particular the O_2 exposure, and the vol. % propene in the reactant mixture. The percent was varied from about 60 to 100 %.

Large increases in selectivity were achieved with these complex cycles relative to the steady state selectivity at the time-average feed composition. Nevertheless, both selectivities and yields appear to be almost an order of magnitude lower than those attained with ethene epoxidation even in steady state operation. The 4-part cycle with an O_2 pulse revealed a 4 % selectivity, compared with almost zero yield of the epoxide under steady state operation. Replacing O_2 exposure by an H_2 pulse, more than doubled the selectivity to 11 %. Figure 12.14a displays the time variation of the selectivity, the epoxide, O_2 and CO_2 within a cycle for the Ba promoted catalyst. Figure 12.14b is for the Cd/Se promoted catalyst which gave a time-average selectivity of 5.4 % rather than 11 % attained with the Ba promoted catalyst. Notable in both cycles is the lag before products appear after the $C_3H_6:O_2$ mixture is introduced. The lag reaches 20 s when H_2 is used to clean the catalyst surface. The quantities of epoxide and CO_2 formed in this cycle was much less than in the 4-part cycles using an O_2 pulse. CO_2 was evolved during the H_2 pulse. Selectivity to the epoxide was much higher during the reactant mixture pulse than for the overall cycle for

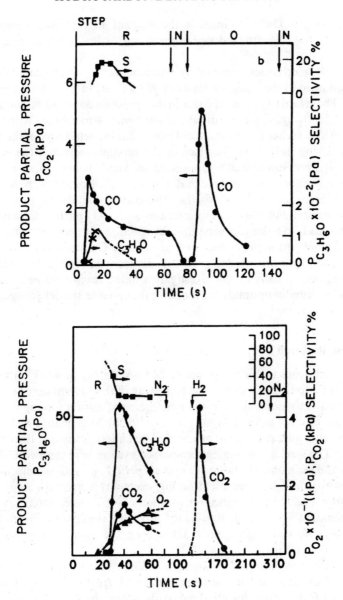

FIGURE 12.14 Variation of reactant and product partial pressures and selectivity during a 4-part composition modulation cycle: (a) Ba promoted AgO/SiO_2 catalyst at 413°C with a $C_3H_6:O_2/N_2/O_2/N_2$ cycle and 90 vol.% C_3H_6 in the reactant mixture portion of the cycle, 60 s duration of each cycle part except for the first N_2 pulse which is 20 s, (b) Cd/Se promoted AgO/SiO_2 catalyst at 435°C with a $C_3H_6:O_2/N_2/H_2/N_2$ cycle and 85 vol.% C_3H_6 in the reactant mixture portion of the cycle. Durations of each cycle part were 70, 60, 120 and 60 s in the order reactant mixture, N_2, H_2 and N_2 (Figure adapted from Balzhinimaev *et al.* (1984) with permission of the authors)

both strategies. Dichloroethane in the reactant mixture always improved selectivity, but it sharply reduced propene conversion and thus the epoxide yield.

Highest selectivities observed by Balzhinimaev et al. reached 29% and were attained in the 8-part cycles using 96 % C_3H_6 in the reactant mixture pulse. The selectivity to the epoxide in that pulse exceeded 80 %. Increasing the vol. % O_2 in the pulse reduces the selectivity. 4-part cycles with the O_2 portion at a higher temperature did not attain the selectivity of the 8-part cycles. Using only C_3H_6 in place of the mixture increased the epoxide selectivity, but substantially suppressed the yield. These observations led Balzhinimaev et al. to conclude that molecularly adsorbed O_2 participates in epoxide formation. Dichloroethane functions to inhibit O_2 dissociation on adsorption, but dissociation increases as the O_2 partial pressure rises and this leads to a dehydrogenation side reaction with possibly cyclization and oligimerization to produce surface residues. On further increase of the O_2 partial pressure, strongly adsorbed aldehydes form. Acrolein was detected occasionally by the investigators in the reactor off gas. Unfortunately, even under optimal conditions for the epoxide, the dehydrogenation reaction is favored.

Chinese Research

A research team under the direction of C.-Y. Li began work on various types of periodic operation at the beginning of the current decade. One of their first activities was to re-examine the composition forcing of ethene epoxidation over a Ag catalyst (Li et al., 1992b). Step change and cycling experiments were carried out in a packed bed micro reactor at 100 kPa and 250°C. The forcing experiments employed cycling between two $C_2H_4:O_2$ ratios and examined the influence of cycle period and split on selectivity to ethylene oxide. For symmetrical forcing, selectivity increased from about 65 % under steady state operation to about 70 % under forcing. Global enhancement was found. Resonance was also observed and maximum selectivity occurred at $\tau = 60$ s. Decreasing the cycle split brought the selectivity to about 72 % at s = 0.3.

A model was prepared assuming a redox mechanism and two classes of reactive sites on the surface: AgO and Ag_2O_2. The former was assumed to be active for ethylene oxide, while the Ag_2O_2 site strongly bound ethene and led stepwise to total oxidation. A CSTR reactor was chosen for the model. Parameters were estimated from step change data (Li et al., 1992a) and from the forcing results. This model qualitatively reproduced the experimental observations, but the predicted resonance, the maximum in the selectivity vs. cycle period, was at a τ well below the measured value.

D. PROPENE PARTIAL OXIDATION AND AMMOXIDATION

Sohio Research

The Sohio organization, now BP America, undertook their work on propene partial oxidation to find petroleum based routes to acrylonitrile and methacrylonitrile. A paper by Lewis *et al.* (1949) proposed using the lattice oxygen of a readily reducible metal oxide as the oxygen source in place of air by reasoning that the limited supply from the surface would reduce the extent of the competing total combustion reaction. Lewis *et al.* suggested that partial oxidation using their concept could be conducted using two fluidized beds, one for propene oxidation and the other for re-oxidizing the catalyst, and moving catalyst between the beds. Only concept and implementation were proposed in this early paper. Experimental investigations were not performed. The early effort at Sohio was directed at identifying catalysts that could be used in the two bed process and this established that complex oxides of molybdenum or phosphomolybdic acids on silica where effective when promoted with copper, antimony, tin or bismuth (Callahan *et al.*, 1970). Yields of up to 60 % of acrolein were obtained by passing propene through a fluidized bed of bismuth phosphomolybdate oxidant. However, the Sohio researchers found that a 50 % yield could be obtained by co-feeding air and propene to a single fluidized bed. They rejected the cyclic operation with two beds because of the solids circulation required. Callahan *et al.* estimated that up to 300 kg oxidant/kg of propene would have to circulate between the beds if the bismuth phosphomolybdate was to maintain its most active oxidation state.

Early Composition Forcing Experiments

Using the term periodic pulsing for forcing, Niwa and Murakami (1972) used this operation together with electrical conductivity measurements to investigate catalytic mechanisms in the partial oxidation of propene to acrolein. The catalysts listed in Table 12.1 were used in their study. These were unsupported materials of relatively low surface area except for one of the bismuth molybdate catalysts which was supported on SiO_2. All catalysts were calcined before use. A large vessel was placed after the reactor and acted as an integrator so that all product measurements are time-average ones. These were made once the cycling system became stationary.

The forcing used increased neither the selectivity or yield, except to a minor extent, as Table 12.1 demonstrates. Analyses in each part of the cycle showed very little acrolein was formed during the O_2 exposure, but

Table 12.1 Acrolein Production and Selectivity for Redox Catalysts Under Composition Modulation[a] and Steady State Operation[b] at Atmospheric Pressure and 386 C (Niwa and Murakami, 1972)

Catalyst	Atomic Ratio of Metals	Operation Cycling/Steady State	Production State Rate (mols/h)	Selectivity
Bi–Mo	1/1	Cyc.	1.86	0.84
		S.S.	5.42	0.92
Bi–W	1/1	Cyc.	0.306	0.27
		S.S.	1.11	0.55
Sb–Mo	2/3	Cyc.	0.518	0.75
		S.S.	0.925	0.61
Sn–Sb	10/1	Cyc.	3.48	0.92
		S.S.	5.25	0.92
Sn–Sb	4/1	Cyc.	3.46	0.93
		S.S.	5.46	0.93
Sn–P	10/1	Cyc.	0.368	0.49
		S.S.	0.388	0.48
MoO_3		Cyc.	0.362	0.85
		S.S.	0.342	0.88

[a] Symmetric cycling between 25 Vol. % C_3H_6 in N_2 and air with $\tau = 30$ s
[b] Steady state operation at 12.5 Vol. % C_3H_6 and 10.5 Vol. % O_2.

that with the exception of the Sb–Sn and Sb–Mo and MoO_3 catalysts, more than 70 % of the carbon oxides originated in this part of the cycle. Catalysts that formed 50 % or more of the carbon oxides in the C_3H_6 part of the cycle were those that did not show large differences in acrolein production between the forcing and steady state modes of operation. Selectivity did not differ either. Production rate was sensitive to cycle period. Figure 12.15 shows the influence of cycle period on the acrolein and carbon oxide production rates. Niwa and Murakami conclude from the continuity of the curves to a period of zero duration (steady state) that the oxidation mechanism does not change when composition modulation is applied.

Electrical conductivity was measured by a d.c. method using particles pressed into a disk. The measurement was checked by an a.c. observation. Variation of the conductivity with the composition of the gas phase passing through the catalyst bed is given in Figure 12.16a for the Bi–W material. Plateaus are approached just prior to switching. The gas phase corresponding to each plateau appears in the figure. The resistance for the plateaus observed with the different compositions are summarized in Figure 12.16b

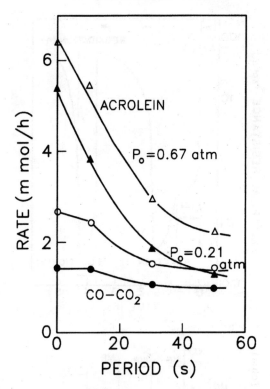

FIGURE 12.15 Influence of cycle period on the rates of acrolein and carbon oxide formation at 386°C in switching between 25 vol.% C_3H_6 in N_2 and air or 67 vol.% O_2 in N_2 at $s = 0.5$ over a Bi-Mo catalyst. Steady state rates at the time average feed composition are the points at $\tau = 0$ (Figure taken from Niwa and Murakami (1972) with permission, © 1972 Academic Press, Inc.)

for the catalyst studied. Conductivity measures the free electron density so a high resistance means a lower free electron density. For these experiments, the catalysts were held in air at 400°C for several hours and then exposed to N_2, the reactant mixture used at steady state (Table 12.1), air, and finally the $C_3H_6-N_2$ mixture used in the cycling experiments. For most of the catalysts, the conductivity changed 3 decades between their oxidized and reduced states. Direction of change indicated all but the Sn–Sb catalyst were n-type semiconductors. The large change in conductivity between air and N_2 exposures suggests oxygen loss during flushing with N_2. If this is loss from the lattice, it means high oxygen mobility in these solids. Lack of change for the Sn–Sb material suggests oxygen is strongly adsorbed on this catalyst. The small change on exposure to the reactant mixture or to propene imply low hydrocarbon adsorption or little reduction of the

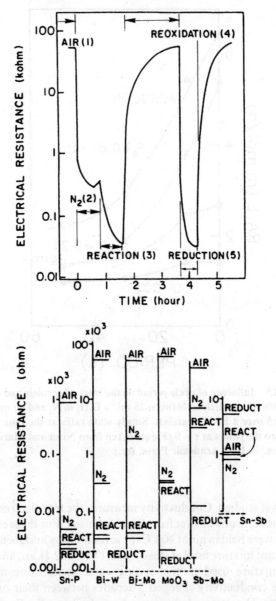

FIGURE 12.16 Variation of the catalyst electrical resistance with the gaseous atmosphere at 386°C: (a) measurements for a Bi-W catalyst for (1) air, (2) N_2, (3) reactant mixture with 12.5 vol.% C_3H_6 and 10.5 vol.% O_2, (4) air, and (5) 25 vol.% C_3H_6 in N_2, (b) measured plateaus for the oxide catalysts for the same sequence of conditions as in (a) (Figures taken from Niwa and Murakami (1972) with permission, © 1972 Academic Press, Inc.)

catalyst. Niwa and Murakami (1972) interpret these measurements as strong C_3H_6 adsorption on the Bi-Mo type of catalysts during the reducing portion of a cycle and combustion of these residues in the oxidizing portion. Oxygen retained on the surface in the Sb-Sn catalyst results in higher carbon oxide production in the reducing portion of the cycle. They conclude that Bi increases the re-oxidation rate of Mo and W and that catalysts with these components have high oxygen mobility compared with the Sn-Sb, Sb-Mo catalysts.

Successive Pulsing of the Ammoxidation Reaction

In this technique, as applied by Brazdil et al.(1980) to a study of the redox behavior of bismuth molybdate catalysts, a sequence of brief pulses of a propene:ammonia mixture alternate with a sequence of brief oxygen: helium pulses. The off gas from the catalyst is analyzed in each pulse so that for the system studied, the change in C_3H_6 conversion, in acrylonitrile, methyl cyanide, and carbon oxide formation, and in oxygen removal from the catalyst can be followed during the reducing pulse sequence. In the oxidizing sequence, O_2 uptake by the catalyst can be tracked. This procedure models a composition modulation experiment in which the number of brief pulses in the sequence determine the cycle period and split. The application, however, is to separate the reduction and re-oxidation steps in the redox system and to investigate mechanisms occurring in the catalyst and on its surface. The procedure is an extension of the alternating pulse method examined in the previous section.

Brazdil et al. (1980) examined calcinated α-$Bi_2Mo_3O_{12}$, β-$Bi_2Mo_2O_9$, γ-Bi_2MoO_6, $Bi_2O_3 - x(MoO_3)$ co-precipitated with other metal oxides, and ordered and disordered scheelite ($Bi_3FeMo_2O_{12}$) at a range of temperatures around 430°C. For catalysts that had C_3H_6 conversions greater than 10 % in the 1st pulse, conversion remained constant for the co-precipitated oxide, but dropped with succeeding pulses for the α-and β-bismuth molybdates; while the acrylonitrile selectivity rose for the co-precipitated oxide and went through a minimum in the 2nd or 3rd pulse for the bismuth molybdates before rising again with the following C_3H_6:NH_3 pulses. Utilization of catalyst oxygen dropped for almost all catalysts with succeeding pulses. The researchers concluded that oxygen drawn from the catalyst can either partially oxidize propene to acrylonitrile or cause total oxidation. With the co-precipitated oxide, ammoxidation sites were those with a partial degree of oxygen coordination of the metal ions in the solid so that reduction of the catalyst causes a transformation of combustion sites into ammoxidation ones prior to the loss of activity. For the α- and β-bismuth molybdates, strength of the M $-$ O bond of surface sites controls the ammoxidation selectivity. Reduction of the catalyst causes structural change which destroys reaction sites, but increases M $-$ O bond strength.

Catalyst re-oxidation was found to be 1st order in O_2 partial pressure and to be the slow step in the redox cycle. For the most active co-precipitated oxide, reduction was estimated to be almost 600 × faster then re-oxidation, whereas for the less active α-and β-bismuth molybdates, there was about a 50-fold difference. Activation energy was observed to be a function of the degree of reduction of the catalyst and this was thought to reflect two controlling steps: surface oxidation and diffusion of the oxygen anion in the solid phase.

Results of Brazdil et al. have implications for periodic operation. Separating the reactants should lead to higher conversion rates, because the slow re-oxidation step can be then carried out under optimal conditions. The selectivity behavior in the reducing sequence for the most active catalysts, however, shows some reduction is necessary to achieve highest selectivity. This suggests a CO or perhaps a diluent flush prior to exposure to the $C_3H_6:NH_3$ mixture to condition the catalyst. Without this, the Brazdil findings indicate a selectivity improvement may not occur.

Acrolein From Propene Partial Oxidation

Silveston (1980) and Silveston and Forrisier (1985) studied the composition modulation of propene partial oxidation at 400°C on a $SbSnO_2$ catalyst, containing Sb and Sn in 10:90 proportions, and an α-$Bi_2Mo_3O_{12}$ catalyst. Their objective was to test the application of composition modulation to partial oxidation reactions using redox catalysts with different semi conductor character. The Sb-Sn catalyst appears to a p-type semiconductor whereas the bismuth molybdates are n-type. Experiments were undertaken with a differential reactor operating at 1 bar using symmetrical forcing. Rather long cycle periods were used on the basis of step change from an oxidizing mixture (13.2 Vol. % O_2 in N_2) to a reducing mixture (13.2 vol. % C_3H_6 in N_2) in which acrolein formation was still evident 90 minutes after the change.

Figure 12.17 compares the acrolein concentration in the product gas and selectivity for the two catalysts as a function of cycle period and amplitude. The time average composition was 6.6 vol. % C_3H_6 and 6.6 vol. % O_2 for all experiments. Data for the α-$Bi_2Mo_3O_{12}$ catalyst were collected at 5 cycle periods, but at only two amplitudes, the larger of which consisted of cycling between C_3H_6 in N_2 and O_2 in this diluent. For the Sb—Sn catalyst, measurements were made at 6 cycle periods and 4 amplitudes. Despite the data scatter, which the investigators attribute to imprecise CO_2 measurements, low frequency, symmetric forcing increases selectivity to acrolein when cycling between mixtures is used. The improvement is about 20 % for the Bi-Mo catalyst (Figure 12.17b), but is considerably larger, about 80 %, for the Sb-Sn catalyst (Figure 12.17a). Switching between C_3H_6 and O_2, however, decreases the selectivity to acrolein to less than half that attained

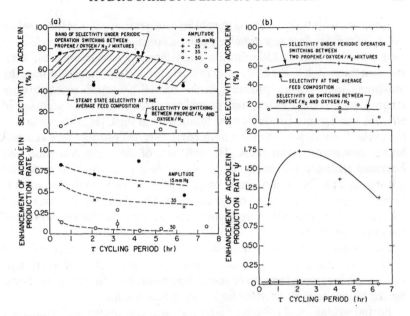

FIGURE 12.17 Influence of cycle period and amplitude on selectivity to acrolein and acrolein production for symmetric cycling at 400°C and 1 bar: (a) Sb–Sn oxide catalyst, (b) Bi_2O_3–$3MoO_3$ catalyst. Acrolein production has been normalized with the production rate under steady state at the time-average feed concentration.

in steady state operation and thus is an unsatisfactory strategy. There appears to be an optimal frequency corresponding to cycle periods between 2 and 4 hours. This is surprisingly low.

Composition modulation suppresses the acrolein concentration in the off gas when normalized with respect to steady state, evidently by decreasing C_3H_6 conversion for the Sb – Sn catalyst as the bottom portion of Figure 12.17a shows. The normalized vol. % acrolein decreases with increasing cycle period and with the amplitude of the switch. Amplitude is expressed in the figure as vol. % and represents half the change in C_3H_6 vol. % in the composition switch. At the maximum possible amplitude for switching between 13.2 vol. % C_3H_6 in N_2 and 13.2 vol. % O_2 in N_2, the time average acrolein vol. % approaches zero for both catalysts. The small amount detected contributes to the scatter seen in the selectivity plot for this amplitude, at least for the Sb-Sn catalyst. The opposite behavior can be observed for the bismuth molybdate in Figure 12.17b. For this catalyst, the normalized acrolein production is significantly higher. Thus, propene partial oxidation has increased. The only products detected with this catalyst were acrolein and the carbon oxides.

Sampling within a cycle using the Sb-Sn catalyst showed that the acrolein vol. % in the product gas remained almost constant through a cycle even for a period as long as 250 minutes. The volume % rose by no more than 10 % on the switch to the feed richer in C_3H_6 for all amplitudes less than the maximum used. Minor products in the partial oxidation reaction were ethanol, propanol, acetone, and propylene oxide. With the exception of ethanol, these minor products were found only in the C_3H_6 rich portion of the cycle. Ethanol was found in both parts, but this product was at much higher concentration in the C_3H_6 rich part of the cycle. Silveston and Forrisier attribute the small change in acrolein concentration to strong adsorption of C_3H_6 and perhaps acrolein precursors. Suppression of CO_2 formation by reduction of the catalyst would then explain improved selectivity. The reduced catalyst would also account for the appearance of the minor products.

Results with the bismuth molybdate suggest much weaker adsorption of C_3H_6. Periodic exposure of the surface to higher O_2 concentrations apparently increases the rate of re-oxidation and it is this which accounts for the higher rates of product formation under forcing.

Partial oxidation of propene under transient conditions was studied on a copper molybdate, $Cu_2Mo_3O_{10}$, by Labastida-Bardales et al.(1989) and by Labastida-Bardales (1991) in a Ph.D. thesis. This catalyst was chosen for study because a reversible change from Cu(I) to Cu(II) occurs on alternating exposure to air and to propene. However, recrystallization also takes place separating out MoO_3 and restructuring the catalyst surface. These alterations are partially irreversible (Haber, 1980; Haber et al., 1980a,b). Haber (1980) employed a successive pulsing procedure similar to the procedure used by Brazdil et al.(1980) and discussed in the previous section. Haber's observations indicated that the acrolein yield could be substantially increased by periodically introducing O_2 pulses into the reactor feed. The objective of the Labastida-Bardales study was to determine the performance under composition modulation for a catalyst which responded to the gas environment with changes in the metal ion coordination, phase separation and surface restructuring.

Labastida-Bardales' steady state and step change measurements undertaken in preparation for the forcing study showed a curious steady state hysteresis similar to that observed by Greger et al.(1984) for the partial oxidation of propene on a 1.2 wt.% Cu on Al_2O_3 catalyst: acrolein formation rates measured under decreasing C_3H_6 partial pressure were less than rates measured when the pressure increased provided a critical $O_2:C_3H_6$ ratio was reached. Observations of the transients on switching between $O_2:C_3H_6$ ratios of 3:1 and 1:6 showed a fast transient that disappeared in seconds on the step to increased C_3H_6 and a slow transient that required about 3 hours. On the step in the other direction, the fast transient spanned about 5 minutes and resulted in an undershoot. Correction of the undershoot occurred over 30 minutes. It took about 6 step up and step down cycles before the transients became reproducible and attained a constant steady state. Labastida et al. concluded the results were consistent with the recrystallization and restructuring arguments of Haber and co-workers. The hysteresis was attributed to stabilization of a copper oxidation state, ordinarily reducible in the absence of reaction, by the concurrent propene oxidation reaction. This was also the explanation advanced by Greger et al. for their observations.

Composition forcing experiments were performed at 623 K on O_2 pretreated $Cu_2Mo_3O_{10}$ by switching between $C_3H_6:O_2$ feed mixtures of composition 6:1 and 3:1, 1:1 and 1:3. Only the last is an oxidizing mixture. Cycle splits, defined by the duration of the propene rich mixture, ranged from 0.5 to 0.9, while periods were between 2 and 60 minutes. Time average rates of both acrolein and acetaldehyde were as much as 170 % greater than the maximum steady state rates, measured at $C_3H_6:O_2 = 6:1$. Selectivity to acrolein, however, was not changed by forcing. The cycling variables, split and period, did not exert a significant effect on the time average rate, except for switching between the 6:1 and 1:3 (oxidizing) mixture. In that experiment, time-average rates of formation of all products decreased with increasing cycle period. Amplitude was significant: increasing amplitude as measured by the change in the $C_3H_6:O_2$ ratio suppressed time-average rates. XRD, ESCA and BET measurements undertaken for an accompanying set of step experiments indicated changes in the mineral phases present and were in general agreement with the observations made by Haber and co-workers. Because the cycling experiments were incomplete, Labastida-Bardales (1991) did not carry out an exhaustive analysis or interpretation.

Acrylic Acid From Propene Partial Oxidation

Acrylic acid, an important monomer in the production of acrylic polymers, is now made from propene in a two step, partial oxidation process using different catalysts and operating conditions in each step. Increasing severity by raising temperature or reducing the space velocity to make acrylic acid in a single step leads to combustion. Limiting the oxygen contact by using the catalyst as an O_2 carrier would appear to offer a way of increasing acrylic acid yields for a single step process. Furthermore, addition of steam to the reactor feed could raise yields. Steam in partial oxidation feed stocks apparently inhibits total combustion and for this reason steam addition is widely practised and often mentioned in process and catalyst patents. The objectives of University of Waterloo researchers were to explore a one step propene partial oxidation process to acrylic acid (Salah-AlHahamed et al., 1992). Of the range of periodic forcing strategies possible, three were examined: 1) modulation of the steam content of the feed, 2) modulation of the $C_3H_6:O_2$ ratio in the feed, and 3) periodic switching between feeds containing C_3H_6 in air and pure air. Two catalysts were chosen so as to be able to examine the interaction of catalyst and forcing strategy. The first, an Sb/Sn/V oxide in the atomic ratios 2:1:1, is a proprietary catalyst of the Distillers Company Ltd. and was chosen because some acrylic acid is formed from C_3H_6 when it is used. The second was $Bi_2Mo_3O_{12}$ (Bi_2O_3-3 MoO_3), a frequently used partial oxidation catalyst. The Sb/Sn/V catalyst produces acetaldehyde, acetic acid, and, if water is present, acetone along with acrolein and acrylic acid. Organic acids did not seem to form when the bismuth molybdate was used. Only limited cycling measurements were made with the bismuth catalyst as step change observations showed monotonic response which usually means a reaction system which is not excitable by composition forcing.

Cycling experiments were undertaken using a differential fixed bed of catalyst at 613 K for Sb/Sn/V oxide and at 633°K for the Bi-Mo oxide to compensate for its lower activity. Space velocity was held at 0.25 L/ min g.cat for all experiments. Pressure was atmospheric. Cycle periods were chosen from observation of step change dynamics. For water addition or removal, there was an initial rapid response on the order of 10 to 20 s and a slower response that required from 10 to 50 minutes. Thus, periods ranged from 30 s to 50 minutes. Symmetrical cycles were used (Salah-Alhahamed et al., 1992).

Various water amplitudes were used as indicated in figure captions around a time average water vol. % in the feed of 7.25. Cycle splits were between 0.1 and 0.8 where split is based on the relative duration of the portion of the cycle in which steam is in the feed. Time-average rates and selectivity to acrolein and acrylic acid are plotted vs. cycle split for different cycle periods in Figure 12.18. The dashed line labelled with percent water gives the steady state rate or selectivity. For the time-average 7.25 vol. % water in the feed, the best rate and selectivity data points are above (below for acrolein selectivity) the dashed line for 10 vol. % H_2O. Thus, water cycling can enhance acrolein and acrylic acid rates significantly, roughly 15 to 30 %. However, both time average formation rates and selectivities under forcing are below or at best just equal to the steady state rates and selectivities at the optimal water content in the feed. Thus, there is no global enhancement. Cycle split appears to be the primary variable with oxygenate formation rates. Selectivities increase with split. A split of 1 means steady state operation at 10 vol. % H_2O in the feed. Thus, forcing of the promoter, in this case water, is not attractive.

In a companion study, Salah-Alhahamed et al. (1993) attempted to explain the role of water in partial oxidation and why the water cycling strategy was unsuccessful. For the Sb/Sn/V oxide, water accelerated the formation of both acrolein and acrylic acid apparently by suppressing propene inhibition of the partial oxidation reaction. Under steady state operation, water suppresses entirely the combustion side reaction, prob-

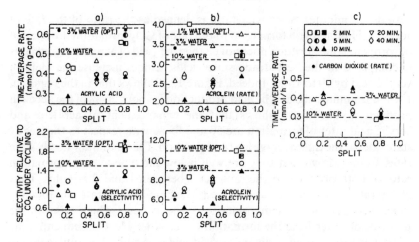

FIGURE 12.18 Influence of cycle split, period and amplitude on the time-average rate of product formation and the time-average selectivity ratio for modulation of water vol.% in the catalyst feed: (a) acrylic acid, (b) acrolein, (c) CO_2. Measurements on 40/60 U.S. mesh particles at 1 bar and 613 K with SV = 0.25 L/g.cat min. Open symbols represent switching between 0 and 10 vol.% H_2O in the feed, half filled symbols represent switching between 0.5 and 14 vol.%, while the solid symbols are for switching between 4.5 and 10 vol.%. Dashed, horizontal lines labelled with % water represent steady state results (Figure taken from Salah-Alhahamed et al., 1992, with permission, © 1992 Pergamon Press, Inc.)

ably by blocking the strong adsorption sites on the catalyst surface. All CO_2 produced comes from cleavage of C_3H_6 forming acetaldehyde and acetic acid. Water also forced desorption of strongly bonded acrylic and acetic acid and increased thereby the re-oxidation of the catalyst. Water adsorption on the mixed oxide surface was found to be rapid and reversible. This appears to be the main reason for the failure of the water cycling strategy. A further reason was that above about 5 vol.% H_2O in the feed, water inhibits acrylic acid formation. With respect to acrolein, inhibition occurs above 2 vol.%. These findings suggest that shorter cycle periods and much lower water concentrations should have been used in the modulation experiments.

The second strategy investigated by Salah-Alhahamed et al. (1992), modulating the concentration of propene in the air stream passing through the catalyst bed, had a remarkable effect on rates of acrylic acid and acrolein formation. Experimental results are shown in Figures 12.19 and 12.20. The former compares rates of formation and selectivity ratios for forcing and steady state operation as a function of cycle period for symmetrical cycling at an amplitude of 5 vol.% C_3H_6, while the latter figure plots the effect of amplitude on rates and selectivity ratios for a cycle period of 2 minutes. Selectivity ratio is defined as the ratio of the rate of acrylic acid or acrolein formation to the rate of CO_2 formation. Forcing of the propene concentration does not change the yield of acrolein significantly compared to the best steady state yields; it does increase the yield of acrylic acid by 10 to 15 %. Indeed, Figure 12.19 shows that in the region of $1 < \tau < 2$ minutes, the acrylic acid production is about 20 % of the acrolein production. The figure shows that the enhancement of acrolein and acrylic acid rates of formation are 50 and 85 % with respect to the steady state rates at the time average feed composition. The maximum steady state rates are not shown but they are well below the time-average formation rates of both products in the range of τ between 1 and 2 minutes. Thus global rate enhancement was observed. The time-average rate plot for CO_2 shows just a small rate enhancement for this product which arises in the presence of water from the formation of acetaldehyde. Consequently, the selectivity ratios for acrolein and acrylic acid in the 1 to 2 minute region in Figure 12.19 are 40 and 75 % greater than the ratios under steady state operation. Global enhancement of the selectivity ratio is also observed.

Salah-Alhahamed et al. (1992) explain the enhancement achieved by modulating the C_3H_6 concentration in the feed through a reduction in flooding of the catalyst surface by propene and oxidation products. Access of oxygen to the surface is greater and this increases the coordination of the surface and near surface metal ions. More oxygen is now available in the following half cycle when the C_3H_6 concentration again increases.

Results for the third strategy investigated appear in Figure 12.20. Operating with an amplitude of 10 vol. % C_3H_6 corresponds to switching between an air feed and a 1:1 C_3H_6-O_2 feed mixture. This strategy may be advantageous for acrolein formation, but it is not as attractive as modulating the propene concentration for acrylic acid. Figure 12.20 shows that with this strategy, the time-average selectivity ratios fall to the steady state levels.

Ketones From Olefins

A French patent (Stamicarbon, 1974) claims an Sn-Mo oxide catalyst and a fluidized bed process for the production of ketones from propene, butene and cyclic olefins. The patent states the catayst is alternately exposed to the hydrocarbon and to an oxygen containing stream with a very short contact time for the latter stream. This suggests a riser-fluidized bed system of the type described in Chapter 1 and mentioned with respect to ammoxidation earlier in this chapter. This reactor type will be discussed in the next section. Using propene, a selectivity to acetone of up to 98 % is mentioned at yields of up to 91 %. Similar results are cited for a butene feed where the main product is methylethyl ketone. Unfortunately, process details are lacking.

HYDROCARBON DEHYDROGENATION

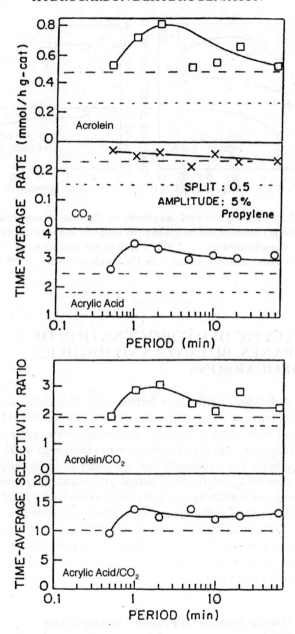

FIGURE 12.19 Influence of cycle period on the time-average rate of product formation and selectivity ratio for symmetrical modulation of the propene concentration at an amplitude of 5 vol.% C_3H_6. Measurement for a Sb/Sn/V catalyst, 40/60 U.S. mesh at 1 bar, 613 K and SV = 0.25 L/g.cat. min (Figure taken from Salah-Alhahamed et al., 1992, with permission, © 1992 Pergamon Press, Inc.)

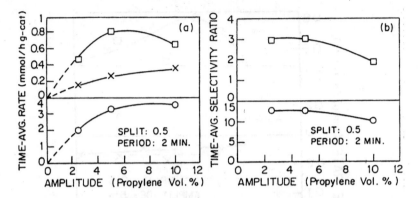

FIGURE 12.20 Influence of cycle amplitude on the time-average rate of acrylic acid and acrolein formation and their selectivity ratios for symmetrical modulation of the propene concentration at $\tau = 2$ min. Catalyst and conditions as in Fig. 12.19 (Figure taken from Salah-Alhahamed *et al.*, 1992, with permission, © 1992 Pergamon Press, Inc.)

E. CATALYTIC DEHYDROGENATION OF PROPANES, BUTANES AND HIGHER HYDROCARBONS

Olefins and diolefins, particularly butenes and butadiene, are currently produced in large volumes in both refineries and petrochemical plants. The conventional technology is catalytic dehydrogenation using a paraffin-steam mixture. Although dehydrogenation has been used widely for many years, the process has a relatively low selectivity to olefins and diolefins. With a butane feed, selectivity to butene and butadiene is about 60 %. Dehydrogenation is endothermic and occurs at moderately high temperatures so the reaction is carried out in a tube furnace. It is a hazardous operation. The process has a high steam demand, but nevertheless coking of the catalyst remains a problem so shutdowns for regeneration are common. All of these considerations has led to investigations of an oxidative dehydrogenation alternative.

Oxidative Dehydrogenation Of Propane On Magnesium Vanadate Catalysts

Step change experiments on propane dehydrogenation over a magnesium vanadate catalyst at 1 bar and 510°C gave the same propene response when carried out in the presence or absence of O_2 in the gas flowing over the catalyst. Carbon oxide responses, however, changed. This led Creaser

(1996) to investigate modulation of this dehydrogenation. His experiments used a bang-bang strategy, switching between streams containing 6 vol. % C_3H_8 in He and 6 vol. % O_2 in He. Creaser observed that a stationary cycling state was attained in about 5 cycles. Figure 12.21 shows that with symmetrical forcing at $\tau \leqslant 30$ s there is a 50 % increase in the time-average selectivity to the olefin relative to the value for steady state operation at the time-average feed composition, but the selectivity drops off as τ becomes larger. Conversion at the short cycle periods is suppressed. The conversion-selectivity trade off results in a sharp maximum in the propene yield. At this maximum, the yields is 30 % higher than the comparable yield at steady state. Cycle split was studied, but turns out to have just a small effect on selectivity and yield.

A mass spectrometer was utilized for analysis so the variation of product concentration within a cycle could be closely followed. Figure 12.22a shows the variations as τ decreases from 400 to 40 s. The bottom concentration trace for $\tau = 40$ s gives a blow up of the peak following the switch to C_3H_8. Maximum C_3H_8 and CO_2 concentrations are reached in just 2 s after the switch. There is no O_2 in the feed so propane is drawing oxygen from the catalyst and this proceeds rapidly. The plateaus seen at $\tau = 400$ and 130 s imply that the oxygen source is in the catalyst bulk and the dehydrogenation rate becomes diffusion limited. On the switch to O_2, the

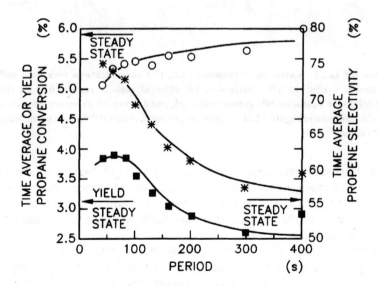

FIGURE 12.21 Time-average propane conversion, selectivity to propene and propene yield as a function of cycle period for the symmetrical cycling between 6 vol.% C_3H_8 in He and 6 vol.% O_2 in He with a MgO/V_2O_5 catalyst at 1 bar and 510°C (Figure taken from Creaser (1996) with permission of the author)

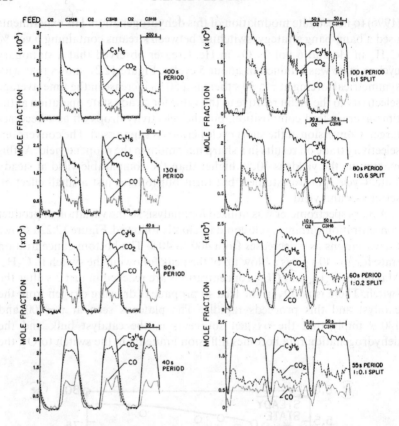

FIGURE 12.22 Variation of propene, CO_2, CO concentrations with time within successive cycles: on left - variations for different periods and $s = 0.5$, on right - variations for a constant 50 s exposure to C_3H_8 and different O_2 exposure. Modulation conditions given in Figure 12.21 (Figure taken from Creaser (1996) with permission of the author)

dehydrogenation reaction requires about 4 to 5 s to disappear indicating there is not much hydrocarbon stored on the surface during exposure of the MgO/V_2O_5 to C_3H_8. Crossover of the CO_2 and C_3H_6 traces indicates some of the adsorbed hydrocarbons are burnt off the surface by gas phase or adsorbed O_2.

Concentration traces as the duration of O_2 exposure is reduced are shown in Figure 12.22b. The initial peak on switching to C_3H_8 disappears between a 10 and 20 s exposure, however, only at a 5 s O_2 exposure is a significant reduction of product concentrations seen. Thus, re-oxidation of the catalyst is rapid. The loss of the initial peak suggests that the oxygen-vacancy interaction at the surface may be slower than oxygen diffusion into the bulk.

Oxidative Dehydrogenation of Butane on MoO_3/MgO Catalysts

Dow Chemical Company has been developing a new process for styrene starting from butane and proceeding through butadiene which is dimerized to 4-vinylcyclohexene. Styrene follows by oxidative dehydrogenation. Dow is considering making butadiene by oxidative dehydrogenation employing a circulating fluidized bed (Murchison et al., 1993; Khazai et al., 1995; Vrieland and Murchison, 1995 and Vrieland et al., 1995). The challenge faced by the Dow researchers was to develop a catalyst with high selectivity and butane conversion as well as high attrition resistance. They also had to design a circulating fluidized bed to achieve, if possible, laboratory scale performance and low catalyst breakage and loss. The starting point for the Dow work was several publications from the Ukraine describing a MoO_3/MgO catalyst (Doroshenko et al., 1986; Shapovalova et al., 1987; Luk'yanenko et al., 1987). The Dow development effort utilized a cycled packed bed reactor, a single tube, dense phase riser reactor and a continuously operating riser reactor, fluidized bed system.

The Ukrainian catalyst development used composition modulation of a fluidized bed reactor. Performance experiments were run at 1 bar between 843 and 873 K using a symmetrical cycle encompassing various mixtures. Cycle period was held at 60 s. Doroshenko et al. (1986) cycled between a $C_4H_{10}:O_2$:Ar mixture in the proportions 1:0.1:10 and air. The space velocity was 1 s^{-1}. Luk'yanenko et al. (1987) used a C_4H_{10}: diluent mixture in the ratio 1:10, and switched periodically to an O_2:diluent mixture in the ratio 1:2.5. Diluents were argon and steam, however, the choice of diluent did not effect selectivity or conversion. The experiments reported by Shapovalova et al. (1987) resembled those undertaken by Luk'yanenko except that the diluent was always steam. Their research was undertaken for catalyst development so that neither the mode of operation or the operating conditions were changed. Doroshenko et al. dealt with catalyst characterization, while Luk'yanenko et al. investigated the effect on performance of adding vanadia to the catalyst. Shapovalova et al. were concerned with the effect of calcination conditions. The best selectivity and conversion results from each study are summarized in Table 12.2.

An exhaustive examination of potential catalysts based on MoO_3/MgO was undertaken by the Dow Chemical Co. Most experiments employed a 10 cc packed bed of catalyst at 833 to 853 K and used a complex cycle: 5 – 30 s catalyst exposure to butane, 60 s flush with He, 60 s exposure to an O_2 – He mixture, followed by a 60 s He flush. Thus, the cycle period varied from 185 to 210 s which is about twice the period used by the Ukrainian workers, although the exposure to reactants persisted for 65 to 90 s. This is much closer to the Ukrainian experiments which were performed at about the same temperatures. The butane portion of the cycle used a diluent

Table 12.2 Comparison of Composition Modulation Results for MoO$_3$/MgO catalysts

Catalyst	Temperature K	Selectivity* C$_4$H$_6$ %	C$_4$H$_6$+C$_4$H$_8$ %	Conversion %	Reference
30% Mo	873	46	—	53	Doroshenko et al. (1986)
25% Mo	873	34	—	77	Doroshenko et al. (1986)
2%V – 20%Mo	863	68	70	72	Luk'yanenko et al. (1987)
2%V – 20%Mo	843	71	75	60	Shapovalova et al. (1987)
0.25%K – 30%Mo	833	58	87	23	Murchison et al. (1993)
0.25%K – 20%V – 30%Mo	833	61	74	48	Murchison et al. (1993)
0.15%K – 18% Mo	853	54	69	38	Murchison et al. (1993)
0.25%K – 2%V – 18%Mo	853	59	79	38	Murchison et al. (1993)
0.31%Cs – 22%Mo	838	70	86	36	Vrieland and Murchison (1995)
0.3%K – 22%Mo	853	73	80	61	Vrieland et al. (1995)
Spinel (0.7%Cs – 25%Mo – 21%Mg)	843	55	69	40	Vrieland et al. (1995)
Spinel-MgO Composite 0.35% K – 18% Mo – 36%Mg	853	69	83	40	Vrieland et al. (1995)

For the catalysts, V = V$_2$O$_5$, Mo = MoO$_3$, Mg = MgO (When MgO is given the remaining component is Al$_2$O$_3$)
*Feed composition and cycle structure differ so selectivity and conversion are not fully comparable

mixture containing from about 10 to 50 % C_4H_{10}. Results for packed bed composition forcing are given in Table 12.2. Selectivities measured by the different researchers are consistent although conversion in the Ukrainian work is sometimes twice as great as in the Murchison study. Conversion is, of course, a strong function of space velocity. Vriesland and Murchison (1995) and Vriesland et al. (1995) show high conversions reduced C_4H_6 selectivity.

The Dow team explored different alkalis as well as the wt % in the MoO_3/MgO catalyst and concluded the choice of alkali made little difference, but that the amount was important (Vriesland and Murchison, 1995). Best results were obtained with about 3 wt. %, which are the entries in Table 12.2. Although the best selectivities were obtained with the alkali promoted MoO_3/MgO materials, these had poor strength. Spinels provided adequate strength and consequently abrasion resistance, but as can be seen in the table above, selectivity to C_4H_6 was poorer. Impregnation of the spinels to increase the MgO content to 30 wt. % or more gave catalysts with almost the same activity and selectivity of the MoO_3/MgO material without sacrificing the strength from the spinel structure as can be seen from the last entry in Table 12.2.

Dow Riser Reactor Studies

The recirculating solids system for oxidative dehydrogenation proposed by the Dow Chemical Co. uses a dense phase riser reactor for the critical butane conversion step. Re-oxidation of the catalyst appears to be straight forward and is apparently carried out in a fluidized bed. Details of this step are not given by Murchison et al. (1993) or by the Dow European Patent (Khazai et al., 1995). Scale up of the riser was addressed in stages: 1) a bench scale unit with diameter from 4 to 12 mm and a transport length of 1 to 2 m operated with a single catalyst pass, 2) a somewhat larger version holding 6 L of catalyst and operating continuously, and 3) a demonstration scale unit operating continuously and holding between 120 and 200 L of catalyst.

Figure 12.23 shows the single pass system and two of the bench scale riser reactor designs. The left-hand design in Figure 12.23b is evidently the reactor shown in Figure 12.23a. The downcomer is operated just above minimum fluidization by diluent entering through the side lines attached to the downcomer through quartz frit plates. Butane enters the reactor at the bottom of the elbow and carries catalyst upward through the right-hand leg. The gas-catalyst emulsion separates in the disengager located just above the fresh catalyst hopper. This is shown in Figure 12.23a. In continuous operation, the hopper and the disengager would be connected. The hopper is pressurized and catalyst is transported pneumatically to a fluidized bed which acts as the catalyst source for the riser as can be seen in

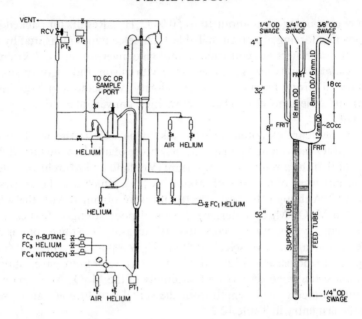

FIGURE 12.23 Schematic of a dense phase upflow transport reactor investigated by Dow Chemical Co.: (a) system schematic, (b) riser design studied (Figure from Murchison *et al.* (1993) with permission of the authors)

the figure. Air can be introduced into this bed so that the re-oxidation step could be undertaken there. The solenoid valve shown in the bottom of Figure 12.23a indicates re-oxidation could also be accomplished in the dense phase riser for test purposes.

The single pass unit was used to test the effect of particle size as well as the performance and attrition resistance of different catalyst formulations. Catalyst strength was an important consideration and the solution was a composite catalyst produced as indicated in Figure 12.24. Drying and calcining temperatures are indicated in the figure. Performance of the single pass riser is compared with the periodic operated packed bed in Table 12.3. The 4-part cycle using He flushes between the C_4H_{10} and O_2 feeds approximates the single pass riser system which used He to transport the catalyst and fluidize the downcomer leg (Figure 12.23). Results are similar so the performance measured in a periodically operated packed bed properly predicts what can be achieved with a recirculating solids reactor. An experiment using a mixed butane-butene feed is also given in the table. This indicates butene recycle is possible; it also suggests that butadiene forms from a further extraction of H_2 from C_4H_8 and not directly from butane. The periodically operated fixed bed reactor was used to test catalyst deactivation. Figure 12.25 shows the composite catalyst exhibited no loss of

Table 12.3 Comparison of Cycled Packed Bed and Single Pass Riser Reactor Performance

Reactor	Feed*	Selectivity				Conversion(%)
		C_4H_6	$C_4H_6+C_4H_6$	C_5^+	C_2+C_3	
Cycled Packed Bed	45%C_4H_{10}	60	77	14	4	34
Riser	45%C_4H_{10}	57	81	9	6	29
Riser	71%C_4H_{10}	49	76	–	–	22
Riser	68%C_4 ($C_4:C_4^=$ = 88:12)	74	74	–	–	21

*Butane in the reactor feed is mixed with a diluent

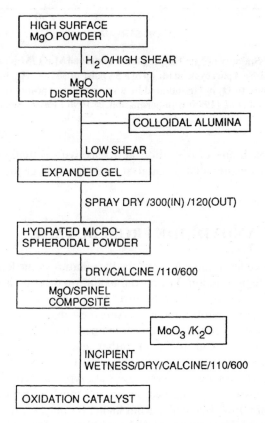

FIGURE 12.24 Preparation procedure for a high strength, attrition resistant promoted MoO_3/MgO catalyst for use in a recirculating solids reactor system. Numbers in the diagram are drying and calcining temperatures in °C (Figure from Murchison et al. (1993) with permission of the authors)

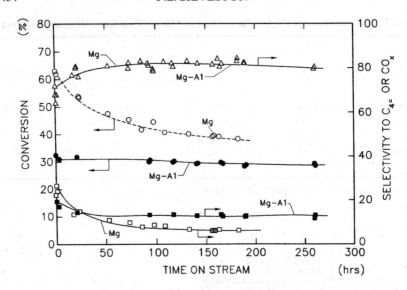

FIGURE 12.25 Stability of MgO impregnated spinel and MoO_3/MgO with time on stream at 853 K with a 4 part cycle made up of a 5 s pulse of 50% C_4H_{10}, 60 s exposure to He, 60 s exposure to O_2 in He followed by a second 60 s exposure to He (Figure taken from Vrieland et al. (1995) with permission, © 1996 Elsevier Science B.V)

activity over 250 hours of operation and selectivity actually improved. The more active magnesia based catalyst lost activity over the same time span.

F. MALEIC ANHYDRIDE FROM BUTANE

Over the last decades, butane has replaced aromatics as the feedstock for maleic anhydride production. The partial oxidation reaction is conducted commercially in fixed beds pierced by an array of tubes carrying coolant or by placing the catalyst in tubes and immersing these in the coolant medium. Temperature control is important. Lower temperatures favor selectivity to maleic anhydride, but conversion or rather space velocity is reduced, whereas at higher temperatures parametric sensitivity arises and runaway becomes a concern. Butane to oxygen ratios are typically 1:10. Butane and air are mixed upstream of the bed so butane must be kept below the flammability limit. Fluidized beds permit higher butane:O_2 ratios and offer better heat transfer. Their use for partial oxidation has been described, but a fluidized bed process has not yet been commercialized. Interest in accomplishing the maleic anhydride reaction with composition modulation, particularly in a recirculating solids system, is that most attractive catalysts function in a redox cycle and, just as for propene partial oxidation, they seem capable of acting as oxygen carriers. Composition modulation may offer higher selectivity. Even if selectivity is not improved, segregating the reactants probably will reduce the cost of product recovery and recycle of unused reactant. Another potential advantage of this route is that heat generation is distributed. With a recirculating solids system, both the butane partial oxidation and catalyst re-oxidation steps are exothermic. The heat of reaction, thus, can be removed in two vessels rather than one. Parametric sensitivity may be substantially reduced in this way.

Investigations At The University Of Waterloo

A tellerium promoted vanadium molybdate supported on TiO_2 was used at 623 K by Lang *et al.* (1989b, 1991) in a packed bed system employing solenoid valves and flow controllers on the hydrocarbon, air or O_2, and diluent streams for modulation. Because oxidative dehydrogenation of butane to butadiene was quantitative and rapid over this catalyst, butadiene rather than butane was chosen as the hydrocarbon feedstock. The Lang study was part of a cooperative research program with the University of Erlangen in Germany which involved using the same catalyst but different feedstocks for maleic anhydride production. The companion work using benzene as the hydrocarbon will be discussed in Section G of this chapter. Other objectives of the Waterloo project were to measure rate and selectivity improvement through forcing, identify the important cycling variables, and explore the effect of different cycling strategies on performance. The strategies considered were 1) modulation of butadiene concentration at a constant O_2 mol fraction, 2) modulation of oxygen concentration at a constant C_4H_6 mol fraction, 3) on-off modulation of both O_2 and C_4H_6 concentrations, and 4) on-off reactant concentration modulation with inert flushes between the reactant pulses. The idealized cycles for the last two strategies are shown in Figure 12.26. Unavoidable mixing in the packed bed reactor, feed lines and valving distorted the intended square waves as illustrated in Figure 12.27.

Modulation of the individual reactants was tested for a time average feed composition of 0.4 vol. % C_4H_6 and 10 vol. % O_2 with N_2 as the diluent for $s = 0.5$ and τ from 1 to 40 minutes. Step change experiments indicated a rapid response occurring over a minute's duration and a slower concentration adjustment which needed about 20 to 40 minutes (Lang *et al.*, 1989a). It was found that butadiene modulation suppressed rates of formation for all products, but resulted in about a 10 % increase in selectivity to the furan intermediate expressed as a selectivity ratio (ratio of furan formed to CO_2 formed) for $\tau = 1$ minute. Modulating the O_2 concentration resulted in a small increase in the formation of maleic anhydride at a period of a minute, but all other rates were reduced. Maleic anhydride selectivity

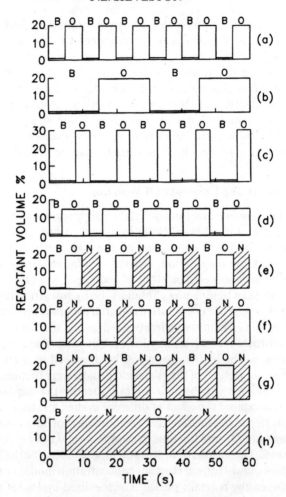

FIGURE 12.26 Multi-step cycles investigated to determine the benefit of flushing with an inert in on-off composition modulation of a two reactant system: B = 0.8 vol.% butadiene in N_2, O = air, N = N_2 (Figure taken from Lang *et al.* (1991) with permission of the authors)

ratio increased by about 10 % at $\tau = 1$ minute, while the ratio for furan improved by just 5 %. These shifts are small. The reason for this seems to be in the symmetry of the response to concentration change. Figure 12.28 shows the variation of the butadiene and product volume fractions during a cycle for both modulation strategies. Although the overshoot-undershoot pattern for butadiene cycling indicates a non-linear response, the undershoot in the C_4H_6 step down in Figure 12.28a is either greater or the same as the overshoot in the step up so that the rates of formation do not increase

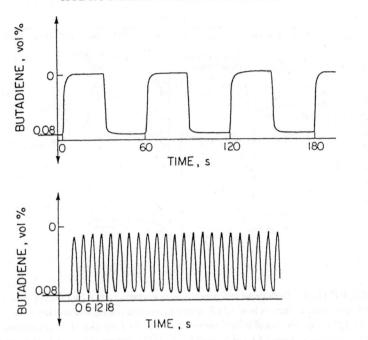

FIGURE 12.27 Shape of the butadiene steps leaving the bed at ambient temperature: (a) $\tau = 60$ s, (b) $\tau = 12$ s (Figure taken from Lang *et al.* (1989b) with permission of the authors)

under forcing. Lang *et al.* attribute the overshoot-undershoot to adsorption competition between butadiene and the oxygenates. The pattern for O_2 modulation, in Figure 12.28b, shows monotonic change after a composition change. This appears to be the consequence of the weak adsorption of O_2 by the catalyst. On a step up in O_2, displacement of the oxygenates occurs slowly. Re-oxidation of the surface, on the other hand increases butadiene conversion. Lang *et al.* (1989b) conclude that modulating the concentration of a single reactant is not an effective strategy for maleic anhydride production. They observe further that rate and selectivity varied little with τ; amplitude of the modulation, however, was a significant variable.

Bang-bang modulation of both hydrocarbon and oxygen proved to be a more effective means of exciting this reaction system, but only for production of the furan intermediate. Cycle periods from 6 s to 60 s were used. For the same time average feed composition as in the single reactant forcing experiment, a 10 % increase in the rate of furan formation was found for $\tau = 12$ s. Selectivity ratio for this intermediate was 35 % greater for periodic operation and this was independent of cycle period. Removal of O_2 from the reactor feed seems the cause of the improved yield of furan. This is

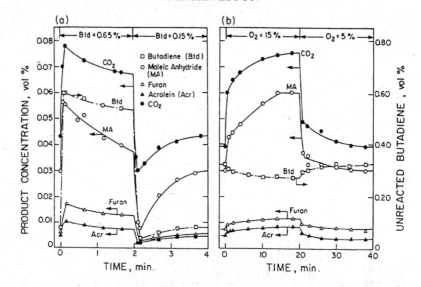

FIGURE 12.28 Butadiene and product concentrations as a function of time within a composition forcing cycle at 823 K over a promoted vanadium molybdate supported on TiO_2: (a) modulation of butadiene volume % in the feed with O_2 held constant at 10 vol.%, (b) modulation of O_2 vol.% with C_4H_6 held constant at 0.4 vol.% (Figure taken from Lang et al. (1989b) with permission of the authors)

shown in Figure 12.29a. At the end of the exposure of the catalyst to 0.8 vol. % C_4H_6, the furan concentration leaving the catalyst bed is almost equal to the maleic anhydride concentration. Indeed, condensing the products from the butadiene half cycle would yield a product that is about 1/3 furan. According to Lang et al., the rapid rise in oxygenate formation at the beginning of this half cycle indicates the O_2 half cycle has depleted the adsorbates on the surface and recharged the oxygen held by the catalyst. The relatively slow rise of butadiene points to its adsorption and conversion. Fully oxidized surface sites form the anhydride or split butadiene forming acrolein, while partially oxidized sites yield furan. Reduction of the surface causes the fall in maleic anhydride. The increase in partially reduced sites compensates for sites further reduced by forming furan so that furan concentrations in the stream leaving the reactor remain constant. When O_2 is re-introduced, adsorbed C_4H_6 is oxidized to maleic anhydride accounting for its continuous appearance in the product stream. The dotted steady state line in Figure 12.29a demonstrates that bang-bang modulation of both reactants decreases maleic anhydride production and selectivity to that product.

The motivation for the strategy of separating reactant exposure by inert flushing was the discovery that on oxide surfaces oxygen can exist as a O^-

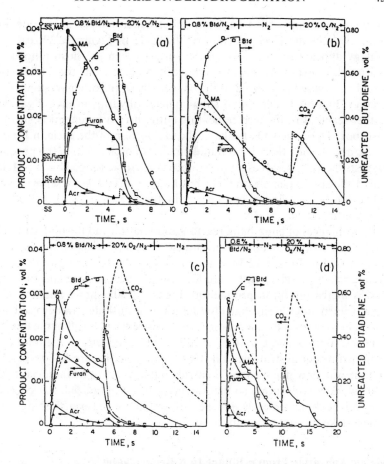

FIGURE 12.29 Butadiene and product concentrations as a function of time within a composition forcing cycle at 823 K over a promoted vanadium molybdate supported on TiO_2: (a) on-off modulation of both butadiene and O_2, (b) as in (a) with a 5 s flush of N_2 following the C_4H_6 pulse, (c) as in (a) with a 5 s flush of N_2 following the O_2 pulse, (d) as in (a) with a 5 s flush of N_2 following both reactant pulses (Figure taken from Lang et al. (1989b) with permission of the authors)

species bonded to metal cations or as a $O^=$ species, perhaps as a surface peroxide. The electron rich species is believed responsible for combustion. That species should occur as an adsorbate. Flushing after O_2 exposure, Lang et al. (1991) conjectured, could reduce the $O^=$ concentration on the surface, improving thereby the selectivity to the oxygenates. Flushing after the hydrocarbon exposure, on the other hand, would decrease the hydrocarbon adsorbate concentration susceptible to total oxidation once O_2 is re-introduced. This strategy was not successful for the vanadium molybdate as may be seen from the smaller maleic anhydride and furan areas in the C_4H_6 portion of the cycle in Figure 12.29b,c,d. The strategy appears to over oxidize the surface when the flush follows the hydrocarbon exposure, and to remove butadiene from the surface that would otherwise be oxidized to the oxygenates in the other versions of the strategy. Figures 12.29b and d show maleic anhydride in the off gas with neither reactant in the feed indicating either it or a precursor is strongly adsorbed. The large CO_2 peaks in the O_2 portion of the cycle in Figures 12.29c and d when there is no butadiene, furan or acrolein in the products suggests there are residues on the surface that are scavenged by O_2. Maleic anhydride also appears with CO_2, but Lang et al. (1989a) observe that the anhydride is just slowly combusted in the presence of oxygen at the temperatures used.

Performances of bang-bang reactant modulation and flushing strategies for partial oxidation on the promoted vanadium molybdate/TiO_2 are summarized in Figure 12.30. The letters on each bar correspond to the cycling structures given in Figure 12.23. Both strategies are effective if the goal is increasing furan production. Figure 12.30b shows about a 75 % improvement in yield through modulation. The flushing strategy used does not seem to matter. Furan production rate is reduced by all of the flushing schemes. With respect to maleic anhydride production, Figure 12.30a demonstrates both strategies fail. The loss in yield (Figure 12.30b) is small and the way in which flushing is carried out appears to be important. The Lang work is exploratory so that forcing was not optimized. Indeed, there may well be choices of cycle split, period, amplitude and even flush duration which make this operation mode attractive.

Maleic Anhydride From n-Butane In A Recirculating Solids Reactor System

Experiments undertaken by Du Pont and brought to commercialization are described in a patent (Contractor, 1987), an overview paper (Contractor et al., 1987) and several contributions dealing with specific aspects of the process (Contractor et al., 1988; Contractor and Sleight, 1988; Contractor et al., 1990). Use of modulation by means of a recirculating solids reactor was motivated by speculation that adsorbed oxygen is responsible for total combustion so that segregating the reactants could raise selectivity to maleic anhydride. A further incentive was to employ butane partial pressures higher than possible in a steady state operation and reduce the oxygen requirement. The catalyst was a vanadium phosphate ($VO_2P_2O_7$) promoted by silica and small amounts of either indium, antimony or tantalum. The Du Pont design uses a riser or transport reactor for butane oxidation and a fluidized bed for re-oxidation of the catalyst. Figure 12.31 shows a schematic of the system. This figure was shown in Chapter 1 as Figure

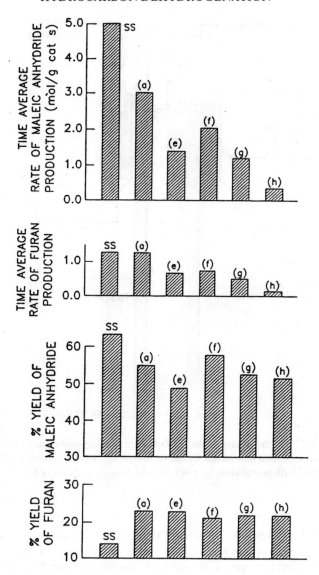

FIGURE 12.30 Influence of forcing strategy on time average maleic anhydride and furan production rates, and time average yields. Letters above each bar indicate the cycle used (Figure 12.26); SS = steady state (Figure taken from Lang *et al.* (1991) with permission of the authors)

2.16. Some preliminary experiments used riser reactors for both steps as reoxidation proceeds rapidly in the 623 to 643 K range employed (Contractor, 1987). Unfortunately, details of neither the Du Pont pilot plant or their full scale reactor have been published.

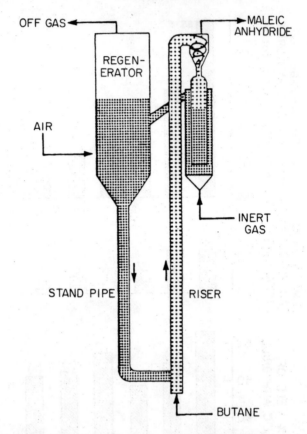

FIGURE 12.31 Schematic of the Du Pont recirculating solids reactor pilot plant for the partial oxidation of n-butane to maleic anhydride (Figure taken from Contractor et al. (1987) with permission, © 1987 Elsevier Science Publishers B.V.)

Table 12.4 compares the performance of a recirculating solids system consisting of a riser and a fluidized bed with the that of a single fluidized bed. The catalyst was a mechanical mixture of vanadium phosphate promoted by Si, Sb and In and a phosphate promoted by Si and Tl. The mixture was calcined at 663 K and activated by exposure to 1.4 vol.% n-C_4H_{10} in air at 723 K. Variables in the experiment were C_4H_{10}:O_2 ratio and gas residence time in the riser.

In our discussion of butane dehydrogenation in Section *E*, the problem of catalyst attrition was mentioned. This was also a major concern in the Du Pont process development. Both the spray dried promoted and unpromoted Vanadium phosphate have low strength and show attrition loss in a fluidized bed and even higher losses in a recirculating solids reactor.

Table 12.4 Comparison of n-Butane Conversion and Selectivity to Maleic Anhydride for a Recirculating Solids Reactor and a Fluidized bed (Contractor, 1987)

Reactor	N-C_4H_{10} vol.%	C_4H_{10}:O_2	Temperature °C	Gas Residence Time Seconds	Conversion %	Selectivity %
RSR	1.2	1:16.7[1]	357/355[2]	5[3]	46.6	90.1
RSR	1.2	1:16.7[1]	358/355	5[3]	47.7	88.2
FB	1.3	1:15.9	356	6.4	60.4	66.0
FB	1.0	1:20.8	363	6.4	68.0	66.7
FB	1.0	1:20.8	357	4.7	49.7	63.2
RSR	2.7	1:7.4	363/360	5	30.2	85.7
FB	1.3	1:7.7	369	6.4	54.9	65.3
RSR	5.4	1:3.7	356/355	5	16.0	83.2
FB	1.3	1:3.85	355	6.4	36.8	76.0
RSR	11.0	1:1.82	355/355	5	9.0	90.8
FB	2.8	1:1.79	354	6.4	22.8	75.6
RSR	3.5	1:5.7	350/350	3.2	16.3	93.2

RSR = Recirculating Solids Reactors (Riser/Fluidized Bed)
FB = Fluidized Bed

[1] Diluent in the riser was N_2; Feed to the fluidzed bed was 20 vol.% O_2 in He. Ratio is based on the time averaged feed composition for the recirculating solids reactor system
[2] 1st number is the riser temperature; 2nd number is the fluidized bed temperature
[3] Gas residence time is for the riser; gas residence time in the fludized bed was constant at 4 s.

The conventional method of strengthening the catalyst particle, addition of 30% by weight of colloidal silica, decreased selectivity substantially (Contractor et al., 1987). Instead polysilicic acid was added to the vanadium–phosphate co-precipitation to give after calcining a solid containing 5 to 10 wt. % silica. Examination of the particle showed the silica was concentrated in a porous shell after calcining. The shell provided mechanical strength without a large effect on selectivity. Performance of this catalyst in the recirculating solids reactor system is shown in Figure 12.32. The catalyst used to obtain the data in the figure was calcined at 663°C and activated at 450°C in a N_2 stream containing 1.5 mol. % n-C_4H_{10}, 12 mol. % O_2. This was the same mixture used in the three fluidized bed experiments given in the figure.

Figure 12.32 confirms the superiority of the recirculating solids system shown in the above table. At the same conversion, selectivity to the anhydride with recirculating solids is about 20 % higher. This represents an increase of 75 to 100 %. Some of the selectivities for the recirculating solids in the figure were for complete segregation of the reactants, as for the data in Table 12.4; but about half the data are for switching between a feed mixture containing 3.0 mol.% n-C_4H_{10} and 8.0 mol.% O_2 (balance N_2) and a 20 mol.% O_2 in He. Although not indicated in the figure, Contractor (1987) reports that complete segregation gave somewhat higher selectivities at the same conversion. It can be seen from the figure that the highest conversions and selectivities are at 360°C, the lowest temperature employed. Selectivity

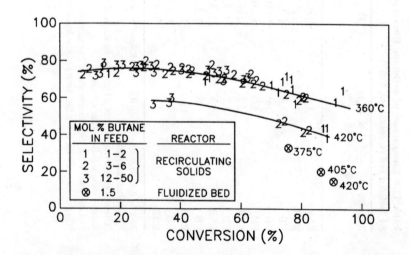

FIGURE 12.32 Selectivity to maleic anhydride vs. butane conversion for the partial oxidation of n-butane over a promoted vanadate phosphate at different temperatures using recirculating solids reactors of various design and a single fluidized bed (Figure from Contractor, 1987).

remains constant up to about 40 % conversion and then decreases slowly with rising conversion. If conversion is below 40 to 50 %, butane concentration in the riser feed, has only a minor influence on selectivity. At conversions approaching 90 %, butane must be under 6 mol.% to avoid over reduction of the catalyst. The selectivity advantage of the riser-fluidized bed recirculating solids system appear to be sufficient to justify commercialization.

Further experiments given as examples in the Contractor (1987) patent deal with the effect on selectivity of 1) adding oxygen to the butane feed and 2) flushing the catalyst with helium between the hydrocarbon and air portions of the catalyst cycle. Table 12.5 summarizes the results for these experiments. Addition of O_2 is attractive, if selectivity is not changed, because it reduces the catalyst recirculation rate. The table shows that 6 mol.% O_2 which is 28 % of the stoichiometric O_2 demand for maleic anhydride does not effect the selectivity. Increasing oxygen to 71 % of the demand causes a significant loss of selectivity, although yield remains almost constant. The experiments with and without flushing between reactant exposures demonstrate that stripping O_2 from the catalyst is important.

Oxidation-reduction studies of vanadium phosphate catalysts indicated oxygen drawn from the catalyst lattice was the primary source for selective oxidation and was involved in the oxidative dehydrogenation of butane as well as in the oxygen insertion steps (Contractor et al., 1990). Lattice oxygen also participated in the parallel combustion reaction. High oxygen availability on the surface appeared to be important for selectivity to maleic anhydride.

Catalyst preparation was found to be crucial for the redox cycle. Contractor et al. do not indicate if preparation is also important for

Table 12.5 Selectivity to Maleic Anhydride and n-Butane Conversion in a Riser-Fluidized Bed Recirculating Solids Reactor System Using a High Strength Vanadium Phosphate Catalyst

n-C_4H_{10} mol.%	O_2 mol.%	He Flush	Temperature °C	Gas Residence Time	Conversion %	Selectivity %
12	0[1]	Yes	360	4.6	47.7	75.2
12	6	Yes	360	4.6	51.5	74.8
12	16	Yes	360	4.7	53.2	69.5
3.5	0	No	360	3.5	51.5	69.8
3.5	0	Yes	360	3.5	49.1	77.0

O_2 in the riser feed; the fluidized bed is fed with 20% O_2 in helium, temperature is 360°C and the gas residence time is 5 s.

selectivity and conversion. Preparation with an isobutyl alcohol/benzyl alcohol mixture, used in the Du Pont patent (Contractor, 1987) and in subsequent pilot plant development, resulted in a single crystalline phase, $(VO)_2P_2O_7$. This catalyst was capable of providing oxygen for selective oxidation at temperatures as low as 573 K in about 7 fold excess of that estimated to exist on the surface according to a successive pulsing experiment at 673 K (Contractor et al., 1990). Re-oxidation was also examined at the same temperature using successive pulsing and took about 15 × the pulses, that is the contact time, needed for reduction. Assuming a recirculating solids system can be designed to provide the contact time for re-oxidation, the critical feature of the design is the weight of catalyst which must be circulated per unit of maleic anhydride produced. This amount depends upon the butane concentration in the riser of the recirculating solids system and is indicated in Figure 12.33. Although the figure suggests as much as a gram of maleic anhydride can be produced per kg of circulating catalyst if the butane volume percent is kept above 8 %, conversion is reduced at this level for the experimental unit run by Du Pont.

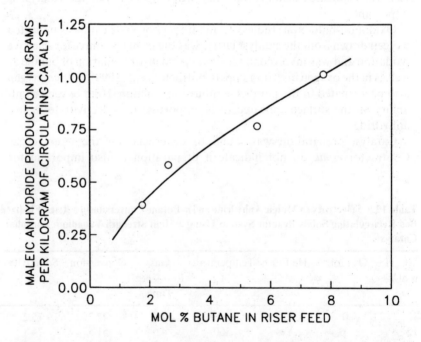

FIGURE 12.33 Weight of maleic anhydride produced per kilogram of circulating VPO catalyst in a riser reactor as a function of the *n*-butane concentration in the riser feed for a bench scale recirculating solids system operating at 673 K (Figure adapted from Contractor *et al.* (1990) with permission, © 1990 Elsevier Science Publishers B.V.)

Figures 12.32 and 12.34 indicate high selectivity can be maintained at conversion up to about 60 %. To obtain such conversion, a lower n-butane volume percent must be used so it is likely that only 0.5 to 0.7 gms maleic anhydride can be achieved per kg of circulating catalyst.

Stability of the attrition resistant catalyst is also discussed by Contractor et al. (1990). Composition modulation experiments performed in a labora-

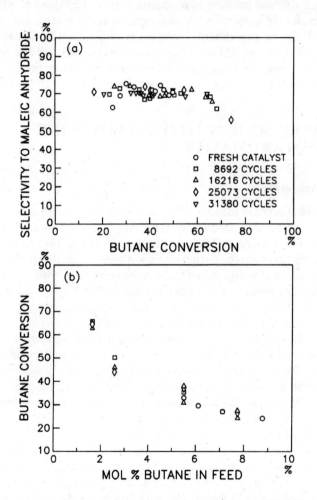

FIGURE 1234 Catalyst performance as a function of the number of composition cycles or time on stream expressed as (a) the variation of selectivity with conversion and (b) the conversion as a function of n-butane concentration in the reactor feed for the Du Pont attrition resistant vanadium phosphate catalyst in a cyclically operated fluidized bed at 1 bar and 693 K with alternating exposures to 7.5 mol % C_4H_{10} in N_2 for 30 s and to 16 mol % O_2 in N_2 for 120 s (Figure taken from Contractor et al. (1990) with permission, © 1990 Elsevier Science Publishers B.V.)

tory scale fluidized bed were carried out for over 31,000 cycles, corresponding to 3–4 months of continuous operation. Switching was between 7.5 mol % n-butane and 16 mol % O_2, both diluted with N_2, using unsymmetrical cycles ($s = 0.2$ based on C_4H_{10}). Compositions and the split were chosen to avoid over reduction of the catalyst. XRD measurements on the catalyst showed no change and both parts of Figure 12.34 confirms that the system performance is also unchanged (compare Figures 12.34 and 12.32).

The results obtained with the recirculating solids system are strikingly different from the experiments using a vanadium molybdate catalyst described earlier in this section. Cycle structure, periods and the operating conditions were about the same as in the Du Pont work so the comparison demonstrates the importance of catalyst choice.

G. MALEIC AND PHTHALIC ANHYDRIDE FROM AROMATICS

Maleic Anhydride

Two publications appeared in 1983 dealing with the partial oxidation of benzene (Bz) to this anhydride; both employed vanadium molybdate catalysts. The two contributions are also similar in that their primary objective was to probe the partial oxidation mechanism rather than explore this application of composition forcing. Nevertheless, both contributions do take up the application question. The catalyst used by Cordova and Gau (1983) was a co-precipitate of ammonium vanadate, ammonium molybdate, manganese sulfate, sodium borate, sodium hypophosphate and nickel chloride on a spherosil (silica) carrier. The minor salts were about 10 % by weight or less of the molybdate. The raw catalyst was calcined at 673 K under N_2. Steady state and composition forcing experiments were undertaken by the investigators. The forcing experiments were modelled on the Niwa and Murakami work discussed in the Section D of this chapter.

Figure 12.35 illustrates the variation of benzene, CO_2, and maleic anhydride (MA) concentrations during a 4-part cycle. Two experiments are shown: (a) a reducing pulse containing 0.27 vol.% Bz in air and (b) a reducing pulse containing 0.47 vol.% Bz in N_2. The cycle period is 240 s and consists of a 60 s exposure to the reducing pulse, a 60 s exposure to O_2 separated by flushing of the catalyst with flowing N_2 for equal durations of 60 s. These data represent stationary, reproducible cycling. Cordova and Gau comment that 12 to 24 h are needed to attain this stationary condition after a change in composition, temperature or from steady state operation is made. A partial oxidation intermediate, p-benzoquinone was also observed in the reducing pulse, but at very low concentration compared to those for maleic anhydride. The steady state indicators in Figure 12.35a refer to the

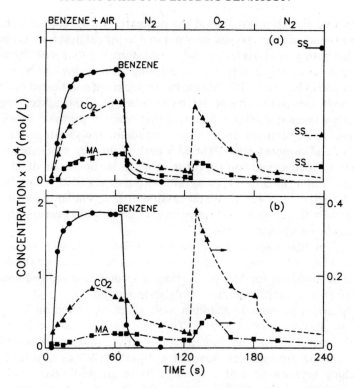

FIGURE 12.35 Reactant and product concentration variation during a 4-part cycle for benzene partial oxidation on a mixed oxide promoted vanadium molybdate supported on a porous silica at 1 bar, 573 K and a 3 L/h g.cat. space velocity with symmetrical cycling and $\tau = 240\ s$: (a) 0.27 vol. benzene in air in the reducing pulse, (b) segregation of the benzene $-O_2$ reactants with 0.47 vol.% benzene in N_2 in the reducing pulse (Figure taken from Cordova and Gau (1983) with permission of the authors)

benzene in air pulse. In this portion of the cycle, maleic anhydride production increases by 36 %, but over the full cycle production is substantially lower. Selectivity to MA is also some 5 % higher than the steady state selectivity during the reducing exposure, but for the full cycle, it drops slightly. In the bottom part of the figure, in which the reactants are completely separated, the selectivity in the reducing portion of the cycle is less than the selectivity at steady state in Figure 12.35a. Thus, fully segregating the reactants is not a suitable forcing strategy. An experiment using a 6 s reducing pulse with 2.5 vol.% Bz in air while retaining 60 s duration of the remaining portions of the 4-part cycle increased the selectivity to MA to 95 %, but suppressed MA production.

Cordova and Gau (1983) used the concentration variations in Figure 12.35 to conclude that oxygen incorporated in the catalyst bulk is responsible for both partial oxidation and combustion, but that mobility of this oxygen is low as only 1 % participated during the reducing and N_2 portion of the cycle in Figure 12.35b. Maleic anhydride is not combusted under the conditions used so the abrupt rise in CO_2 when O_2 is introduced suggests partially oxygenated intermediates or side reaction products formed by benzene adsorption on an oxidized vanadium molybdate surface are reactive and combust easily. The MA peak during O_2 exposure and the drop in CO_2 concentration appear to result from conversion of adsorbed benzene. Tailing after the O_2 pulse results from strong adsorption of maleic anhydride and possibly CO_2 on the catalyst surface. Similar shapes in parts (a) and (b) of the figure indicate preferential adsorption of benzene and high coverage even when air is co-fed.

Based on their observations, Cordova and Gau suggest that a circulating fluidized bed with a distributed feed of air and benzene along the reactor would be attractive for MA production. A second fluidized bed could be used to re-oxidize the vanadium molybdate catalyst.

A tellerium promoted vanadium molybdate, supported on TiO_2, was used by Fiolitakis *et al.* (1983). They too used successive pulsing to investigate mechanism and changes to the catalyst on exposure to oxidizing and reducing atmospheres, however a 2-part cycle was also used with switching between air and a 5 vol.% Bz in air. Fiolitakis *et al.* were interested in the changes occurring in the catalyst as it was exposed to reacting mixtures so they examined the first 12–15 cycles. Their observations are shown in Figure 12.36 for a cycle period of 325 *s*. Benzene in the first pulse is almost completely consumed, but without a large draw down of O_2. Products are formed, but in amounts much smaller than in subsequent pulses. Most of the benzene appears to be irreversibly adsorbed on the catalyst. The temperature drops even though adsorption is exothermic, so changes must be occurring in the catalyst as well. Maleic anhydride and the total oxidation products, H_2O and CO_2, increase in the next 2–3 pulses and then decline. By the 8th pulse the combustion activity of the catalyst has greatly decreased. This indicates catalyst reduction. In subsequent pulses, the primary reaction is MA formation and the pulses become reproducible despite increasing temperature. Low initial selectivity followed by a large increase was found to be typical of fresh catalysts. Selectivity to MA generally reached 70 to 100 %. With extended use, activity and selectivity declined; the latter could go as low as 20 to 35 %. Exposure to air in the absence of benzene restored much of the lost selectivity.

Fiolitakis *et al.* (1983) applied wave front analysis to the 5th Bz rich pulse. This novel technique rests upon thermal inertia so that for the first few seconds following a step up or step down in a reactant concentration

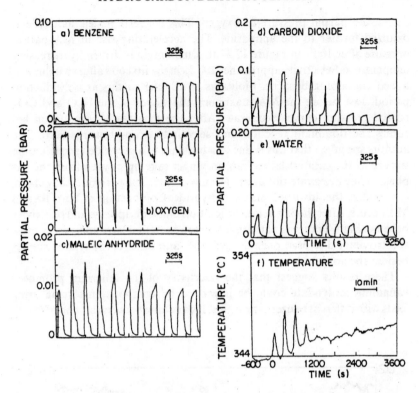

FIGURE 12.36 Mass spectrometer measurements of reactant and product partial pressure variations in successive pulsing of 5 vol.% Bz into air passing over a 28 mm deep bed of a TeO_2-V_2O_5-MoO_3/TiO_2 catalyst at 1 bar and 1.14 s^{-1} space velocity using a cycle period of 325 s: (a) benzene, (b) O_2, (c) maleic anhydride, (d) CO_2, (e) H_2O, and (f) temperature (Figure taken from Fiolitakis et al. (1983) with permission of the authors)

temperature remains constant. Slopes of the concentration changes can indicate the reaction steps occurring because they are not confounded by changing temperature (Fiolitakis and Hofmann, 1983). Figure 12.37 expands the time scale and places on the same plot the responses of the reactants and the main products to the switch from air to 5 vol.% Bz in air. CO_2 peaks and overshoot before Bz and MA reach higher levels suggest CO_2 is being displaced from the surface. The magnitudes of the initial rise in maleic anhydride and the drop in the oxygen partial pressures after benzene is introduced are in the ratio 1:3.5 indicating maleic anhydride is formed through maleic aldehyde. Since it is not seen in the reaction products, the aldehyde must be strongly adsorbed. The slope of MA partial pressure after the sharp rise is constant for about 25 seconds and equals the slope of the decreasing partial pressure of oxygen. This suggests maleic aldehyde is oxidized further to the anhydride. The accelerating rise in MA partial pressure after 100 s in Figure 12.37 to a maximum is driven by increasing temperature. Selectivity approaches 100 % in the first 60 s after switching to a feed containing benzene. Fiolitakis *et al.* refer to this as an induction period. Just before the MA maximum, the slopes of the H_2O and CO_2 partial pressure curves accelerate. After the maximum, combustion becomes the dominant reaction and selectivity drops towards zero. The authors speculate that this is due to a build up of a peroxide species on the surface as the catalyst lattice can no longer take up oxygen from the gas phase. They designate this as an ignition period. The peroxide, they claim, can oxidize the adsorbed maleic anhydride. Combustion of both Bz and MA occurs. After the 8th pulse, the ignition phase disappears and selectivity becomes high. Trapping of the reaction products and later analysis by mass spectrometry disclosed phenols, quinones and other carboxylated species besides the anhydride.

These results suggest that the selectivity of the tellerium promoted vanadium molybdate could be improved by composition forcing using feeds with different benzene-air ratios. If switching between air and a Bz rich

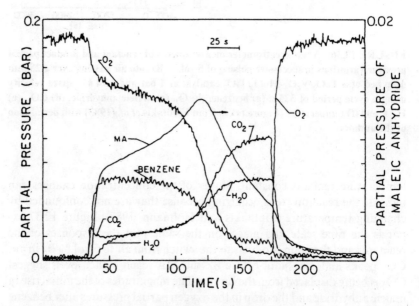

FIGURE 12.37 Replotting of the fifth pulse in Fig. 12.36 with expansion of the time axis to show the partial pressure variation with time of products and reactants in the catalytic oxidation of benzene to maleic anhydride (Figure taken from Fiolitakis *et al.* (1983) with permission of the authors)

stream is used, short periods must be employed to avoid peroxide build up. The first strategy was tested by Fiolitakis *et al.*, but at 693 K. Feed to a fixed bed of the tellerium promoted catalyst was modulated between O_2:Bz volumetric ratios of 2.4 and 4.6. The latter ratio is just above the stoichiometric ratio of 4.5 so the catalyst was forced between reducing and stoichiometric conditions. Cycling was symmetrical with periods ranging from about 10 to 360 s. Results of the modulation experiments are given in Figure 12.38. The broken horizontal lines mark selectivity and conversion under steady state operation at the time-average O_2:Bz ratio of the cycling feed. A 100 % increase in selectivity is evident in Figure 12.38a, but part (b) of the figure shows a 17 % decrease in conversion. Highest selectivities occur between periods of 30 and 120 s. The wave front analysis (Figure 12.37) suggested high selectivity persisted until about 60 s after the switch

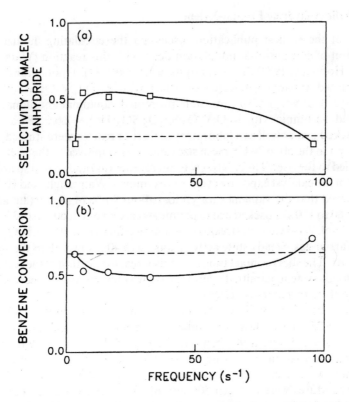

FIGURE 12.38 Reactor performance under symmetrical composition modulation between O_2:Bz = 2.4 and 4.6 as a function of cycle frequency over a TeO_2-V_2O_5-MoO_3/TiO_2 catalyst at 1 bar and 420°C: (a) selectivity to maleic anhydride, (b) benzene conversion (Figure taken from Fiolitakis *et al.* (1983) with permission of the authors)

from air to a Bz rich stream so the results are consistent. The cycling conditions, however, were quite different in the two experiments.

The catalyst used in the study of Fiolitakis et al. was the same as that used by Lang et al. (1989b, 1991) for maleic anhydride from the partial oxidation of butadiene. The Lang work was discussed in the previous section. The choice by Lang et al. was intentional in order to compare the influence of composition modulation when the catalyst and product were the same but the chemical route was altered. Lang et al. observed no improvement in selectivity to MA through forcing. The production rate was also depressed by this operating mode. Although a generalization cannot be made on the basis of just one test, it seems likely that the chemical route and surface mechanism are important factors in a decision to explore periodic operation.

Phthalic Anhydride From o-Xylene

One of the earliest publication discussing the circulating fluidized bed version of composition modulation dealt with this reaction (Wainwright and Hoffman, 1974). Two catalysts were used: (1) a potassium sulfate promoted vanadia supported on silica with $K_2SO_4:V_2O_5:SO_3:SiO_2 = 29:9:12:50$ as wt.%, (2) a potassium promoted vanadium antimonate supported on titania with $K_2O:V_2O_5:Sb_2O_3:SO_3:TiO_2 = 2:6:6:2:84$ as wt.%. A packed bed as well as a laboratory scale riser reactor were studied. Fresh catalyst in the 50 to 70 US mesh size range was employed in the packed bed and fed to the riser. The catalyst was not regenerated and reused so only one part of the catalyst exposure cycle was examined. Wainwright and Hoffman observed that the vanadia catalyst lost about 90 % of its start up activity after 60 to 120 s in packed bed experiments carried out at 603 to 663 K using 1 to 2 vol.% o-xylene in streams containing either 10 or 30 vol.% O_2 in the N_2 carrier gas. Steady state activity was 30 to 40 × lower than the start up activity. The short contact time in the riser reactor portion of the recirculating solids system permitted operation at lower temperatures where selectivity to the oxygenates is better.

For the riser, Wainwright and Hoffman (1974) used a 17.3 mm (i.d.) × 8.23 m stainless steel tube, thermostated to maintain constant temperature. Using a 90° bend at the top of the riser enabled them to operate at void fractions from 0.6 to 0.86. Catalyst flow rates of about 16 kg/min were used with superficial gas velocities from 4.6 to 7.6 m/s. Temperatures were between 506 and 615 K for the vanadia catalyst and 643° for the vanadium antimonate material. Figure 12.39 compares the xylene conversions in the riser experiments with those for packed bed operation and with packed bed results from other investigators using vanadia catalysts. There is better than an order of magnitude increase in conversion. Indeed, even operating at temperatures some 50°C lower than

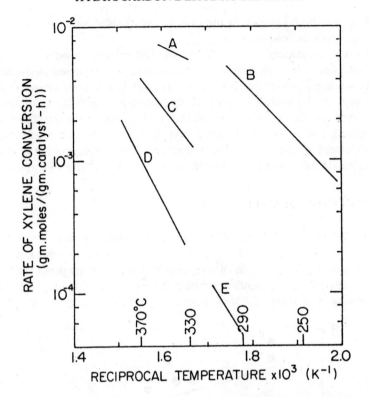

FIGURE 12.39 Rate of o-xylene conversion as a function of reciprocal temperature for a feed of 1 vol.% o-xylene and 0.5 vol.% SO_2 in air at 1 bar using a potassium promoted vanadia catalyst supported on silica. Curve A = 10 s time-on-stream using fresh catalyst in a packed bed, Curve B = riser reactor data, Curve C = literature data using a different vanadia catalyst in a packed bed, Curve D = steady state data for the packed bed reactor, Curve E = literature data using a different vanadia catalyst in a packed bed (Figure taken from Wainwright and Hoffman (1974) with permission, © 1974 American Chemical Society)

those used for their packed bed experiments, Wainwright and Hoffman achieved higher xylene conversions with their riser reactor system. The vanadia catalyst used produced o-tolualdehyde rather than phthalic anhydride in the 506 to 615 K temperature range. Selectivity to this aldehyde was about 95 % at 10 % xylene conversion and dropped only to 75 to 85 %, depending on temperature, for conversions reaching 90 %. The authors comment that industrial reactors attain at best 70 % selectivity to the aldehyde.

Only step change experiments were undertaken with the vanadium antimonate catalyst. These experiments at 643 K, a 200 cm^3/g.cat min SV, using 1.3 vol.% xylene in air showed that 73 to 80 % selectivity to phthalic anhydride could be obtained with short catalyst exposures at xylene conversions about twice those obtained in steady state operation. Selectivity to oxygenates reached 85%

Russian researchers quoted by Matros (1985) worked with a vanadia catalyst using a recirculating solids reactor in which o-xylene was oxidized in a down flow, counter current reactor. The catalyst was re-oxidized in the riser part of the system. Figure 12.40 compares selectivity to phthalic anhydride for the recirculating solids system in which the reactants are segregated and a co-fed, fluidized bed. It can be seen that selectivity improvement in the 80 to 95% xylene conversion range is substantial. Unfortunately details of the Russian work are not available in English.

H. AROMATIC NITRILES

In the 1960's, the Lummus Company, now ABB Lummus Crest Inc., launched an effort to develop a process for producing terephthalic acid from p-xylene via hydrolysis of terephthalonitrile. Starting point for their work was a group of patents describing the formation of aromatic nitriles by passing a mixture of either aromatics and ammonia, or aromatics,

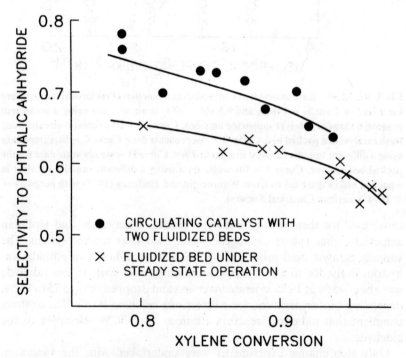

FIGURE 12.40 Selectivity to phthalic anhydride *vs* *o*-xylene conversion on a vanadia catalyst (Figure taken from Matros (1985) with permission, © 1985 Elsevier Science Publishers B.V.)

ammonia and oxygen over various metal oxide catalysts. In the former experiments, catalyst deactivation was traced to reduction of the metal oxides and it was found that the initial catalyst activity could be restored by air blowing of the catalyst at just above the reaction temperature. These observations led the Lummus researchers to pursue the proposal of Lewis et al. (1949) to use the metal oxide as the oxygen carrier. Thus, they proceeded to develop a nitrile process based on shifting metal oxide catalysts between two fluidized beds: an ammoxidation reactor fed with an aromatic feedstock, ammonia and molecular oxygen, and a regenerator blown with air. The pilot plant stage was reached in this development and was described in a paper presented at the March, 1973, Meeting of the AIChE in New Orleans. Nevertheless, the Lummus terephthalic acid process does not seem to have been commercialized.

Experiments with other feedstocks were undertaken by Lummus during their development effort and it was observed that their vanadium based catalyst produced different aromatic nitriles in good yields depending on the feedstock used. Unfortunately, only meager descriptions of the Lummus technology exist in the literature (Sze and Gelbein, 1975; 1976, and Schwendeman et al., 1983) and experimental data are not available.

Lummus Isophthalonitrile Process

This process seems to be the only Lummus vanadium catalyzed aromatic nitrile process to have been commercialized. A plant probably producing about 20×10^6 lbs/yr of this nitrile was built in 1976 near Houston, Texas, for what was then the Diamond Shamrock Corporation. At that time isophthalonitrile was used as an intermediate in the production of tetrachloro-1,3-dicyanobenzene, a herbicide/fungicide. It also has uses as an intermediate in fiber/plastic manufacture.

Isophthalonitrile is formed from m-xylene and ammonia in a sequential reaction with m-toluonitrile as an intermediate. The reaction sequence is given in Figure 12.41. The m-toluonitrile is recovered from the ammoxidation reactor and recycled so that benzonitrile is the only significant by-product. Figure 12.42 is a schematic of the isophthalonitrile plant built in Houston. A similar sequence of operations is used for the other aromatic nitrile processes developed by Lummus (Sze and Gelbein, 1975; 1976). As described above, the nitrile reaction occurs at 672 K in a bed of an attrition resistant, vanadium catalyst fluidized by the m-xylene, NH_3, O_2 and recycle mixture. The m-xylene, hydrocarbon and intermediate recycle from the stripping unit following the quench cooler go to a partial vaporizer. The vapor fraction is used as part of the fluidizing gas, which runs between 10 and 20 vol.% hydrocarbon, while the liquid fraction is injected directly into the bed. The liquid fraction includes gums which are trapped by the catalyst and eventually burned in the fluidized bed regenerator. To maintain high

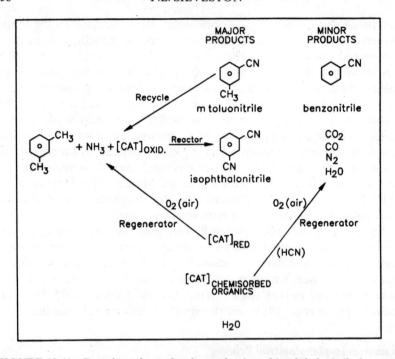

FIGURE 12.41 Reaction scheme for the production of isophthalonitrile from m-xylene. The arrow labelled reactor gives the overall reaction taking place in the ammoxidation step and indicates the main products formed. The arrows labelled regenerator show catalyst regeneration by re-oxidation and combustion of waste products and gums trapped on the vanadium catalyst (Figure taken from Sze and Gelbein (1975) with permission of the authors)

selectivity, conversion is restrained to about 50 % of the xylene fed to the fluidized bed. Xylene, intermediates and the nitrile products are adsorbed by the catalyst which must be steamed stripped before being conveyed pneumatically to the regenerator. The regenerator operates at 700 to 800 K and is fluidized by air. Its function is to re-oxidize the catalyst, combust residues that are not removed by stripping, and to burn CO and HCN found in the ammonia recovery system. Regenerated catalyst returns to the ammoxidation bed through a gravity leg. It is not stripped.

All of the Lummus aromatic nitrile processes operate with excess NH_3 so an ammonia recovery process based on absorption in water is necessary. This process also takes the water phase from the nitrile crystallizer. A second step shown in the Figure 12.42 schematic is to separate NH_3 and CO_2. Nitrile is recovered from the process by centrifugation and crystallization. Crystallization is also utilized for further purification. Assays reveal 99.5 % product purity. The ultimate yield of isophthalonitrile based on m-xylene was 80–85 % for the Lummus pilot plant.

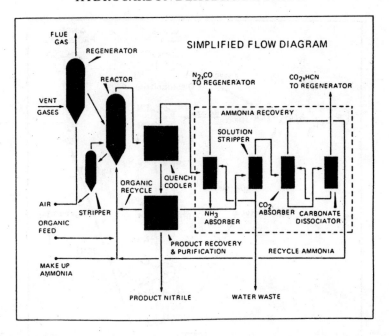

FIGURE 12.42 Schematic of the ABB Lummus Crest circulating catalyst process for the production of aromatic nitriles from aromatic feedstocks, namely isophthalonitrile from *m*-xylene (Figure taken from Sze and Gelbein (1975) with permission of the authors)

If *o*-xylene is substituted for m-xylene, the process described above and shown schematically in Figure 12.42 yields phthalonitrile. Hydration of this nitrile forms phthalamide and ammonolysis of the amide yields phthalimide. Figure 12.43 shows the reaction scheme. After recrystallization, the product assays at 99.5 %. Ultimate yields in the Lummus pilot plant were between 75 and 80 % based on the *o*-xylene fed.

Replacement of *m*-xylene by *p*-xylene produces terephthalonitrile, but different reactor conditions must be used. As the by-product yield from *p*-xylene is negligible, terephthalonitrile can be recovered by centrifuging the aqueous phase from the quench cooler. Sze and Gelbein (1975; 1976) report 99.9 % product purity in their pilot plant runs and ultimate yields of 90 % based on p-xylene. If toluene is used as the feed, the product is benzonitrile. Purities and yields, the same as those given for terephthalonitrile, are mentioned by Sze and Gelbein, but no process details are given.

Lummus Nicotinonitrile Process

This product, a key feedstock for making nicotinic acid, is formed by a multi-step sequence, crudely indicated in the reaction scheme of Figure

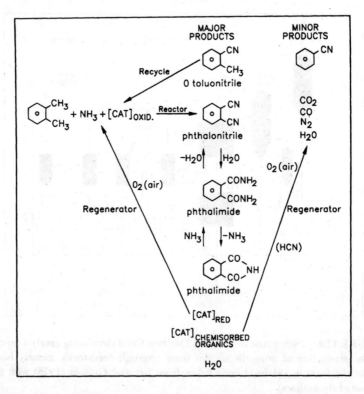

FIGURE 12.43 Reaction scheme for the production of phthalonitrile from *o*-xylene. The arrow labelled reactor gives the overall reaction taking place in the ammoxidation step and indicates the main products formed. The arrows labelled regenerator show catalyst regeneration by re-oxidation and combustion of waste products and gums trapped on the vanadium catalyst. The two way arrows show the further reactions of phthalonitrile (Figure taken from Sze and Gelbein (1975) with permission of the authors)

12.44. All of the intermediates shown in the scheme are formed in the ammoxidation reactor and must be recovered and recycled. Recovery utilizes differences in the water solubility of the pyridines, picolines and nitriles. Purification is complicated. Unfortunately, Sze and Gelbein report neither process details or the yields and purities they achieved in their pilot plant and bench scale work.

I. OVERVIEW AND ASSESSMENT

C_1 Conversion

Claims made by Keller and Bhasin (1982) in their pioneering study of methane oxidative coupling that a segregated feed mode of operation

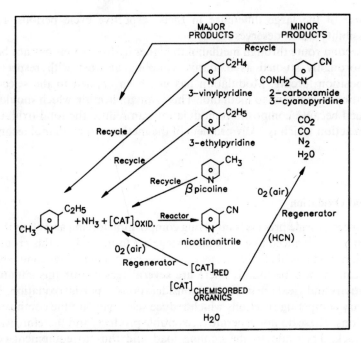

FIGURE 12.44 Reaction scheme for the production of nicotinonitrile from 2-methyl, 5-ethyl pyridine. The arrow labelled reactor gives the overall reaction taking place in the ammoxidation step and indicates the main products formed. The arrows labelled regenerator show catalyst regeneration by re-oxidation and combustion of waste products and gums trapped on the vanadium catalyst (Figure taken from Sze and Gelbein (1975) with permission of the authors)

increases selectivity to the C_{2+} products by up to 10 × have not been borne out by subsequent studies. Segregating CH_4 and O_2 through composition modulation does appear to improve selectivity to the higher hydrocarbons and thereby raises yields somewhat. Still further improvement can be made by flushing between exposures or by partially overlapping them. Nevertheless, it now seems that these yield increases are not sufficient to overcome the yield barrier so cogently discussed by Amenomiya et al. (1990). It seems then that composition modulation cannot contribute significantly to this technology. Only if the yield for economic feasibility can be lowered will there be room for composition modulation. In the discussion closing Section C of this chapter, the application of periodic flow reversal to oxidative coupling was discussed. This could provide significant energy savings thereby reducing the C_{2+} yield for economic feasibility. It would appear possible to combine flow reversal and composition modulation because of the large difference in their respective cycle periods. It is a possibility which deserves study.

A second route through methanol to higher hydrocarbons has not been as thoroughly studied as oxidative coupling at least with respect to composition forcing. Modulation has not been applied to the selective oxidation of methane to methanol. This is an application which should be pursued because composition modulation can reduce the total oxidation side reaction which is a key problem of the methane to methanol technology.

Partial Oxidation

Commercial scale processes involving composition forcing as well as those which have advanced to a pilot plant stage are found only in this group of reactions and it is likely that it is within this group that further commercial applications will be found. There are several reasons for this situation. Selectivity and yield are important considerations in partial oxidation. The primary competing reactions which reduce selectivity are the combustion reactions. These do not generate recoverable products and therefore waste feedstock. They add to the cooling load and thus to equipment cost. Furthermore, competing combustion reactions reduce system stability and thereby bear directly on the complexity and cost of process control. However, what is evident from our discussion of oxidative dehydrogenation and partial oxidation in Sections D to G of this chapter is that composition modulation of these reactions suppresses the total oxidation reactions so that for virtually all systems studied higher selectivities to olefins or to the desirable oxygenates are achieved with respect to steady state operation. Heat release at comparable conversions will be lower and when recirculating catalysts systems are employed, the heat load can be distributed between the two reactor vessels that will be needed. Cooling cost may be reduced thereby. Certainly process control is made simpler. A further advantage is that segregating hydrocarbons and air substantially simplifies product recovery and recycle of the unreacted feed or any intermediates produced.

Composition modulation, on the other hand, will not be universally applicable. Highly selective catalysts have been developed for most of the commercially important oxidative dehydrogenation and partial oxidation processes. For such catalysts, composition modulation augments selectivity just modestly, say, between 10 and 30 %. Examples of such modest increases are the results of Contractor *et al.* (1987) with the selective oxidation of butane to maleic anhydride, discussed in Section F, or the Silveston and Forrisier (1985) study of propylene partial oxidation to acrolein over a Bi-Mo oxide which was reviewed in Section D. Much larger selectivity increases have been observed, of course, but these were generally made with catalysts which exhibited poor selectivity. The Silveston and Forrisier work just mentioned used a mixed Sb-Sn oxide catalyst which exhibited a rather poor selectivity to acrolein. Composition forcing about doubled the selectivity. Because improvements from forcing are expected to be modest with highly selective catalysts, adoption of composition modulation rests upon a comparison of the economic benefits of higher selectivity, lower heat transfer and process control costs, with the higher capital and operating costs of two reactors and circulating catalyst.

It is our view that composition modulation becomes an attractive alternative when feedstock costs are high and at least one of the reaction products has high value. The application of the two reactor, recirculating catalyst technology to the production of aromatic nitriles (Section H) by ABB Lummus Crest appears to be an example of such a situation. The second situation for which composition modulation may become attractive is when the desired reaction product is readily decomposed or oxidized further in the reaction environment. Oxidative dehydrogenation represents such a situation; formation of furan in the selective oxidation of n-butane is another. It seems likely that the use of short contact time, riser reactors for the dehydrogenation or partial oxidation steps followed by catalyst recovery and recycle through air oxidation in a fluidized bed would offer attractive options when such products are sought. Composition modulation should be less interesting when stable reaction products form. Maleic and phthalic anhydride react only slowly in oxidative environments so it is surprising that Du Pont is pursuing the two reactor, recirculating catalyst technology for their maleic anhydride from butane process.

CHAPTER 13

Fischer-Tropsch Synthesis[1]

A. INTRODUCTION

Within the last 15 years, enough studies of the Fischer-Tropsch synthesis (FTS) of hydrocarbons under periodic composition forcing have accumulated to evaluate the success of this operating mode on a potentially important, commercial reaction system. The synthesis was thought to be an attractive target for periodic operation because the steady-state mode suffers from either low activity or poor selectivity to desirable products, or, for some catalysts, from both of these problems. In this chapter, we will (1) summarize the experimental results, (2) search for common features of the FTS under composition forcing, (3) suggest mechanistic explanations for some of the features identified, and (4) identify opportunities for future work. This contribution is only peripherally concerned with the mechanism of the synthesis, which has already received considerable attention in the literature (Vannice, 1976; Bell,1981; Ponec, 1978, and Anderson, 1984).

The Fischer-Tropsch synthesis can be carried out as a solid catalyzed, gas phase reaction using a fluidized or packed bed or as a three phase reaction with the liquid phase functioning as a solvent for the products and as a heat sink. Slurry reactors are most commonly used when three phases are present, but trickle beds or packed bed bubble columns are alternatives. Normally, there are just two reactants, CO and H_2, although there is some interest in replacing part of the CO by CO_2. Only the solid catalyzed, gas phase reaction has been studied under composition modulation.

Strategies available for modulation have been discussed in earlier chapters. For the Fischer-Tropsch synthesis, the strategies investigated have been limited to either modulation of just one of the reactants or to a 180° phase lag when two reactants were used. However, another aspect of strategy, the coupled choice of time-average feed composition and amplitude, has not been dealt with previously. Figure 13.1 illustrates possible

---

[1]This chapter is adapted from a paper, "The Fischer-Tropsch Synthesis Under Periodic Operation", *Catalysis Today*, **25(2)**, 127–145 (1995). The assistance of Dr. A.A. Adesina, Chemical Engineering, Univ. of New Sourth Wales, Australia, and Prof. R. R. Hudgins, Chemical Engineering, Univ. of Waterloo, Canada, with the paper is acknowledged gratefully.

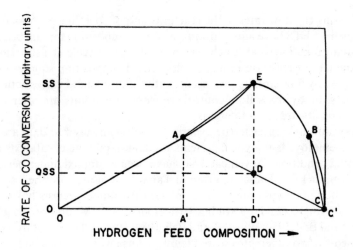

FIGURE 13.1 Different forcing strategies relative to the composition giving the maximum rate of conversion at steady state and the relaxed and quasi-steady states

choices. The figure shows schematically variation of the rate of CO conversion in hydrocarbon synthesis at steady state as a function of the H_2 mole fraction. There is a maximum rate at E. One strategy is to switch between a synthesis mixture, represented by A, and pure H_2, represented by C, while a second strategy is to switch between two mixtures, represented by A and B. Other strategies are cycling between compositions on either side of E. Figure 13.1 also shows two asymptotic situations that can arise in composition forcing which were discussed in Chapter 1. If composition is rapidly cycled between A and C, the catalyst environment, as a result of unavoidable local mixing, cannot follow the change and will remain at a constant mean. This high frequency asymptote is the relaxed steady state and could be the steady state at the time-average feed composition or it could be a new condition with a different rate of reactant conversion. On the other hand, if the period is long and the cycle is symmetrical, steady state will be attained in each half-cycle and the time-average rate of CO usage will be given by D in the figure. This asymptote is called the quasi-steady state. It is clearly not attractive if the rate of CO conversion is important.

B. EXPERIMENTAL SYSTEMS EMPLOYED

As we have described in Chapter 1, laboratory scale periodic composition-forcing employs conventional laboratory reactor systems utilized for the study of kinetics, but adds provision for quickly changing feed composition by adjusting the flow rates of the reactants and often a diluent. All studies published on FTS have adopted square-wave composition forcing because this forcing function is easily achieved on a laboratory scale. It also seems to have the largest effect on rate and selectivity. Composition square-waves are created by flow circuits with needle valves and solenoid valves actuated by timers, or by mass flow controllers whose set points are altered by computer software at set intervals.

Important differences in the experimental systems used to date are the reactors employed, analytical systems used, and, of course, catalysts and operating conditions. Seven catalysts have been examined under periodic operation for FTS. These are given in Table 13.1, along with their source and method of conditioning. The iron and ruthenium catalysts have been used often in recent FTS studies (Anderson, 1984; Madon et al., 1977; Kellner and Bell, 1982; Winslow and Bell, 1984, 1985). Of these catalysts, Ru was used by two different research teams, whose work is discussed in this contribution. Co was also used in two separate studies.

Reactor types, dimensions, operating conditions, temperature control and analytical systems used are summarized in Table 13.2. Barshad and Gulari (1986) used a novel, honeycomb monolith reactor with catalyst wash coated on the surface which, when combined with an IR detector, permitted the evolution of reaction products to be followed instantaneously. A clever system of valve drives and a rapidly responding mass flow meter permitted composition cycling at 2 Hz with just a small distortion of the concentration square-wave. However, the data of Gulari and co-workers (Barshad and Gulari, 1986; Dun and Gulari, 1986) are sometimes problematic because higher hydrocarbons were poorly resolved by the GC system employed. Khodadadi et al. (1996b) were able to obtain good time resolution using an FTIR and a light pipe, but this system was only used to follow dynamic response to step changes.

C. EXPERIMENTAL OBSERVATIONS

Promoted Iron Catalyst

Study of this catalyst under composition modulation is described in two research contributions and a summary paper (Feimer et al., 1985a; 1985b; Silveston et al., 1986). The motivation for Feimer's work was the possibility that periodic H_2-pulsing might curtail chain growth on the catalyst surface, thereby enhancing the light naphtha yield at the expense of wax production. Feimer et al. (1985a) investigated, therefore, strategies represented by A to C or E to C in Figure 13.1. Forcing was also explored around either side of the maximum E, that is, from A to E or B to C. CO-pulsing was not examined because Feimer et al. (1984) observed that in the absence of H_2,

Table 13.1 Catalysts Used and Conditioning Prior to Use

Catalyst	Ruthenium	Ruthenium	Molybdenum	Molybdenum	Iron	Cobalt	Cobalt
Literature Reference	Barshad and Gulari (1986)	Ross et al. (1987)	Dun and Gulari (1986)	Dun and Gulari (1986)	Feimer et al. (1984)	Adesina et al. (1986)	Khodadadi et al. (1996a)
Composition	1 wt.% Ru	0.5 wt.% Ru	6 wt.% Mo, 2 wt.% K	6 wt.% Mo	Fe:Cu:K_2O = 100:20:1	62 wt.% CoO	23 wt.% Co
Support	Alumina-Engelhard X17 monolith	γ-Alumina	Charcoal	Charcoal	None	Kieselguhr	Silica
Source	In-house preparation	Engelhard	In-house preparation	In-house preparation	Exxon Corp.	United Catalysts	In-house preparation
Conditioning Prior to Use	Reduction in H_2 at 493 K for 3 h	Reduction in H_2 at 673 K for 4 h followed by H_2 flushing at reaction temperature for 1.5 h	Not given	Not given	Reduction in H_2 at 543 K for 3 days, activated with 27:1 H_2/CO at 519 K, then flushed with H_2 for 3 h	Reduction in H_2 at 473 K for 30 min, then at 673 K for 2 h. H_2 flushing at reaction temperature for 1.5 h	Reduction in H_2 at 723 K for several hours, then exposed to 2.5:1 H_2/CO at 503 K for 5 h

Table 13.2 Reactor Characteristics and Operating Conditions

Catalyst	Ruthenium	Ruthenium	Molybdenum	Molybdenum	Iron	Cobalt	Cobalt
Literature Reference	Barshad and Gulari (1986)	Ross et al. (1987)	Dun and Gulari (1986)	Dun and Gulari (1986)	Feimer et al. (1984)	Adesina et al. (1986)	Khodadadi et al. (1996a)
Type of Reactor	washcoated catalyst on honeycomb monolith	Tubular, granular packed bed (Differential)	Tubular, granular packed bed	Tubular, granular packed bed	Differential, granular packed bed	Differential, granular packed bed	Tubular, granular packed bed (Differential)
Reactor Dimensions	Cylindrical 32mm diam × 34mm with 50.5 channels per cm^2	Copper tube: 5mm i.d. × 80mm with 0.3g catalyst	Stainless tube: 12.7mm i.d. × 178mm with 140mm catalyst	Stainless tube: 12.7mm i.d. × 178mm with 140mm catalyst	Stainless tube: 7.7mm i.d. with 0.5g catalyst	Copper tube: 5mm i.d. × 80mm with 0.3g catalyst	Stainless tube: 2.5mm i.d. × 50mm with 0.25g catalyst
Catalyst Particle Size		60/100 US mesh	< 1.3 mm	< 1.3 mm	60/100 US mesh	100/400 US mesh	60/100 US mesh
Catalyst Dilution					5:1 with glass beads	1:1 with glass beads	
Space Velocity	1800 h^{-1}	22000 h^{-1}	3000 h^{-1}*	3000 h^{-1}*	5000 h^{-1}*	15000 h^{-1}	3500 h^{-1}
Temperature	435–473 K	484 K	543–673 K	543–673 K	519 K	473 K	423–523 K
Pressure	110 kPa	445 kPa	110 kPa	110 kPa	384 kPa	115 kPa	117 kPa
Method of Temperature Control	Heating jacket computer controlled	Oil bath analog control	Heating oven, analog control	Heating oven, analog control	Oil bath analog control	Oil bath analog control	Oil bath analog control
Analytical System	FTIR	GC (Carle 211 A)	IR & HP 5790 GC	IR & HP 5790 GC	GC (Carle 8700A & 211A)	GC (Carle 211A)	FTIR & Carle 400A GC, Perkin-Elmer 3B GC

*based on an assumed bed porosity of 0.4

CO poisoned the catalyst. In transient experiments, overshoot occurred within minutes of a step-change; thus, cycle periods only up to 15 min were explored because Feimer et al. assumed that cycling using these short periods could capture the production benefits of the overshoots. The slow response of the mass flow controllers and mixing in the IR detectors used to follow CO and CO_2 limited the shortest cycle period to about 1.5 min. The cycle split(s), defined as the fraction of the period the catalyst was exposed to pure H_2, was kept between 0.4 and 0.875.

The observations of Feimer et al. (1985a) are reproduced in Figures 13.2 and 13.3. The first of these figures plots the time-average rate as a function of period for the C_1 to C_3 paraffins. The steady-state rates corresponding to the time-average feed composition (SS) and the time-average quasi-steady state (QSS) are shown as horizontal broken lines and are given for

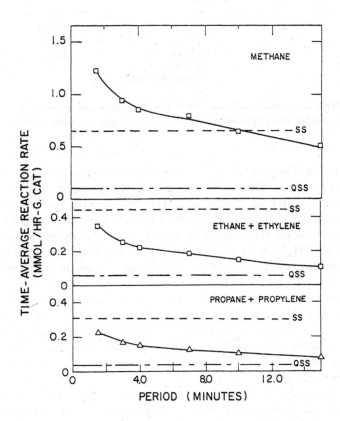

FIGURE 13.2 Effect of cycle period on time-average iron-catalyzed formation rates of methane, C_2 and C_3 species at $A_1 = 1$, $A_2 = 0.476$, $s = 0.7$, 246 °C, 384 kPa (Figure adapted from Feimer et al. (1985a) with permission of the authors)

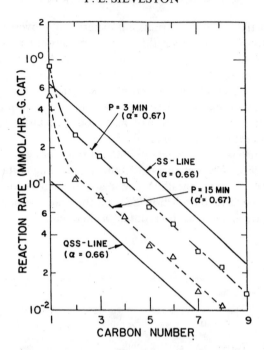

FIGURE 13.3 Anderson-Schulz-Flory plot for conditions given in Figure 13.2 (Figure adapted from Feimer *et al.* (1985a) with permission of the authors)

comparison. These results are for forced cycling of pure H_2 and a CO/H_2 mixture across the maximum steady state, a strategy represented by the line between A and C in Figure 13.1. The time-average feed composition corresponds to D' in the figure, so the steady-state rate is the maximum rate possible in steady state operation. Figure 13.2 indicates that this strategy stimulates CH_4 formation for a period (τ) less than 8 min. At $\tau = 1.5$ min, the time-average rate is about 90% greater than the maximum steady-state rate. Thus, at least for CH_4 production, composition forcing can achieve rates of formation in the FTS that exceed the maximum possible under steady state. The rising slopes of the time-average rates of the C_2 and C_3 products with increasing frequency in Figure 13.2 suggest that the formation of higher hydrocarbons could be stimulated by higher frequency forcing. Feimer *et al.* observed that cycle split had little effect on the time-average rate.

The relatively slow forcing used by Feimer *et al.* (1985a) for the experiments shown in Figure 13.2 had a negligible effect on the distribution of hydrocarbon chain length (carbon number) as may be seen in Figure 13.3. This figure is the conventional Anderson-Schulz-Flory (ASF) plot widely used to display product distributions for the FTS. The symbol α in the figure is the chain growth factor in the ASF model for FTS polymerization. Only CH_4 formation is altered by feed composition modulation. As Figure 13.2 shows, the period affects the hydrocarbon formation rate. Increasing the period in Figure 13.3 reduces the formation rates of all products until the quasi-steady-state limit is attained. Although product distribution, apart from CH_4, is not altered, the H_2-pulsing strategy affects the olefin/paraffin ratio. The formation rates of all olefins above C_2H_4 are suppressed by composition forcing. Ethylene was not separable from C_2H_6 in the GC system used by Feimer. Formation rates of the paraffins, on the other hand, are stimulated by this strategy.

Composition forcing on either side of point E in Figure 13.1 produced time-average rates vs. period similar to those shown in Figures 13.2 and 13.3. C_1 formation was stimulated at periods of less than 2 min, but the rate was well below the maximum at composition D′ in Figure 13.1. The chain growth factor, α, was not changed; however, unlike pulsing with H_2, the olefins/paraffins ratio is not significantly altered. The ratio is also unchanged by cycling on the H_2-rich side of Figure 13.1, that is, from B to C. Both of these forcing strategies promoted CO_2 formation at periods shorter than 4 to 10 min. This has some benefit as it reduces H_2 wasted in removing oxygen from the synthesis reaction.

Ruthenium Supported on Al_2O_3

As may be seen from Table 13.1, this catalyst was studied by two separate research teams. However, operating conditions, Ru loading on the catalyst, and the reactors used for the two studies were quite different (Table 13.2). Ross et al. (1987) examined a strategy of hydrogen-pulsing similar to that investigated by Feimer et al. (1985a), while Barshad and Gulari (1986) tested two strategies: (1) H_2-pulsing using a synthesis gas mixture with 11% CO, and (2) bang-bang switching between CO and H_2. With their rapidly responding flow controllers, Barshad and Gulari were able to employ cycle periods as short as 10 s, whereas the shortest periods used by Ross et al. (1987) were 30 times longer.

Both teams found the H_2-pulsing strategy promoted the formation of light paraffins and suppressed olefin production. This is illustrated by results for symmetrical forcing (split = 0.5) given in Figure 13.4 (Ross et al., 1987). Time-average rates for C_n's have been normalized by dividing by the steady-state rate at the time-average feed composition. Results for paraffins and olefins have been separated; however, the total hydrocarbon production rate is plotted in part (b) of the figure along with data for olefins. For this type of plot, QSS rates will vary for each hydrocarbon; thus, this limit of forcing is shown by bands in the figure. Periods as long as 40 min resulted in higher rates of formation for the C_1 to C_3 paraffins. At the shortest periods used, the rates were 30 to 50% greater than the corresponding steady-state

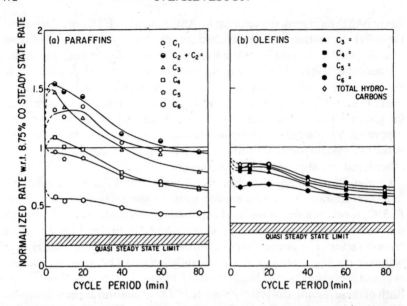

FIGURE 13.4 Normalized time-average formation rates over a supported ruthenium catalyst under composition forcing *vs.* cycle period (τ) at a mean CO volume fraction in the feed of 8.75% for $s = 0.5$ (Figure adapted from Ross *et al.* (1987) with permission of the authors)

rates, which were also close to the maximum attainable steady-state rates. The figure shows that the rate enhancement decreases with paraffin carbon numbers above C_4. Evidently, the polymerization process damps out the effects of composition modulation. Part (b) of Figure 13.4 shows that olefin formation is strongly suppressed by periodic pulsing of hydrogen. Just as for paraffins, the normalized olefin production rate decreases with increasing carbon number. Despite the strong improvements in paraffin production, the plot of the normalized rates for total hydrocarbons is below unity for all periods investigated.

The effects of split were explored by Barshad and Gulari (1986) with forcing at a period of 25 s. Figure 13.5 plots the time-average rate of CO consumption normalized by the rate under steady-state operation with a synthesis mixture containing 11% CO. At high splits, the H_2-pulsing strategy increases the rate by a factor greater than 2. At a split of 0.5, the condition used by Ross *et al.* (1987), the increase in CO consumption is about 45%. This seems much higher than the increases measured by Ross *et al.* because it covers all the hydrocarbons formed rather than just the C_2 and C_3 hydrocarbons. Consumption also includes CO used to scavenge oxygen. Figure 13.6 shows that at $\tau = 25$ s, H_2-pulsing promotes C_3H_8 formation at low values of s; at high values, Barshad and Gulari (1986) report that CH_4 is the dominant product. Figure 13.4 also shows a substan-

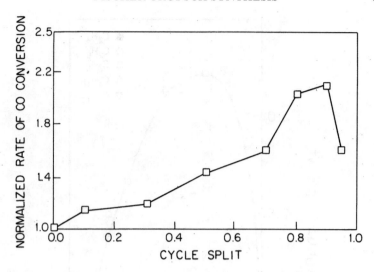

FIGURE 13.5 Normalized time-average rates of CO usage over a Ru/Al$_2$O$_3$ catalyst with switching at $\tau = 25$ s between H$_2$ and a synthesis gas containing 11%CO, 435 K, 1 bar (Figure adapted from Barshad and Gulari (1986) with permission, © 1986 Gordon & Breach Publishers)

tial increase in C$_3$H$_8$ production at 5 min, the shortest period used by Ross et al. (1987). Barshad and Gulari report that C$_2$H$_4$ is the only olefin formed in measurable amounts. Figure 13.4 indicates other olefins are formed at higher temperatures and pressures but feed composition forcing sharply reduces their production. There appears, therefore, to be a measure of agreement between these two studies using Ru catalysts.

In contrast to iron catalysts, the product distribution for the lower hydrocarbons in FTS over ruthenium at steady state does not follow the ASF distribution (King, 1978; Pannell et al., 1981; Kellner and Bell, 1982). This is explained by readsorption of C$_2$H$_4$ and C$_3$H$_6$ and their incorporation into growing hydrocarbon chains (Dywer and Somorjai, 1979). The steady-state distribution is given in Figure 13.7 and compared with the time-average data obtained by Ross et al. (1987). The hydrogen-pulsing strategy alters the distribution of products at higher cycling frequencies so as to approach the ASF distribution. Readsorption of the lower olefins appears to be reduced. It is likely that this resulted from the hydrogenation of these olefins prior to their desorption. Barshad and Gulari (1986) did not report data in a form that could be plotted on an ASF diagram.

Employing a different strategy, alternating the reactor feed from CO to H$_2$ and back again, leads primarily to CH$_4$ at $20 < \tau < 60$ s, at a split of 0.7, as Figure 13.8 shows. This plot shows selectivity to C$_1$ and CO conversion. The conversion drops off after $\tau = 40$ s which is explained by CO saturation of the surface; thus, CO leaves the reactor in the off-gas during the CO pulse

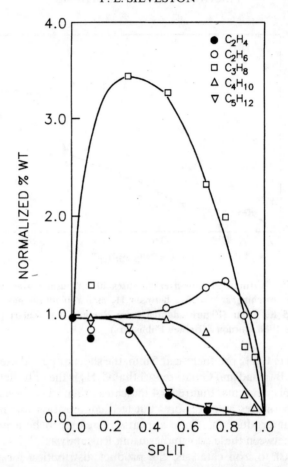

FIGURE 13.6 Normalized product distribution with H_2-pulsing, conditions as given in Figure 13.5 (Figure adapted from Barshad and Gulari (1986) with permission, © 1986 Gordon & Breach Publishers)

(Barshad and Gulari, 1986). For these longer periods and high split, the extended exposure of the surface to H_2 appears to hydrocrack adsorbed carbon chains or convert monomer to methane, which explains the high C_1 selectivity shown in the figure. At the other extreme, alternating short pulses of CO and H_2 provides a surface conducive to monomer formation and subsequent polymerization so C_1 selectivity drops off. For cycle periods below 40 s, and a constant split, CO conversion falls because decreasing CO contact time results in incomplete CO adsorption. Step-change experiments with a honeycomb catalyst showed that higher hydrocarbons appear even before CH_4 when the catalyst is first exposed to CO and then to H_2. But after that initial burst, higher hydrocarbons quickly disappear from the

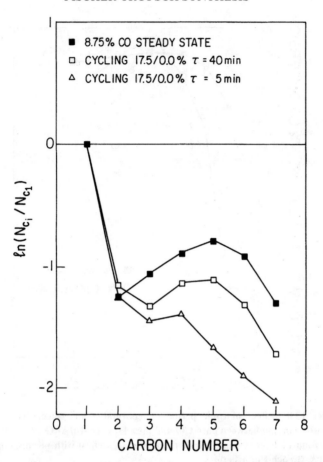

FIGURE 13.7 Anderson-Schulz-Flory plot of normalized rate of olefin + paraffin formation over ruthenium in steady and periodic modes. Supported ruthenium catalyst using operating conditions given in Figure 13.4 (Figure adapted from Ross *et al.* (1987) with permission of the authors)

gaseous product, even though CH_4 continues to be observed. The higher hydrocarbons appear only if the catalyst is exposed for several minutes to CO prior to the H_2 switch. In similar experiments, Ross *et al.* (1987) switched between a synthesis gas mixture and pure H_2 and observed a rapid rise and fall of the C_2 to C_4 products; CH_4 continued to appear many minutes after the switch to H_2. These observations suggest rapid dissociative CO adsorption and formation of skeletal carbon chains on the Ru surface.

Molybdenum Supported on Charcoal

An unpromoted and a potassium-promoted Mo/charcoal catalyst were investigated by Dun and Gulari (1986) because these catalysts are not

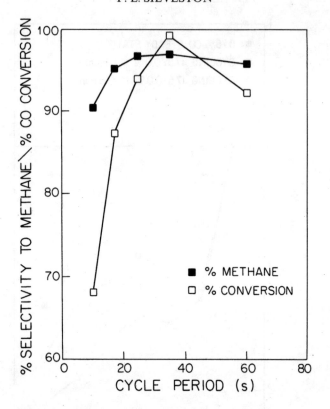

FIGURE 13.8 Effect of cycle period on conversion and selectivity to CH_4 for composition forcing between pure CO and pure H_2 over Ru/Al_2O_3 at $s = 0.7$ (473 K, 1 bar) (Figure adapted from Barshad and Gulari (1986) with permission, © 1986 Gordon & Breach Publishers)

deactivated by pure CO and do not produce higher paraffins, thereby avoiding analytical problems. The second catalyst was included because adding a potassium promoter depresses the catalyst activity at steady state, but sharply increases the selectivity to C_{2+} hydrocarbons (Murchison, 1982; Dun et al., 1985). Thus, an objective of these investigators was to see if composition modulation affects the two molybdenum catalysts differently. In these studies, feeds were either H_2 or CO.

Figure 13.9 plots the normalized time-average rate of CO consumption against forcing period and split for both catalysts. Rates were normalized with respect to the rate under steady state at the same time-average feed composition. QSS rates are zero for this bang-bang strategy and do not appear in the figure. Both period and split influence the time-average rate. For the unpromoted catalyst, normalized rates at a split of 0.3, based on CO, show a 45% increase at $\tau = 5$ s. The steady-state rate at the time-

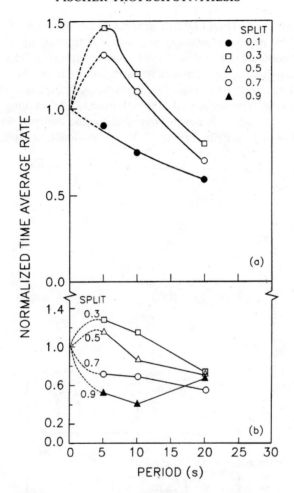

FIGURE 13.9 Effect of cycle period on the normalized time-average rate of CO consumption at different cycle splits; temperature = 400 °C. (a) 6% Mo on charcoal; (b) 6% Mo plus 2% K on charcoal (Figure adapted from Dun and Gulari (1986) with the permission of the authors)

average feed composition corresponding to this split is the maximum steady state rate of CO utilization so global enhancement was observed. With the promoted catalyst (Fig. 13.9b), the highest rates are obtained also at a split of 0.3 and $\tau = 5$ s. The increase above the steady-state rate is about 25%. There is also rate enhancement at a split of 0.5 provided the period is 5 s. This split corresponds to the optimal steady state feed composition. As with ruthenium, both promoted and unpromoted molybdenum under composition forcing catalyze CO conversion at rates that significantly exceed the maximum steady state rate. Changing the split from 0.3 for the unpromoted catalyst (Fig. 13.9a) decreases CO consumption. If the relative duration of H_2 exposure becomes large ($s > 0.3$), this forcing strategy fails. For the promoted catalyst (Fig. 13.9b), splits greater than 0.5 result in an unsatisfactory forcing performance regardless of cycle period.

The relative selectivity to higher hydrocarbons, defined as the C_{2+}/CH_4 ratio, is sensitive to period as well as split. Figure 13.10a is a normalized plot of this ratio against cycle period and split for the unpromoted catalyst, similar to Figure 13.9. From Figure 13.10a, it is evident that rapid switching

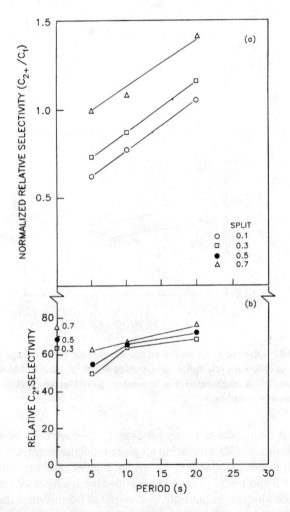

FIGURE 13.10 Normalized relative selectivity ($C_2 + /C_1$) as a function of period at different splits: (a) normalized time-average relative selectivity; (b) relative time-average selectivity to C_{2+}. Conditions and catalysts as in Figure 13.9 (Figure adapted from Dun and Gulari (1986) with permission of the authors)

($\tau < 10$ s) suppresses, relative to steady operation, the buildup of growing hydrocarbon chains on the Mo surface and thus the selectivity to higher hydrocarbons. For the promoted catalyst (Fig. 13.10b), selectivities to C_{2+} rather than relative selectivities are plotted. The results are not normalized and the steady state data are shown at $\tau = 0$. Evidently, cycle periods must be longer than 20 s for selectivities to exceed those at steady state. This period is larger than the period needed to affect improvement seen in Figure 13.10a. The cycle split, based on the duration of CO exposure, must be 0.7 or greater for bang-bang cycling to be advantageous. It appears, then, that this composition forcing strategy is much more effective for the unpromoted catalyst than the promoted one.

According to Dun and Gulari (1986), both the normalized relative selectivity and CO consumption rate depend on temperature. As temperature increases, methane is favoured, thereby reducing the normalized selectivity. On the other hand, lowering temperature decreases the enhancement of the CO utilization rate due to forcing. The improvement shown at 400 °C for $s = 0.3$ and $\tau = 5$ s (Fig. 13.9a) decreases to 10% at 270 °C. The period for this maximum enhancement moves from 5 s to 10 s as well (Dun and Gulari, 1986). For the promoted catalyst, the enhancement decreases as temperature decreases, just as observed for the unpromoted Mo catalyst.

Under forced cycling, Dun and Gulari (1986) found that there was a large decrease in the C_2H_4/C_2H_6 ratio that was independent of period, but strongly dependent on split. At small splits, corresponding to relatively long H_2 exposures, olefins are hydrogenated on the Mo surface, but it can be seen from the C_{2+} selectivity data in Figure 13.10b that there is little variation in the C_{2+} selectivity due to period. Thus, olefin hydrogenation occurs without a large increase in CH_4 production. Dun and Gulari concluded from their investigation that bang-bang composition forcing can elevate selectivity to higher hydrocarbons to values attainable using a K promoter without the penalty of decreased catalyst activity.

Cobalt Supported on Kieselguhr or Silica

An H_2-pulsing strategy was examined for a cobalt catalyst (Table 13.1) by Adesina *et al.* (1986, 1988, 1990). Cobalt catalysts are employed for wax production and should exhibit significant effects for H_2 pulsing if the premise is correct that chain-growth can be controlled through abrupt termination induced by H_2 exposure. Additional strategies are discussed and evaluated by Khodadadi *et al.* (1996a) using a similar cobalt catalyst.

Adesina *et al.* carried out two separate studies. In the first, forcing periods up to 80 min were used because composition step-change experiments indicated a slow relaxation after a composition change. Much shorter cycles were used in the second study. Figure 13.11 shows the forcing results for symmetrical cycling for both studies. Data points for the first study are

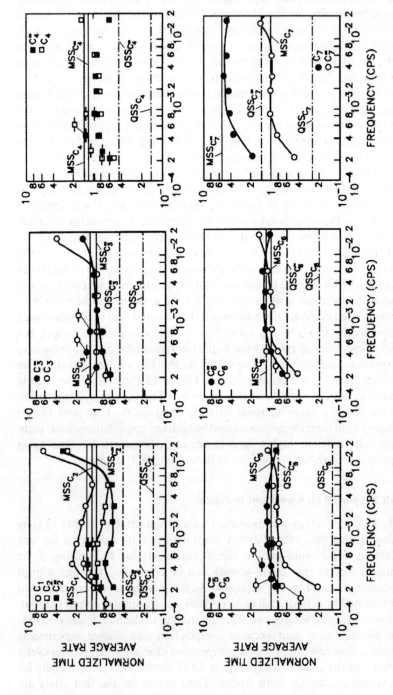

FIGURE 13.11 Effect of cycle period on the normalized rate of product formation at $s = 0.5$, $A = 0.125$ for a cobalt catalyst (473 K, 115 kPa): (a) alkanes, (b) alkenes (Figure taken from Adesina et al. (1986) with permission of the authors)

indicated by a horizontal line through the point. In Figure 13.11, the normalized time-average rate is plotted against frequency for C_1 and for paraffins and olefins up to C_7. Steady-state rates measured at the time-average composition of the reactor feed were used for normalization. QSS limits for each hydrocarbon are also shown. The MSS lines in the figures are the maximum steady-state rates for the temperature, pressure, and space velocity used by the investigators. Figure 13.11 suggests that the H_2-pulsing strategy significantly enhances the time-average rate of formation of most paraffins and olefins up to C_7 at some frequency. Generally, enhancement decreases with increasing carbon number, although heptane is strongly enhanced and seems to be an exception. Forcing was carried out across the composition corresponding to the maximum rate. Thus, for many products, the steady-state rates used to normalize the time-average values were equal or close to the maximum steady-state rates. Figure 13.11 demonstrates that for a cobalt catalyst composition modulation can provide global rate enhancement.

The mass of data shown in the figure and the plotting method makes interpretation difficult. The effect of modulation on the product distribution is better seen in Figure 13.12. This figure is an ASF plot showing

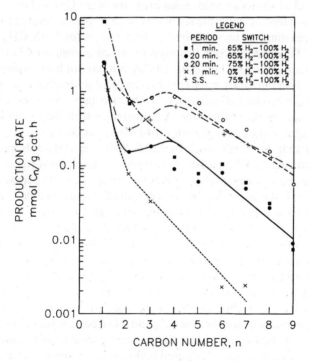

FIGURE 13.12 ASF plot of rate of product formation for composition forcing at $s = 0.5$ and steady-state operation. Conditions and catalyst as given in Figure 13.11 (Figure adapted from Adesina *et al.* (1986) with permission of the authors)

product rates as a function of carbon number for a near optimum steady state and four periodically forced experiments that are also near optimal for the C_2 and C_3 hydrocarbons within the limits of Adesina's investigation. It shows only a small change in the C_{3+} hydrocarbon product distribution through forcing, although rates of formation are suppressed by modulation at most frequencies with the exception of C_1. The slope, corresponding to the inverse of the propagation probability, increases under forcing meaning that the shorter chain products are favored by H_2 pulsing. Like Figure 13.11, the ASF plot shows that rate improvement can be achieved by the proper selection of amplitude and cycle period. In the figure, the best cycle period is 20 minutes. Figure 13.12 illustrates that at steady state and for cycle periods of 20 minutes the C_2 and C_3 products do not follow the Anderson-Schultz-Flory distribution suggesting that ethene and to a lesser extent propene are consumed in chain building. Relatively fast cycling with $\tau = 1$ min appears to suppress incorporation of the light olefins into the carbon chains so that the straight lines extend to a carbon number of 2 for this cycle period. Figure 13.7 showed a similar result for the Ru catalyst. Figure 13.12 indicates an H_2-pulsing strategy enhances the formation of methane.

Figure 13.11 shows an unforeseen property of the Co catalyst under feed composition forcing: multiple resonance with respect to forcing frequency, i.e., rate maxima, in two different period ranges for both CH_4 and C_2 products. This phenomenon has been observed as well for CO oxidation over a vanadia catalyst (Jain et al., 1982a), but has not been explained. The resonance is in the catalyst activity; Figure 13.12 shows that changes in the time-average product distributions at the two periods corresponding to rate maxima are about the same. A comparison with the steady state line indicates that α, the chain growth probability, has changed.

Noting that Adesina's work indicates that a fast H_2-pulsing strategy might be attractive, Khodadadi et al. (1996a) repeated the Adesina study by modifying the experimental equipment to bring the cycle period down to 7 s. The catalyst was similar, but was prepared in house (Table 13.1). These investigators also examined 3 additional strategies: 1) bang-bang cycling of H_2 and CO, 2) CO pulsing and 3) cycling between different syngas mixtures. The intent of the CO pulsing was to suppress C_1 formation and to force the production towards larger hydrocarbon chains. Provision was made in the analytical system to measure products up to C_{20} but at the low conversions used, only hydrocarbons up to C_{14} were found.

Results for fast cycling are best illustrated by bang-bang modulation, that is, periodic switching between H_2 and CO feeds, shown in Figure 13.13. All data in the figures are normalized by steady state rates measured at the time average feed compositions used in the cycling experiments. Parts (a) to (c) in the figure are for a 5 second exposure of catalyst to H_2 followed by a CO exposure of variable duration, while in (d) to (f) a 5 s exposure to CO is

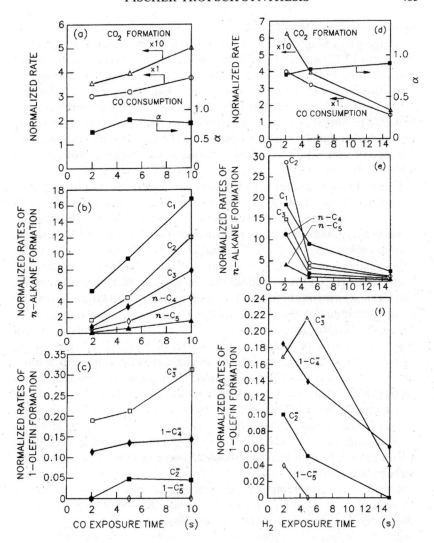

FIGURE 13.13 CO consumption, chain growth probability and product formation rates for bang-bang cycling between H_2 and CO feeds over a Co/SiO$_2$ catalyst at 483°C, 110 kPa and a space velocity of 10775 h^{-1} (STP). All data have been normalized with steady state measurements at the same conditions and for the time-average feed composition (Figure adapted from Khodadadi et al. (1996a) with permission of the authors) with permission of the authors)

followed by a H_2 exposure of variable duration. The remarkable feature of these experiments is the enormous increase in CO_2 production: 60 × the steady state rate for a 7 s cycle with a 2 s H_2 pulse duration. Oxygen is removed from the FTS by CO rather than H_2 which is desirable as hydrogen is the expensive component of syngas. A consequence of O_2 scavenging by CO is the much more rapid consumption of CO also evident in Figures 13.13a and 13.13d. There is actually a net increase in the rate of hydrocarbons formed, but most of the increase is methane production. C_1 formation increases $10 \times$ for $s = 0.5$ in a bang-bang mode. With just a $2\,s\,H_2$ pulse, ethane formation increases about $28 \times$. Large increases in C_4 and C_5 paraffins appear to arise entirely from hydrogenation of the olefin. Olefins higher than C_5 were not observed in the products. Parts (c) and (f) of Figure 13.13 show that $C_3^=$ and $C_4^=$ persist in the synthesis products even with 10 s exposure to H_2.

Figure 13.14 compares a CO-pulsing strategy (Figs. 13.14d and e) to H_2-pulsing (Figs. 13.14a to c). Unlike H_2-pulsing, switching between CO and syngas seems beneficial. With just a $2\,s$ syngas pulse, C_{2+} production remains the same as under steady state operation, but Figure 13.14e shows a large enhancement of C_2 to C_4 paraffins. Thus, CO-pulsing of a syngas feed stream seems to be a better strategy than H_2-pulsing if operating at a low mean H_2/CO ratio is desirable. Bang-bang switching between H_2 and CO and switching between H_2 and syngas or CO and syngas is a matter of the amplitude of the change in the H_2 or CO concentrations. Thus, Figures 13.14 a,b,c can be compared to Figures 13.13 a,b,c, while (d) to (e) of Figure 13.14 may be likened to (d) and (e) of Figure 13.13. Decreasing amplitude appears to increase the Flory chain growth probability, α, but it decreases the rate of CO consumption. Decreasing amplitude increases olefins in the synthesis product, but works oppositely on the paraffins. These generalizations were confirmed by experiments with cycling between syngas mixtures. Some enhancement was seen, but syngas cycling gives just small changes over steady state performance (Khodadadi et al., 1996b). Chain growth probabilities are about the same as those measured under steady state operation.

Choice of a best forcing strategy depends on objectives. At the present time the Fischer-Tropsch synthesis is seen as a possible source for jet and diesel fuels, for waxes and possibly for low molecular weight hydrocarbons, preferably olefins. Consequently, one objective for forcing could be increasing the yields of C_{8+} paraffins, while another might be increasing the C_2 to C_4 yields or, indeed the yield of light olefins. There are other more general objectives for applying composition modulation to the FTS: 1) improving the utilization of hydrogen, and 2) reducing methane production. Table 13.3 summarises the extensive data obtained by Khodadadi et al. (1996b). From this table, it can be seen that if the objective is to increase C_{8+} formation, cycling between syngas mixtures should be used but the best measured production rate for this strategy is 24.3 μmol/gcat·min while the optimal steady state rate is 40 μmol/gcat·min. Or, if yield as mol of C_{8+}/mol of CO is the objective, syngas cycling again is the best strategy for C_{8+}/CO yield, but the best yield is 0.13 vs the best steady state yield of 0.27.

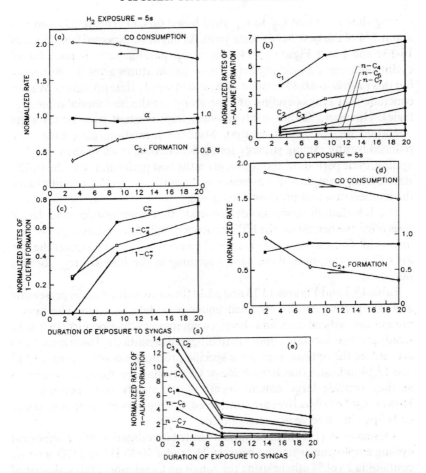

FIGURE 13.14 CO consumption, chain growth probability and paraffin formation rates for a) H_2-syngas cycling and b) CO-syngas cycling over a Co/SiO$_2$ catalyst at 483°C, 110 kPa and a space velocity of 10775 h^{-1} (STP). All data have been normalized with steady state measurements at the same conditions and for the time-average feed composition (Figure adapted from Khodadadi *et al.* (1996a) with permission of the authors)

Syngas cycling or H_2-pulsing of syngas results in the highest chain growth probability, but $\alpha = 0.64$ is less than the comparable steady state $\alpha = 0.70$. On the other hand, if the objective is to improve the C_{2+} production rate or yield, switching between syngas mixtures or H_2-pulsing continue to be the best strategies, even though the enhancement over the comparable steady state performance may be less than for other strategies studied. Thus, with syngas cycling, the C_{2+} formation rate is 123 μmol/gcat·min. However, the steady state rate is 140μmol/gcat·min. With respect to olefins, syngas-H_2

cycling shows the best $C_2^=$ to $C_4^=$ yield based on CO. This is 0.33, but it is about 3/4 of the best steady state yield. As might be expected from Figures 13-13c and f and Figure 13-14c, the olefin/paraffin ratio is poor for all cycling strategies. Switching between syngas mixtures gives the best ratio, but this ratio is just 0.39 versus a 1.10 ratio in steady state operation. We can conclude that syngas cycling or H_2-pulsing are the best modulation strategies. However, neither strategy offers improvement over steady state operation for the cobalt catalyst. Methane formation, also, cannot be reduced by any cycling strategy tested. Khodadadi et al. showed that fast cycling, cycle periods of about 10 s, gives the best performance so the steady state-cycling comparison was made under optimum cycling conditions for the temperature and pressure considered.

The Khodadadi study did demonstrate that composition modulation does offer one benefit for the FTS: hydrogen utilization can be significantly increased. This can be seen in Figure 13.13a and in Table 13.3 under the row for CO consumption. Bang-bang switching is the best strategy for this purpose.

Table 13.3 and Figures 13.13 and 13.14 illustrate a dilemma in presenting and interpreting results obtained under periodic operation. The figures present normalized data and therefore compare cycling performance with steady state performance under matching feed conditions. These conditions are seldom the optimal ones for a specific process objective. Figures 13.13 and 13.14 indicates that bang-bang or CO-cycling are attractive strategies as they provide large enhancements of the steady state performance. However, when objectives are considered, cycling between syngas mixtures or H_2-pulsing emerge as better choices.

Adesina et al. (1991b) undertook a limited investigation of 3 component cycling employing switching between H_2 and a 50:50 H_2 and CO mixture containing 2 vol.% ethene using the cobalt on kieselguhr catalyst described in Table 13.1 and the experimental system summarized in Table 13.2. Temperature was 200°C, pressure 111 kPa and the space velocity was 15,000 h^{-1}. Substantial enhancement of rates of formation was observed up to C_9 compared to steady state rates in a feed containing 1 vol.% $C_2^=$. These were as high as 100 %. In addition, olefin production in the C_3 to C_8 range was also enhanced by this cycling strategy. Two cycle splits were considered: $s = 0.5$ and 0.8. The latter represents H_2-pulsing. Improvements were found for both splits. Generally, higher cycling frequencies resulted in greater enhancement. Methane production also increased under modulation so at longer cycle periods, the ratio of higher hydrocarbon to C_1 was less than at steady state. However, at the highest forcing frequencies used (corresponding to a period of a minute), the selectivity ratio, that is, the rate of hydrocarbon production to the rate of C_1 production exceeded the ratio at steady state up to C_9 hydrocarbons while operating at $s = 0.8$. In this experiment, thus, both rate and selectivity increased through modulation.

Table 13.3 Comparison of the Best Performance Using Different Modulation Strategies

Operating Conditions or Performance Criteria	Bang-Bang		CO Pulsing		H_2 Pulsing		Mixture Cycling		Steady State	
	for CO conversion	for C_{8+} Olefins	for CO conversion	for C_{8+} Olefins	for CO conversion	for C_{8+} Olefins	for CO conversion	for C_{8+} Olefins	for CO conversion	for C_{8+} Olefins
CO-rich% CO	100	100	100	100	70	70	70	70		
Duration, s	2	5	5	5	3	20	9	15		
H_2-rich% CO	0	0	7	7	0	0	7	7		
Duration, s	5	2	20	2	5	5	20	2		
Time-average CO/H_2 (mole ratio)	0.40	2.50	0.34	2.76	0.36	1.27	0.36	1.67	0.40[#]	1.67[#]
CO Conversion (%)	51*	14	30	7	43	10	33	6	20	5
Co Conversion Rate, μmol/g cat·min	804*	490	389	225	535	293	384	188.1	260	150
C_{2+} Formation Rate, μmol/g cat·min	67.1	102.3	68.1	92.1	77.6	106.8	82.3	123.1*	160	140
C_{8+} Formation Rate, μmol/g cat·min	Nil	7.7	1.6	8.9	4.1	22.1	6.0	24.3*	16	40
α	0.33	0.57	0.448	0.564	0.513	0.635*	0.563	0.625*	0.48	0.70
C_1/CO Selectivity	0.718	0.507	0.728	0.381	0.783	0.594	0.751	0.311*	0.36	0.31
C_2/CO Selectivity	0.198	0.284*	0.097	0.210	0.072	0.042	0.034	0.034	0.016	0.02
C_8^*/CO Selectivity	Nil	0.002	0.004	0.040	0.008	0.075	0.016	0.129*	0.060	0.27
$C_2^=$/CO to $C_8^=$/CO Selectivity	0.008	0.016	0.053	0.192	0.061	0.333*	0.04	0.27	0.23	0.43
Olefin/Paraffin Mole Ratio	0.01	0.023	0.021	0.102	0.025	0.186	0.04	0.39	0.31	1.10

*best performance under cycling
[#]steady-state value in comparison to time-average value.

However, for symmetrical cycling, selectivity, measured as a production rate ratio was not augmented.

Explanation of the improvement according to Adesina et al. is that ethene initiates chains rather than adds to existing chains. Also the effect of ethene on hydrocarbon formation rate is strongly non-linear. Mechanisms are discussed in Adesina et al., 1990.

D. COMPARISONS

All the experiments discussed in the previous section show enhancement of catalyst activity with respect to the formation of at least one product through composition forcing. With an iron catalyst, this product was CH_4. Indeed, methane production is enhanced with most of the catalysts investigated. With other catalysts, rates of formation of the lighter paraffins, from C_2 to C_4, can be made to exceed steady-state maximum rates by an appropriate choice of period and/or split. Increased catalyst activity through periodic operation is evident from Figure 13.15 which plots the normalized rate of CO consumption against period for the cobalt and molybdenum catalyst. With symmetrical cycling, both the cobalt and the molybdenum data show that decreasing the cycle period enhances CO consumption in most cases. The exceptions are for cobalt and cycling between syngas mixtures where improvements appear at cycle periods of about 5000 s. Furthermore, for bang-bang cycling, some of the data suggest decreasing cycle period decreases enhancement. Just three such points are shown in Figure 13.15a. However, for the molybdate catalyst, bang-bang modulation of both CO and H_2 increases CO consumption as the cycle period diminishes. Decreasing the split appears to change the behaviour quite significantly. For the cobalt catalyst, there is enhancement over a wide range of cycle periods. However, enhancement is less under the bang-bang mode and even drops below unity for the shortest cycle period used. A resonance phenomenon is evident in Figure 13.15b. With a molybdate catalyst, enhancements are small and the presence of a promoter seems to influence the enhancement achieved for the cycle periods explored.

Enhancement of the rates of formation of the lower hydrocarbons can also be seen in Figure 13.16 which plots normalized rates for the lighter hydrocarbons against period. All normalization is with respect to the corresponding steady state as discussed in several places in the previous section. Unfortunately, just limited product distribution data were compiled by Gulari and co-workers, and these could not be converted into terms compatible with this figure.

Compared to the 8- to 10-fold increases in the rate of NH_3 formation under periodic operation (Rambeau and Amariglio, 1981b) discussed in Chapter 5, or the 20-to 40-fold increase in CO oxidation (Zhou et al., 1986)

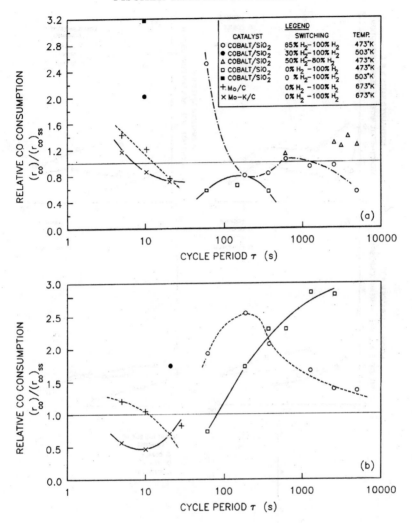

FIGURE 13.15 Normalized time-average rate of CO consumption as a function of cycle period for $s = 0.5$, temperature and pressure differ for each catalyst; (a) $s = 0.5$, (b) $s = 0.2$

shown in Chapter 3, the largest rate increases evident in Figures 13.15 and 13.16 are in the relatively modest range of 20 to 300%, depending on the catalyst. Examination of Figure 13.16 shows cycling effects diminish with increasing carbon number. Rate enhancements are largest for C_1, but for C_4 formation in (e) the enhancement has disappeared.

A striking feature of both figures is the differences in resonance periods for the different catalysts. The charcoal supported Mo catalyst is excited by

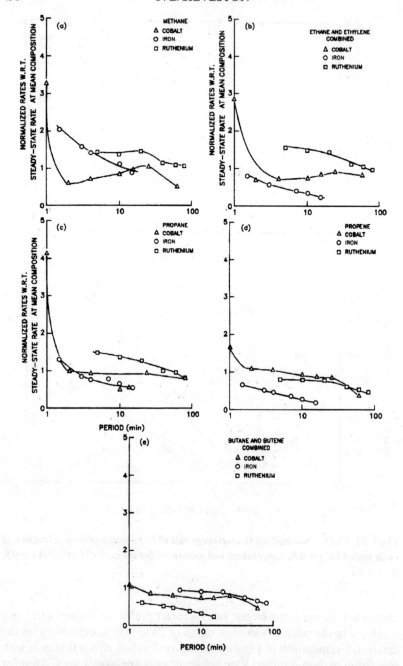

FIGURE 13.16 Normalized time-average rate of product formation as a function of cycle period. Split, temperature and pressure differ for each catalyst (Figure taken from Silveston *et al.* (1986) with permission, © 1986 Elsevier Science Publishers)

forcing in the 5-to 10-s range (Fig. 13.15). Ruthenium on Al_2O_3 appears to be excited by a wide range of periods from 10 s to about 40 min (Fig. 13.16); cobalt on silica exhibits resonance in two ranges of period: below 1 min and about 40 min.

The periods associated with resonance appear to be good indicators of the relaxation times of the rate controlling surface steps for hydrocarbon formation following an abrupt composition change. This relationship is borne out by Figure 13.17 which compares, for three of the catalysts, the instantaneous rates of product formation, normalized with respect to the steady state rate after a step change to pure H_2 occurs. C_3H_6 disappears from the iron within 10 min after H_2 is introduced, whereas with Co and Ru catalysts, about 20 min are required. On the other hand, methane is still being formed on the iron catalyst at about 20% of the steady-state rate 2 h after the switch to a pure H_2 feed. Similarly, small amounts of CH_4 are being formed on the Co and Ru catalysts 2 h after the switch.

Composition modulation changes the product distribution observed at steady state, but the changes are small or, as for iron, undesirable, since CH_4 formation is stimulated (Fig. 13.16). H_2-pulsing, in general, shifts the products towards lower carbon numbers. For Ru and Mo, and to a lesser extent the Fe and Co catalysts, the H_2-pulsing strategy decreases the olefin/paraffin ratio. This can be seen in Figure 13.16 for C_3H_8 and C_3H_6. Figure 13.18 plots the time-average rate of product formation vs carbon number. The time-average rate is measured at the period giving the maximum methane production in Figure 13.16 and normalized with respect to the steady-state rate at the corresponding time-average feed composition. A horizontal line means that the product distribution is not altered by forcing, whereas a line with a negative slope means that forcing shifts the distribution in favour of shorter-chain hydrocarbons. Thus, composition modulation with an iron catalyst does not alter the distribution of the C_{2+} hydrocarbons. A change towards shorter hydrocarbons is observed for the Co and Ru catalysts. For Ru catalyst, the bulge at a carbon number of 2 indicates ethane/ethene formation is strongly favoured over the higher hydrocarbons. On the other hand, Co catalyst seems to favour C_1 and C_2 and also C_6 to C_8.

E. MECHANISM CONJECTURES

The inability of H_2-pulsing to alter selectivity substantially and its success in raising the rates of formation for some hydrocarbons seems best explained for all catalysts by fast dissociative adsorption of CO to yield an active carbon. The active carbon causes rapid surface polymerization to create skeletal carbon of various chain lengths. These skeletal species are H_2-deficient and scour any adsorbed H_2 from the catalyst surface to which

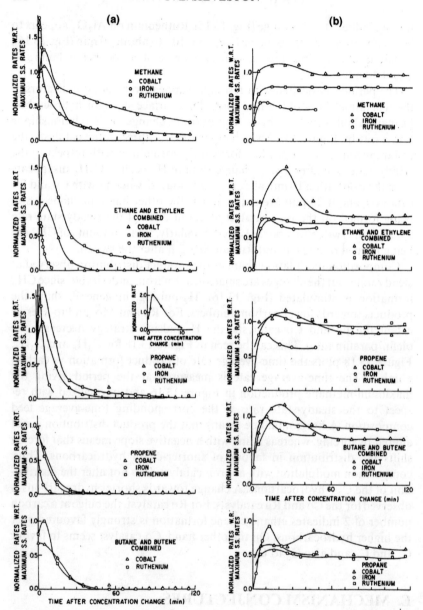

FIGURE 13.17 Normalized product concentration *vs.* time: (a) step-change from synthesis gas mixture to H_2, (b) step-change from H_2 to a synthesis gas mixture. Temperature and pressure differ for each catalyst.

they are strongly bonded. Introducing H_2 leads to hydrogenation of the skeletal species, weakening of the adsorptive bond, and eventual desorption. Chain growth does not occur during the H_2-pulse portion of a cycle. Thus, the appearance of different carbon-number products reflects what takes place in the portion of the cycle devoted to CO or syngas exposure. Since selectivity does change for Co, the polymerization step may be slower for this catalyst.

Evidence for a polymerization-hydrogenation sequence is the abrupt overshoot of synthesis products seen in Figure 13.17 when H_2 replaces a synthesis gas mixture. An overshoot of a particular hydrocarbon when the carbon source is withdrawn indicates a reservoir of reactive carbon must be present on the surface. Similarly, FTIR measurements on the honeycomb support discussed earlier (Barshad & Gulari, 1986), disclose that higher hydrocarbons form at the instant H_2 replaces synthesis gas or CO. The time-resolution of the FTIR-honeycomb system is of the order of a fraction of a second. Even with the poorer time-resolution of Figure 13.17a, it is evident that CH_4 and the C_{2+} appear together after the switch to H_2, and C_4's do not lag the C_2's. On the other hand, the transients for the C_2's to C_4's are much longer for the step from H_2 to the synthesis gas mixture (Fig. 13.15b). The CH_4 transients are different and will be considered below. Furthermore, there is a small lag between the combined $C_2H_6-C_2H_4$ and the combined $C_4H_{10}-C_4H_8$ maxima for the Fe and Ru catalysts. This lag is enlarged as the carbon number rises (data not shown). These results suggest it takes time for chain growth from carbon laid down by CO dissociation on a surface largely covered with adsorbed H_2.

The CH_4 transients for the three catalysts seen in Figure 13.17a are much longer than those for the switch from H_2 to synthesis gas in Figure 13.17b and are also longer than the transients for the higher hydrocarbons. This has been widely observed (e.g., Matsumoto and Bennett, 1978; Raupp and Delgass, 1979; Winslow and Bell, 1984; Zhou and Gulari 1987). Both Fe and Co catalysts appear to be capable of forming metal carbide on exposure to CO and Matsumoto and Bennett (1978) and Raupp and Delgass (1979) have thus proposed that the transients result from hydrogenation of the bulk carbide that requires H_2 diffusion into the catalyst. Because the FTS follows the ASF distribution for both Fe and Co, living-chain addition-polymerization must occur involving a C_1 species. This monomer should be capable of hydrogenation and the transients for this should be similar to the transients for the higher carbon number products. However, Figure 13.17 shows that this is not so. From modelling studies, Feimer et al. (1985b) conclude that there are at least two pathways to methane. For the Fe and Co catalysts, we propose that there are parallel routes to CH_4: (1) hydrogenation of the surface monomer, and (2) hydrogenation of a bulk carbide to methane. The route via carbide was excited by the relatively low H_2-pulsing frequencies used by Feimer et al. (1985a). This resulted in the large increase in CH_4 formation and the suppression of the higher hydrocarbons for the Fe catalyst that is evident in Figures 13.2, 13.3, 13.16 and 13.18. Stimulation of both C_1 and C_2 was observed for the Co catalyst. The explanation for this is that the composition-forcing experiments used a synthesis gas mixture with 25% CO, whereas the composition for the maximum steady-state rate of hydrocarbon formation is about 40%. Thus, probably just small amounts of carbide were formed in the H_2-pulsing experiment. We speculate that the resonance in the rate of C_1 formation at two different periods (Fig. 13.11) may be associated with the two routes for formation of this product.

Ruthenium does not form a carbide under the conditions used in the forcing studies. The long C_1 transient in Figure 13.17a is explained by the presence of a reservoir of inactive, possibly graphitic, carbon on the surface, which, in the presence of H_2 is converted into an active form (Winslow and Bell, 1984; 1985). We suggest that under the conditions of H_2-pulsing, the conversion of this inactive carbon occurs on a Ru surface flooded with H_2 so that hydrogenolysis dominates and most of the carbon desorbs from the surface as CH_4.

The FTS is excited for Ru over a surprisingly large range of frequencies as Figure 13.16 suggests. The explanation for this appears to be that the Ru surface is H_2-starved under steady-state operation. Zhou and Gulari (1987) observe that the synthesis is limited by chain termination (by H_2) on this catalyst. Furthermore, examining the effect of composition on the steady-state rate of hydrocarbon formation shows that the rate maximum occurs for a feed of about 95% H_2 (Ross et al., 1987), compared to 80 to 90% for Fe and about 60% for Co, indicating that CO is more strongly adsorbed than H_2 on the Ru surface. H_2-pulsing serves to increase the hydrogen adatom concentration on the surface by scavenging CO and some of the carbon reservoirs. A minimum exposure is necessary: about 20 s seems adequate from the Barshad and Gulari (1986) results in Figure 13.8. There appears to be a sufficient reservoir of carbon on the surface that H_2 exposures for up to 20 min still lead to rate improvement over steady state. This is a curious feature of this catalyst.

In Figures 13.16 and 13.18 the large increase in C_2 for the Ru and Co catalysts and in C_3 for the Co catalyst at $\tau = 1$ min appears to result from the hydrogenation of C_2H_4 and C_3H_6 to the corresponding paraffins through H_2-pulsing. The paraffin may be readsorbed but cannot participate in chain-building, whereas at least some of the C_2 and C_3 olefins are consumed in this way. Evidence for the hydrogenation explanation is the substantial shift in the olefin/paraffin ratio under composition forcing (Fig. 13.4).

With the Fe catalyst, CO_2 formation is enhanced under H_2-pulsing. Thus, it appears that this cycling mode excites the water gas shift reaction. Water is the primary reaction product in the FTS, but FT catalysts are often good shift catalysts; thus, the appearance of CO_2 in the product gases is

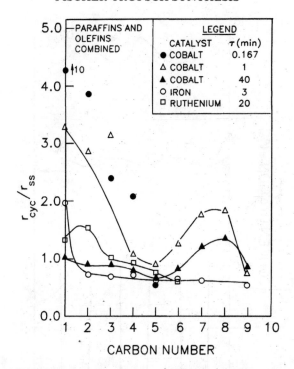

FIGURE 13.18 Normalized time-average product distribution. Conditions correspond to Figure 13.16 (Figure taken from Silveston *et al.* (1986) with permission, © 1986 Elsevier Science Publishers)

both expected and widely observed. The shift reaction appears to take place in several steps. Barshad and Gulari (1986) followed CO_2 formation by FTIR on the Ru catalyst they employed. Their observations, given in Figure 13.19, show that CO_2 and hydrocarbon formation are 180° out of phase. This result suggests a lag of about 6 s between the FTS reaction and the water gas shift.

F. CONCLUSIONS AND DIRECTIONS FOR FUTURE WORK

The experimental evidence demonstrates that composition forcing using an H_2-pulsing strategy can significantly increase catalyst activity and, for all the catalysts investigated to date, the proper choice of cycle period and split provides rates of formation for the lower carbon number paraffins which exceed the maximum rates attainable through steady-state operation at a specified temperature and pressure. For the Co catalyst, rates of formation of the lower olefins are enhanced as well. Product distributions are

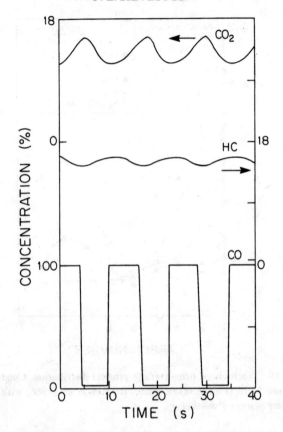

FIGURE 13.19 Variation of CO_2 and hydrocarbon concentrations during forcing with switching between CO and H_2 at $s = 0.5$, $\tau = 12$ s for Ru/Al_2O_3 (573 K, 1 bar) (Figure adapted from Barshad and Gulari (1986) with permission, © 1986 Gordon & Breach Publishers)

altered by composition forcing, but the relative change is smaller than for activity. The change is also catalyst-dependent. For Ru and Co, there is a reduction in the mean carbon number and a shift towards a higher paraffin content in the product. Also, there is a significant increase in the C_2 and C_3 products. Insufficient data were collected to establish product distribution changes for the Mo catalysts. The mean carbon number can be increased by composition forcing, but only under conditions that sharply decrease the rates of hydrocarbon formation. For the Fe catalyst, CH_4 formation is promoted but otherwise the product distribution is unchanged.

Selectivity is probably the most important consideration for FTS. Composition forcing so far has shown changes that are too small to elicit interest in applying composition modulation on a larger scale. Further exploratory research is needed. In particular, the H_2-pulsing strategy that has attracted the most study seems ineffectual because it forces up CH_4 production.

A CO-pulsing strategy should be examined along with further investigation of switching between different synthesis gas mixtures. For both the Fe and Co catalysts, any further study should use higher forcing frequencies and examine the effect of split on performance. For the Ru catalyst, a more thorough study of forcing frequency is needed to seek resonance effects that can strongly alter product distributions within the ranges explored by Barshad and Gulari (1986) and Ross *et al.* (1987).

CHAPTER 14

Polymerization Reactions

A. INTRODUCTION

The early simulations of the periodic operation of reactors by Douglas and Rippin (1966) and Douglas (1967) discussed in Chapter 1 were quickly followed up by researchers interested in polymerization. Ray (1968) and Laurence and Vasudevan (1968) discovered apparently simultaneously that the molecular weight distribution of polymers may be modified by periodically varying either the monomer, initiator, or transfer agent concentrations fed to a polymerization reactor. Since these studies, a number of further simulations have been undertaken and several experimental studies have been performed so that quite a bit is known about the effect of periodic operation on molecular weight distribution (MWD). Most simulations and experiments have manipulated the initiator or monomer concentrations. The activity in this research area has fallen off in the last 5 years, based on the number of studies appearing in the literature. Table 14.1 summarizes the published studies. It can be seen that a wide variety of polymerizations, mechanisms, reactor types have been examined. The types of forcing that have been considered and the wave forms used are also given in the table. The long chain material formed in polymerization can be described in different ways and several of these have been used in the modulation studies to date as Table 14.1 indicates.

Polymerization can proceed either homogeneously or heterogeneously. We will consider both cases in this chapter. Justification for including homogeneous polymerization is that an initiator is usually employed and/or the reaction is triggered by an external source of energy usually in the form of visible or UV radiation. The presence of a material active in the reaction along with the reactant which accelerates the reaction without (usually) being incorporated in the product makes homogeneous polymerization resemble a catalytic system. The industrially important emulsion polymerization is heterogeneous, but non-catalytic. However, the micelles play a key role in creating product properties and in this, emulsion polymerization is analogous to a catalytic process. Consequently, we will discuss this non-catalytic but heterogeneous case in this chapter. Light modulation has been used to investigate the mechanism and kinetics of free

Table 14.1 Literature Overview

Reference	Type of Study		Polymerization Type			Mechanism Chain Growth				Reactor Type			Concentration or Flow Rate of			Periodic Operation Forced Variable			Wave Form	MWB Defined Through
	Simulation	Experimental	Solution Homopolymerization	Other	Step Growth	Free-radicals	"Living" anionic	Cationic	Coordination	CSTR	Tubular	Batch	Initiator	Monomer	Transfer agent	Effluent flow rate in CSTRs	UV-or-radiation	Temperature	Square or sinusoid Pulses	Other	Averages and polydispersity	Complete curve
Roy (1968)	x		x		x					x			x								x	
Laurence and Vasudevan (1968)	x		x		x					x			x									
Claybaugh et al. (1969)		x	x			x						x									x	
Bandermann (1971)	x		x			x				x			x						x		x	
Lee & Bailey (1974)	x		x			x				x			x						x		x	
Konopnicki and Kuester (1974)	x		x							x			x						x		x	
Spitz et al. (1976)	x	x	x									x	x							x	x	
Hashimoto et al. (1976a)	x	x	x									x	x						x		x	x
Hashimoto et al. (1976b)	x	x	x									x	x						x		x	x
Meira et al. (1979)		x	x				x			x						x			x		x	x
Klohn & Bandermann (1979)	x			x			x			x						x			x		x	
Crone and Renken (1979)	x				x					x				x							x	
Marconi and Pagni (1979)		x		x		x						x	x				x				x	
Chen et al. (1980)	x		x								x						x				x	x
Meira & Johnson (1981)	x		x				x			x						x			x		x	
Couso & Meira (1984)	x		x			x				x			x					x	x		x	x
Thiele (1984)	x	x	x			x				x			x					x	x		x	x
Frontini et al. (1986, 1987)	x	x	x				x			x					x	x					x	x
Couso et al. (1985)	x	x	x				x			x					x	x			x		x	x
Gugliotta & Meira (1988)	x	x	x						x			x	x					x	x		x	x
Gupta et al. (1985)	x	x	x		x							x	x								x	
Elicabe and Meira (1989)	x	x	x			x				x			x						x		x	x
Hungenberg (1992)	x			x					x			x	x								x	
Gosden et al. (1995)	x	x	x			x				x			x	x						x	x	
Vega et al. (1991)	x		x			x				x			x								x	
Gosden et al. (1996)		x	x			x				x			x					x	x		x	x

radical polymerization. Because our interest in this monograph is improvement of reactor performance through composition modulation, this application of modulation will not be examined here. It will be considered briefly in Chapter 15.

Description of MWD poses some problems. Although moments of the number chain length distribution are calculated directly in most simulations and describe the distribution adequately, these moments are not used in practice. The number average chain length, μ_n, or molecular weights, M_n, M_w, and the polydispersity, D, are normally used instead. Polydispersity is the ratio of the weight average and number average molecular weights, $D = M_w/M_n$.

The subject of this chapter was treated in a 1981 review (Meira, 1981) which covered more than half the publications now in print. With this in mind, our discussion in this chapter up dates this important contribution.

B. SIMULATION OF POLYMERIZATION UNDER FLOW OR COMPOSITION FORCING

Much of the research on the application of periodic operation to polymerization has employed mathematical models of polymerization dynamics and predicted reactor behaviour through simulation. Such studies use part or all of the set of elementary reactions shown in Table 14.2 for "living" chain (anionic) and free radical homopolymerization. These two mechanisms, employing only monomer addition, have been investigated most often. The table does not include step polymerization where the monomer has two functional groups capable of participating in chain enlargement. The elementary steps, essentially, are those shown in Table 14.2. Complications arise in formulating the rate terms (Table 14.3). Heterogeneously catalyzed polymerization can follow either a "living" chain or free radical mechanism. Like emulsion polymerization, it involves a second phase so that transport steps must be added to a reactor model.

In all mechanisms, polymers with different chain lengths will be present in a back mixed reactor. Material balances and, if temperature is manipulated, a heat balance lead to a large set of simultaneous, ordinary differential equations as Table 14.3 illustrates for the free radical mechanism (neglecting the use of chain transfer agents). The calculational problem of solving a large number of equations is avoided by introducing population balances which replace polymer concentrations of different chain length by moments of the polymer number chain length distribution. Moment generating functions or the z-transform (Table 14.4) can be used to provide a set of equations for moments of the polymer number chain length distribution. Table 14.5 gives these equations for the simplified free radical model given in Table 14.3. Details of the transform technique are given by Ray (1967,

1972). The number average chain length and the polydispersity may be calculated directly from the moments:

$$\mu_n = \sigma_1/\sigma_o \tag{14.1}$$

$$D = \sigma_o\sigma_2/(\sigma_1)^2 \tag{14.2}$$

Simulation studies listed in Table 14.1 have been concerned with the influence of periodic operation on the number average chain length, μ_n, the monomer conversion, X_m and the polydispersity, D. Chain length and monomer conversion, as in any chemical reaction, depends on residence time, operating conditions, contacting pattern in the reactor and, of course, on the polymerization mechanism. Increasing residence time, raising monomer concentration, introducing plug flow characteristics, for example, all increase μ_n and X_m. These parameters of polymerization reactions are limited only by viscosity of the monomer-polymer solution. Increasing viscosity as polymerization proceeds leads to switching of rate control from kinetic to diffusional processes and, ultimately, gel or cage effects appear. With polydispersity, however, limited values may be observed even at low M_n where viscosity and gel phenomena are absent. For example, with "living" anionic polymerization in batch or plug flow reactors, D will be unity. A polydispersity of 2, indicating a Schultz-Flory distribution of polymer chain lengths, arises with step growth polymerization in a plug flow or batch reactor, with "living" anionic polymerization in a fully back mixed reactor, or with free radical polymerization in the latter reactor when termination occurs only through disproportionation (and in the absence of high viscosity or gel formation). D becomes 1.5, however, if recombination is the only termination mechanism. In all these cases, steady state operation has been assumed.

With any given co- or homopolymer, physical properties and thus value of the polymer depends on μ_n and D. The monomer conversion, X_m, of course, influences process economics. Perhaps because of the economic significance as well as the limits on D just discussed, investigators of periodic operation have focused on the polydispersity.

The influence of mechanism on D makes it desirable to discuss simulation results for the studies listed in Table 14.1 separately. Because experimental work is quite limited, it will be examined after discussion of the simulation results. Simulation contributions are grouped under polymerization type or mechanism and then subdivided to bring papers considering the same or similar operating conditions together. As a consequence, chronology is not maintained.

C. FREE RADICAL POLYMERIZATION

Examples of polymerizations proceeding via this mechanism are the production of polymethylmethacrylate or polystyrene. These are industrially important processes.

Table 14.2 Mechanistic Models for Polymerization Reactions

Living, Anionic Polymerization

Initiation Step:
$$I + M \xrightarrow{k_1} P_1(+I) \tag{14.3}$$

Propagation Step:
$$P_1 + M \xrightarrow{k_p} P_2 \tag{14.4}$$
$$P_n + M \xrightarrow{k_p} P_{n+1} \tag{14.5}$$

Free Radical, Homo Polymerization

Initiation Steps:
$$I \xrightarrow{k_1} 2R_c^* \tag{14.6}$$
$$R_c^* + M \xrightarrow{k_2} R_1^* \tag{14.7}$$

or
$$I + 2M \xrightarrow{k_i} 2R_1^*$$

Propagation Steps:
$$R_1^* + M \xrightarrow{k_p} R_2^* \tag{14.8}$$
$$R_n^* + M \xrightarrow{k_p} R_{n+1}^* \quad n > 1 \tag{14.9}$$

Termination Steps:
$$R_n^* + R_m^* \xrightarrow{k_{td}} P_n + P_m \tag{14.10}$$
$$R_n^* + R_m^* \xrightarrow{k_t} P_{n+m} \tag{14.11}$$

Transfer Steps:
$$R_n^* + M \xrightarrow{k_{fm}} P_n + R_1^* \tag{14.12}$$
$$R_n^* + S \xrightarrow{k_{fs}} P_n + S^* \tag{14.13}$$
$$R_n^* + T \xrightarrow{k_{ft}} P_n + T^* \tag{14.14}$$
$$S^* + M \xrightarrow{k_{sm}} R_1^* + S \tag{14.15}$$

Where I = initiator, R_c = initiator radical, M = monomer, R_1^* = activated monomer or unit chain length polymer, R_n^*, R_m^* = polymer radical of chain length n, m; P_n, P_m = polymer of chain length n, m; S, S* = solvent and solvent radical, T, T* = transfer agent and transfer agent radical.

Simulation

For this mechanism, the earliest observations were that periodic, usually sinusoidal, variation of initiator or monomer feed rate increases the polydispersity with little change in n or X_m. Ray (1968) and Laurence and Vasudevan (1968) found that the molecular weight distribution for free radical polymerization in an isothermal CSTR broadens as cycle period increases so that the maximum D occurs under quasi steady state operation.

POLYMERIZATION REACTIONS

Table 14.3 Material and Heat Balances for a Simplified Free Radical Polymerization Mechanism [Table adapted from Konopnicki and Kuester (1974) with permission, © 1974 Marcel Dekker Inc.]

Component mass balances:*^{t}

$$dI/dt = I°/\theta - k_i I - I/\theta \tag{14.16}$$

$$dM/dt = M°/\theta - M/\theta - k_p MR_c - 2k_i fI - k_{fm}MR_c \tag{14.17}$$

$$dR_1/dt = -R_1/\theta + 2k_i fI - k_p R_1 M - k_t R_1 R_c + k_{fs} SR_c$$
$$+ k_{fm}MR_c - k_{fs}R_1 S - k_{fm}MR_1 \tag{14.18}$$

$$dR_n/dt = -R_n/\theta - k_p R_n M + k_p R_{n-1}M - k_t R_n R_c - k_{fs}SR_n - k_{fn}R_n M \tag{14.19}$$

$$dP_n/dt = -P_n/\theta + \tfrac{1}{2} k_t \sum_{n=1}^{n-1} R_{n-m} + k_{fs}R_n S + k_{fm}MR_n \tag{14.20}$$

where n, m = 1, 2, 3, ⋯, ∞

Energy balance:

$$dT/dt = [v_T(S°C_{p,s} + M°C_{p,m})(T° - T) - UA(T - T_j)$$
$$- V\Delta H_{rxn,p} k_p M \sum_{n=1}^{\infty} R_n]/[v_T(S°C_{p,s} + M°C_{p,m})] \tag{14.21}$$

---

* Assumes chain transfer agent not present
^{t} CSTR assumed; other assumptions are standard. See Konopnicki and Kuester (1974) for details

Table 14.4 Transformations used to Convert Polymer Mass Balances to Moment Form [Table adapted from Konopnicki and Kuester (1974) with permission, © 1974 Marcel Dekker Inc.]

Z-Transform:	$\bar{R}(z,t) = \sum_{n=1}^{\infty} R_n(t)z^{-n}$	(14.22)
Moment Generation Theorem:	$\lambda_k = [-z\, d/dz]^k \bar{R}(z,t)\|_{z=1}$	(14.23)
Shift Theorem:	$z[R(n-k)] = z^{-k}\bar{R}(z)$	(14.24)
Real Convolution Theorem:	$z\left[\sum_{n=1}^{k} R_1(n) R_2(n-k)\right] = \bar{R}_1(z)\bar{R}_2(z)$	(14.25)

Experimental values of rate constants and realistic concentrations, flow rates and reactor volume were used by these authors. Imperfect mixing was treated by Lee and Bailey (1974) who found that this broadened the MWD and increased the polydispersity for both periodic and steady state operation.

Couso et al. (1987) also assumed an isothermal CSTR but allowed for transfer agent variation as well as switching of monomer and initiator concentrations. They observed that D could be increased without changing n with monomer or transfer agent concentration switching, however, this was not possible with sinusoidal variation of the initiator concentration. Results obtained were similar to those of earlier workers. Variations of the transfer agent concentration gave the largest increases in D. These reached 3 to 4 times the steady state value in quasi steady state operation. Simultaneous variation of any two of the three manipulated feed concentrations, 180° out phase, always increased D. A "relaxed" steady state was not observed as $f \to \infty$.

Konopnicki and Kuester (1974) modelled styrene polymerization in a toluene solution with azobisisobutyronitrile as initiator. A heat balance was added to the model so that nonisothermal operation could be considered. The effect of viscosity on the termination rate constant and the initiator efficiency were also added to the model. Viscosity was treated as a function of temperature and polymer weight fraction. The simulation allowed for sinusoidal and square wave variation of initiator feed concentration and cooling jacket temperature separately and simultaneously. Amplitude and phase lag was considered but not comprehensively. Figure 14.1a shows the

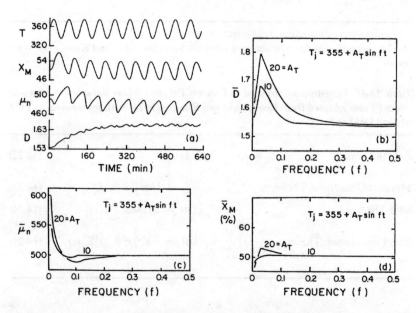

FIGURE 14.1 Simulated behaviour of styrene polymerization with sinusoidal forcing of cooling jacket temperature (T_j), (a) start up behaviour and stationary cycling response to forcing of T_j, (b) effect of cycling frequency on number average chain length (μ_n), (c) polydispersity (D), (d) monomer conversion (X_m) (Figure adapted from Konopnicki and Kuester (1974) with permission, © 1974 *J. Macromol. Sci. Chem.*)

simulation of jacket temperature variation around a mean of 355°C for a 60 minute period. The effect of frequency and amplitude on D, μ_n and X_m may be seen in the other parts of the figure. A resonance phenomena arises at low frequencies with D. It is also present for μ_n (which decreases) and conversion. Similar results were obtained for variation of the initiator feed concentration and when both variables were cycled. Square wave manipulations gave greater changes than sinusoidal variations and varying both initiator concentration and jacket temperature produced the largest increase in D. Only one phase lag was examined and its effect on D was negligible. Tables 14.3 shows the model used by Konopnicki and Kuester while Table 14.5 summarize the moment equation numerically solved by these investigator.

Thiele (1984) modelled the co-polymerization of styrene–acrylonitrile and simulated periodic variation of the cooling jacket temperature. An amplitude dependent increase in D of about 10% was found. Experimental observations for the conditions simulated showed about a 20% increase in D at an amplitude of 10°C, whereas a 5% decrease was seen at an amplitude of 2.5°C. The simulation indicated no change in D.

Experimental

The free radical polymerization of styrene in a jacketed CSTR using an on-off variation of the azobisisobutyronitrile (AIBN) initiator was investigated by Spitz et al. (1976). Measurements were made at 110, 126 and 136°C so as to consider different initiator half lives. Monomer conversions were in the 40 to 60% range. Thus, the experiments were performed in a range which is of commercial interest. Refractometry and size exclusion gel chromatography were used to assess monomer conversion and the MWD. To analyze the results, the ratios of mean residence time ($\bar{t}$) to period (τ) and initiator half life ($t_{1/2}$) were useful. The former indicates the damping occurring in the CSTR, while the latter indicated the variation of initiator concentration during a cycle. Thus, undesirable oscillating monomer conversions and number average chain lengths, μ_n, were observed only if $\bar{t}/\tau < 2.5$. Broadening of the MWD occurred when $\tau/t_{1/2} > 5$. This is indicated by the polydispersity, D, given as a function of τ in Table 14.6. Two sets of experiments are shown and D increases with τ. Large increases in D are observed when $\tau/t_{1/2}$ exceeds 4.3 in both data sets. Figure 14.2 shows the experimental MWD broadening under well and poorly damped conditions for the second set in the table. D increases through flow modulation, but more strongly under poor damping, i.e., when $\theta/\tau < 2.6$. Thus, cycle period has a strong effect as the simulations predicted. The curves at $\tau = 40$ in Figure 14.2b are samples taken at different times in the cycle and indicate an important variation of polymer properties with time. The entry at $\tau = 40$ min in Table 14.6 illustrate the variations. Although the

Table 14.5 Moment Equations from Simplified Free Radical Polymerization Model given in Table 14-3 [Table adapted from Konopnicki and Kuester (1974) with permission, © 1974 Marcel Dekker Inc.]

Live Polymer Moments:

$$d\lambda_0/dt = -\lambda_0/\theta + 2k_i fI - k_t\lambda_0^2 \qquad (14.26)$$

$$d\lambda_1/dt = \lambda_1/\theta + 2k_i fI + k_p M\lambda_o - k_t\lambda_o\lambda_1 - k_{fs}S\lambda_1$$
$$- k_{fm}M\lambda_1 + k_{fs}S\lambda_o + k_{fm}M\lambda_o \qquad (14.27)$$

$$d\lambda_2/dt = \lambda_1/\theta - \lambda_2/\theta + 2k_i fI + 2k_p M\lambda_1 + k_p M\lambda_o$$
$$+ k_t\lambda_o\lambda_1 - k_t\lambda_o\lambda_2 + k_{fs}S\lambda_1 - k_{fs}S\lambda_2 + k_{fm}M\lambda_1$$
$$- k_{fm}M\lambda_2 + k_{fs}S\lambda_o + k_{fm}M\lambda_o \qquad (14.28)$$

Dead moments:

$$d\mu_o/dt = -\mu_o/\theta + \lambda_o[\tfrac{1}{2}k_t\lambda_o + k_{fs}S + k_{fm}M] \qquad (14.29)$$

$$d\mu_1/dt = -\mu_1/\theta + \lambda_1[k_t\lambda_0 + k_{fs}S + k_{fm}M] \qquad (14.30)$$

$$d\mu_2/dt = -\mu_2/\theta + \mu_1/\theta - \lambda_1[k_t\lambda_o - k_t\lambda_1 + k_{fs}S$$
$$+ k_{fm}M] + \lambda_2[k_t\lambda_o + k_{fs}S + k_{fm}M] \qquad (14.31)$$

Conversion of Heat Balance:

$$\sum_{n=1}^{\infty} R_n = \lambda_o \qquad (14.32)$$

Total moments:

$$\sigma_o = \lambda_o + \mu_o \qquad (14.33)$$
$$\sigma_1 = \lambda_1 + \mu_1 \qquad (14.34)$$
$$\sigma_2 = \lambda_2 + \mu_2 \qquad (14.35)$$

Pseudo steady state assumption:

$$d\lambda_o/dt = d\lambda_1/dt = d\lambda_2/dt = 0 \qquad (14.36)$$

forcing of initiator feed to the polymerization tank can alter polydispersity significantly, Table 14.6 shows that polymer molecular weight, M_w, and monomer conversion, $X_{monomer}$, decrease as the cycle period augments.

The physical explanation for the effect of τ on D, the sensitivity to damping and the importance of the cycle period to initiator half life ratio is that the molecular weight of the polymer is controlled by the termination reactions and is sensitive as well to the initiator concentration. Thus, to obtain a broad MWD, the initiator concentration must change dramatically during both a cycle and the residence time of a growing polymer chain in the back mixed reactor. This requires large ratios of $\tau/t_{1/2}$ and $\bar{t}/t_{1/2}$. It

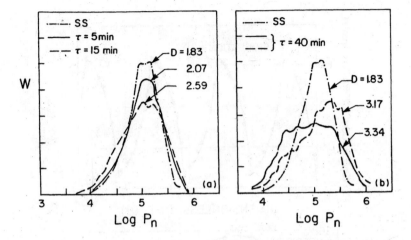

FIGURE 14.2 Influence of cycle period on the polystyrene molecular weight distribution for symmetrical on-off AIBN initiator flow rate variation for $T = 126°$: a) $\bar{t}/\tau = 8.26$ (Damping), $\tau/t_{1/2} = 10.30$, b) $\bar{t}/\tau = 1.0$, $\tau/t_{1/2} = 80$ (Figure adapted from Spitz et al. (1976) with permission, © 1976 American Chemical Society)

Table 14.6 Effect of Initiator Modulation on Polystyrene Properties*

Experimental Set	Cycle Period (τ) min.	$\tau/t_{1/2}$	$\tau/\bar{t}$	$M_N \times 10^4$	D	$X_{monomer}$
2	1.25	0.543	32	4.88	1.74	0.54
	5	2.17	8	4.94	1.76	0.51
	10	4.30	4	4.36	1.95	0.47
	16.7	7.25	2.4	3.68	2.20	0.42
	Steady State	–	–	5.06	1.68	0.55
3	5	10	8	7.44	2.07	0.45
	15	30	2.6	5.60	2.59	0.42
	40	80	1.0	4.13 to 5.90	3.17 to 3.34	0.30 to 0.42
	Steady State	–	–	7.38	1.83	0.48

*Adapted from Spitz et al. (1976) with permission, © 1976 AIChE

suggests then that $\bar{t}/\tau \geqslant 1.0$. A limit on $\bar{t}/\tau$ is imposed by mixing; ratios below 2.6, Spitz et al. suggest, lead to observable time variations in polymer proerties which would require post reactor mixing to smooth out.

Crone and Renken (1979) worked with a similar styrene polymerization system, but used higher initiator concentrations and a lower temperature (80°C). They simultaneously varied initiator and styrene concentrations in the CSTR feed (0° phase lag) by increasing or decreasing the toluene solvent. Although square wave forcing in the feed was used, Figure 14.3a

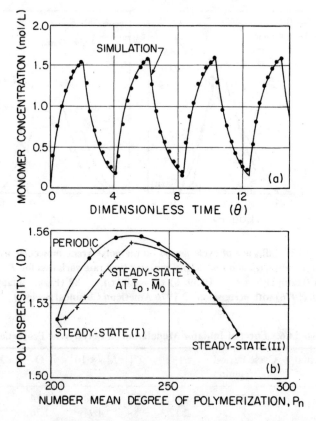

FIGURE 14.3 Influence of periodic operation on polydispersity for styrene polymerization in an isothermal CSTR with in phase, square wave cycling of AIBN initiator and styrene monomer concentration in the reactor feed: (a) styrene concentration leaving the reactor for $\bar{t} = 4.1$ hrs., (b) polydispersity as a function of number average polystyrene chain length. $T = 80°C$, $\bar{C}_i = 0.005$ mol/L, $\bar{C}_1 = 2.1$ mol/L. toluene solvent (Figure adapted from Crone and Renken (1979) with permission; © 1979 Verlag Chemie GmbH)

shows that mixing and reaction distort the monomer concentration time pattern in the reactor. Nevertheless, broadening of the MWD was obtained as the D values in Figure 14.3b demonstrate. These investigators observed that the space time yield of polymer could be increased by up to 44% above that obtained at steady state through periodic operation. D and μ_n were unchanged, but this required doubling the mean residence time.

Simultaneous sinusoidal and square wave variation of AIBN initiator and styrene monomer concentrations in the feed to a CSTR were also studied by Meira et al. (1979). These investigators switched the concentration out of phase (180° lag). Attention was paid to isothermal operation and maintaining a constant mean residence time that matched the time used in steady state experiments. They observed that 4 to 5 cycles were needed to attain a stationary cyclic invariant state. Periodic variation of concentrations led to higher polydispersities and number average molecular weights as illustrated in Figure 14.4.

Free radical co-polymerization of styrene and acrylonitrile was studied by Thiele (1984) in the gel region using a helix stirrer equipped, water jacketed 5L reactor. Among the various dynamic experiments performed were two which employed square wave cycling of the reactor jacket temperature at $\tau = 1$ hr inducing a 2° to 20° oscillation of the reactor temperature. With the larger amplitude, the polydispersity increased from 2 to 2.1, whereas with a 1° amplitude, a decrease in D of 0.05 units was found.

Hashimoto et al. (1976b) examined the X-ray induced, free radical homo polymerization of methyl methacrylate in a 50 mL gas stirred reactor equipped with a moveable shield so that the X-ray dosage could be switched between 153 and 5.3 rad/s. The objective of this study and a prior one using simulation (Hashimoto et al., 1976a) were to study reactor control by manipulating irradiation. If the process is undisturbed externally, square wave variation of irradiation caused oscillating initiator concentrations and temperatures in the well stirred reactor as illustrated in Figure 14.5. The figure shows square wave variation of irradiation and the resulting irregular fluctuation of temperature and the MWD broadening that result. The a) portion of the figure shows that the pulse width of irradiation depends on the flow rate through the reactor. Broadening of the MWD is clearly seen when square wave irradiation is used. Other results obtained by Hashimoto

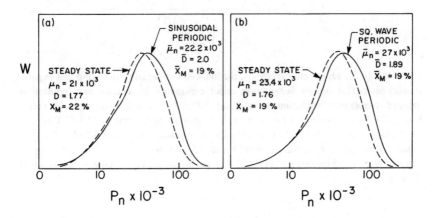

FIGURE 14.4 Effect on polystyrene molecular weight distribution of simultaneously varying AIBN and styrene monomer concentration 180° out of phase: (a) sinusoidal forcing, (b) square wave forcing of the feed concentrations (Figure adapted from Meira et al. (1979) with permission, © 1979 American Chemical Society)

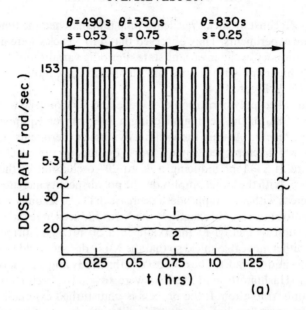

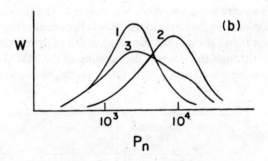

FIGURE 14.5 Dose rate variations, the induced temperature variation and experimental molecular weight distribution under continuous irradiation and irradiation induced temperature variation: a) set point = 23.5°C. Curve 1: reactor temperature; Curve 2: inlet temperature of monomer; b) reaction temperature = 52.5°C. Curve 1: continuous irradiation at 153 rd/s; Curve 2: continuous irradiation at 5.3 rd/s; Curve 3: temperature varied by alternating 153 and 5.3 rd/s irradiation ($s = 0.32$) (Figure adapted from Hashimoto et al. (1976b) with permission; © 1976 American Chemical Society)

and co-workers showed MWD broadening depended on periodicity and pulse length relative to period.

A quite different irradiation experiment was undertaken by Chen et al. (1980) using styrene polymerization initiated by the photo dissociation of benzoin methyl ether. Control was through a thermocouple held in the helix stirred, glass reactor which operated a shutter on the UV light source. Chen's experimental system is shown in Figure 14.6. With this system, the researchers were able to operate their reactor at a higher temperature, metastable state which narrowed the MWD significantly and raised μ_n and X_m by a factor of about 6 ×. This result is illustrated in Figure 14.7.

D. ANIONIC 'LIVING' CHAIN POLYMERIZATION

Anionic polymerization is also industrially important and, because of its simplicity, it has been the most frequent topic of simulation studies. Examples of anionic, living chain polymerization are the synthetic rubbers: styrene-butadiene copolymers, and polyisoprene.

Simulation

Models for this type of polymerization are much simpler than the free radical model given in Table 14.3. In the ideal case there are no termination or transfer reactions so terms containing the parameters k_t, k_{fs} and k_{fm} vanish. There is no dead polymer so the material balances (the set of

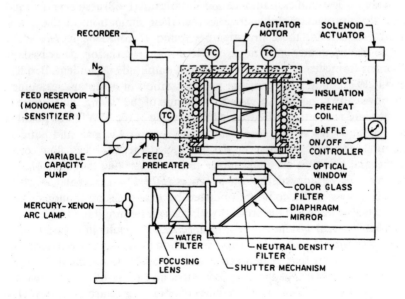

FIGURE 14.6 Experimental unit for temperature controlled, periodic UV irradiation of a stirred styrene polymerization reactor (Figure adapted from Chen *et al.* (1980) with permission, © 1980 AIChE)

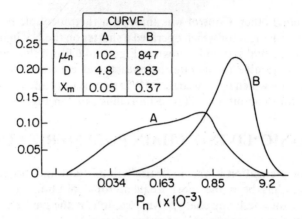

FIGURE 14.7 Molecular weight distribution of polystyrene produced using periodic irradiation to stabilize a metastable reactor operating state A: low temperature stable state at 340 K, B: metastable state at 440°C for well mixed reactor with $t = 200$ s, $(C_M^\circ) = 8.66$ mol/L, $(C_I^\circ) = 1.5 \times 10^{-5}$ mol/L (Figure taken from Chen et al. (1980) with permission, © 1980 AIChE)

P equations in the table) also drop out. The moment equations too are reduced as $\sigma_n = \lambda_n$.

In the earliest studies, Laurence and Vasudevan (1968) used experimental data for caprolactam polymerization for their simulation. In the quasi-steady state limit, these investigators found D could be increased by sinusoidal monomer variations, while initiator variation decreased D. Changes from steady state were small and amplitude dependent. Bandermann (1971), modelling isoprene polymerization in n-heptane, examined the physically impossible sinusoidal variation of the "living" ends concentration and found that broadening or shrinking of the MWD was possible depending on the steady state mean concentration. Langner and Bandermann (1978) used the same model but cycled the total feed flow to an isothermal CSTR in an on-off mode and demonstrated that the polydispersity could be reduced from 2 to as low as 1.375 ($D = 1$ in a batch reactor). Values as low as 1.2 could be obtained by manipulating monomer and initiator flow rates and the product withdrawal rate simultaneously. This is a rather unusual system as the flow modulation results in a variation of reactor volume. An extension of the simulation by Klahn and Bandermann (1979) concluded that D between 1 and ∞ could be obtained by using a symmetric forcing of the feed flow rates and pulse-like withdrawal of product from the CSTR. The possibility of obtaining an arbitrary MWD in a tubular reactor (with axial mixing) was examined theoretically and experimentally by Meira and Johnson (1981). Constant initiator flow was assumed and the monomer flow profile was varied to achieve the desired

MWD. The operation is periodic, but the monomer flow profile within a cycle takes on ramp or ramp plus step discontinuities.

Couso and Meira (1984) returned to the isothermal CSTR and considered independent and simultaneous sinusoidal and square wave variation of initiator, monomer and solvent flow rates. The latter were used to hold the mean residence time constant. These investigators found that large values of $D(>10)$ can be achieved by operating at quasi-steady state with square wave monomer flow rate variation (constant initiator flow) or by varying both monomer and initiator flow rates 180° out of phase. They also observed that D as low as about 1.7 is possible at intermediate frequencies, but μ_n exhibits a maxima which greatly exceeds its steady state value (3-4 fold). High frequency cycling of monomer concentration leaves D and μ_n at steady state values, but increases monomer conversions.

Building on the Bandermann and Meira work, Frontini et al. (1986, 1987) considered periodic operation from the stand point of an optimal control problem in which the manipulated variables are initiator, monomer, and transfer agent feed concentrations or flow rates and the objective is to achieve a desired μ_n, D and production rate. Cycle periods and concentration variations for initiator and monomer that are irregular functions of dimensionless time within a cycle (t/τ) were found which would raise D from 2 to 4 or decrease the polydispersity to 1.58 while maintaining a constant chain length and monomer conversion (Frontini et al., (1987)). Elicabe and Meira (1988) in their review of control of polymerization reactors point out that polymerization is too complex for excitation by simple composition forcing, but they nevertheless conclude periodic operation is a valid control strategy. Optimal periodic control, however, will require complex, unsymmetric cycles such as those proposed by Frontini et al. (1987) and illustrated in Figure 14.8. If reactive impurities are present in the feed to the reactor, however, the wide range of polydispersity possible by resorting to modulation is sharply reduced and the cyclic pattern of monomer or initiator flow variation for the impurity free feed is no longer optimal (Alassia et al., 1988a,b).

The observations of Alassia et al. led Meira and co-workers to reformulate the optimal control problem considered in the previous paragraph. Vega et al. (1991) introduce a poisoning reaction to form "dead" polymer, Q,

$$K + P_n \longrightarrow Q_n \quad (14.37)$$

to the description of "living" chain polymerization given in Table 14.2. Impurities also consume initiator to yield an inactive material, IK,

$$K + I \longrightarrow IK \quad (14.38)$$

Material balances for the impurities and for dead polymer Q_n with chain length from 1 to ∞ must be introduced into the model, and terms must be

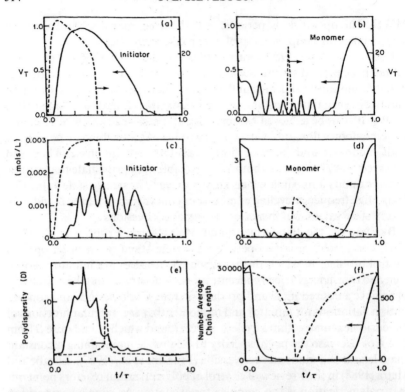

FIGURE 14.8 Optimal initiator and monomer flow policies to maximize polydispersity and resultant reactor concentrations and outputs for simulation of the polymerization of isoprene in n-heptane with a n-butyllithium initiator in an isothermal CSTR ($V = 0.9$ dm) at 25°C with $k_p = 4284$ dm^3/gmol h, $k_i = 21.4$ dm^3/gmol h, $M° = 6$ gmol/dm^3, $I° = 0.003$ gmol/dm^3, $v_M = v_I = 0.5$ dm^3/h: (a) $v_I(t)$, (b) $v_M(t)$, (c) I(t), (d) M(t), (e) D, (f) M_n (Figure adapted from Frontini et al. (1987) with permission, © 1987 John Wiley & Sons)

added to the balances written for I, P_1 and P_n to account for loss of initiator and the conversion of "living" polymer to "dead" polymer. This results in a model which resembles loosely the free radical polymerization model given in Table 14.3, although the initiation term is different and there is only one term for conversion of polymer rather than two. The moment form for this expanded mechanism is given by Vega et al. (1991). It is structured like the set of equations in Table 14.5, but has fewer terms and these are different as the reactions of impurities are not the same as termination and chain transfer steps. Equations (14.1) and (14.2) apply but the moments are the sums of moments for the "living" and "dead" chains. For this modified optimal control problem, Vega et al. show that for steady state operation in an isothermal CSTR, the MWD follows the Schultz-Flory model so $D = 2$.

Results for periodic cycling of feed flows depend on the kinetics of the polymerization. Vega *et al.* investigated the n-butyllithium initiated polymerization of isoprene in a *n*-heptane solution. As might be expected, the presence of impurities at just half the initiator concentration in a CSTR restrict monomer conversion, chain length and polydispersity. For example, if the target conversion and chain length at steady state are achieved, a D of 2.85 can be attained by an optimal periodic flow cycling policy in the presence of impurities, whereas in their absence, $D = 4$ with an optimal, but different, policy. As Frontini *et al.* (1987) show, an optimal flow cycling policy can be found which will reduce D. Vega *et al.*(1991) find that the policy changes little in the presence of impurities and a D of 1.5 can be obtained.

A solution technique for periodic operation of a "living" chain polymerization reactor which does not convert polymer material balances to moments as been published by Gosden *et al.* (1995). It uses the model of Couso and Meira (1984) but adds the restriction that the space velocity in the reactor remains constant. The solvent flow is manipulated to achieve this. Gosden *et al.* note that the mean chain length or polymer molecular weight depends on the residence time in the reactor or the "age" of the polymer. Use of a CSTR rather than a PFR means that although the average chain length may be the same, the polymer from a CSTR will exhibit a Poisson distribution of chain lengths. The distribution is actually a convolution of two Poisson distributions, one of which is in the means of the chain length because of the different ages of polymer leaving the CSTR. Indeed, the method of Gosden *et al.* employs the evaluation of convolution integrals which they claim can be done much faster than the numerical integration of a system of moment equations. From the "age" and chain length distributions, Gosden *et al.* find a simple relationship for μ_n for a "living" chain polymerization:

$$d\mu_n/dt = k_p M \tag{14.39}$$

where t is understood to be "age" or space time. In a steady state CSTR, M, the monomer concentration in the reactor, is constant so $\mu_n \propto t$ and it is easily shown that the MWD for the "living" chain will follow the Schultz-Flory model. The effect of introducing a square wave variation in the monomer flow rate can be understood in terms of Equation (14.39). At high conversions, M in the reactor will change with the feed flow rate so the reactor output will be proportional to a convolution of the distributions resulting for steady state operation at the two flow rates used. When the convolution integral is exactly evaluated, which must be done numerically, the MWD in the reactor outlet is given by Figure 14.9a for a symmetrical cycle. The smooth curve in the figure is the MWD under steady state operation at the same conversion of monomer and mean chain length. Symmetry of the square wave disturbance is important. The effect of cycle split on D for an isothermal CSTR forced at $\tau/\bar{t} = 2$ and $k_p/k_i = 10$ appears

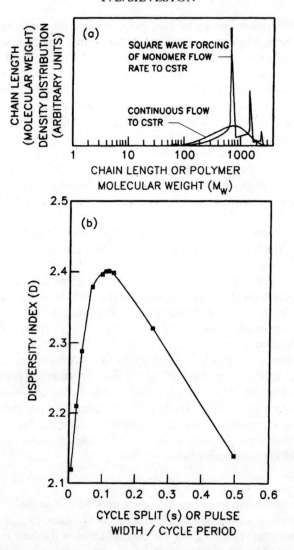

FIGURE 14.9 Simulation of square wave forcing of monomer flow for an anionic, "living" chain polymerization in an isothermal CSTR with $\bar{t} = 40$ min, $= 80$ min, $M° = 5.2$ gmol/L, $I° = 0.0173$ gmol/L, $k_p = 50$ L/gmol s, $k_i = 5$ L/gmol s: (a) molecular weight population density distribution, (b) polydispersity, D, as a function of cycle split, s (Figure adapted from Gosden et al. (1995) with permission, © 1995, Marcel Dekker, Inc.)

in Figure 14.9b. This simple, unsymmetrical flow rate forcing increases D by a maximum of 20 % over that obtained at steady state. Gosden et al. (1995) find that if $k_p/k_i \approx 1$, D can be increased to about 5 in their modulation scheme. Their results were obtained earlier by Couso and Meira (1984).

Experimental

Meira and Johnson (1981) have been the only experimental investigators of "living" anionic polymerization despite the large number of theoretical studies of this polymerization mechanism. Their study pursued the approach of Langner and Bandermann (1978) and Frontini et al. (1986, 1987) who employed periodic operation with irregular concentration patterns to obtain a specified μ_n and D. Figures 14.10a is a schematic of the apparatus used. Unlike the free radical polymerization studies which were performed in a CSTR, a tubular, 1.2 mm i.d. x 10,000 mm reactor was used by Meira and Johnson. Complete and rapid mixing up stream of the reactor is essential with a tubular reactor so a special T joint with a needle injector was used. This is shown in Figure 14.10b. Computer controlled positive displacement metering pumps are needed for the concentration pattern within a cycle. These are shown in the figure. Since time average results were desired, the reactor empties into discharge controlled receiving (mixing) vessels. Further polymerization in the receiving vessels was killed by air injection.

Experiments were performed on styrene polymerization in tetrahydrofuran (THF) using sec-butyllithium in THF or hexane, or a methyl styrene tetramer in THF as the initiator. The latter was also used as a scavenger for impurities. Extreme care was used to prevent contamination by water, oxygen and other impurities that can react with the "living" polymer ends. The experiments were performed at 30°C with a constant initiator concentration: 8.75×10^{-4} mol/L. A single cycling period of 15 minutes was used and the mean residence time was 13 to 15 seconds. Two experiments were performed to produce via flow modulation: a) a rectangular MWD with a 20:1 ratio of highest to lowest molecular weight in the polymer, 2) a discrete number MWD of equimolar mixtures of two polymer species in a 20:1 molecular weight ratio.

Figure 14.11 shows the MWD desired in two ways, the monomer feed rate profile needed and the results achieved. The steady state results were for the time averaged monomer concentration. The desired MWD results were not obtained. In the first experiment, μ_n for periodic operation is close to the desired value (and the steady state value), but D is 2.57 instead of the target 1.29. In the second experiment, μ_n for periodic operation is twice μ_n for steady state, but was less than desired. The MWD is broad rather than discrete with two sharp peaks. Figure 14.11 shows a deconvolution of the single curve for the periodic operation. The ratio of μ_n for each of the curves is 13:1 instead of the intended 20:1. Areas under the curves differ substantially rather than being equal as desired. Meira and Johnson attribute the problem to impurities which kill living ends and mixing effects which result in monomer conversion under 100%. This is demonstrated by the spread of the steady state MWD's.

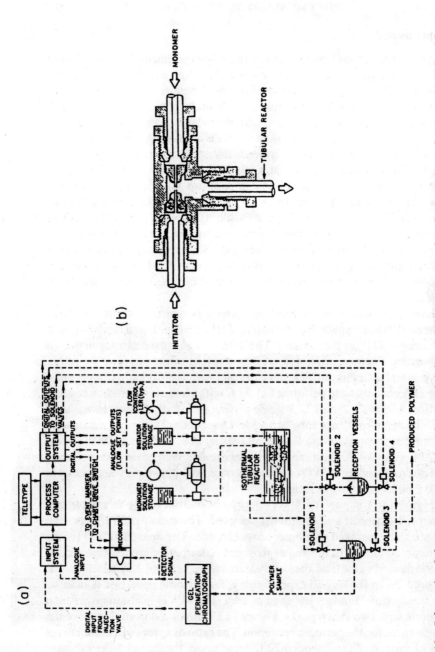

FIGURE 14.10 Schematic of the apparatus used by Meira and Johnson (1981) (Figure taken from Meira and Jonsson (1981) with permission, © 1981 Polymer Eng. Sci.)

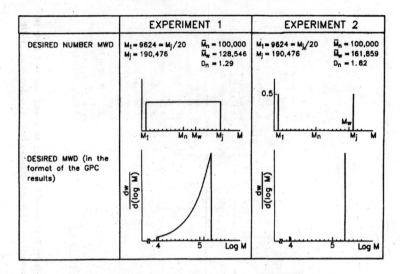

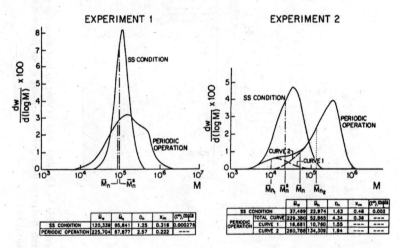

FIGURE 14.11 Flow modulation experiments to control molecular weight distributions (a) and results obtained (b) (Figure taken from Meira and Johnson (1981) with permission, © 1981 Polymer Eng. Sci.)

About 15 years after the experiments just discussed, Johnson and co-workers returned to the forcing of "living" anionic polymerization and performed another set of two experiments using this time a CSTR (Gosden et al., 1996). The objective of this recent work was to verify the use of convolution integrals in solving the model equations, so a comprehensive investigation was not undertaken.

A schematic of the Gosden experimental system appears as Figure 14.12. The reactor was a 1 litre, jacketed glass vessel whose fluid volume was held

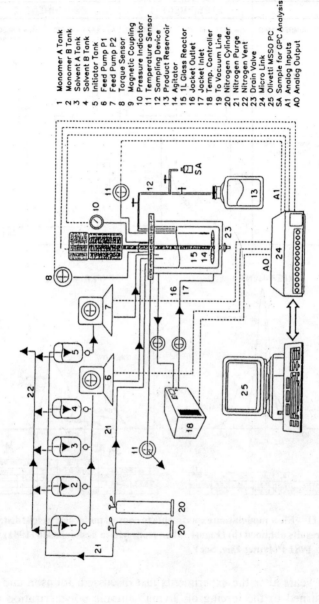

FIGURE 14.12 Experimental system used by Gosden *et al.* (1996) to investigate periodic forcing of monomer flow and initiator flow to a 1 L glass CSTR for the polymerization of styrene in cyclohexane using n-butyllithium as initiator (Figure adapted from Gosden *et al.* (1996) with permission, © 1996 Marcel Dekker, Inc.)

constant through an overflow tube. Two variable delivery, chromatography pumps were computer controlled to produce a square wave in the flow of initiator or monomer to the reactor. The reaction chosen by Gosden et al. was the n-butyllithium initiated polymerization of styrene in a cyclohexane solvent. In the first experiment, initiator flow was manipulated about a mean concentration of 0.007 gmol/L using a 3 min. pulse of initiator in a 25 min. cycle ($s = 0.12$ based on the duration of initiator flow). The monomer concentration was 1 gmol/L and the temperature was held at 50°C. Amplitude was 50 in terms of initiator flow. The monomer flow was cycled in the second experiment about a mean concentration of 1.75 gmol/L with the initiator concentration held constant at 0.0048 gmol/L. The temperature was 40°C. Cycling parameters were $\tau = 20$ min., $s = 0.3$ based on monomer flow and $A = 50$. Molecular weight distribution was determined by gel permeation. The "living" polymer was "killed" by dumping reactor overflow into methanol. Time average as well as instantaneous data were collected.

Figure 14.13 demonstrates the dramatic changes in the instantaneous polymer characterization on pulse introduction of the initiator. The polydispersity index rises abruptly from 2 to about 10, while the number average molecular weight plummets from 40,000 to 5000 g/mol when initiator pulse begins. These changes reverse when the pulse ends and are repeated on each successive pulse. Time averaging of this instantaneous data clearly shows that periodic composition forcing must increase D and reduce M_n.

Gosden et al. (1996) ran into difficulty in meeting the objective of their work. Despite careful attention to purifying monomer, solvents and initiator, they were not able to simulate their experimental results without adding through convolution a Gaussian function with a width of 0.2 decades at half peak height. This broadening of the time response to a monomer or initiator pulse was introduced to allow for imperfect mixing in the CSTR, non-instantaneous "killing" of the polymer when doused in methanol and data smoothing in gel permeation. Figure 14.14 compares experimental and calculated molecular weight distributions for a sample collected at $t/\tau = 0$ when a monomer or initiator pulse was introduced. The model parameters for simulation were taken from the literature ($k_p = 600$ L/gmols and $k_i = 77$) and reproduced the steady state reactor performance adequately at $\bar{t} = 10$ min. In the first experiment, initiator cycling, the agreement is reasonable: peak shapes are alike and there is just a small offset of the peaks suggesting a difference in the measured and predicted M_n (Fig. 14.14b). However, the agreement is poor in Figure 14.14a where the monomer concentration was cycled. Gosden et al. mention that the first molecular weight peak is spurious and arose from contamination of the sampling system. Thus, simulation was not able to produce the shape of the distribution curve and also failed to correctly predict the number average molecular weight. Nevertheless, the measurements of Gosden et al. confirm

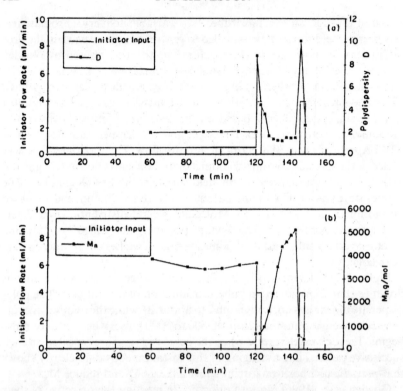

FIGURE 14.13 Variation of polydispersity and number average molecular weight with time in a CSTR on the start up of periodic initiator pulsing. In the steady state portion of the figure, initiator and monomer concentrations are 0.007 and 1 gmol/L, $\bar{t} = 10$ min and T = 50°C. Initiator concentration in the pulse is 0.7 gmol/L (Figure adapted from Gosden et al. (1996) with permission, © 1996 Marcel Dekker, Inc.)

the work of earlier investigators which show that composition modulation can be used to control the polymer molecular weight distribution.

E. STEP GROWTH POLYMERIZATION

If the monomer contains more than one functional group capable of chain formation, say $R - A_f$, where A is the functional group and f is the number of groups per molecule, the growing chains will have available groups so that chain addition can occur without converting a "living" chain to a "dead" one. Condensation polymerization is also used to describe polymerization where multiple functional groups arise. This important but complex polymerization has been considered under sinusoidal forcing of the monomer concentration by Gupta et al. (1985). Only simulations are

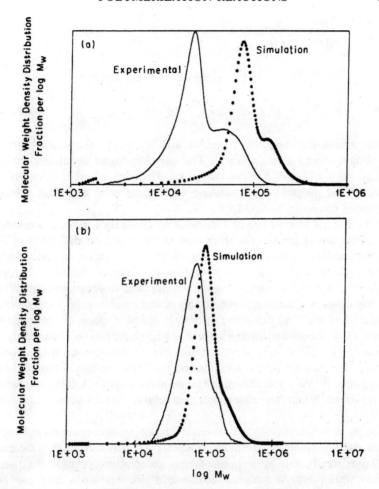

FIGURE 14.14 Comparison of experimental and calculated molecular weight distributions for a) monomer cycling at 40°C and b) initiator cycling at 50°C. Measurement conditions for b) are given in Figure 14.13 and $\tau = 25$ min, $s = 0.12$, while for (a) $\tau = 20$ min, $s = 0.3$. Mean initiator and monomer concentrations were 0.0048 and 1.75 gmol/L (Figure adapted from Gosden *et al.* (1996) with permission, © 1996, Marcel Dekker, Inc.)

available. Gupta *et al.* consider, for simplicity, a monomer that becomes active on entering the reactor. For an isothermal CSTR, the materials balances on the monomer and a polymer of chain length n become:

$$\frac{dP_1}{dt} = \frac{((P_1)^\circ(t) - P_1)}{\bar{t}} - k\frac{f}{2}P_1 \sum_{m=1}^{\infty} [mf + 2 - 2m][P_m] \quad (14.40)$$

$$\frac{dP_n}{dt} = \frac{(-P_n)}{\bar{t}} - \frac{k_p}{2}\sum_{m=1}^{n-1}[mf+2-2m][(n-m)f+2$$
$$-(n-m)]\frac{P_m P_{n-m}}{2} - k(nf+2-2n)[P_n]$$
$$\sum_{m=1}^{\infty}(mf+2-2m)\frac{P_m}{2} \quad n=2,3,4\ldots \tag{14.41}$$

In the above, the monomer concentration is forced sinusoidally with an amplitude A and a frequency ω. The two non-linear equations replace Equations (14.16) to (14.21) in Table 14.3. They are solved numerically by forming the moments, normalizing with space time and time-average monomer concentration (Gupta et al., 1985).

Two distinct behaviors with time were observed for a monomer possessing 3 functional group ($f=3$): in one, the normalized difference of the second and first moments of the chain number distribution remained finite, while in the second the difference continued to increase eventually becoming infinite. Gupta et al. reason that the latter behavior represents gelation. In this case, modulation never attains a stationary cyclic state so that modulation is not an attractive option. In $\bar{t}/\tau$ vs. ς space, the gelling and non-gelling response to forcing appear as a line bifurcation. The parameter ς is $k_p/2\,[f-2]^2(P_1)^\circ \bar{t}$ and it depends only on the number of functional groups, the polymerization kinetics and the time-average monomer concentration. When $f=4$, the same two responses occur. A critical value of ς is observed. When this value is exceeded gelation always occurs regardless of $\bar{t}/\tau$.

Gupta et al. found that the sinusoidal forcing exerts the greatest enhancement at the quasi-steady state limit where $\bar{t}/\tau = 0$. Ray found a 16% increase in D over steady state by monomer feed concentration cycling at the quasi-steady state limit. At a higher frequency, there was a 6% decrease. No decrease was seen by Gupta et al. for the model parameters they explored ($A = 0.6$ to 0.8, $f = 3$ to 4 and ς from 0.01 to 0.075). The increase in D over steady state approached 40% at $\bar{t}/\tau = 0$.

F. ZIEGLER-NATTA POLYMERIZATION

Lee and Bailey (1974) simulated olefin polymerization on Ziegler-Natta catalysts using a hydrogen transfer agent for MWD control. Polymerization can follow either a "living" chain or free radical type of mechanism. Indeed, Lee and Bailey assumed the steps given by Equations (14.3) to (14.5), (14.12) and (14.14) in Table 14.2. They also added termination by disproportionation and wall adhesion steps that are not stated in the table. Square wave cycling of hydrogen concentration in a CSTR and in an imperfectly mixed vessel were treated. The model for the former resembles the free radical model in Table 14.3. Lee and Bailey converted this model to moments for their numerical studies. For the most part, just the limiting quasi-steady state (very slow cycling) and relaxed steady state (very rapid cycling) were evaluated. In the choice of model, parameters and operating conditions, an attempt was made to model the experiments of Claybaugh et al. (1969). Lee and Bailey found that at a constant μ_n, D could be increased from about 6 to 10 at quasi-steady state if the vessel is imperfectly mixed. Much larger increases, however, could be obtained at the relaxed steady state limit. Several simulations were undertaken for intermediate frequencies. These fell between the two limits. Indeed, a 3 × increase in D was reported for hydrogen concentration cycling in a CSTR.

Experimental results reported in the Claybaugh et al. (1969) U.S. patent deal with propylene polymerization in an impeller mixed vessel with a Ziegler-Natta catalyst, $TiCl_3$-0.33 $AlCl_3$ in aluminum triethyl. Hydrogen was used as a chain transfer agent to control molecular weight and the H_2 flow was switched on and off. A bidispersed MWD should have been produced ideally. Due to non ideal contacting a broad distribution was obtained with $D = 10.8$. Under steady state operation, $D = 8.0$. The simulation of Lee and Bailey (1974) substantiated the effect of non ideal contracting on the MWD distribution and also showed the MWD of product emerging from the reactor did not vary greatly with time despite the H_2 pulsing.

Modulation of the H_2 partial pressure as a method for increasing polydispersity with Ziegler-Natta catalysts was also examined by Hungenberg (1992) through simulation by approximation. The question addressed was which one of various methods was best for achieving a higher polydispersity so an exact model was felt to be unnecessary, moreover, kinetics were not available for the modern Ziegler-Natta catalyst that the Hungenberg study considered. Instead, a steady state integral model expressing the effect of H_2 on chain length was used. Although the reactor type is not given, the treatment used by Hungenberg suggest a CSTR was assumed. Two methods of forcing were examined: 1) rising and falling ramps between two levels (saw-tooth variation of H_2), 2) bang-bang switching between 0 and 4 mol.% H_2 in the feed. Abrupt switching in the feed does not mean abrupt changes occur in the reactor. A more realistic analysis makes abrupt switching in the feed become a 4 part cycle in the reactor in which the H_2 partial pressure decays exponentially from 4 to 0%, remains at 0 for an interval and then increases linearly back to 4% where it remains for a short time. Hungenberg found that only the bang-bang operation was able to attain $D > 10$ which was his target polydispersity. A saw-tooth variation of H_2 partial pressure gave a polydispersity of 8, which was not high enough. Figure 14.15 shows the product polymer variations for the saw tooth strategy and one in which the cycle has four parts as just described. The upper part shows that D varies within a cycle from 5 to 12 for the saw tooth

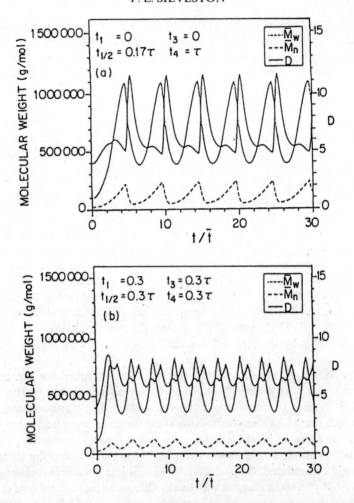

FIGURE 14.15 Predicted polymer properties for two strategies of H_2 partial pressure forcing in propylene polymerization using current Ziegler-Natta catalysts (a) forcing in a sawtooth pattern with $t_1 = t_3 = 0$, (b) forcing in a square wave pattern with $t_1 = t_3 = 0.3$ where t_1, t_3 are durations of constant H_2 concentration and t_2, t_4 are durations of sinking and raising concentration; $t_{1/2} = H_2$ half life in reactor and $\bar{t}$ = space time in reactor. (Figure taken from Hungenberg (1992) with permission; © 1992 VCH Verlagsgesellschaft)

variation, while the lower part shows a smaller variation, between 6 and 8. In both cases the time average D is less than 10. When the 4 part cycle was considered, D was seen to decrease as the duration of the exponential decay lengthened. Nevertheless, H_2 modulation increases D significantly over the 4 to 5 accomplished under steady state operation just as the Claybaugh patent and Lee and Bailey (1974) concluded.

G. EMULSION POLYMERIZATION

It is hardly surprising that the periodic forcing of emulsion polymerization has had little study despite its enormous industrial importance. It is probably the most complicated polymerization process mechanistically and obtaining reproducible experiments has been difficult for even batch runs in a bomb or steady state operation in a CSTR. The chemical mechanism usually consists of steps expressed by Equations (14.6), (14.8) to (14.12) and (14.14). Superimposed on these steps are micelle nucleation and particle growth, transport of monomer and chain transfer agents to the particle, adsorption and transport within the micellar phase. Partition coefficients must be considered and gel effects can be important even for low monomer conversions.

Writing in 1985, Penlidis et al. (1985) mention that progress has been made in modelling the emulsion polymerization system. Models are now sophisticated enough to predict the spontaneous oscillation of emulsion particles, particle diameter and monomer conversion that are observed experimentally for PVC and PVAc systems. Rawlings and Ray (1988a) for example, describe a sophisticated model which allows for radical concentrations in particles, radical entry and adsorption in both micelles and polymer particles, inhibition by foreign matter, and the gel effect. In a second paper, these authors demonstrate that their models successfully predict experimental results such as overshoot on start up and the spontaneous oscillations mentioned above (Rawlings and Ray, 1988b) Thus, tools are available to investigate if benefits can be obtained from one or more of the composition forcing strategies possible for this type of polymerization.

Gugliotta and Meira (1986,1988), nevertheless, avoided considering the complex emulsion polymerization models by examining the sinusoidal variation of the chain transfer agent (t-dodecal mercaptan) and the solvent (water) at the quasi-steady state limit in their study of the forcing of the co polymerization of styrene and butadiene in an aqueous emulsion. An isothermal CSTR was considered. The authors used a steady state model for the number of polymer particles produced per unit volume of reactor which depends on the mean residence time, the rate of decomposition of the initiator, the emulsifier concentration and its properties. The RTD of the reactor behaves as a spreading function, e.g. a linear filter on the expression for the number of particles. The particle size distribution (PSD), monodispersed for a batch or PFR, then is "spread" by the RTD of the reactor used. In a similar way, the transfer agent or solvent flow variation acts as a spreading function for the steady state PSD. Gugliotta and Meira invert the spreading arguments and show how in principle, if a PSD is known, the flow variation policy can be determined. The authors illustrate the procedure using the copolymerization of styrene and butadiene under conditions just mentioned. They demonstrate that forcing the transfer agent increases D from 2 to 8 × at quasi-steady state depending on amplitude with no change in μ_n. Solvent forcing had no effect on D, but reduced X_m at the low frequency limited.

Experimental composition forcing of emulsion polymerization does not seem to have been reported in the literature. Nevertheless, a start towards exploring this type of operation has been made by Gordon and Weidner (1981). These investigators used interruption of emulsifier flow to a batch reactor to control the polymer particle size distribution in the reactor. Interruption was controlled by the monomer conversion as detected by a temperature rise in the reactor jacket. However, control could also have been achieved by periodically manipulating the flow without conversion sensing. The concept followed by Gordon and Weidner was that excess emulsifier over that needed to cover the growing polymer particle leads to formation of more micelles and hence fine particles. Initiation of new particles results in PSD and MWD broadening.

H. ASSESSMENT

Simulation results reviewed in this Chapter considered various types of polymerization but generally assumed ideal operating conditions such as well defined mixing (CSTR or PFR), isothermal operation (usually) and the absence of impurities and initiator side reactions which can cause chain terminations. These simulations predicted large increases in the polydispersity, particularly for the living chain mechanism. For Ziegler-Natta catalyzed polymerization, a better than 3 × increase in D through on-off H_2 cycling was indicated. Simulation predicted a 2 to 8 × increase in D for emulsion polymerization, With free radical polymerization, which has been fairly well studied, composition forcing of monomer and initiator concentrations in phase or forcing of the CSTR jacket temperature was found to increase D up to 2 fold. Indeed, complex variations of total flow, thus changing reactor volume, or of initiator, monomer and transfer agent/solvent flows at constant reactor volume can, in principle, either raise or lower D depending on the cycle.

The experimental results are much less sanguine than those predicted through simulation. For one, only free radical polymerization has been looked at to any extent experimentally. Just one experimental study on living chain polymerization appears in the literature; for Ziegler-Natta systems, a patent gives just incomplete experimental data. Composition modulation remains to be tested for the industrially important emulsion polymerization. With anionic, "living" chain polymerization, periodic flow manipulation increased the polydispersity 2 ×, but this was much less than anticipated. Some control of the MWD and μ_n through flow manipulation was achieved, but again they were much less than models predicted. On-off

H_2 addition provided only a 20% increase in D, rather than the 2 to 3 × change expected. For free radical polymerization, just small increases in polydispersity, ranging from 15 to 20%, were observed with composition forcing. Temperature forcing, using the reactor jacket temperature, resulted in almost no change in D. Furthermore, in many of the experiments, reproducibility problems were indicated so results are suspect.

Certainly, the current challenge for periodic operation of polymerization reactors is to demonstrate experimentally the control of polydispersity predicted by the idealized simulations. This situation may have contributed to Hungenberg's (1992) assessment that periodic operation is not an attractive alternative for broadening the MWD of polypropylene polymers. Hungenberg dealt with the situation introduced by the new generation of Ziegler-Natta catalysts which produce polymer with $D = 4$ to 5, whereas several polymer markets require $D > 10$. Forcing of the H_2 partial pressure was only one of several options considered.

Hungenberg points out a further problem with composition modulation. The largest effects on polydispersity are observed at the quasi-steady state limit or at least under low frequency cycling. In either condition, the properties of polymer leaving the reactor will vary temporally. This cannot be tolerated so that a blender would have to follow the reactor to produce a product of constant properties. This blender could be as great as 4 × the reactor volume and would have to be well mixed. Thus, it is questionable whether or not a low frequency periodic polymerization process could ever be attractive. On the other hand, high frequency modulation is impractical with stirred reactors which dominate polymerization on an industrial scale because of damping due to the coupling of mixing and the high material holdup in the reactor.

These considerations suggest that there is a region, $0.2 \leqslant \tau/\bar{t} \leqslant 2$ in which periodic forcing may be attractive. This operational window needs to be looked at in simulations and even more pressingly through experimentation. The literature examined in this chapter suggests that any experimental effort undertaken should focus on emulsion, step growth, and Ziegler-Natta polymerization.

CHAPTER 15

Application to Mechanistic Studies

A. INTRODUCTION

Primary reasons for employing composition modulation to study catalytic mechanisms and kinetics are: 1) modulation data provide insights into the elementary steps through which a reaction proceeds and offer a sensitive discriminant among rival kinetic models, and 2) models based on steady state or even pulse or step change measurements do not reliably predict the performance of periodically forced catalytic reactors. Models are useful for extrapolating experimental data for design studies or for indicating what further experiments should be undertaken so that modelling is an important part of a research study. As will be seen in this chapter, models intended for modulation applications require measurements made on periodically forced systems.

Composition modulation constrains a catalytic system to operate in an unsteady state so that as a technique for studying mechanism or measuring kinetics, modulation can be classed as a dynamic or transient method. These methods have become a popular means for studying mechanism in the last decades. Three versions are widely used: 1) response to a single pulse of a reactant or promoter, 2) successive pulsing of a reactant (Hattori and Murakami, 1968), 3) response to a step change of a reactant or promoter (Kobayashi and Kobayashi, 1974; Kobayashi, 1982; Müller and Hofmann, 1987). Temperature programmed desorption (TPD) or temperature programmed reaction (TPR) are other variants of transient methods. Kobayashi and Kobayashi (1974) and Bennett (1976) have published comprehensive reviews of the application of these methods to mechanistic studies. More recent reviews by Renken (1990a,b, 1993) discuss the use of periodic operations in this context. Step change methods and their results have been examined in many of the previous chapters. Successive pulsing was discussed in Chapter 12 in conjunction with epoxidation and partial oxidation.

B. MEASUREMENT APPLICATIONS OF COMPOSITION MODULATION

Qualitative Applications

Resonance frequencies can indicate the rate controlling step in a catalytic reaction and identify in this way portions of the reaction mechanism. In an early experimental study of composition modulation, Unni *et al.* (1973) observed that a diffusion step, probably of O_2 or SO_2, in a melt or glassy phase must control the rate of SO_2 oxidation to account for resonance observed at frequencies between 0.000035 and 0.00028 Hz for this reaction over potassium promoted vanadia catalysts. Consequently, the oxidation must occur in a melt phase. At the other end of the frequency spectrum, Silveston (1995) reasoned from the temperature dependency of the magnitude of the resonance effect and the relatively high frequencies at which resonance is observed that CO adsorption and/or surface reaction are the rate controlling steps for CO oxidation over noble metal catalysts.

Qualitative information on catalytic mechanism can be derived as well from observing start up of composition modulation from a steady state. For example, Chiao *et al.* (1987) found that the time-average rate of NH_3 synthesis increased slowly with time after switching on $N_2 - H_2$ modulation, but when modulation was switched off hours later, the steady state rate was much higher than previously. They used this observation to account for the failure of models to predict synthesis performance under periodic forcing, concluding that composition modulation activates the catalyst probably by surface reorganization. Labastida-Bardales *et al.* (1989), investigating the composition forcing of propene partial oxidation over a copper molybdate catalyst, also concluded surface transformation occurs. These investigators proposed that new copper molybdate crystal phases form causing structure change on the surface during start up as the result of periodic forcing. Labastida-Bardales *et al.* used an unsymmetrical cycle and long periods so that their forcing method resembled successive pulsing.

The role of surface residues and the participation of oxygen in ethene epoxidation was explored by Park *et al.* (1983) using 4 and 8 part cycles. The 4 part cycles consisted of successive exposure of the catalyst to C_2H_4 or a $C_2H_4 - O_2$ mixture, followed by a N_2 flush, O_2 exposure and a second flush. The 8 part cycle added H_2 exposure (to hydrogenate residues), another O_2 exposure and two more N_2 flushes. Park *et al.* varied the ethene and O_2 exposure duration. The temperature of the second O_2 exposure in the 8 part cycle was also increased in some runs. Further details of these experiments and a discussion of results are given in Chapter 12 as part of a discussion of forcing epoxidation reactions.

The mechanism of propene partial oxidation was investigated using 2 and 3 part cycles in one of the earliest studies of composition forcing by Niwa and Murakami (1972). They formed the catalyst into disks and measured electrical conductivity as well as products formed during successive pulses of different composition. From the resistance measurements they were able to identify weak and strong adsorption of propene on different partial oxidation catalysts and relate the adsorption strength to catalyst performance under steady state and cyclic operation. Bismuth in bismuth containing oxide catalysts was found to contribute to catalyst reoxidation and to provide high lattice oxygen mobility. A two page discussion of the experiments and results will be found in Chapter 12 together with a brief description of the successive pulsing technique. Successive pulsing studies of Brazdil et al. (1980) were similar to the work of Labastida-Bardales et al.(1989) mentioned on the previous page. They and the Labastida-Bardales study are also discussed in Chapter 12.

The question of intermediates in the reaction pathway under transient conditions was examined by Zhou and Gulari (1986a) using a near transparent wafer of silica impregnated with Pt mounted in a flow through IR cell. The flow of CO and O_2 was modulated in a bang-bang mode and the absorption spectrum measured. Figure 15.1 shows the variation in the % absorbed with wave length under steady state and when the flow is modulated with different cycle periods. The broad peak centered at 2125 cm^{-1} under steady state operation is attributed to CO adsorbed on a Pt oxide surface, while the narrow peak at 2080 cm^{-1} is CO bridge-bonded to Pt atoms. Under composition forcing, the 2080 cm^{-1} peaks splits and overlaps the oxide bonded CO. Zhou and Gulari attribute this transformation to CO now bridge-bonded to a chemisorbed oxygen and Pt. The changing amplitude of the peaks suggest a shift in the reaction intermediate as the cycling frequency decreases from 0.1 to 0.02 Hz from Pt bonded CO to CO bridge-bonded to O and a Pt site. Zhou and Gulari suggest that this switch in mechanism explains the remarkable increase in the time-average rate of CO oxidation under bang-bang modulation (See Chapter 3).

Our short survey was intended to illustrate the different ways of employing composition modulation for qualitative studies of mechanism. It is, of course, incomplete. Many more examples will be found in earlier chapters of this monograph.

Quantitative Applications

An elegant use of modulation for mechanistic insight has been presented by Renken and co-workers (Marwood et al., 1994; 1996). Simultaneous measurements of the time variation of gas phase and adsorbate concentrations were made for the methanation of CO_2 over a 2 wt.% Ru on TiO_2

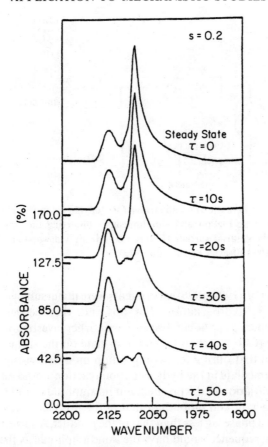

FIGURE 15.1 Variation of the absorption spectra of chemisorbed CO on Pt in the presence of O_2 with cycle period for the bang-bang modulation of feed composition. A cycle period of zero = steady state (Figure adapted from Zhou and Gulari (1986a) with permission, © 1986 Pergamon Press)

catalyst at 383°C and near atmospheric pressure. A fixed bed of the supported Ru catalyst was placed in a recycle loop and a product stream was withdrawn through a diffuse reflectance, Fourier transform infrared spectrometer (DRIFTS) cell containing a small amount of the Ru catalyst maintained at the same temperature as the fixed bed. Figure 15.2 is a schematic of the Marwood experimental system. The recycle loop assures a near uniform composition in the fixed bed, while just a small amount of catalyst in the DRIFTS cell keeps the gas phase composition in the cell the same as the composition in the reactor so that adsorbate concentrations represent those in the reactor. Placing the cell in the exit line rather than in the recycle loop greatly reduced the pressure drop in the DRIFTS cell.

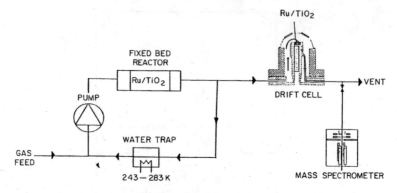

FIGURE 15.2 Schematic of an apparatus for studying composition modulation of a fixed bed catalytic reactor and simultaneously observing the time variation of adsorbates on the catalyst surface (Figure taken from Marwood *et al.* (1994) with permission, © 1994 Elsevier Science Publishers)

Marwood *et al.* (1994, 1996) experiments were in pursuit of a mechanistic model for CO_2 hydrogenation. For this end, they needed to identify intermediate species on the catalyst surface. Earlier investigations observed adsorbed CO (CO_{ad}), carbon C_α and a formate on the surface. Marwood *et al.* reasoned that during feed composition modulation the response of a surface intermediate in the hydrogenation reaction would have the same shape as the variation of the product, but its amplitude would be less than that of the reactant and more than that of the product. The intermediate would exhibit a phase lag less than that of the product. A surface species not in the reaction sequence would have the same amplitude as the modulated reactant. There could be a phase lag, but it would be small. With multiple surface intermediates, the phase lag indicates the reaction sequence. Experiments were conducted using constant flows of H_2 and He by periodically switching the CO_2 feed on and off so as to maintain a time-average composition of $CO_2:H_2:He = 1:4:5$ (10 vol.% CO_2).

Gas phase and adsorbate responses for a 20 minute, symmetrical cycle are presented in Figure 15.3. The curves have been normalized with steady state measurements at the time-average feed composition. Distortion of the CO_2 square wave and reduced amplitudes of surface species and the CH_4 product indicate strong adsorption. Both product and the surface species are present when there is no CO_2 flowing through the reactor or the DRIFTS cell. The flattening of the CO_2 concentration after 5 minutes and its approach to 2 indicates the surface is becoming saturated with CO_2. The shape of the time variation of the surface formate is close to that of CO_2 and there appears to be no lag with respect to the feed. Thus, the formate is an early intermediate in the reaction sequence. On the other hand, the shape of the CO_{ad} response is quite similar to the time variation of the CH_4 product

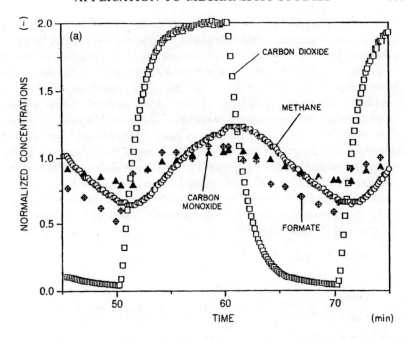

FIGURE 15.3 Response of product and surface species to the symmetrical on-off modulation of CO_2 feed rate to a recycle reactor containing a packed bed of 2 wt.% Ru on TiO_2 catalyst operating at 383 K, about 105 kPa, time-average feed composition as $CO:H_2:He = 1:4:5$ with $\tau = 20$ min. (Figure taken from Marwood *et al.* (1996) with permission of the authors)

and the adsorbate response shows a significant lag. This lag can be measured by blowing up the CH_4 and CO_{ad} responses so that the amplitude of the former matches the amplitude of the CO_2 change. When this was done, it was seen that the CO_{ad} peak lags CO_2 by 0.8 min., while the lag of the CH_4 peak is about 1.2 min. Furthermore, the CO_{ad} lag indicates another intermediate must be present which is either transformed to or decomposes to CO_{ad}. Marwood *et al.* suggest this intermediate is the surface formate $HCOO_{ad}$. The mechanism proposed by these authors is

$$CO_2 + OH - s_s \rightarrow HCO_3^- - s_s \tag{15.1}$$

$$HCO_3^- - s_s + H_2 \rightarrow HCOO^- - s_s + H_2O \tag{15.2}$$

$$HCOO^- - s_s + s_m \rightarrow HCOO^- - s_m + s_s \tag{15.3}$$

$$HCOO^- - s_m + s_m \rightarrow CO - s_m + OH - s_m \tag{15.4}$$

$$CO - s_m + 3H_2 \rightarrow CH_4 + H_2O + s_m \tag{15.5}$$

The IR data suggest that the bicarbonate forms on the TiO_2 surface so Renken and co-workers speculate that the catalytic mechanism involves spillover of formate and hydroxyl surface species or that reaction occurs at the interface of the Ru crystallites and the TiO_2 support. In the equations above, s_s represents a surface TiO_2 site, while s_m is a site on the Ru surface. The formate-support, formate-metal and hydroxyl-metal species are assumed to be mobile and capable of spillover. Molecular hydrogen appears in the equations, but the mechanism certainly involves adsorbed atomic hydrogen. H_2 adsorption and dissociation is rapid on Ru surfaces at the operating conditions used. Equation (15.5) is, of course, an overall reactions involving hydrogenation of successively formed intermediates.

Marwood et al. (1994) constructed a kinetic model from the mechanism shown above assuming the surface is always saturated with hydrogen and that spillover is instantaneous:

$$\frac{dC_1}{dt} = \frac{(C_1)_o(t) - C_1}{\bar{t}} - (k_1(C_1)(\theta_s) - k_{-1}(\theta_1)) \tag{15.6}$$

$$\frac{d\theta_1}{dt} = k_1(C_1)(\theta_s) - k_{-1}(\theta_1) - (k_2\theta_1 - k_{-2}(\theta_2)) = 0 \tag{15.7}$$

$$\frac{d\theta_2}{dt} = k_2(\theta_1) - k_{-2}(\theta_2) - (k_3(\theta_m)(\theta_2) - k_4(\theta_s)(\theta_3)) = 0 \tag{15.8}$$

$$\frac{d\theta_3}{dt} = k_3(\theta_m)(\theta_2) - k_4(\theta_s)(\theta_3) = 0 \tag{15.9}$$

$$\frac{d\theta_4}{dt} = k_5(\theta_3) - k_6(\theta_4) \tag{15.10}$$

$$\frac{dC_2}{dt} = -\frac{C_2}{\bar{t}} - k_6(\theta_4) \tag{15.11}$$

Also, $1 = \theta_s + \theta_1 + \theta_2$ and $1 = \theta_m + \theta_3 + \theta_4$ where $\bar{t}$ is the space time, $C_1 = CO_2$ concentration in the gas phase, $C_2 = CH_4$ concentration also in the gas phase, θ_s = fraction of free sites on the support, θ_1 = fraction of sites occupied by adsorbed CO_2, θ_2 = fraction of support sites occupied by the formate species, θ_m = fraction of free sites on Ru, θ_3 = fraction of metal sites occupied by the formate species and θ_4 = fraction of metal sites occupied by adsorbed CO. Employing the definition of θ and Equations (15.7) to (15.9), the set of differential equations collapses to 3 and the θ_i terms can be eliminated. Nonetheless, 9 parameters have to be evaluated in order to employ the model in a simulation.

In their first modelling attempt, Marwood et al. (1994) handled the problem of parameters by deriving rate expressions for the formation of CH_4, adsorbed CO and the adsorbed formate species from steady state experiments and using these in material balances for these species. This is tantamount to assuming no lag between the CO_2 variation and the response of the product and surface species. Figure 15.4 compares this crude simulation with normalized time-average cycling data obtained for different cycle periods, τ. The simulation was adequate for the formate, but underestimates the normalized concentrations for methane and CO_{ad}.

A later simulation'(Marwood et al., 1996) simplifies the model by assuming the surface formate acts as a reservoir for the CO_{ad} intermediate. This reduces the model to 3 differential equations and just 3 kinetic parameters. These were evaluated from step change and cycling data. Figure 15.5 shows this model reproduces the experimental data closely.

There are several examples in the engineering literature in which modulation has been employed to measure adsorption/desorption rates. Li et al.(1989) used a sinusoidal modulation of CO concentration in a carrier gas to measure sorption rates on Pt/SiO_2. Their catalyst, 50 to 80 mesh, 1 wt.% Pt on SiO_2, was pressed into a wafer which was mounted perpendicular to the IR beam in a temperature-controlled IR flow cell. After reduction in situ in H_2 at 728 K for several hours under steady conditions, the cell and catalyst wafer were cooled to the measurement temperature,

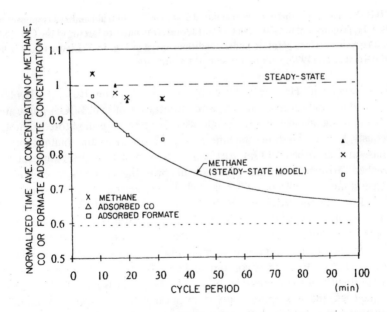

FIGURE 15.4 Comparison of normalized, time-average CH_4, CO_{ad} and adsorbed formate concentration calculated from an approximate reaction model with experimental values obtained at cycle periods from 7.5 to 95 minutes. Experimental conditions as given in Fig. 15.3 except for cycle period (Figure taken from Marwood et al. (1994) with permission, © 1994 Elsevier Science Publishers)

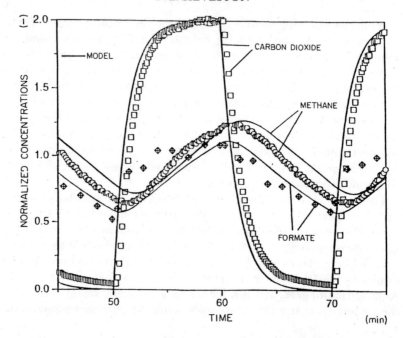

FIGURE 15.5 Comparison of simulated and experimental normalized responses of the CH$_4$ product and formate and CO surface intermediates to forcing of the CO$_2$ flow to a recycle reactor operating under conditions given in Fig. 15.3 (Figure taken from Marwood *et al.* (1996) with permission of the authors)

then flushed with He. The feed used by Li *et al.* was CO in He with a 1 vol.% sinusoidal variation around the mean concentration of 2 vol.%. Gas phase and adsorbate signals at the appropriate wave lengths were monitored simultaneously. Figure 15.6 shows the time varying absorbance data obtained for gas phase and adsorbed CO (at 2080 cm^{-1}) once a stationary cycling state is reached. Adsorbate and gas phase signals have the same frequency but are phase shifted. Amplitudes differ as well. To extract rate constants from this data, Li *et al.* wrote a dynamic model for dissociative adsorption for a differential reactor. They used perturbation theory to linearize their model. After applying a Laplace transform, they solved for the response in the transform domain and obtained expressions for the amplitude ratio and the phase shift in terms of the adsorption rate constant and the adsorption equilibrium constant. Consequently, if the amplitude ratio and phase shift is measured at several frequencies, the adsorption rate, desorption rate, adsorption equilibrium constant and a sticking coefficient can be estimated. Problems encountered in using the modulation data and interpretation of the constants and coefficient extracted are given in the Li *et al.* (1989) paper.

Adsorption rates on α-Al$_2$O$_3$ for a variety of gases (CH$_4$, CO and CO$_2$) were measured by Lynch and Walters (1990) using sinusoidal forcing of the

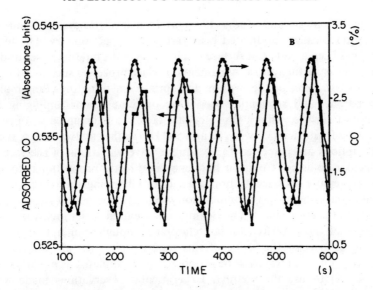

FIGURE 15.6 Experimental response to the sinusoidal forcing of CO flow to a constant temperature IR cell containing a wafer of 1 wt.% Pt on SiO_2 at 343 K and 101.3 kPa for $\omega = 0.3$ rad/s. Absorbance of gas phase CO as volume% is given by the right axis, while absorbance of adsorbed CO at a wave number of 2080 cm^{-1} is given by the left axis (Figure taken from Li *et al.* (1989) with permission, © 1989 *AIChE*)

adsorbent flow rate. Their experimental system employed a packed bed of adsorbent with gas recycle, modelled as a CSTR, and used an IR spectrometer to monitor time changes in gas concentration. Both the phase shift and the amplitude ratio were used to measure reactor volume, gas recycle rate and obtain adsorption rate constants. Lynch and Walters demonstrated that the technique is accurate with respect to reactor volume and recycle rate. They also found that CO and CH_4 do not adsorb on α-Al_2O_3. CO_2 adsorbs and these investigators found that a two site adsorption model was needed to fit the modulation data. Adsorption constants for each group of sites were determined. Lynch and Walters used their measurements to test the validity of the linearization described above. They found it introduced negligible error. Like Li *et al.* (1989), Lynch and Walters devised a model for their experimental system, linearized and then transposed the model into the Laplace or rather Fourier domain. In this domain the model can be fitted directly to measured amplitude ratios (gain) and phase shifts to evaluate model parameters such as adsorption rate constants. Sinusoidal forcing appears to be more reliable than the pulse methods which suffer from parameter correlation.

In the field of chemical physics, composition modulation has been applied to investigate mechanisms and measure the kinetics of gas-solid reactions (Schwarz and Madix, 1974). Molecular beams are used and the technique is called modulated beam relaxation spectrometry. The technique has been used to test surface mechanisms of the type discussed above and to measure the rate of fundamental surface processes such as adsorption, surface diffusion, desorption and surface reactions, including multistep parallel and sequential ones. One or more beams impinging on a catalyst surface, often a single crystal face, are interrupted (chopped), usually at frequencies between 10 and 500 Hz. At the catalyst surface these interruptions are perceived to be square wave variations of the rate of molecular collision. Collision frequency is directly proportional to the partial pressure or concentration of a species. Chopping, therefore, has the same effect as bang-bang composition modulation, except for the higher energy in the molecular beam. Beams, too, are used under high vacuums. Because of these differences, fast chopping, though similar to periodic forcing, is beyond the scope of this monograph. Like all modulation methods, a mechanism must be proposed and formulated as a mathematical model to use the method. Experimental observations made with modulated beam relaxation spectrometry usually deal with the influence of chopping frequency, surface temperature and beam intensity on the amplitude ratio and phase lag observed for the reactor products. Variation of the amplitude ratio and phase lag as these independent variables change may be used to discriminate between rival models and, through adjustments of surface temperature and chopping frequency, rate controlling steps can be isolated and their rate constants determined. Schwartz and Madix (1974) describe the technique. There is a large literature.

Fast chopping has been applied to measuring the rate of carbon deposition on a Pt filament under pressures of about a torr using a system that did not employ molecular beams (Halpern and al Mutaz, 1986). Chopping was accomplished using a rapidly swinging syringe which passed a razor edge separating a bypass to the vacuum pump inlet from the flow channel containing a hot Pt filament. The system is shown schematically in Figure 15.7. In the authors' experiment, two swinging syringes were used one feeding hydrocarbon and the other feeding oxygen. These were mounted so that they swung past the razor edge moving in opposite directions. This caused alternating square wave exposure of the Pt filament to the two reactants 180° out of phase. Thermionic electron emission from a heated filament is very sensitive to carbon on the Pt surface so change of the emission with time measured the deposition rate. The O_2 pulse cleaned the Pt surface prior to the next exposure. Halpern and al Mutaz operated their flow modulator below 20 Hz and so were able to observe the rise time of the emission signal. For faster reactions, the variation of the time-average electron emission rate with oscillating frequency could be used for the measurement. Halpern and al Mutaz obtained from their measurements the probability of carbon deposition per molecular impact and demon-

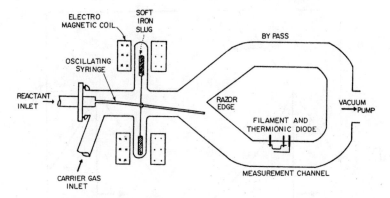

FIGURE 15.7 Fast chopping apparatus for the study of carbon deposition on a hot Pt filament using a swinging gas syringe (Figure adapted from Halpern and al Mutaz (1986) with permission, © 1986 Pergamon Press)

strated that this was a logarithmic function of the carbon number of the paraffin used.

C. MODULATION OF LIGHT INTENSITY

A classic application of modulation to investigations of kinetics and mechanisms has been the modulation of light intensity to gas phase free radical reactions, primarily in a bang-bang mode. The technique employs a brief, intense illumination of the contents of a reactor and the decay of reactants or formation of products is measured in the dark period by a variety of methods. Systems studied are non-catalytic and thus fall outside the scope of this monograph. There is, of course, an extensive literature on the technique or, better stated, the collection of modulation techniques used for measurements of elementary reactions. Our justification for including light modulation in this chapter is its application to free radical polymerization reactions.

Consecutive pulses of light are used and the polymer properties, such as number chain length, μ_n and polydispersity, D, are related to frequency and duration of the dark or light periods. Initially, a rotating sector was used for pulsing, but this has been replaced by pulsed laser illumination of a reaction cell (Olaj et al., 1985; 1987). Light photons generate free radical which react with monomer to initiate chains, but the vastly increased radicals also cause chain termination by the apparent elementary reactions given in Table 14.2 in the previous chapter. Figure 15.8 illustrates the effect of a light flash on

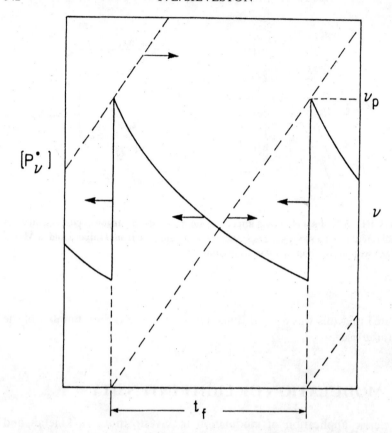

FIGURE 15.8 Schematic variation of the free radical concentration (p*) and number average chain length (v) with time after a laser pulse. v_p is the chain length generated in the time t_f between successive laser pulses (Figure taken from Davis *et al.* (1989a) with permission of the authors)

the radical concentration in a reaction cell and the chain length of the polymer formed. The latter would be obtained from gel permeation chromatography. Olaj *et al.* (1985) demonstrate that the propagation rate constant, k_p, can be obtained from the chain length of the polymer formed between successive light pulses through

$$k_p = v_p/Mt_f \qquad (15.12)$$

where v_p is the number chain length of the polymer formed, M is the monomer concentration and t_f is the duration between pulses. The interpretation of v_p is shown in Figure 15.8. The termination rate constant, k_t, can be obtained from the rate of polymerization measurement through the relation

$$\frac{r_p t_f}{M} = \frac{k_p}{k_t} \ln\left\{1 + \frac{R*k_t t_f}{2}\left[1 + \left(1 + \frac{r}{R*k_t t_f}\right)^{1/2}\right]\right\} \quad (15.13)$$

In this relation, r_p is the polymerization rate and $R*$ is the free radical concentration (Olaj et al., 1987).

Davis et al. (1989a) used light pulsing to measure the propagation constant for the homopolymerization of styrene and methyl methacrylate in various solvents using benzoin or azobisisobutyronitrile as initiators. The constants were found to be independent of chain length, initiator and solvent. This is not true for the co-polymerization of the two monomers; k_p depends on composition in this reaction (Davis et al., 1989b). O'Driscoll and co-workers applied light pulsing to measure k_p for other co-polymerization systems and used their results to explore explanations for the composition effect (Davis et al., 1990, 1991) and eventually to test the suitability of different models to explain the variations observed (Sanayei and O'Driscoll, 1994).

D. APPLICATION TO THE TESTING OF RIVAL MODELS

Composition modulation data, either reactant and product variation during a cycle, the time-average reaction rate or the cycle period associated with resonance, offers a discriminating test for rival models or, indeed, rival mechanisms if they can be formulated as models. This was the conclusion of Li et al. (1985a,b) based on attempts to model the moderate pressure ammonia synthesis data published by Jain et al. (1983b). Renken and co-workers reached the same conclusion at about the same time and have since then championed this application of composition modulation (Renken, 1990a,b, 1993).

At the time of Li's investigation, the mechanism for the synthesis reaction on a promoted iron catalyst was thought to involve the dissociative adsorption of H_2 and N_2 and the stepwise hydrogenation of the adsorbed N moiety. There was controversy over whether adsorption was competitive or H_2 and N_2 adsorbed on separate sites. This question is discussed in Chapter 7. Li et al. (1985a) constructed dynamic models of the synthesis for each of the proposed adsorption routes and evaluated the parameters for the models from steady state data by setting the time derivatives to zero. In doing this, rate constants collapse into equilibrium constants so the number of parameters in the model decreases. The remaining parameters in the dynamic models were obtained by fitting the step response data of Jain et al. (1983b). These models were tested against response data for the step change from a H_2 environment to a N_2-H_2 mixture which were not used for fitting. The competitive adsorption model failed to properly predict the experimentally observed overshoot response. The two site adsorption model provided a better match for the data. This is illustrated in Figure 15.9. The two site model is quite robust. Figure 15.9b shows that it predicts the behaviour closely for a very different step change experiment in which flow through the catalyst bed switched from H_2 to a N_2—H_2 mixture containing 90 % H_2. Model testing shown in this figure illustrates the use of dynamic data to distinguish between rival models or, indeed, rival mechanisms.

When Li et al. (1985b) extended testing to composition forcing data, the two site model failed to predict the resonance seen as the cycling frequency varied (Fig. 15.10). Modifying the two step model to allow for N_2 storage in the bulk of the catalyst (see the discussion in Chapter 7 and the bulk dissolution model in Table 15.1) and using a portion of Jain et al.'s cycling data to fit the two new parameters introduced, Li was able to match Jain's resonance results as Figure 15.10 demonstrates. Allowing for storage in the catalyst bulk improved prediction of the step change response; there was no effect on the fit of the steady state data. Nevertheless, the two site with bulk storage model failed to adequately reproduce the variation of the ammonia synthesis rate during a forcing cycle.

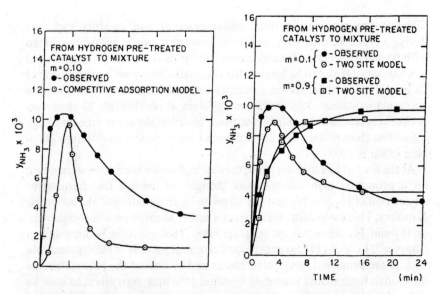

FIGURE 15.9 Comparison of predicted and measured response to a step change from a H_2 feed to feed containing both N_2 and H_2 (m = vol.% H_2 in the mixture) over a promoted iron catalyst at 400°C and 2.38 MPa: (a) competitive adsorption model, (b) non-competitive adsorption model (Figure taken from Li et al. (1985a) with permission of the authors)

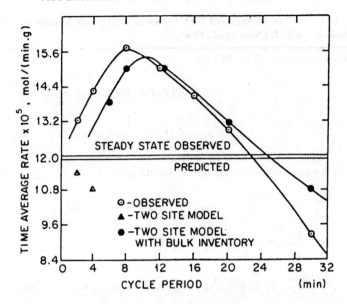

FIGURE 15.10 Predicted time-average rate of ammonia synthesis for two site models as a function cycle period, τ, and comparison with CSTR data collected at 400°C and 2.39 MPa (Figure taken from Li et al. (1985b) with permission of the authors)

Nam et al. (1990) extended Li's work by examining two classes of storage models: 1) a class assuming the surface contained spillover regions where reaction could not occur but where adsorbate storage was possible, and 2) models which stored a reactant below the catalyst surface through dissolution when the reactant was present in the gas phase at high concentration. These reservoirs, either on or below the surface, supplied the active surface with one of the reactants when that reactant was absent or in low concentration in the gas phase. Figure 15.11 shows the two classes as well as variants dealing with how transfer occurs between storage and active sites or the surface. Table 15.1 compares models for the two classes. Following Li et al., the models assume non-competitive adsorption and that the multi-step hydrogenation of atomic nitrogen is lumped into a single rate expression. Although the experiments of Jain et al. (1982c) observed storage of both N_2 and H_2, the storage of H_2 was much less than that of N_2, so that only N_2 storage is considered in the models. In both models the Elovich equation is used for adsorption which is assumed to be irreversible.

Parameters for the models were obtained from steady state and step change data as described earlier. Additional parameters allowing for N_2 storage had to be drawn from part of the cycling data of Jain et al. (1983b). Unfortunately, these data were also used for model discrimination. In all

Table 15.1 Surface Storage and Bulk Dissolution Dynamic Models for Ammonia Synthesis in a CSTR (Nam et al., 1990)

Balances Common to Both Models

$$\frac{dy_{N_2}}{d\xi} = (y_{N_2})_o - y_{N_2} - \Phi[k_1 P y_{N_2} \exp(-b_N \theta_N)]/2 \quad (15.14)$$

$$\frac{dy_{H_2}}{d\xi} = (y_{H_2})_o - y_{H_2} - \Phi[k_2 P y_{H_2} \exp(-b_H \theta_H)]/2 \quad (15.15)$$

Surface Storage-Surface Storage Model (a) in Figure 15.11

$$\frac{dy_{NH_2}}{d\xi} = -y_{NH_2} + \Phi[3k_3 \theta_N \theta_H + (1-f)\chi] \quad (15.16)$$

$$\frac{d\theta_N}{d\xi} = \zeta_N[k_1 P y_{N_2} \exp(-b_N \theta_N) - \{k_3 \theta_N \theta_H + (1-f)\chi\}] \quad (15.17)$$

$$\frac{d\theta_H}{d\xi} = \zeta_H[k_2 P y_{H_2} \exp(-b_H \theta_H) - 3\{k_3 \theta_N \theta_H + (1-f)\chi\}] \quad (15.18)$$

$$\frac{d\theta_N^*}{d\xi} = \zeta_N f \chi \quad (15.19)$$

$$\chi = k_s[\theta_N(1-\theta_N^*) - \theta_N^*(1-\theta_N)) \quad (15.20)$$

This flux can be either positive or negative

Bulk Dissolution - Bulk Model (a) in Figure 15.11 (Single N_2 Adsorption Site)

$$\frac{dy_{NH_2}}{d\xi} = -y_{NH_2} + \Phi[3k_3 \theta_N \theta_H] \quad (15.21)$$

$$\frac{d\theta_N}{d\xi} = \zeta_N[k_1 P y_{N_2} \exp(-b_N \theta_N) - \{k_3 \theta_N \theta_H + k_D(N_N \theta_N - C_N)\}] \quad (15.22)$$

$$\frac{d\theta_H}{d\xi} = \zeta_H[k_2 P y_{H_2} \exp(-b_H \theta_H) - 3k_3 \theta_N \theta_H] \quad (15.23)$$

$$\frac{dC_N}{d\xi} = \bar{t} k_D(N_N \theta_N - C_N) \quad (15.24)$$

where y_i = gas phase mole fraction, θ_i = fraction coverage of i, θ_N^* = fraction coverage of atomic N on spillover surface, N_i = concentration of i on spillover surface, C_N = N atom concentration in catalyst bulk, P = total pressure, $\xi = t/\bar{t}$, $\Phi = \bar{t} W/VC$, $\zeta_i = \bar{t}/N_i$

cases, the Levenberg-Marquardt algorithm was used to extract the parameters. For the transient and cycling data, the model was integrated within each step of the search routine using a 4th order Runge-Kutta procedure.

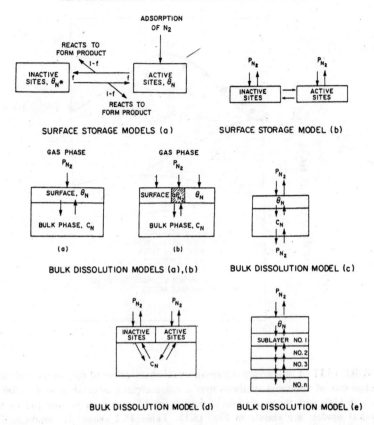

FIGURE 15.11 Storage models tested using forcing data by Nam *et al.* (1990) in their attempt to model ammonia synthesis under composition modulation (Figure adapted from Nam *et al.* (1990) with permission, © 1990 Pergamon Press)

All of the models tested by Nam *et al.* described Jain's (1981) steady state ammonia synthesis data closely and most reproduced the step change measurements adequately. However, all but two of the models were rejected when they were used to predict resonance under variation of the cycling frequency. One of the surviving models was the surface storage model (a) in Figure 15.11. To make this model satisfactory, it was necesary to assume that the inactive surface had greater than $10 \times$ more capacity for N_2 adsorption than the active surface where reaction takes place. Figure 15.12 compares predicted normalized rates of ammonia production for four models with the measured normalized rates. The bulk dissolution model assuming separate sites for N_2 adsorption and reaction and for N_2 dissolution (Bulk dissolution model (b) in Fig. 15.11) predicts the observed resonance for a feed containing 25 vol.% N_2 closely. The single dissociation site model (Bulk dissolution

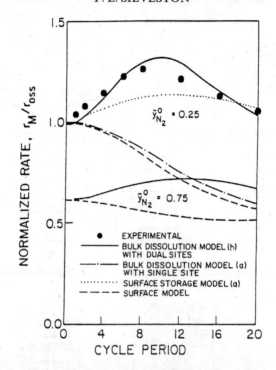

FIGURE 15.12 Comparison of experimental observations of the normalized time-average rate of ammonia synthesis over a commercial iron catalyst at T = 400°C, P = 2.38 MPa with the predictions of models for different cycle periods and s = 0.5. Physical models are shown in Fig. 15.11. Table 15.2 shows the mathematical formulations for each class of model. The surface storage model assumes the ratio of the N_2 adsorption capacities on the inactive and catalytically active surfaces = 10 (Figure adapted from Nam *et al.* (1990) with permission, © 1990 Pergamon Press)

model (a) in Fig. 15.11) does not. Equally, the surface storage model (a) (Fig. 15.11) predicts resonance when the N_2 capacity on the inactive surface is 10 × the capacity on the active surface, while if this capacity ratio is zero, no resonance with respect to the cycle period is seen (surface model in Fig. 15.12). Although a statistical test would suggest that that the bulk dissolution model provides a superior fit to the data shown as well as to other data not presented, there is enough uncertainty in fitting the 9 parameters used in the models to justify collecting further data. Because Nam *et al.* (1990) used literature data for their investigation of suitable models, they were not able to choose a single best model. Certainly, if they had the ability to collect further experimental data, an experimental series could be designed to discriminate between these two suriving models.

A weakness of the modelling studies of Li *et al.* (1985b) and Nam *et al.* (1990) is the use of the same data set for parameters estimation and model discrimination. Renken and co-workers take pains to use separate data for these two tasks. In one of their earliest studies, modelling the Pd/SiO_2 catalyzed acetoxidation of ethene to form ethyl acetate, separate measurement of the adsorption kinetics of acetic acid and ethyl acetate were undertaken (Rabl et al. 1986). These showed adsorption was slow and suggested that the step could be modelled by assuming the activation energy for adsorption was a linear function of the adsorbate coverage. This is a form of the Elovich equation for chemisorption. Several models for the acetoxidation

$$CH_3COOH + C_2H_4 \rightleftharpoons CH_3COOC_2H_5$$

were considered. All models assumed ethyl acetate forms from the reaction of the acetic acid and ethene adsorbates. Material balances terms are the same for a differential reactor considered by Truffer and Renken (1986) and for a continuously stirred tank reactor (CSTR) that are described by Equations (15.6) to (15.11) or in Table 15.1; rate expressions and the species involved, of course, vary between applications and models. Rather than write out the full model, just the rate expression will be discussed now. For a differential reactor, only reactants need to be considered so the model is written for two gas phase species, acetic acid and ethene, and the same two species as adsorbates. The rate equations used by Truffer and Renken for the adsorption/desorption and reaction steps are

$$(R_{ad})_1 = (k_{ad})_1 p_1 (1-\theta) - (k_{de})_1 \theta_1 \tag{15.25}$$

$$(R_{ad})_2 = (k_{ad})_2 p_2 (1-\theta) - (k_{de})_2 \theta_2 \tag{15.26}$$

$$R_{re} = k \theta_1 \theta_2 \tag{15.27}$$

where 1 = acetic acid, 2 = ethene.

The first model tested assumed that the adsorption steps were fast so that the equilibrium adsorbate coverage could be used. This results in a Langmuir-Hinshelwood model which adequately fitted the author's steady state data, but could not reproduce the step chance response. Their second model treated the adsorption and desorption rate constants e.g. $(k_{ad})_1$, as invariant. The model represented the steady state data and qualitatively, at least, the step change behavior. Using the results of Rabl et al. (1986) and noting that the acetoxidation is inhibited by acetic acid, the adsorption/desorption constants in the third model were expressed as

$$(k_{ad})_1 = (k_{ad})_1^\circ \exp(-[(E_{ad})_1^\circ + \beta\theta_1]/RT) \tag{15.28}$$

$$(k_{ad})_2 = (k_{ad})_2^\circ \exp(-[(E_{ad})_2^\circ + \beta\theta_1]/RT) \tag{15.29}$$

$$(k_{de})_1 = (k_{de})_1^\circ \exp(-[(E_{de})_1^\circ + \beta'\theta_1]/RT) \tag{15.30}$$

$$(k_{de})_2 = (k_{de})_2^\circ \exp(-[(E_{de})_2^\circ + \beta'\theta_1]/RT) \tag{15.31}$$

The 11 parameters in this model make it impractical so Truffer and Renken used their observation that the inhibition effect worked mainly on ethene and simplified the rate terms to give a fourth model in which Equations (15.25) and (15.27) were retained, but Equations (15.29) and (15.31) were substituted into Equation (15.26) to give after simplification

$$(R_{ad})_2 = (k_{ad})_2 \, p_2 \exp(-b\theta_1) - (k_{de})_2 \exp(-b\theta_1)\theta_2 \quad (15.32)$$

Model parameters were reduced to 6. These were fitted to the step change response data, including measurements for adsorption. Model 4 represented the step change response data well as might be anticipated. Figure 15.13 shows that the model satisfactorily depicts the start up of a cycling experiment. None of the cycling data were used for parameter estimation. Chapter 9 discusses the acetoxidation reaction further.

The "stop" effect in which interrupting the flow of reactants leads to a rapid but brief acceleration of the reaction has been used perceptively by Thullie and Renken (1993) to demonstrate the use of modulation for model discrimination. The effect is examined in Chapter 9. There are in the literature two competing explanations or models for the effect (Koubek et al., 1980a, b). The first (Model 1) proposes that two different surface sites on the alumina catalyst participate:

$$A + s_1 \underset{k_{-1}}{\overset{k_1}{\rightleftharpoons}} A\text{-}s_1 \quad (15.33)$$

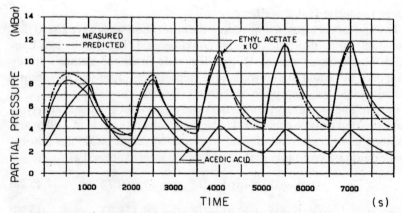

FIGURE 15.13 Comparison of measured (—) and predicted (----) partial pressure of ethyl acetate leaving a differential reactor on start up of cycling of the acetic acid flow to the reactor for $s = 0.5$ and $\tau = 1500$ s (Figure adapted from Truffer and Renken (1986) with permission, © 1986 AIChE)

APPLICATION TO MECHANISTIC STUDIES

$$A + s_2 \underset{k_{-2}}{\overset{k_2}{\rightleftharpoons}} A-s_2 \quad (15.34)$$

$$A - s_1 + s_2 \overset{k_3}{\rightarrow} C + s_1 + s_2 \quad (15.35)$$

The s_1 site is acidic and strongly adsorbs the alcohol or amine at its basic OH or NH_2 end. Both reactants can absorb as well on basic sites on the alumina which block this site for H extraction to give the olefin. Reducing the reactant concentration frees the basic sites allowing the reaction to proceed. Nowobilski and Takoudis (1986) developed and analyzed a model based on this proposal.

The second explanation (Model 2) treats the acid and basic sites on alumina as an ensemble:

$$A + s \underset{k'_{-1}}{\overset{k'_1}{\rightleftharpoons}} A-s \quad (15.36)$$

$$A + A - s \underset{k'_{-2}}{\overset{k'_2}{\rightleftharpoons}} A - s - A \quad (15.37)$$

$$A - s \overset{k'_3}{\rightarrow} C + s \quad (15.38)$$

The adsorption of a second alcohol or amine on the site blocks its dissociation capacity. This is a form of educt inhibition and assuming adsorptive equilibrium leads to a substrate inhibition model that is widely used to describe enzyme kinetics. A third model has been proposed by Moravek (1992) based on the Ipatiev mechanism for bimolecular substitution. Some simplification gives this model the same mathematical formulation as Model 2.

Thullie and Renken (1993) demonstrate that in steady state operation when adsorptive equilibrium can be assumed the Langmuir-Hinshelwood models that result are virtually indistinguishable. By examining the predicted rise in reaction rate after the flow of reactant is interrupted, these authors show that the rise is proportional to the steady state rate predicted by the model. Consequently, if the models cannot be distinguished using steady state data, it will be difficult to choose between them using step change response data. This is demonstrated in Figure 15.14 which shows virtually identical predicted response to stop flow for identical sets of model parameters.

Employing data obtained in composition modulation, however, can discriminate between the two models. At the asymptotic limit of the relaxed

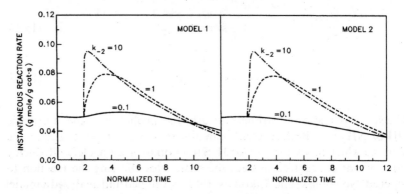

FIGURE 15.14 Calculated response of the dehydration rate to the cessation of reactant flow to a differential reactor for the two competing stop flow models with $k_1 = k'_1 = 0.001$, $k_2/k_{-2} = k'_2/k'_{-2} = 100$ and $k_3 = 0.1$ (Figure adapted from Thullie and Renken (1991) with permission, © 1991 Pergamon Press)

steady state, Thullie and Renken show that

$$\text{Model 1: } R_{\text{Rss}} = R_{ss}\left\{(1-s) + Q\,\frac{E(1-s)s}{E(1-s)+F}\right\} \quad (15.39)$$

$$\text{Model 2: } R'_{\text{Rss}} = R'_{ss}\left\{(1-s) + Q'\,\frac{D(1-s)s}{D(1-s)+G}\right\} \quad (15.40)$$

where R_{ss} and R'_{ss} are the Langmuir-Hinshelwood steady state models for the two reaction mechanisms, s is the cycle split and D, E, F, G and Q are constants incorporating different rate constants. If these rate constants where evaluated from steady state and step change data so that, for example, $k_1 = k'_1$ then the relaxed steady state models must predict different relaxed steady state rates. Figure 15.15a illustrates that at this high frequency modulation limit for on-off flow of the reactant the models become distinguishable. In the figure, the rate under steady state for either model is also shown using the same scale as the predicted difference in rates between competing models. From this figure, it can be seen that the cycling experiment should be performed at a time-average feed concentration of alcohol or amine for which the steady state rate is maximal. The cycle split, defined as the duration of reactant flow to cycle period, should also have a large value.

Away from the relaxed steady state limit as the cycle period increases, the difference between the calculated time-average rates decrease, as may be seen in Figure 15.15b, and it becomes more difficult to discriminate between the models. As our discussion of the contributions of Li *et al.* (1985b) and Nam *et al.* (1990) showed, variation of rate or a product concentration with time within a cycle is a more sensitive test of a model than the prediction of the time-average behavior or of resonance. Thus, it should be possible to choose between the models using realistic cycle periods, but instantaneous measurements of the dehydration rate of the olefin or the olefin concentration must be made.

Randall et al. (1996) employed modulation as proof of model acceptability rather than for model discrimination. Discrimination among 3 alternative models for NO reduction by CO over a silica supported iron catalyst was handled satisfactorily using step response methods. The models considered by Randall et al. were

I. homogeneous reactions:

$$NO + CO \rightarrow N_2O + CO_2 \qquad (15.41)$$

$$N_2O + CO \rightarrow N_2 + CO_2 \qquad (15.42)$$

II. heterogeneous reactions:

$$NO + s \rightarrow NO - s \qquad (15.43)$$

$$NO + NO\text{-}s \rightarrow N_2O + O - s \qquad (15.44)$$

$$N_2O + s \rightarrow N_2 + O - s \qquad (15.45)$$

$$CO + O - s \rightarrow CO_2 + s \qquad (15.46)$$

III. redox reactions:

$$2NO + v \rightarrow N_2O \qquad (15.47)$$

$$N_2O + v \rightarrow N_2 \qquad (15.48)$$

$$CO \rightarrow CO_2 + v \qquad (15.49)$$

In the set of redox reactions, v represents an oxygen vacancy in the iron oxide. Oxygen is extracted filling the vacancies. Passing CO over the surface then re-creates oxygen vacancies. In the mechanisms above, each step is considered elementary and leads to a rate expression of the type given in our earlier examples. In the case of the redox mechanism, the vacancy is treated as the degree of reduction of the oxide.

When tested against steady state data, all of the three rate models were found to be adequate. Step change experiments in which a N_2O/Ar or NO/Ar stream was introduced onto a reduced catalyst resulted in a burst of N_2 production which then fell off slowly. N_2 formation in the absence of CO disqualified Model I. Also, the N_2 formed exactly equalled the N_2O consumed so that models need not allow for N_2O adsorption. With the N_2O step, the initial rise in N_2 was independent of the reduction of the iron oxide, but the total production increased with oxide reduction. This is consistent with both Models II and III. With an NO step from Ar to 0.4

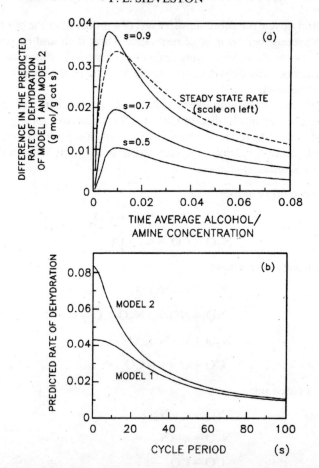

FIGURE 15.15 (a) Difference in the calculated relaxed steady state rates for competing models as a function of the reactant concentration and the cycle split for on-off reactant flow and $k_1 = k'_1 = 10$, $k_{-1} = k'_{-1} = 0.001$, $k_2/k_{-2} = k'_2/k'_{-2} = 100$ and $k_3 = 0.1$. Steady state rate for either model with the same set of rate constants is given by the ---- curve. (b) Calculated time-average rate of dehydration for the two models versus cycle period for a time-average reactant concentration = 0.054, the same set of rate constants and s = 0.9 (Figures adapted from Thullie and Renken (1993) with permission, © 1993 Pergamon Press)

vol % NO in Ar, product appeared in the order N_2, then N_2O and finally NO. The latter appeared in the off gas 6 s after NO was introduced. NO adsorption was indicated which means Model II is the better model. Experiments with the catalyst exposed to CO prior to the introduction of the NO step showed that the initial rate of N_2 formation was unaffected by the CO exposure duration, but that the total N_2 formed varied directly with this duration. The appearance of N_2O was delayed as the CO exposure duration increased, but the total N_2O formed was independent of duration. These latter observations are not predicted by either model so Randall et al. were forced to modify their model to allow for CO adsorption and the reaction of adsorbed CO with NO to form N_2.

Choice of the modified Model II was confirmed by a modulation experiment utilizing a 4 part cycle consisting of a 15 minute exposure to CO, a 2 minute Ar flush followed by exposure to 0.4 vol.% NO in Ar for 20 minutes and a second 2 minute Ar flush. The long cycle period allowed data collection during NO flow across the catalyst. Observations of a lag in NO appearance, a sharp N_2 peak 3–4 seconds after the switch to NO followed about 2 to 3 s later by a small N_2O peak and then a slow drop off in N_2 and N_2O formation were consistent with modified Model II.

Competing mechanism for N_2O reduction by CO were also investigated by Sadhankar and Lynch (1994) using composition modulation. Their experiments examined a Pt/Al_2O_3 catalyst and were performed at 499°C. Chapter 4 describes the equipment used and the operating conditions employed. The steps considered were Equations (15.45) and (15.46) and the adsorption of CO

$$CO + s \rightarrow CO - s \qquad (15.50)$$

which Randall et al. (1996) found had to be added to their NO reduction model. Reduction of the Pt surface involved adsorbed CO rather than gas phase CO (Eqn. 15.46). It was observed that CO strongly inhibits the Pt catalyzed reduction rendering the activity of this catalyst low. Two models considered for steady state experiments were (1) adsorption equilibrium using the Langmuir adsorption model which gives the familiar Langmuir Hinshelwood model, and (2) CO adsorption with self exclusion. In the later, it is assumed that a molecularly adsorbed CO molecule excludes adsorption of CO on a surrounding ensemble of N_{CO} Pt sites. N_2O, however, can adsorb and react on those sites. Model parameters were obtained by adsorption and steady state rate measurements.

In testing these models with independent data, Sadhankar and Lynch found various sets of model parameters which fitted the steady state data equally well. These researchers chose a novel mode of forcing to resolve the modeling problem. They employed a form of bang-bang cycling varying the concentration of each reactant between 0 and 2.4 vol.% and introduced phase lead (or lag) as an additional cycling parameter. This type of cycling was examined in Chapter 4. It is illustrated in Figure 15.16 for 90° and 270° phase leads with respect to the introduction of the N_2O feed. In terms of phase lead, most applications of periodic forcing have utilized a 180° lead. When Sadhankar and Lynch used a 180° phase lead they found that the Langmuir-Hinshelwood and CO exclusion model fitted the low frequency data equally well, but that only the latter model matched their higher frequency data. In between these two frequencies both models failed.

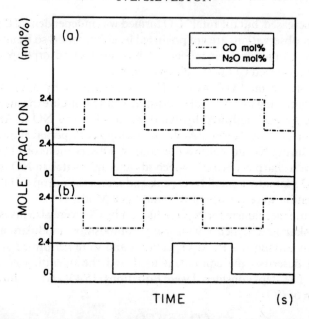

FIGURE 15.16 Out of phase cycles of reactant flow or composition used for model discrimination: (a) 90° phase lead of the N_2O feed, (b) 270° phase lead of the N_2O feed (Figure adapted from Sadhankar and Lynch (1994) with permission, © 1994 Academic Press)

Figure 15.17 shows their observations. To resolve the poor fit, Sadhankar and Lynch followed Chiao and Rinker (1989) and introduced a factor for transformation of the surface under cycling. They referred to this as an adsorbate induced phase transformation. The factor as a property of cycling had to be evaluated from the forcing data. Figure 15.17 shows that including the factor in the CO exclusion model now fits the experimental CO_2 data satisfactorily. CO_2 rather than N_2O or N_2 data are used because more accurate measurements can be made of CO_2 concentrations.

Sadhankar and Lynch (1994) undertook forcing with different phase leads to further test their model. Results are illustrated in Figure 15.18 which compares model prediction and data for N_2O phase leads of 90° and 270°. Use of $\pi/2$ or $3\pi/2$ phase leads with $s = 0.5$ results in 4 part cycles with equal part durations. These parts are traced out in the figure by the N_2O and CO concentrations. Predicting rate or concentration behavior within a cycle is an extremely sensitive test of model performance. It is evident from Figure 15.18 that the CO exclusion, phase transformation model for N_2O reduction by CO over a Pt catalyst is a very good model indeed.

Systematizing model discrimination using instantaneous rates or concentrations measured at points within a packed bed of catalyst during

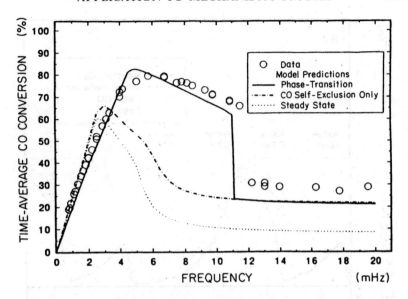

FIGURE 15.17 Comparison of calculated and experimental CO_2 conversion as a function of ω for the symmetrical bang-bang cycling of N_2O and CO from 0 to 2.4 vol.% over a Pt/Al_2O_3 catalyst at 499°C. Three models are equilibrium adsorption of CO, adsorption of CO with CO exclusion and CO exclusion modified for phase transformation of the Pt surface (Figure adapted from Sadhankar and Lynch (1994) with permission, © 1994 Academic Press)

a cycle has been undertaken by Yuan and co-workers using the partial oxidation of benzene to maleic anhydride over a vanadia catalyst as a test system. As mentioned above, this appears to be the most sensitive use of modulation data. Tan *et al.* (1994) propose applying the Karhunen-Loeve expansion. Let C(z, t) be the predicted concentration of one of the competing models at a point z in the catalyst bed at time t. Subtracting the time average value gives a deviation **V** (z, t). Assuming the predicted deviations are part of a stationary time series, deviations at n different points for m different times form a matrix **V** (z, t). $\mathbf{C} = \mathbf{VV}^T$ and $\mathbf{D} = \mathbf{V}^T\mathbf{V}$ can be shown to be non-negative, real symmetrical matrices sharing a set of common coefficient λ_i. The Karhunen-Loeve expansion sets

$$\mathbf{V}(z, t) = \sum_{i=1}^{r} \lambda_i \phi_i(z) \xi_i^T(t) \qquad (15.51)$$

where $\phi_i(z)$ and $\xi_i^T(t)$ are terms in the eigenfunctions of **C** and **D** respectively and are orthogonal. The sum in Equation (15.51) can be truncated to h terms where h is related to some target error. The eigenfunction terms $\phi_i(z)$ can be evaluated from the model once its parameters are determined

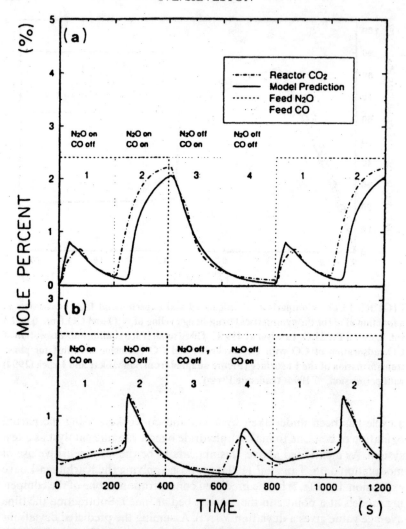

FIGURE 15.18 Comparison of experimental CO_2 concentrations leaving a differential fixed bed reactor with predictions from a modified CO exclusion model for N_2O reduction by CO for the system and conditions in Fig. 15.17, however, the cycles use (a) a 90° phase lead of the N_2O feed and (b) a 270° phase lead (See Fig. 15.17) (Figure adapted from Sadhankar and Lynch (1994) with permission, © 1994 Academic Press)

and measured eigenfunction terms $\phi'_i(z)$ can be calculated from the experimental observations through Equation (15.51). This involves recording instantaneous rates or concentrations at n points in the bed at m different times. These are used to establish time-average values at the n points which are used to find $\mathbf{V}(z, t)$ for the catalyst bed.

To demonstrate the discrimination procedure, Tan et al. collected data in a 12 mm (i.d.) × 200 mm bed of 40–60 mesh commercial V_2O_5 for the well studied partial oxidation of benzene to maleic anhydride at atmospheric pressure and three temperatures (633, 653 and 673 K). The catalyst bed was maintained isothermal by a thick copper sheath outside of the reactor shell. This sheath and the reactor shell were pierced by 4 sampling tubes located 42, 100, 158 and 196 mm from the top of the bed. Feed composition modulation was achieved by solenoid valves which switched flow from an air-N_2 stream containing 1.4 vol.% Bz to an air-N_2 stream without Bz. Cycles were symmetrical and τ was 60 s. A quadrupole mass spectrometer monitored the gas composition of the samples which were collected and used to construct the deviation matrix $V(z, t)$. Step change measurements were also made for parameter estimation.

Tan et al. considered the mechanisms for the partial oxidation given in Table 15.2. Rate expressions for each mechanism are given in the table and were used with an isothermal, one dimensional plug flow model:

$$\frac{\partial C_k}{\partial t} = -\frac{u}{\varepsilon_B}\frac{\partial C_k}{\partial z} - \frac{\rho_B}{\varepsilon_B} R_{ik} \tag{15.52}$$

$$\frac{\partial \theta_s}{\partial t} = -\frac{R_{is}}{Q_{is}} \tag{15.53}$$

$$\sum \theta_s = 1 \tag{15.54}$$

In the these material balances, i designates the model, k the components in the gas phase and s the surface species. Q_{is} is the surface capacity at saturation for the species s. R_{ik} are the first 4 terms for each model in Table 15.2, while R_{is} are the last 3 or 4 terms. Boundary equations depend on the start up condition: either an empty reactor and an empty catalyst surface or from steady state. The initial condition for Bz and O_2 depend on whether the experiment is periodic or employs a step change.

Parameters for each model were estimated by discretizing the z dimension and expanding Equations (15.52) to (15.54) using Eqn. (15.51) to obtain a set of ordinary differential equations. These were integrated numerically and fitted to the step change data using a simplex routine to minimize the sum of the errors at all of the measurement points. Four of the 6 models were discarded at this stage because parameters were either negative or the residual errors could not be reduced to the target level.

The remaining models, Models 2 and 3, were tested against the modulation data using the Karhunen-Loeve expansion. Figure 15.19 shows the test results. In (a) using the maleic anhydride data, model discrimination is poor, although it is evident that Model 3 is preferred. However, if the benzene data are employed, the model predictions are dramatically different and there is no question that Model 3 is superior.

Tabel 15.2 Competing Mechanistic Models and Rate Expressions for the Oxidation of Benzene to Maleic Anhydride (Tan *et al.*, 1997)

Model 1

$$O_2(g) + s \xrightarrow{k_1} O_2 - s \qquad R_{O_2} = k_1 p_{O_2} \theta_s$$

$$O_2 - s \underset{k_3}{\overset{k_2}{\rightleftharpoons}} 2O_l \qquad R_{Bz} = k_4 p_{Bz} \theta_{O_2}$$

$$Bz(g) + s \xrightarrow{k_4} Bz - s \qquad R_{MA} = k_5 \theta_{Bz} \theta_O$$

$$Bz - s + 8O_l \xrightarrow{k_5} MA + s + CO + CO_2 + 2H_2O \qquad R_{CO_2} = k_5 \theta_{Bz} \theta_O + 6k_6 \theta_{Bz} \theta_{O_2} + 6k_7 p_{Bz} \theta_{O_2}$$

$$Bz - s + 7O_2 - s \xrightarrow{k_6} 5CO_2 + CO + 3H_2O + 8s \qquad R_{\theta O_2} = k_6 \theta_{Bz} \theta_{O_2} + k_7 p_{Bz} \theta_{O_2} + k_2 \theta_{O_2}$$
$$- k_3 \theta_O - k_1 p_{O_2} \theta_s$$

$$Bz(g) + 7O_2 - s \xrightarrow{k_7} 5CO_2 + CO + 3H_2O + 7_s \qquad R_{\theta Bz} = k_5 \theta_{Bz} \theta_O + k_6 \theta_{Bz} \theta_{O_2}$$
$$- k_4 p_{Bz} \theta_s$$
$$R_{\theta_O} = k_3 \theta_O + k_5 \theta_{Bz} \theta_O - 2k_2 \theta_{O_2}$$
$$\theta_s = 1 - \theta_{O_2} - \theta_{Bz}$$

Model 2

$$O_2(g) + s \xrightarrow{k_1} O_2 - s \qquad R_{O_2} = k_1 p_{O_2} \theta_s$$

$$O_2 - s \underset{k_3}{\overset{k_2}{\rightleftharpoons}} 2O_l \qquad R_{Bz} = k_4 p_{Bz} \theta_s + k_5 p_{Bz} \theta_O$$

$$Bz(g) + s \xrightarrow{k_4} Bz - s \qquad R_{MA} = k_7 \theta_{MA}$$

$$Bz(g) + 8O_l \xrightarrow{k_5} MA + CO + CO_2 + 2H_2O \qquad R_{CO_2} = k_7 \theta_{MA} + k_6 \theta_{Bz} \theta_{O_2}$$

$$Bz - s + 7O_2 - s \xrightarrow{k_6} 5CO_2 + CO + 3H_2O + 8s \qquad R_{\theta O_2} = k_6 \theta_{Bz} \theta_{O_2} + k_2 \theta_{O_2}$$
$$- k_3 \theta_O - k_1 p_{O_2} \theta_s$$

$$MA - s \xrightarrow{k_7} MA + s \qquad R_{\theta Bz} = k_6 \theta_{Bz} \theta_{O_2} - k_4 p_{Bz} \theta_s$$
$$R_{\theta_O} = k_3 \theta_O + k_5 \theta_{Bz} \theta_O - 2k_2 \theta_{O_2}$$
$$R_{\theta_{MA}} = k_7 \theta_{MA} - k_5 \theta_{Bz} \theta_O$$
$$\theta_s = 1 - \theta_{O_2} - \theta_{Bz} - \theta_{MA}$$

Model 3

$$O_2(g) + s \xrightarrow{k_1} O_2 - s \qquad R_{O_2} = k_1 p_{O_2} \theta_s$$

$$O_2 - s \underset{k_4}{\overset{k_3}{\rightleftharpoons}} 2O_l \qquad R_{Bz} = k_6 p_{Bz} \theta_s + k_2 p_{Bz} \theta_{O_2}$$

$$Bz(g) + 4O_2 - s \xrightarrow{k_2} Bz(O_2 - s)_4 \qquad R_{MA} = k_5 \theta_{in} \theta_O$$

APPLICATION TO MECHANISTIC STUDIES 561

Tabel 15.2 (Continued)

$Bz(O_2-s)_4 + \xrightarrow{k_5} MA + 3s + 2CO_2 + 2H_2O$
$\quad R_{CO_2} = 2k_5 \theta_{in} \theta_O + 6k_7 \theta_{Bz} \theta_{O_2}$

$Bz-s + 7O_2-s \xrightarrow{k_7} 5CO_2 + CO + 3H_2O + 8s$
$\quad R_{\theta_{O_2}} = k_2 p_{Bz} \theta_{O_2} + k_3 \theta_{O_2} +$
$\quad k_7 \theta_{Bz} \theta_{O_2} - k_4 \theta_O$
$\quad - k_1 p_{O_2} \theta_s$

$Bz(g) + s \xrightarrow{k_6} Bz-s$
$\quad R_{\theta_{Bz}} = k_7 \theta_{Bz} \theta_{O_2} + k_6 p_{Bz} \theta_s$
$\quad R_{\theta_O} = k_5 \theta_{in} \theta_O + k_4 \theta_O - 2k_3 \theta_{O_2}$

$\quad R_{\theta_{in}} = k_5 \theta_{in} \theta_O - k_2 p_{Bz} \theta_{O_2}$
$\quad \theta_s = 1 - \theta_{O_2} - \theta_{Bz} - \theta_{in}$

Model 4

$O_2(g) + s \xrightleftharpoons{k_1} O_2-s$
$\quad R_{O_2} = k_1 p_{O_2} \theta_s - k_2 \theta_{O_2}$

$O_2-s \xrightleftharpoons[k_4]{k_3} 2O_l$
$\quad R_{Bz} = k_5 p_{Bz} \theta_s$

$Bz(g) + s \xrightarrow{k_5} Bz-s$
$\quad R_{MA} = k_6 \theta_{Bz} \theta_O$

$Bz-s + 7O_2-s \xrightarrow{k_7} 5CO_2 + CO + 3H_2O + 8s$
$\quad R_{CO_2} = k_6 \theta_{Bz} \theta_O + k_7 \theta_{Bz} \theta_{O_2}$

$Bs-s + 8O_l \xrightarrow{k_7} MA + s + CO + CO_2 + 2H_2O$
$\quad R_{\theta_{O_2}} = k_2 \theta_{O_2} + k_3 \theta_{O_2}$
$\quad - k_7 \theta_{B_2} \theta_{O_2} - k_1 p_{O_2} \theta_s - k_4 \theta_O$
$\quad R_{\theta_{Bz}} = k_6 \theta_{Bz} \theta_O - k_5 p_{Bz} \theta_s$
$\quad R_{\theta_O} = k_4 \theta_O + k_6 \theta_{Bz} \theta_O$
$\quad - 2k_3 \theta_{O_2}$
$\quad \theta_s = 1 - \theta_{O_2} - \theta_{Bz}$

Model 5

$O_2(g) + s \xrightarrow{k_1} O_2-s$
$\quad R_{O_2} = k_1 p_{O_2} \theta_S - k_5 p_{O_2}^3 \theta_{in}$

$Bz(g) + s \xrightarrow{k_2} Bz-s$
$\quad R_{Bz} = k_3 p_{Bz} \theta_{O_2} + k_2 p_{Bz} \theta_s$

$Bz(g) + O_2-s \xrightarrow{k_3} BzO_2-s$
$\quad R_{MA} = k_5 p_{O_2}^3 \theta_{in}$

$Bz-s + 7O_2-s \xrightarrow{k_7} 5CO_2 + CO + 3H_2O + 8s$
$\quad R_{CO_2} = 6k_4 \theta_{Bz} \theta_{O_2} + k_5 p_{O_2}^3 \theta_{in}$

$BzO_2-s + 3O_2(g) \xrightarrow{k_5} MA + s$
$+ CO + CO_2 + 2H_2O$
$\quad R_{\theta_{O_2}} = k_3 p_{Bz} \theta_{O_2} + k_4 \theta_{Bz} \theta_{O_2}$
$\quad - k_1 p_{O_2} \theta_s$
$\quad R_{\theta_{Bz}} = k_4 \theta_{Bz} \theta_{O_2} - k_2 p_{Bz} \theta_s$
$\quad R_{\theta_{in}} = k_5 p_{O_2}^3 \theta_{in} - k_3 p_{Bz} \theta_{O_2}$
$\quad \theta_s = 1 - \theta_{O_2} - \theta_{Bz} - \theta_{in}$

Tabel 15.2 (Continued)

Model 6

$$O_2(g) + s_1 \xrightarrow{k_1} O_2 s_1 \qquad R_{O_2} = k_1 p_{O_2} \theta_{s_1}$$

$$Bz(g) + s_2 \xrightarrow{k_4} B_z - s_2 \qquad R_{Bz} = k_7 p_{Bz} \theta_{O_2} + k_4 p_{Bz} \theta_{s_2}$$

$$O_2 - s_1 \underset{k_3}{\overset{k_2}{\Leftrightarrow}} 2O_l \qquad R_{MA} = k_5 \theta_{Bz} \theta_O$$

$$Bz - s_2 + 8O_l \xrightarrow{k_5} MA + s_2 + CO \qquad R_{CO_2} = k_5 \theta_{Bz} \theta_O + 5k_7 p_{Bz} \theta_{O_2}$$
$$+ CO_2 + 2H_2O \qquad\qquad\qquad\qquad + 5k_6 \theta_{Bz} \theta_{O_2}$$

$$Bz - s_2 + 7O_2 - s_1 \xrightarrow{k_6} 5CO_2 \qquad R_{\theta_{O_2}} = k_5 p_{Bz} \theta_O - k_1 p_{O_2} \theta_{s_1}$$
$$+ CO + 3H_2O + s_2 + 7s_1 \qquad\qquad + k_2 \theta_{O_2} - k_3 \theta_O$$

$$Bz(g) + 7O_2 - s_1 \xrightarrow{k_7} 5CO_2 \qquad R_{\theta_{Bz}} = k_5 \theta_{Bz} \theta_O + k_6 \theta_{Bz} \theta_{O_2}$$
$$+ CO + 3H_2O + 7s_1 \qquad\qquad\quad - k_4 p_{Bz} \theta_{s_2}$$
$$\qquad\qquad\qquad\qquad\qquad R_{\theta_{Bz}} = k_3 \theta_O - 2k_2 \theta_{O_2}$$
$$\qquad\qquad\qquad\qquad\qquad \theta_{s_1} = 1 - \theta_{O_2}$$
$$\qquad\qquad\qquad\qquad\qquad \theta_{s_2} = 1 - \theta_{Bz}$$

OVERVIEW AND RESEARCH CHALLENGES

Our intent in this chapter has been to demonstrate that composition modulation has a place in research apart from investigating whether or not dynamic operation or stationary systems with circulating catalyst beds can improve selectivity, yield or throughput. Although studies of modulation for mechanism identification are only about a decade old, it is clear that this is a promising use of modulation. Furthermore, modulation equipment is relatively simple, at least for square wave cycling, and measurements require no more effort than following concentrations after a step change. There is no experimental barrier for the application of modulation to mechanism studies.

Qualitative applications for ammonia synthesis and the use of a DRIFT cell by Marwood *et al.* (1994, 1996) have demonstrated the successful use of modulation in mechanistic studies. In the later case, they were used under conditions where the relaxation time of the catalytic processes were relatively slow so that time changes within a cycle were observable. In principle, modulation was not necessary, step change or pulse response could have been used instead. Response techniques, however, become difficult to apply when the relaxation processes are rapid. Modulation could be advantage-

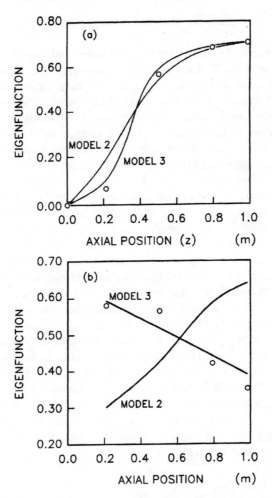

FIGURE 15.19 Comparison of calculated and experimental eigenfunctions for Models 2 and 3 in Table 15.2 for composition modulation of the partial oxidation of benzene over V_2O_5 at atmospheric pressure and 673 K with u = 0.677 m/s, s = 0.5 and τ = 60 s: (a) maleic anhydride, (b) benzene (Figures adapted from Tan *et al.* (1994) with permission, © 1995 Elsevier Science Ltd.)

ous in such cases. High frequency cycling should be able to stabilize a surface species that might otherwise be unobservable so that the species' response to changing mean concentrations, temperature or other operating conditions can be determined and related to product behavior thereby identifying the species as a reaction intermediate or not as one. This is an important and difficult research challenge. The first task will be to identify a suitable catalytic system for study, one which offers other means of identifying intermediates so that the modulation observations can be tested.

The use of sinusoidal forcing for measurement of adsorption/desorption rates is well established. Experimentally, however, sinusoidal forcing is more difficult than square wave forcing. Amplitude ratio and phase lag relationships for sinusoidal inputs need to be extended to square waves, or better still arbitrary forcing to provide simple methods for extracting adsorption or desorption rate constants. Fitting in the Fourier or other transform domain, of course, is possible, but it is difficult and often leads to high correlation between parameters obtained from fitting.

Several examples in this chapter have shown that modulation experiments lead to model discrimination in cases where step change experiments have failed. It is not clear why this should be. Can it be demonstrated that this is always true? The investigations of Li et al. (1985b) and Nam et al. (1990) suggest that a model which closely predicts the resonance frequency and the rate or selectivity enhancement at this frequency may not be able to predict the variation of reactant or product concentration with time within a cycle. We concluded from this observation that reproducing concentration variation is more sensitive for model discrimination than reproducing resonance behavior. Is this correct? If it is, why is this true? Model discrimination may prove to be an important laboratory application of composition modulation. The challenge is determine the best way of using modulation. The work of Tan et al. (1994) discussed in the previous section provides a method, but further study of this approach seems called for. Truncation of an infinite sum of terms is necessary, but it is not known what the effect of truncation is on discrimination. Probably, there is no general answer to this question. Truncation effects are likely to be model dependent. Further research is needed.

Author Index

(*Roman numerals indicate bibliography pages*)

Abdel'gani, A. Kh.   225   xxiii
Abdul-Kareem, K. H.,   15, 72–73, 92–95   xii, xxxiii
Adesina, A. A.   22, 466–468, 479–481, 486–488, 490, 495   xii, xxxiv
Ai, D.   412   xxv
Akehata, T.   499, 509–510   xx
Alassia, L. M.   499, 513,   xii, xvi
Al-Taie, A. S.   350–352, 357   xii
al Mutaz, I.   540–541   xix
Amariglio, A.   369, 373–384   xii, xiv, xxiv, xxvii
Amariglio, H.   2, 22, 211, 225–236, 249–253, 260, 265–266, 369, 373–384, 488   xii, xiv, xxiv, xxvii, xxx
Amenomiya, Y.   386, 390, 393, 461   xiii
Anderson, R. B.   464, 466   xiii
Andersson, B.   23   xxi
Andersson, S.   177   xxv
Annapragada, A.   44, 48   xxxvi
Aris, R.   51, 66, 68   xxii, xxvi
Arpe, H.-J.   295, 312   xxiv
Auguste, S.   499, 515–516   xviii–xix
Ausikaitis, J.   8   xiii

Baiker, A.   97, 163–168, 183, 189–190   xiii, xv, xxix
Bailey, J. E.   6–7, 22–23, 25, 58, 77, 90, 100, 107–110, 113, 183, 186–189, 243, 335, 339–342, 350, 353–357, 359, 499, 503, 524–526   xiii–xiv, xvi, xxi, xxiv, xxix, xxxiii
Baldi, G.   190   xxxvi
Balzhinimaev, B. S.   267, 280–283, 409–412   xiii, xxi, xxiii, xxvi
Banderman, F.   499, 512   xiii, xxiii–xxiv
Baras, F.   24   xix
Baron, K.   62–64, 150   xx, xxviii
Barshad, Y.   16, 22, 40–48, 55, 97–99, 466–468, 471–474, 476, 488, 493–497   xiii–xiv, xxxviii
Bartsch, S.   393   xiv
Bassett, M. R.   110, 122–123, 131   xviii
Baudais, F. L.   13, 15, 95   xxix
Beachell, H. C.   406   xviii
Beck, D. D.   175–177   xxvii
Belgued., M.   369, 373–383   xii, xiv
Bell, A. T.   370–371, 381, 464, 466, 473, 493–494   xiv, xxii, xxxvii
Belles, F. E.   26   xxx
Belot, G.   177–178   xxxiv
Belyeava, N. P.   267   xiii
Bennett, C. O.   493, 530   xiv, xxvi
Bergna, H. E.   20–21, 440–444, 462   xvi
Bergougnan, M.   189–190   xiii
Bernhardt, P.   153, 155–156   xxxvii
Bhasin, M. M.   24, 386, 388, 397, 460   xxii

Bi, Y.-L.   393   xiv
Bics, V. I.   386, 390, 393, 461   xiii
Bijsterbosch, J.   177–178   xxxiv
Bilimoria, M. R.   350, 353–357   xiv
Bitai, I.   541–543   xxviii
Björnkvist, L.   177   xxv
Blackstone, C. M.   20–21, 440–444, 462   xvi
Bogdashev, S.   282, 286   xxxiv
Borckmans, P.   24   xix
Boreskov, G. K.   22–23, 267, 291   xiv, xxiii, xxvii
Boudart, M.   244–245, 260   xxix
Boylan, D. R.   318   xxxvii
Brazdil, J. F.   417–418, 420, 532   xiv
Briggs, J. P.   8, 18, 267–268, 271–280, 283, 285–286, 288   xxii, xiv
Bucker, E. R.   466   xxv
Budman, H.   401   xv
Bulko, J. B.   363   xx
Bunimovich, G. A.   287–291   xiv–xv, xxxiv

Callahan, J. L.   19, 24, 413   xv
Capsaskis, S. C.   66–67   xv
Casanova, R.   97   xiii, xv
Cavendish, J. C.   62–64, 99   xxviii
Chakrabarty, T.   73   xv
Chanchlani, K. G.   9–11, 14, 199–206, 363–364, 369   xv
Chang, C. C.   143–146   xx
Chappelear, D. L.   499, 505–507   xxxiv
Charon, O.   27–29   xv
Chartier, P. A.   499, 510–512   xv
Chen, H. T.   499, 510–512   xv
Chiao, L.   18, 23, 211, 213, 218, 221–224, 226, 245–249, 255, 259–265, 531, 556   xv, xxxv–xxxvi
Cho, B. K.   58, 63, 77, 83–89, 99, 103, 107–110, 129–130, 265   xv–xvi, xxix
Chouichi, H.   326–329   xxxv
Chowdhry, U.   20–21, 440–444, 462   xvi
Claybaugh, B. E.   499, 525–526   xvi
Contractor, R. M.   20–21, 440–447, 462   xvi
Cordova, V. R.   17, 24, 448–450   xvi
Couso, D. A.   499, 504, 513–517   xvi, xviii
Creaser, D.   427–428   xvi
Crone, G.   499, 507–508   xvi
Cserényi, J.   369, 373, 381–382   xvii, xxxiv
Cutlip, M. B.,   39–41, 58–59, 67   xvi, xviii

Dai, Y.-C.   557–564   xxxv
D'Aniello, M. J. Jr.   175–177   xxvii
Dath, J.-P.   110, 122–123, 131   xvi, xviii
Davis, T. P.   543   xvi–xvii
Deelen, M. J. A. G.   369–374   xxiii
De Jong, W. A.   192   xxxvi
Dekker, N.   177–178   xxxiv
Delgass, W. N.   493   xxx
Dettmer, M.   298–302, 368   xvii, xxxi
Dialer, K.   406–407   xxix
Doepper, R.   194–195, 305, 368, 406, 532–538, 549, 553–555, 562   xxvi, xxx–xxxi, xxxvii
Dolya, L. P.   429–430   xvii, xxxiii
Dong, Y.   324   xvii

AUTHOR INDEX

Doroshenko, V. A.
   429–430   xvii
Douglas, J. M.   22, 25, 498   xvii
Dun, J.-W.   466–468, 475–479
   xvii
Dywer, D. J.   473   xvii

Ebner, J.   440, 445–447   xvi
Edvinsson, R.   303–305   xxi
Edwards, H. G. M.   499, 515–516
   xviii
Eiswirth, M.   68–71
   xvii, xxiii, xxxii
Elicabe, G. E.   499, 504, 513–515
   xvii–xviii
El Masry, H. A.   303–305   xvii
Emig, G.   58–59   xxv
Engel, A. J.   8   xiii
Erdöhelyi, A.   369, 373, 381–382
   xvii, xxxiv
Ertl, G.   24, 68–71, 110, 122–123
   xvi–xviii, xxiii, xxxii

Fabbricino, L.   15, 72–73,
   196–198, 363   xxxiii, xxviii
Falkowski, J.   393   xiv
Fan, C. W.   318   xxxvii
Farhadpour, F. A.   23, 339
   xvii–xviii
Feimer, J. L.   13, 22, 466–471,
   490, 493–495   xviii, xxxiv
Felvgi, A.   373, 381   xxxiv
Fink, Th.   110, 122–123, 131
   xvi, xviii
Finn, B. P.   363   xx
Fiolitakis, E.,   22, 450–454   xviii
Fjeld, M.   25   xviii
Flank, W. H.   406   xviii
Floy, K. R.   318   xxxvii
Forrissier, M.   2, 22, 418–420,
   462–463   xxxiii
Frestad, A.   177   xxv
Frontini, G. L.   499, 504, 513–515,
   517   xii, xviii, xxxvi

Fu, J.   324   xxiv
Fujitani, Y.   29, 33–37, 44, 55,
   73–76, 103–107, 137–138,
   141–143, 145–149, 151, 155,
   157–158, 160–163, 168–174,
   179, 181, 208–210   xxvii–xxviii,
   xxxii–xxxiii, xxxvii

Galuszka, J.   386, 390, 393, 461
   xiii
Garin, F.   153, 155–156
   xxxvii
Gau, G.   11, 16–17, 21, 24, 402,
   407–412, 448–450, 531
   xiii, xvi, xviii, xxix
Gelbein, A. P.   21, 24, 457–460
   xxxv
Genies, B.   27–29   xv
Ghazali, S.   11, 407–408, 531
   xviii, xxix
Gibilaro, L. G.   23, 339
   xvii–xviii
Gilliland, E. R.   19–20, 24, 413,
   457   xxiv
Gleixner, G.   541–543   xxviii
Goldman, O. V.   291   xv
Goledzinowski, M.   386, 390, 393,
   461   xiii
Goodman, M. G.,   58–59, 72–73
   xviii
Gonzalez, R. D.   537–539   xxv
Gordon, D. L.   529   xviii
Gosden, R. G.   499, 515–516,
   519–523   xviii–xix
Grabmueller, H.   23   xix
Graham, W. R. C.   44, 51–62,
   98–100, 112   xix
Grasselli, R. K.   19, 24, 413,
   417–418, 420, 532   xiv–xv
Gray, P.   24   xix
Greger, M.   420–421   xix
Griffiths, B.   20–21, 440–444,
   462   xvi
Griffon, J. R.   499, 525–526   xvi

AUTHOR INDEX

Gugliotta, L. M. 499, 527 xix
Gulari, E. 6, 22, 40–48, 55, 97–99, 466–468, 471–479, 488, 493–497, 532–533 xiii–xiv, xvii, xxxvi, xxxviii
Gupta, S. K. 499, 522–524 xix

Haber, J. 420–421 xix
Halpern, B. 540–541 xix
Hamielec, A. E. 527 xxix
Hashimoto, S. 499, 509–510 xx
Hattori, T. 7, 530 xx
Haure, P. M., 23 xx
Hawkins, C. J. 58–59, 67 xvi
Hegedus, L. L. 62–64, 77, 86, 99, 143–146, 150 xx, xxviii
Helmrich, H. 22, 183–186, 364–367 xx, xxxi
Henry, B. E. 24, 71 xxxii
Herman, R. G. 363 xx
Herz, R. K. 134–138, 140–141 xx
Hinkelmann, F. 541–543 xxviii
Hirako, O. 153 xxii
Hoffman, T. W. 454–455 xxxvi
Hoffmann, U. 23 xix, xxxii
Hofmann, H. 22–23, 393, 450–454, 530 xiv, xviii, xxvii, xxxii
Hogan, L. 306 xx
Horn, F. J. M. 22, 25, 243, 267, 335 xiii, xx–xxi
Horowitz, H. S. 20–21, 440–444, 462 xvi
Huang, J. 371–372 xxxvii
Huber, S. L. 72–73 xxxiii
Hudgins, R. R. 8–16, 18, 22–23, 58, 61, 72–73, 90, 92–95, 100, 120, 196–208, 211–221, 224, 226, 237–239, 243, 252–258, 263, 265, 267–280, 282–283, 285–286, 288, 303–305, 363–364, 369, 387, 393–401, 420–426, 435–441, 454, 466–473, 475, 479–488, 490, 493–495, 497, 531–532, 543–548, 552, 564 xii, xiv–xv, xviii, xx–xxv, xxvii–xxix, xxxi–xxxiv, xxxvi
Hugo, A. J. 16, 90–92, 120 xxi
Hungenberg, K. D. 499, 525–526 xxi

Ihme, B. 420–421 xix
Imbihl, R. 70–71, 110, 122–123, 131 xvi–xviii
Inoue, H. 353, 355–356 xxiii
Inoue, K. 23, 326, 332 xxxv
Irandoust, S. 23 xxi
Irish, D. E. 13, 15, 95 xxix–xxx
Isaacs, M. 21, 24, 457 xxxii
Ito, T. 393 xxi
Ivanov, A. A. 267, 280–283 xiii, xxi, xxiii, xxvi–xxvii

Jaeger, N. I. 24, 80–83, 192, 194, 210 xxi, xxv, xxxiv–xxxv
Jagtap, S. B. 315, 323–324 xxi
Jain, A. K. 8, 15, 58, 61, 72–73, 92–95, 100, 120, 211–221, 224, 226, 237–243, 252–253, 257, 265, 482, 543–545, 547 xxi–xxii, xxxiii
Jakelski, D. M. 16, 90, 92, 120 xxi
Jiang, Q. 412 xxv
Jiang, Y.-T. 393 xiv
Johnson, A. F. 499, 508–509, 512, 515–523 xviii–xxix, xxvi
Jones, C. A. 386–387, 389–392, 397 xxii, xxxiv
Jorti, A. 22, 231–236 xxx
Jouvard, D. 27–29 xv
Joy, G. J. 175, 177 xxii

Kadlec, R. H. 23 xxxvi
Kaliaguine, S. 406 xxxvi

AUTHOR INDEX

Kane, D. M. 8 xxii
Kaneko, Y., 153 xxii
Kapteijn, F. 177–178, 325 xxvii, xxxiv
Karnatovaskaya, L. M. 267 xxvi
Kasemo, B. 177 xxv
Kawakami, W. 499, 509–510 xx
Keil, W. 78–80 xxii
Keimer, H. 406–407 xxix
Keller, G. E. 24, 386, 388, 397, 460 xxii
Kellner, C. S. 466, 473 xxii
Kenney, C. N. 58–59, 66–67 xv–xvi, xviii, xxvii
Kentfield, J. A. C. 26 xxx
Kerschenbaum, L. S. 350–352, 357 xii
Kevrekidis, I. G. 51 xxii
Khazai, B. 429–434 xxii, xxviii, xxxvi
Khodadadi, A. A. 466–468, 479, 482–486 xxii–xxiii
Kibby, C. L. 473 xxix
Kilty, P. A. 406 xxiii
King, D. L. 473 xxiii
Kiselov, O. V. 291 xiv
Klahn, J. 499, 512 xxiii
Kleta, J. B. 135–136, 140 xx
Klier, K. 363, 401 xx, xxix
Klumperman, B. 543 xxxii
Klusacek, K. 190–193 xxxiv
Kobayashi, H. 153, 406–407, 530 xxii, xxvii
Kobayashi, M. 410, 530 xxiii
Kobylinski, T. P. 363, 473 xx, xxix
Koerts, T. 369–374 xxiii
Komagoma, I. 153 xxii
Komiyama, H. 353, 355–356 xxiii
Konopnicki, D. 499, 503–506 xxiii

Kotter, M. 420–421 xix
Koubek, J. 306–309, 314, 550 xxiii
Kozyrev, S. V. 267 xxiii
Krischer, K. 69 xxiii
Krueger, M. H. 175–177 xxvii
Krylova, A. V. 225–226 xxiii
Kuan, C. N. 499, 510–512 xv
Kuester, J. K. 499, 503–506 xxiii
Kumar, A. 499, 522–524 xix
Kutnyaya, M. Yu. 429–430 xxv, xxxiii
Kzyonsek, M. 401 xv

Labastida–Bardales, J. R. 420–421, 531–532 xxiii
Lai, H. 324 xvii, xxiv
Lang, X.-S 11, 12, 22, 435–441, 454 xxiv
Langner, B. 512, 517 xxiv
Lapinaa, O. B. 267, 280–283 xiii, xxvi
Laurence, R. L. 498–499, 502, 505–507, 512 xxiv, xxxiv
Lee, C. K. 25, 339–342, 353, 355–356, 359, 499, 503, 524–526 xxiv
Lefort, L. 383–384 xxiv
Lenthe, M. 364–367 xxxi
Leonard, J. J. 386–387, 389–392, 397 xxii, xxxiv
Lester, G. R. 175, 177 xxii
Leupold, E. I. 295–298, 312, 367–368 xxiv
Lewis, W. K., 19–20, 24, 413, 457 xxiv
Li, C. 324 xxiv
Li, C.-Y. 120, 214, 252–255, 257, 263, 412, 543–545, 547, 552, 564 xxiv–xxv
Li, Y.-E. 537–539 xxv
Liauw, M. A. 82–83 xxv
Lin, C.-H. 393 xxi

Lin, R. C. 267 xxi
Lööf, P. 177 xxv
Lübke, M. 69 xxiii
Luk'yanenko, V. P. 429–430 xxv, xxxiii
Lunsford, J. H. 393 xxi
Luu, H. D. 15, 94–95 xxv
Lynch, D. T. 44, 51–62, 98–100, 110–121, 124–129, 131, 180, 538–539, 555–558 xix, xxv, xxxii
Lysukho, T. V. 429–430 xxxiii

MacGregor, J. F. 527 xxix
Madix, R. J. 540 xxxii
Madon, R. J. 466 xxv
Maire, G. 153, 155–156 xxxvii
Malone, B. 20–21, 440–444, 462 xvi
Manna, L. 190 xxxvi
Marconi, P. F. 499 xxvi
Marek, M. 80 xxvi
Marwood, M. 194–195, 532–538, 562 xxvi
Mastikhin, M. 267, 280–283 xiii, xxiii, xxvi
Matros, Yu. Sh. 22–23, 282, 286–291, 456, xiv, xxii, xxvi, xxxiv
Matsubars, M 23, xxxvii
Matsumoto, H. 493 xxvi
Matsunaga, S.-I. 137–138, 168–174, 181 xxvi
McBride, R. 21, 24, 457 xxxii
McCarty, J. G. 370–371, 381 xxvi
McEwen, D. J. 143–146 xx
McKarnin, M. A., 51, 66, 68 xxvi
McNeil, M. A. 195, 204–208 xxvi
Meira, G. R. 22–23, 499–500, 504, 508–509, 512–519, 527 xii, xvi–xix, xxvi, xxxvi

Meshcheryakov, V.D. 267 xxvii
Meszena, Z. G. 499, 519–523 xix
Mikami, J. 406–407 xxvii
Mielczarski, E. 369 xxvii
Milberger, E. C. 19, 24, 413 xv
Mitchell, P. J. 22, 137, 139–140, 153–154, 160, 170–171, 175 xxxii
Möller, P. 70–71 xvii, xxxii
Mohsin, M. A. 499, 519–523 xix
Molinaro, F. S. 175, 177 xxii
Monroe, D. R., 175–177 xxvii
Monteverdi, S. 369, 373–380, 383 xiv, xxvii
Moravek, V. 551 xxvii
Morris, C. E. 315–321 xxvii
Mortazavi, Y. 387, 393–401 xxvii
Morton, W. 58–59, 67 xvi, xviii, xxvii
Moulijn, J. A. 177–178, 325 xxvii, xxxiv
Mueller, M. 402–405, 530 xxvii, xxxi
Mukesh, D. 58–59, 67 xvi, xviii, xxvii
Mummery, M. J. 440, 445–447 xvi
Murakami, Y. 7, 413–417, 530, 532 xx, xxviii
Muraki, H. 29, 33–37, 44, 55, 73–76, 103–107, 137–138, 141–143, 145–152, 155, 157–158, 160–163, 168–174, 179, 181, 208–210 xxvi–xxviii, xxxii–xxxiii, xxxvii
Murchison, C. B. 429–434, 476 xxii, xxviii, xxxvi

Naito, K. 326–329 xxxv
Nakayama, O. 153 xxii

Nam, Y. W.   252–258, 263, 545–548, 552, 564   xxviii
Nappi, A.   15, 196–198, 363   xxviii
Nath, S.   499, 522–524   xix
Nicolis, G.   24   xix
Niwa, M.   413–417, 532   xxviii
Ng, K. Y. S.   476   xvii
Nowobilski, P. J.   23, 99, 309–312, 551   xxvii–xxviii

O'Driscoll, K. F.   543   xvi–xvii, xxxii
Oh, S. H.   62–64, 77, 86, 99   xxviii–xxix
Ohme, H.   326, 328–329   xxxv
Ohtsuka, Y.   326   xxviii
Olaj, O. F.   541–543   xxviii
Olsson, P.   292–294   xxix
Onken, H. U.   83–84   xxix
Onogi, K.   23   xxxvii
Ozaki, A.   244–245, 260   xxix

Padeste, L.   163–168   xxix
Pagni, R.   499   xxvi
Pannell, R. B.   473   xxix
Pareja, P.   369, 373–383   xii, xiv
Park, D. W.   11, 16, 21, 24, 407–412, 531   xiii, xviii, xxix
Pasek, J.   306–309, 314, 550   xxiii
Penlidis, A.   527   xxix
Pierson, W. R.   292   xxix
Pitchai, R.   401   xxix
Piton, M. C.   543   xvi–xvii
Plath, P. J.   80–83, 192, 194, 210   xxi, xxv, xxxv
Ponec, V.   464   xxix
Prairie, M. R.   77, 183, 186–189, 305   xxix, xxxv
Prasad, K. B. S.   190   xxx
Prauser, G.   406–407   xxix

Prigent, M.   153, 155–156   xxxvii
Prokopowicz, R. A.   13, 15, 95   xxix–xxx
Putnam, A. A.   26   xxx

Qin, F.   51, 65   xxx

Rabl, A.   368, 549   xxx
Rambeau, G.   2, 22, 211, 225–236, 249–253, 260, 265–265, 488   xii, xxx
Ramsay, J.   499, 508–509   xxvi
Randall, H.   305, 553–555   xxx, xxxvii
Rao, A. S.   190   xxx
Rao, M. B.   190   xxx
Raupp, G. B.   493   xxx
Rawlings, J. B.   527   xxxi
Ray, W. H.   498–502, 524, 527   xxxi
Razon, L. F.   24   xxxi
Reed, W. A.   19–20, 24, 413, 457   xxiv
Renken, A.   2, 22, 25, 29–33, 183–186, 194–195, 295–302, 305, 309–313, 335, 342–346, 357–362, 364–368, 402–406, 499, 507–508, 530, 532–538, 543, 549–555, 562   xvi–xvii, xx, xxiv, xxvi, xxx–xxxi, xxxv–xxxvii
Reuter, D.   21, 24, 457   xxxii
Richarz, W.   97, 183   xiii, xv
Riekert, L.   420–421   xix
Riel, T.   315–316, 322   xxxvii
Rinker, R. G.   18, 23, 99, 195, 204–208, 211–213, 218–224, 226, 242–249, 255, 259–265, 531, 556   xv, xxvi, xxxv–xxxvii
Rippin, D. W. T.   22, 25, 498   xvii
Roche, R.   177–178   xxxiv
Ross, G. S.   466–468, 471–473, 475, 490, 494–495, 497   xxxi, xxxiv

Rothschild, W. G. 175 xxxv
Roussel, J. 72–73 xxxiii
Ruzicka, V. 306–309, 314, 550 xxiii

Sachtler, W. M. H. 406 xxiii
Sadhankar, R. R. 110–121, 124–129, 131, 555–558 xxxii
Saint-Just, J. 369, 373–381, 383 xiv
Sakashita, Y. 326, 328–329 xxxv
Salah-Al Hahamed, Y. A. 421–426 xxxii
Sanayei, R. A. 543 xxxii
Sanger, A. 386, 390, 393, 461 xiii
Satoh, S. 406–407 xxvii
Schack, C. J. 208 xxvi
Schädlich, K. 23 xix, xxxii
Schlatter, J. C. 22, 137, 139–140, 150, 153–154, 160, 170–171, 175 xx, xxxii
Schlosser, E. G. 295, 312 xxiv
Schmid, M. 22, 110, 122, 368, 406, 450–454 xviii, xxxi
Schmidt, L. D. 24, 51, 66, 68, 71 xxii, xxvi, xxxii
Schmitz, R. A. 24 xxxi, xxxiii
Schöön, N.-H. 292–294 xxix
Schürgerl, K. 22, 183–186 xx
Schüth, F., 24, 71 xxxii
Schwankner, R. J. 71 xxxii
Schwartz, S. B. 110, 122 xxxii
Schwarz, J. A. 540 xxxii
Schwendeman, M. N. 21, 24, 457 xxxii
Scott, D. S. 328–333 xxxvii–xxxviii
Scott, S. K. 24 xix
Sell, J. A. 135–136, 140–141 xx
Setthachayanon, S. 499, 510–512 xv

Severyanin, V. S. 26 xxxiii
Shanks, B. H. 58, 90, 100, 107–110, 113 xvi, xxxiii
Shaporalova, L. P. 429–430 xvii, xxv, xxxiii
Sheintuch, M. 24 xxxiii
Sheplev, V. S. 267 xxvii
Shinjoh, H 29, 33–37, 44, 103–107, 137, 143, 145, 147–149, 155, 157–158, 160–162, 168–173, 208–210 xxviii, xxxii–xxxiii
Shinouskis, E. J. 77, 135 xx, xxix
Sicardi, S. 190 xxxvi
Silveston, P. L. 2, 8–16, 18, 22–23, 58, 61, 72–73, 90, 92–95, 100, 120, 196–206, 211–221, 224, 226, 237–239, 243, 252–258, 263, 265, 267–280, 282–283, 285–286, 288, 303–305, 328–333, 363–364, 369, 387, 393–401, 418–426, 435–441, 450–454, 462–463, 466–473, 475, 479–488, 490, 493–495, 497, 531–532, 543–548, 552, 564 xii, xiv–xv, xviii, xx–xxv, xxvii–xxx, xxxi–xxxiv, xxxvi–xxxviii
Simmons, G.W. 363 xx
Simonova, L. G. 267 xxvi
Sinkevitch, R. M. 22, 137, 153–154, 158–161, 170–171, 173–174 xxxii, xxxv
Sleight, A. W. 20–21, 440–444, 462 xvi
Slin'ko, M. G. 24 xxxiv
Slin'ko, M. M. 24 xxxiv
Sloan, E. M. 143–146 xx
Smith, L. L. 315–316 xxvii
Snyder, J. D. 291 xxxiv
Sobukawa, H. 44, 55, 73–76, 103–107, 137, 141–143, 146, 155, 157–158, 160–163, 168–173, 179 xxvii–xxviii
Sofranko, J. A. 386–387, 389–392, 397 xxii, xxxiv

AUTHOR INDEX

Solodkaya, V. S. 429–430 xxv, xxxiii
Solymosi, F. 369, 373, 381–382 xvii, xxxiv
Somorjai, G. A. 473 xvii
Spitz, J. J. 499, 505–507 xxxiv
Stamicarbon 424 xxxiv
Stanitsas, G. 16, 90–92, 120 xxi, xxxiv
Stankiewicz, A. 190 xxxvi
Stegenga, S. 177–178 xxxiv
Sterman, L. E. 23 xxxiv
Stoch, J. 420–421 xix
Strecker, H. A. 19, 24, 413 xv
Strots, V. O. 287–291 xv, xxxiv
Stuchly, V. 190–193 xxxiv
Su, E. C. 175 xxxv
Su, D. 412 xxv
Su, S. 305 xxxv
Subramaniam, B. 291 xxxiv
Sullivan, G. R. 16, 90–92, 120 xxi
Summers, J. C. 150 xx
Suresh, D. D. 417–418, 420, 532 xiv
Suzuki, T. 23, 326–329, 332 xxxv
Svensson, P. 80–88, 192, 194, 210 xxi, xxxv
Swift, W. M. 315–321 xxxv
Sze, M. C. 21, 24, 457–460 xxxv
Szke, A. 373 xxxiv

Takoudis, C. G. 23, 99, 309–312, 551 xxvii–xxviii
Tamai, Y. 326 xxviii
Tan, H. 557–564 xxxv
Tao, L. 371–372 xxxvii
Taradi, C. C. 20–21, 440–444, 462 xvi
Taylor, H. S. 244–245, 260 xxix

Taylor, K. C. 22, 133, 158–161, 170, 173–174 xxxv
Taylor, W. F. 466 xxv
Teng, C.-W. 393 xiv
Thiele, R. 499, 505, 509 xxxv
Thullie, J. 18, 23, 211, 213, 218, 221–224, 226, 245, 259–260, 265, 309–313, 531, 550–552, 554 xv, xxv, xxxv–xxxvi
Toledo, E. 27–29 xv
Tomita, A. 326 xxviii
Truffer, M.-A. 299, 301–302, 368, 549–550 xxxi, xxxvi

Unger, B. D. 211–213 xxxvi
Unni, M. P. 22, 267–271, 288 xxxvi

Vamling, L. 14, 16–17, 25, 347–350 xxxvi
van Dierendonck, L. L. 190 xxxvi
Van Doesburg, H. 192 xxxvi
van Santen, R. A. 369–374 xxiii
Van Vyve, F. 194–195, 532–537, 562 xxvi
Vannice, M. A. 464 xxxvi
Vaporciyan, G. G. 23, 44, 48 xxxvi
Vasudevan, G. 498–499, 502, 512 xxiv
Vega, J. R. 499, 513–515 xxxvi
Verma, A. 406 xxxvi
Vernijakovskaja, N. 282, 286 xxxiv
Visser, J. B. M. 190 xxxvi
Voznyuk, V. I. 429–430 xxxiii
Vrieland, G. E. 429–434 xxii, xxviii, xxxvi

Wagler, R. 72–73 xxxiii
Wainwright, M. S. 454–455 xxxvi

Walker, B. 299 xxxvi
Walters, N. P. 538–539 xxv
Wandrey, C. 22, 29–33, 357–362, 402–405 xxxi, xxxvii
Wang, J. 412 xxv
Wang, J.-X. 393 xxi
Wang, L. 371–372 xxxvii
Wanke, S. E. 58–59 xxv
Watanabe, N. 23 xxxvii
Watanabe, Y. 23, 326–329, 332 xxxv
Watson, A. T. 499, 525–526 xvi
Weibel, B. 153, 155–156 xxxvii
Weidner, K. R. 529 xviii
Weihl, E. 429–433 xxviii
West, L. A. 83–89, 107 xv
Wetzl, K. 70–71 xvii, xxxii
Wheelock, T. D. 315–324 xxi, xxvii, xxxv, xxxvii
Wicke, E. 78–80, 83–84 xxii, xxix
Willcox, D. 537–539 xxv
Wilson, H. D. 211, 213, 218–221, 226, 242–265 xxxvii
Wiltowski, T. 420–421 xix
Winnik, M. A. 543 xvi–xvii
Winslow, P. 370–371, 381, 466, 493–494 xxxvii
Wise, H. 370–371, 381 xxvi

Wolf, E. E. 51, 65 xxx
Wolf, W. 69 xxiii

Xie, M. 371–372 xxxvii
Xu, G. 371–372 xxxvii
Xu, Y. 371–372 xxxvii

Yadav, R 99 xxxvii
Yang, X.-G. 393 xiv
Ydstie, B. E. 23 xxxiv
Yeramian, A. 267, 271 xxxviii
Yeung, S. Y. S. 25, 339–342, 359 xxiv
Yokota, K. 44, 137–138, 150, 155, 157–158, 160–162, 168–174, 181 xxvi, xxviii, xxxvii
Yuan, W.-K. 557–564 xxxv

Zack, F. K. 18, 211, 213, 218, 221–224, 226, 259–260, 265, 531 xv
Zamaraev, K. I. 267, 280–283 xiii
Zasa, Ph. 305 xxxv, xxxvii
Zhang, Z.-G. 328–333 xxxvii–xxxviii
Zhen, K.-J. 393 xiv
Zheng, D. 324 xxiv
Zhou, X. 41–48, 98–99, 488, 493–494, 532–533 xiv, xxxviii

Subject Index

ABB-Lummus-Crest process
  for aromatic nitriles
  Background for composition
    modulation 456
  Circulating fluidized bed 457
  Plant description 457
  Process flow sheet 457, 459
Acetylene hydrogenation
  Adsorbate storage in 356
  Experimental system 353
  Modeling 355
  Period effects 353
  Stoichiometric ratio effects 353
Adiabatic operation
  Ammonia synthesis 239
  Global enhancement with 239
  Temperature distribution
    in 240
  Temperature fluctuations
    in 240
Adsorbents mixed with catalysts
  in a single bed in ethene
  epoxidation 408
Adsorption
  Ammonia synthesis, effects
    in 227
  H_2 inhibition 227, 231
  Modeling of H_2 inhibition 249
  Rate of N_2 adsorption
    in ammonia synthesis 227
Adsorption rate measurement
  CO adsorption on platinum
    metals 537
  CO_2 adsorption on
    alumina 538
  Experimental system 537

Mathematical
    manipulation 538
Air Flushing
  Asymmetrical flushing 278, 289
  Breakthrough behavior 276
  Disymmetry 290
  Flow sheet 278
  Modeling 282
  SO_2 oxidation 272
  Temperature behavior 277
Air-fuel ratio
  Carbon monoxide oxidation
    and 140
  Catalyst aging effects and
    153, 170
  Control system for 135
  Coupled oxidation-reduction
    reactions and 142, 145
  Exhaust composition
    relationship and 133
  Fluidized bed decomposition of
    calcium sulfate 319
  Future challenges for
    composition modulation 182
  Light off and 160, 172
  Modulation of 132, 135, 163, 168, 171
  N_2O formation and 165
  Period effects in decomposition
    of calcium sulfate 319
  Performance improvement
    and 181
  Performance with automotive
    engine exhaust and 148, 168
  Performance with simulated

engine exhaust and 148, 153, 155, 160
Performance with sulfur poisoning and 175
Performance with three-way catalysts 150, 153, 156, 163, 168, 177
Stoichiometry and 140, 142
Variation in driving cycles 133
Alumina as catalyst or as catalyst support
 CO_2 adsorption rate measurements on 538
Ammonia synthesis
 Adiabatic operation 237
 Adsorption effects 227
 Bang-bang modulation 218, 221
 Best model 548
 Chromatographic studies 212
 Cycle split effects 214, 218
 Diffusion interference 235
 Discrimination between rival models 544, 547
 Dynamics of 216, 227, 265
 Experimental equipment 211
 Exposure duration effects 227
 Future challenges for composition modulation 263
 Global enhancement 215, 218, 220, 224, 239
 Introduction 211
 Iron catalysts 213, 218, 225
 H_2 conversion in 226
 Mechanism for improvement under modulation 216, 222, 225
 Modeling 242, 249, 252, 259, 543, 545
 N_2 adsorption rate in 227
 Osmium catalysts 231
 Overview of composition modulation 263
 Pressure effect 219
 Pulsing of H_2 reactant 225
 Resonance 214
 Ruthenium catalysts 226
 Stoichiometric ratio effect 215, 221, 227, 231
 Temperature effects 219, 227, 231
Amplification of periodic disturbances 83
Amplitude of modulation-see Modulation amplitude
Analysis
 Ammonia synthesis 242
Anderson-Schultz-Flory carbon number distribution in the Fischer-Tropsch synthesis
 Cobalt catalyst 481
 Iron catalyst 470
 Ruthenium catalyst 473, 475
Anionic living chain polymerization
 Comparison of simulation and experiment 521
 Control of molecular weight distribution 512
 Control of polydispersity 512, 516
 Cycle split effects 515
 Experimental problems 519
 Experimental system 517, 519
 Impurities effect 513
 Kinetic models 502, 513
 Modulation strategies 517
 Optimal control policy 513, 515
 Simulation of 511
 Styrene polymerization 521
 Ziegler-Natta polymerization by anionic mechanism 524
ARCO study of oxidative coupling of methane 388
Aromatic nitriles via xylene ammoxidation
 Assessment of composition modulation for 463
 Background for composition

SUBJECT INDEX

modulation 456
Circulating fluidized bed 457
Isophthalonitrile 457, 459
Lummus Co. process 457
Nicotinonitrile 459
Phthalonitrile/Phthalimide 459
Product distribution 457, 458, 460, 461
Terephthalonitrile 459
Attrition resistant catalysts for circulating fluidized beds
  Redox catalysts for butane dehydrogenation 432
  Redox catalysts for butane partial oxidation 442, 447
Automotive engine exhaust
  - see also Automotive exhaust catalysis
  Air-fuel ratio effects 132, 148, 168
  Characterization of 135
  Engine operation effects 132
  Light off of 172
  Periodic air purge 163
  Simulated - see Simulated automotive engine exhaust
Automobile exhaust catalysis - see Automotive exhaust catalysis or Three-way catalysts
Automotive exhaust catalysis
  Air-fuel ratio 132, 168, 171
  Amplitude effect 168, 171
  Background 132
  Catalyst development 148, 153, 163, 170, 177
  Characterization of composition modulation 135
  CO oxidation 29, 33, 39, 60, 73
  Coupled oxidation-reduction reactions 142
  Diffusion interference 83
  Dual oxidation reactions 141
  Engine operation and exhaust composition 132
  Experimental systems 137
  Future challenges for composition modulation 182
  Improvement of through composition modulation 181
  Light off on three-way catalysts 160, 172
  Monoliths as catalyst support 163, 173
  NO_x Reduction 103
  Overview of composition modulation 179
  Periodic air exposure effects 163
  Research objectives for composition modulation 136
  Simulated engine exhaust studies 148
  Single reactions 139
  SO_3 reduction by CO 292
  Source of modulation 135
  Stoichiometric number 74, 142, 171
  Sulfur poisoning 175
  Stoichiometric ratio effects 86
  Three-way catalysts 150, 153, 168
  Without noble metals 177
Autonomous oscillations 24, 51, 66, 78, 122

Bang-bang cycling
  Ammonia synthesis 218, 221, 233
  Benzene partial oxidation 448
  Butane dehydrogenation 437, 440
  Fischer-Tropsch synthesis 473, 476, 479, 482
  Oxidative coupling of methane 394, 398
  Propane dehydrogenation 427

Ziegler-Natta
  polymerization 525
Benzene partial oxidation
  Application of the Karhunen-
    Loeve expansion to 559
  Assessment of composition
    modulation 463
  Bang-bang cycling 448
  Best model 559
  Chemical mechanism 452
  Circulating fluidized bed 450
  Discriminating between rival
    models 559
  Estimation of model
    parameters 559
  Experimental system 559
  Frequency effects 453
  Kinetic models for 560
  Mechanism of improvement
    through modulation 450
  Modeling 559
  Multi-part cycles 448
  Redox catalysts 448, 450
  Successive pulsing for
    mechanistic studies 450
  Wave front analysis 450
Butadiene hydrogenation
  Amplitude effects 351
  Experimental system 350
  Modeling 351
  Period effects 351
Butadiene partial oxidation
  - see also butane partial oxidation
  Background 434
  Bang-bang cycling 437, 440
  Flushing with inert 438, 440
  Furan selectivity 435
  Mechanism of improvement
    through modulation 437
  Modulation strategies 435
  Multi-part cycles 435
  Period effects 435
  University of Waterloo
    studies 435

Butane oxidative dehydrogenation
  Attrition resistance 432
  Catalyst development 429
  Dow Chemical Co. studies 429
  Equipment for 431
  Experimental system 429, 439
  Reactor development 431
  Redox catalysts 429
Butane partial oxidation
  Attrition resistance 442, 447
  Catalyst development 442
  Du Pont studies 440
  Flushing with inert 445
  Mechanism of improvement
    through modulation 445
  Reactor development 441
  Recirculating solids reactor
    440, 446
  Selectivity dependence on
    conversion 444
  Stoichiometric ratio effect 444
  Temperature effects 441, 444
  Time-on-stream testing 447

Calcium oxide regeneration
  Air-fuel ratio cycling in
    a fluidized bed 323
Carbon deposition rate
  measurement through
  modulation 540
Carbon gasification
  Choice of catalyst 325
  CO_2 gasification of low
    rank coal 326
  Oxide state 326
  Redox mechanism 325
Carbon monoxide oxidation
  Autonomous oscillations
    51, 65, 78
  Diffusion interference 83
  Future challenges for
    composition modulation 101
  High temperature studies
    on noble metals 71

SUBJECT INDEX

Introduction to composition
modulation of 38
Low temperature studies on
noble metals 39, 56
Other noble metals 73
Overview of composition
modulation 98
Copper oxide catalysts 95
Hopcalite catalysts 97
Nickel oxide catalysts 90
Palladium catalysts 46, 73
Phase lag effects 55
Platinum catalysts - experimental
39, 51, 60, 72
Platinum catalysts - mechanism
43, 45, 47, 52, 76, 82, 99
Platinum catalysts -
modeling 56
Platinum/tin catalysts 48
Quasi-periodic behavior 49, 94
Silver catalysts 90
Single crystal studies - excitation
diagrams 66
Single crystal studies
- experimental 68
Single crystal studies
- mechanism 69
Single crystal studies
- model 69
Space velocity effects 53
Stoichiometric ratio effects
74, 86
Three-way catalysts 139
Vanadia catalysts 92
Wrong-way behavior 95
Catalyst aging in three-way
catalysts
Automotive exhaust and
153, 172
Simulated engine exhaust
and 165
Catalyst deactivation
Acetoxidation of ethene 368
Ethyl acetate from ethene and
acetic acid 298, 367
Methane homologation 377
Methanol synthesis 369
Catalyst development for
composition modulation
Attrition resistance 432
Butane dehydrogenation 431
Catalyst screening for composition
modulation
Butane dehydrogenation
429, 431
Oxidative coupling of
methane 388
Propene ammoxidation 413
Catalytic
Carbon monoxide oxidation - see
Carbon monoxide
oxidation
Coal Gasification - see Carbon
gasification
Destruction of automotive
exhaust - see automotive
exhaust catalysis and three-
way catalysts
N_2O reduction - see N_2O
reduction
NO_x reduction - see NO_x
reduction
Catalytic combustion
Propene 33
Propane 35
Catalytic converter
Composition of engine
exhaust 133
Operation of 134
see Automotive exhaust catalysis
and Three-way catalysts
Chain growth probability in the
Fischer-Tropsch synthesis 484
Chemical mechanism
Ammonia synthesis on iron
catalysts 244, 260
Benzene partial oxidation 452
CO_2 methanation 535

SUBJECT INDEX

Ethene epoxidation   406, 407
Fischer-Tropsch synthesis   491
Propene partial oxidation   417
SO_2 oxidation over vanadium catalysts   281
Chromatographic operation
  Ammonia synthesis   212
Circulating fluidized beds - see Fluidized beds as well
  Ammoxidation of aromatics to aromatic nitriles   457
  As implementation of composition modulation   19
  Benzene partial oxidation   450
  Butane partial oxidation   440, 446
  Ethene epoxidation   408
  Propene ammoxidation   413
  Xylene partial oxidation   456
Circulating solids reactors - see Circulating fluidized beds
Claus reaction
  Experimental system   303
  Period effect   304
  Stoichiometric ratio effect   303
CO_2 gasification of low ranked coal
  Cycle period effects   327, 329
  Experimental systems   327, 328
  Exposure duration effects   329
  Interpretation of off gas composition profile   329
  Rate controlling step   334
  Temperature effects   329, 334
CO_2 methanation
  Adsorption rate measurements   537
  Chemical mechanism   535
  DRIFT cell   533
  Experimental system   533
  Modeling   536
  Phase lag measurements   534
  Ruthenium catalysts   532
  Similarity of product response curves   534

Cobalt catalysts
  Fischer-Tropsch synthesis   467, 479
  Methane homologation   374
Combined catalyst and adsorbent beds   408
Combustion
  Catalytic   29
  Homogeneous   25
Competitive reactions   337, 363, 364, 367
Complex cycles - see Multi-part cycles
Composition forcing - see Composition modulation
Composition modulation
  Acetylene hydrogenation   353
  Adiabatic operation   239
  Adsorption rate measurements by   537
  Advantages of   385
  Aging for automotive engine exhaust and   160, 172, 175
  Air-fuel ratio and   132, 319
  Ammonia synthesis   211, 213, 218
  Amplitude effects with automotive engine exhaust   150, 155, 168
  Analysis for ammonia synthesis   242
  Anionic living chain polymerization   511, 517, 519
  Attainment of a reproducible cycling state   92, 118
  Aromatic nitriles processes   459
  Automotive engine exhaust   148, 153, 168, 170
  Automotive exhaust catalysis   29, 33, 73, 103, 136
  Autonomous oscillations   51, 65, 78
  Background   1, 22
  Bang-bang   219, 394

SUBJECT INDEX

Benzene partial oxidation 448
Butadiene hydrogenation 350
Butadiene partial
 oxidation 434
Butane dehydrogenation 429
Carbon deposition
 measurements on metals 540
Catalyst aging 153
Catalyst deactivation
 367, 368, 369
Catalytic combustion 29
Catalytic combustion of
 hydrocarbons 32, 357
Char gasification 325
Characterization of, with
 automotive exhaust 135
Circulating fluidized beds and
 408, 413
Claus reaction 303
CO oxidation 39, 56, 90
CO oxidation on single crystals
 66, 68, 69
CO_2 methanation 532
Combustion 25, 357
Deamination reactions 305
Decomposition of calcium sulfate
 or gypsum 315
Dehydration reactions 305
Dehydrogenation of
 methanol 305
Development of 22
Diethanolamine formation 342
Diffusion interference
 63, 83, 235
Discrimination between rival
 kinetic models 543
Emulsion polymerization 527
Equipment 12
Ethanol dehydration 364
Ethanolamine formation 347
Ethene oligomerization 383
Ethene partial oxidation 402
Ethyl acetate from ethylene and
 acetic acid 295

Ethyl adipate saponification
 339
Fischer-Tropsch synthesis 464,
 466, 471, 475, 479, 486
Free radical polymerization
 502, 505, 507
Flushing with inerts 397
Gas-solid reactions 315, 324
Global enhancement 23, 61, 99,
 296, 302, 310
High temperature furnaces 27
History of 22
Homogeneous combustion 25
Homogeneous reactions
 25, 338
Hysteresis 30
Isophthalonitrile from xylene
 ammoxidation 456
Ketones from partial oxidation of
 olefins 424
Light off with automotive
 exhausts and 160, 172
Limitation of, with ammonia
 synthesis 263
Mechanistic studies using
 530
Methane homologation
 369, 373
Methane partial oxidation
 to methanol 401
Methanol synthesis 363
Model verification 553
Modeling - see Modeling
Multi-part cycles 397
Multiple reactions 335
Multi-variable 513
Nature of 2
N_2O reduction by CO 103, 106,
 113, 124, 127, 129
NO_x reduction by CO on single
 crystals 121
NO_x suppression 27
Objectives of 8, 335
Overlapping cycles 398

Oxidative coupling of methane 386, 388, 395
Parametric sensitivity and 403
Parasitic reactions 367, 368
Performance indices 4
Periodic air purge in automotive exhaust catalysts 163
Periodic switching of flow direction and 401
Polymerization reactions 498
Poisoning in automotive exhaust with 175
Propane dehydrogenation 426
Propene ammoxidation 413
Propene epoxidation 409
Propene partial oxidation 418, 421
Regeneration of calcium oxide 323
Resonance at single frequency 39, 73, 110, 124, 129, 150, 155, 157, 214, 218, 269
SO_2 oxidation over vanadium catalysts 268, 271
SO_3 reduction by CO over platinum 292
Source of in automotive exhausts 135
Stabilizing an unsteady state by 29
Step growth polymerization 522
Stoichiometric ratio effects 74, 86
Stoichiometric number effects 158, 170
Stop effect 306
Strategy in 9, 347, 359
Total oxidation of hydrocarbons 357
Two variable 504, 512
Types of 5
Unstable steady states 29
Variables in 4

Wrong-way behavior 95
Xylene partial oxidation 454, 456
Ziegler-Natta polymerization 524
Consecutive-competitive reactions 339, 342, 347, 350, 353, 357
Copper-zinc oxide catalysts
 Methanol synthesis 363
Coupled oxidation-reduction reactions in automotive exhaust
 CO partitioning 145
 NO/CO 142
 Three-way catalysts 143
Deamination reactions
 Future challenges for composition modulation in 312
 Oxide catalysts 305
 Stop effect 306
 Temperature effects 306
Decomposition of calcium sulfate or gypsum
 Air-fuel ratio cycling in a fluidized bed 319, 323
 Experimental 317
 Lime recovery 315
 Temperature limits 316
 Two zone fluidized bed 316
Dehydration reactions
 Oxide catalysts 305
 Stop effect 306
Dehydrogenation of methanol on sodium carbonate
 Dynamic studies 305
 Mechanism studies 305
Diethanolamine formation
 Experimental system 343
 Global enhancement 343
 Kinetic effects on enhancement 346
 Modeling 345
Diffusion interference
 Ammonia synthesis 235

SUBJECT INDEX

CO oxidation 63, 83
Effect on enhancement in
 ammonia synthesis 237
Effectiveness factor 237
Gradientless reactor for
 measurements of 237
Modeling of, in CO oxidation
 63, 83

Dow Chemical Co. research on
 oxidative dehydrogenation
 Catalyst screening 429
 Reactor development 431
DRIFT cell, use of in CO_2
 methanation study 533
Du Pont research on butane partial
 oxidation
 Catalyst development 440, 442
 Reactor development 440, 441
Dual oxidation reactions in
 automotive exhaust
 CO/Propene 141
 Three-way catalysts 141
Duty fraction - see Cycle split
Dynamic (step change) methods for
 study of mechanism 530
Dynamics
 Adsorption modeling 253
 Ammonia synthesis 216
Dynamic measurements for
 mechanistic studies 305
Dynamic models - see Modeling

Effectiveness factors in composition
 modulation
 Ammonia synthesis 237
 Apparent effectiveness
 factor 235
Emulsion polymerization
 Co-polymerization 527
 Emulsifier flow
 interruption 528
 Future challenges for
 composition modulation 529

Mixing effects 527
Modeling 527
Particle size distribution effects
 in 527
Quasi-steady state model 527
Enhancement
 Definition of 4, 337
 Determination of 342
Enhancement factor
 Definition of 4, 337
Equilibrium conversion
 SO_2 oxidation 273
Equipment for composition
 modulation
 Ammonia synthesis 211, 239
 Automotive exhaust
 studies 137
 Circulating fluidized beds 19
 Automotive engine exhaust 137
 Industrial units 19
 Laboratory reactors
 13, 268, 273 295, 303
 Riser reactor for butane
 dehydrogenation 431
Ethanol dehydration
 Cycle split effects 365
 Exposure duration effects 365
 Mechanism for improvement
 under modulation 365
 Modeling 366
 Period effects 365
Ethanolamine formation
 Enhancement under modulation
 349
 Experimental system 348
Ethene acetoxidation - see Ethyl
 acetate from ethylene
 and acetic acid
Ethene epoxidation
 Circulating fluidized beds 408
 Experimental system 402, 408
 Gau studies 407
 Kinetic mechanism 406
 Mechanism of improvement

through modulation 405
Mixed beds of catalyst and
  adsorbent 408
Modeling 406, 412
Multi-part cycles 407
Parametric sensitivity 403
Period effects 402
Renken studies 402
Silver catalysts 402, 407, 412
Successive pulsing for
  mechanistic studies 406
Temperature distribution in
  packed beds 403
Ethene oligomerization
  Cycle structure 383
  Platinum catalyst 383
  Product distribution 383
  Temperature effects 383
Ethyl acetate from ethylene
  and acetic acid
  Acid catalysts 295
  Best model 550
  Catalyst deactivation in 298
  Cycle split effects 296
  Discrimination between rival
    models 550
  Experimental systems 295
  Future challenges for
    composition modulation 312
  Global enhancement 296, 302
  Mechanism for improvement
    under modulation 298
  Modeling 298, 549
Ethyl adipate saponification
  Experimental system 339
  Modeling 341
  Stoichiometric ratio effects
    340
Excitation diagrams 66, 122
Experimental methods
  Ammonia synthesis 211
Experimental systems - see also
  Equipment for modulation
  Acetylene hydrogenation 353

Benzene partial oxidation 559
Butadiene hydrogenation 350
Butane dehydrogenation
  429, 431
Claus reaction 303
CO_2 methanation 533
Diethanolamine formation 343
Ethanolamine formation 348
Ethyl acetate from ethene
  and acetic acid 295
Ethyl adipate saponification
  339
Fischer-Tropsch synthesis 465
Gas adsorption rate
  measurements 537
Hydrocarbon catalytic
  combustion 358
Methane homologation 373
Oxidative coupling of methane
  387, 388, 389
Polymerization 505, 509, 511,
  517, 519,
Propene partial oxidation 413
SO_2 oxidation 268, 273
Exposure duration effects
  Ammonia synthesis 227, 233
  Ethyl acetate from ethene and
    acetic acid 302
  Fischer-Tropsch synthesis 483
  Oxidative coupling of
    methane 391
  Stop effect 307

Feedback in automotive exhaust
  catalysis
  Air-fuel ratio control 135
  Characterization of exhaust
    composition oscillations 135
Fischer-Tropsch synthesis
  Anderson-Schultz-Flory plot
    470, 473, 475, 481
  Assessment 488
  Bang-bang cycling 473, 476,
    479, 482

SUBJECT INDEX

Chain growth probability 484
Chemical mechanism 491
Cobalt catalysts 479
Cycle split effects 472, 476, 479
Ethene cycling in 486
Experimental systems 465
Exposure duration effects 483
Frequency effects 486
Future challenges for
  composition modulation 495
Global enhancement 481
H_2 pulsing 473
Hydrogen utilization 486
Introduction 464
Iron catalysts 466
Mechanism of improvement
  through modulation
  472, 475, 491
Methane formation mechanisms 493
Modulation strategies
  464, 471, 479, 482, 489
Molybdenum catalysts 475
Multiple resonance 482
Optimal modulation strategies 484
Period effects 469, 471, 476, 489
Product distribution 470, 471, 473, 491
Resonance 477, 488, 491
Ruthenium catalysts 471
Selectivity 475, 478
Step change measurements 493
Stoichiometric ratio effects 471
Surface carbides 494
Temperature effects 479
Water gas shift reaction 494
Fluidized beds - see also circulating fluidized beds
Two zone for gypsum
  decomposition 316
Oxidizing and reducing zones
  in 316
Flushing with inerts
  Butadiene partial oxidation
  438, 440
  Butane partial oxidation 445
  Ethene epoxidation 407
  Oxidative coupling of methane 397
Forcing - see Composition
  modulation or Step changes
Forcing of autonomous oscillation
  by composition modulation
  CO oxidation - experimental
    51, 65, 78
  CO oxidation - modeling 66
Free radical polymerization
  Broadening molecular weight
    distribution 505, 506, 508, 509
  Experimental system
    505, 509, 511
  Induced oscillations in 505
  Kinetic model 502
  Measurement of propagation
    rate constant 542
  Mechanism of improvement
    through modulation 506
  Modulation strategy in 505, 507, 508, 509
  Molecular weight distribution
    in 505
  Operation at a meta stable state
    through modulation 511
  Period effects 505
  Radiation modulation 509, 510
  Reactor models 503
  Simulation of 502
  Space time yield in 508
  Styrene polymerization 505, 507
  Transform of model to moment
    form 503
  Ziegler-Natta polymerization
    through free radicals 524
Furan selectivity in butadiene
  partial oxidation 435, 440

SUBJECT INDEX

Gas-solid reactions
  Catalytic coal gasification 325
  Challenges for composition
    modulation 324
  Heat pumps using 324
  Lime recovery 315, 323
Global enhancement
  Definition 4, 337
  Determination of 342
  Factor 4
Global enhancement through
  composition modulation
    23, 61, 99, 215, 218, 220, 224,
    239, 296, 302, 343, 481
Gradientless reactor
  Ammonia synthesis over iron
    catalysts 237
Gypsum 315
Independent reactions 337

Hydrocarbon catalytic combustion
  Experimental system 358
  Mechanism for improvement
    under modulation 359, 362
  Period effects 359
High temperature furnaces 27
Homogeneous reactions 25
Hot spots
  Ammonia synthesis
    reactors 241
Hydrogen pulsing
  Fischer-Tropsch synthesis 473
Hydrogen utilization in the
  Fischer-Tropsch synthesis 486
Hydrogenation
  Adsorbed N_2 on iron 213, 249
  Ammonia synthesis 211
  Iron catalysts - ammonia
    synthesis 213
Hysteresis
  Propene partial oxidation 420

Industrial reactors for composition
  modulation 19

Iron catalysts
  Ammonia synthesis 213, 218,
    225, 239
  Chromatographic operation
    212
  Diffusion interference 235
  Fischer-Tropsch synthesis 466
  Mechanisms for improvement
    under modulation 216, 222,
    225, 227, 231, 260, 265
  Modeling ammonia synthesis on
    242, 252, 259
  NO reduction by CO 553

Ketones from partial oxidation
  of olefins 424
Kinetic mechanism - see Chemical
  mechanism
Kinetic measurements through
  composition modulation
  Bang-bang light modulation for
    gas phase reactions 541
  Bang-bang light modulation for
    free radical polymerization
    reactions 541
  Modulated beam relaxation
    spectrometry 540
Kinetic models
  SO_2 oxidation 281
Kinetics of gas-solid reactions
  through modulation 539

Laboratory reactors for
  composition modulation 13
LiMgO catalysts
  Oxidative coupling of
    methane 393
Light off of automotive engine
  exhaust
  Amplitude effects 172
  Catalyst aging with engine
    exhaust 172
  Three-way catalysts with
    automotive engine

SUBJECT INDEX

exhaust   172
Three-way catalysts with simulated automotive exhaust   160

Mass transfer interference - see Diffusion interference
Mean composition effects - see Stoichiometric ratio effects
Mechanism - see Chemical mechanism
Mechanisms for performance improvement under composition modulation
  Ammonia synthesis   216, 222, 225, 227, 231, 260, 265
  Benzene partial oxidation   450
  Butane dehydrogenation   437, 445
  CO oxidation   43, 45, 47, 52, 76, 80, 99
  Deamination reactions (stop effect)   308
  Ethanol dehydration   365
  Ethene epoxidation   405, 408
  Ethyl acetate from ethene and acetic acid   298
  Fischer-Tropsch synthesis   472, 475, 491
  Hydrocarbon catalytic combustion   359, 362
  Methane homologation   371, 378, 381
  N_2O reduction by CO   127, 129
  NO_x reduction by CO   106, 113
  Propane dehydrogenation   428
  Propene epoxidation   412
  Propene partial oxidation   420, 422, 424
  Stop effect   308
Mechanisms investigated using composition modulation
  Ammonia synthesis on iron catalysts   531

CO oxidation on platinum surfaces   532
CO_2 methanation   532
Ethene epoxidation on silver catalysts   531
Propene partial oxidation on redox catalysts   531, 532
SO_2 oxidation on vanadia catalysts   531
Mechanistic studies
  Assessment of composition modulation for   562
  Dynamic (step change) methods   530
  DRIFT cell, use of for   533
  Future challenges for composition modulation   562
  Multi-part cycles, use of for   531
  Phase lag, use of for   534
  Qualitative information from composition modulation   531
  Resonance frequency, use of for   531
  Similarity of response, use of for   534
  Start up behavior, use of for   531
Methane homologation
  Assessment of composition modulation   380, 382
  CO exposure in   377
  Catalyst deactivation   377
  Chemical mechanism   371, 378, 381
  Cycle structure   370, 373, 382
  Experimental system   373
  Exposure duration effects   376
  High temperature studies   369
  Inert flush, use of in   377, 381
  Low temperature studies   373
  Molybdate catalysts   373
  Product distribution   371, 374
  Ruthenium catalysts   370
  Temperature effects   371, 374, 376, 381

Thermodynamic barrier 369
Transition metal catalysts 373
Zeolite catalysts 371
Methanol synthesis 363
Methanolation/demethanolation reactions in heat pumps 324
Mixed beds of catalysts and adsorbents in ethene epoxidation 408
Mixing effects in polymerization Emulsion polymerization 527
Models - see Modeling
Model discrimination through composition modulation
   Ammonia synthesis 244, 253, 543
   Application of Karhunen-Loeve expansion 557
   Assessment of the use of composition modulation 564
   Ethene acetoxidation 549
   N_2O reduction by CO 555
   Partial oxidation of benzene 557
   Research questions 564
   Stop effect 310, 550
   Use of point values within a cycle 556
Model Verification through composition modulation
   Ammonia synthesis 253, 257, 263
   Ethyl acetate from ethene and acetic acid 300
   NO reduction by CO 553
Modeling of composition modulation
   Acetylene hydrogenation 355
   Allowance for diffusion in porous catalysts 63, 83
   Ammonia synthesis over iron catalysts 242, 249, 252, 543, 545
   Anionic living chain polymerization 502, 511, 513
   Benzene partial oxidation 559
   Butadiene hydrogenation 351
   CO oxidation over platinum catalysts 56
   CO oxidation over silver catalysts 90
   CO_2 methanation 536
   Diethanolamine formation 345
   Emulsion polymerization 527
   Ethanol dehydration 366
   Ethene epoxidation 406
   Ethyl acetate from ethene and acetic acid 298, 549
   Ethyl adipate saponification 341
   Free radical polymerization 502, 503, 506
   Global enhancement of ammonia synthesis rate 246
   H_2 inhibition of ammonia synthesis 249
   N_2 adsorption on ruthenium 249
   N_2 storage in catalysts 252, 256
   N_2O reduction by CO 127, 555
   Relaxed steady state in ammonia synthesis 242
   SO_2 converter with periodic air flushing 282
   SO_2 oxidation in an isothermal CSTR 287
   Step growth polymerization 502, 523
   Stop effect 304, 310, 550
   Super active state of catalyst in ammonia synthesis 260, 261
   Surface activation/restructuring in ammonia synthesis 259
   Ziegler-Natta polymerization 524
Modeling of dynamic behavior of non-isothermal fixed catalyst

SUBJECT INDEX

beds 97
Modulated beam relaxation
  spectrometry 540
Modulation, effect of
  Amplitude 150, 155, 163, 168,
    171, 172, 269, 394, 418
  Cycle split 214, 218, 296, 343,
    365, 394, 422, 472, 476, 479
  Frequency 39, 46, 48, 57, 73,
    93, 103, 110, 114, 150, 155, 157,
    168, 214, 394, 413, 480, 525
  Period 296, 302, 319, 327, 329,
    343, 351, 353, 359, 365, 402,
    412, 418, 422, 424, 427, 435,
    469, 471, 476, 489
  Phase Lag 55, 110, 114, 124
Molybdenum Catalysts
  Fischer-Tropsch synthesis
    467, 475
  Methane homologation 373
Multi-part cycles
  Benzene partial oxidation 448
  Butadiene partial
    oxidation 435
  Ethene epoxidation 407
  Oxidative coupling of
    methane 397
  Propene epoxidation 410
Multiple reactions
  Acetylene hydrogenation 353
  Butadiene hydrogenation 350
  Classes of 337
  Competitive 337, 363
  Consecutive 338
  Consecutive-competitive
    338, 347, 362
  Diethanolamine formation 342
  Ethene oligomerization 383
  Ethanol dehydration 364
  Ethanolamine formation 347
  Ethyl adipate saponification
    339
  Independent 337

Methane homologation
  369, 373
Methanol synthesis 363
Model systems 335, 339
Parasitic 347, 363, 367
Total oxidation of hydrocarbons
  357
Multiple resonant frequencies
  93, 287, 482

Nickel catalysts
  Acetylene hydrogenation 353
  Butadiene hydrogenation 350
Nitrogen dissolution
  Iron catalysts in ammonia
    synthesis 252
Nitrogen storage
  Iron catalysts in ammonia
    synthesis 252
  Modeling of, on iron catalysts
    253, 256
NO reduction by CO
  Discrimination of models by step
    change response 553
  Iron catalysts 553
  Kinetic models 553
  Model verification 555
N_2O formation in simulated
  automotive engine exhaust 165
N_2O reduction by CO
  Bang-bang cycling 555
  Experimental on platinum
    catalysts 124, 129, 555
  Mechanism 127, 129
  Model verification 556
  Modeling 127, 555, 556
  Phase lag effects 124, 556
Noble metal catalysts - see also
  Three way catalysts
  SO_3 reduction by CO 292
NO_x reduction by CO
  Autonomous oscillations
    110, 122

Noble metal catalysts 103
Platinum catalysts
    110, 113, 114
Future challenges for
    composition modulation 130
Introduction 101
Mechanism 106, 113
N_2O formation 165
Phase lag effects 110, 114
Reactions 101
Single crystals 121, 122
NO_x suppression 27

Optimal modulation strategies
    Fischer-Tropsch synthesis 484
Osmium catalysts
    Ammonia synthesis 231
    Exposure duration effects 233
Overlapping cycle parts in oxidative
    coupling of methane 398
Oxidative coupling of methane
    ARCO studies 388
    Amplitude effects 394
    Assessment of composition
        modulation 460
    Background for composition
        modulation 386
    Bang-bang cycling 394
    Catalyst screening 388
    Cycle split effects 394
    Experimental systems
        387, 388, 389
    Exposure duration effects 391
    Flushing with inerts 397
    Frequency effects 394
    Future challenges for
        composition modulation
        401, 460
    Li-MgO catalysts 393
    Modulation strategies 393, 400
    Multi-part cycles 397
    Overlapping cycle parts
        398
    Periodic reversal of flow direction

    combined with 401
    Stoichiometric ratio effects 394
    Union Carbide study 388
    University of Waterloo
        study 393
    Yield barrier for 386, 390, 398,
        401
Oxidative dehydrogenation - see
    Butane oxidative
    dehydrogenation or propane
    oxidative dehydrogenation
Oxide catalysts
    Copper oxide with CO
        oxidation 95
    Deamination and dehydration
        reactions with 305
    Hopcalite with CO
        oxidation 97
    Nickel oxide with CO
        oxidation 90
    Stop effect with 306
    Vanadia with CO oxidation 92

Palladium catalysts
    Acetoxidation of ethene 368
Parallel reactions - see
    Competitive reactions
Parameter estimation
    N_2 storage models 254
    Surface restructuring
        models 260
Parametric sensitivity
    Ethene epoxidation 403
Parasitic reactions
    Acetoxidation of ethene 368
    Ethyl acetate formation 367
    Methanol synthesis 369
Partial oxidation
    Assessment of composition
        modulation 462
    Methane to methanol 401
Period effects - see also Modulation
    and Frequency
    Ammonia synthesis 214

SUBJECT INDEX

Periodic air purge in automotive
exhaust catalysis 163
Periodic reversal of flow direction
Combined with composition
modulation 401
SO_2 oxidation 291
Phase lag effects
CO oxidation on platinum
catalysts 55
N_2O reduction on platinum
catalysts 124
NO_x reduction on platinum
catalysts 110, 114, 124
Phase lag, use of in CO_2
methanation study 534
Platinum catalysts
CO oxidation 39, 51, 60, 72,
82, 99
Coupled oxidation reduction
- CO/NO system 143
Ethene oligomerization 383
Hydrocarbon catalytic
combustion on gauze 359
Methane homologation 374
Model for CO oxidation 56,
63, 83
Model for N_2O reduction 127
N_2O reduction 124, 127,
129, 555
NO_x reduction 110, 113, 114,
122
SO_3 reduction by CO 292
Polymerization reactions
Anderson-Schultz-Flory
distribution of chain
lengths 501
Anionic living chain
polymerization 501, 511,
517, 519
Assessment of composition
modulation for 528
Background for composition
modulation 498
Emulsion polymerization 527

Free radical polymerization
500, 502, 505, 507
Future challenges for
composition modulation
with 529
Mean chain lengths and
molecular weights 500
Molecular weight
distributions 500
Moment methods 500
Polydispersity 500
Simulations of 500
Step growth polymerization
501, 522
Summary of the literature 499
Ziegler-Natta polymerization
524
Pressure effects
Ammonia synthesis 219
Product distribution
Ethene oligomerization 383
Fischer-Tropsch synthesis 470,
471, 473, 481, 491
Methane homologation
371, 374
Propene partial oxidation 419
Propane oxidative dehydrogenation
Bang-bang cycling 427
Mechanism of improvement
through modulation 428
Period effects 427
Redox catalysts 426
Propene ammoxidation
Background for composition
modulation 413
Chemical mechanism 417
Circulating fluidized beds
413
Sohio studies 413
Successive pulsing for
mechanistic studies 417
Propene epoxidation
Mechanism of improvement with
modulation 412

Modulation strategies 410
Multi-part cycles 410
Silver catalysts 409
Propene partial oxidation
  Amplitude effects 418
  Cycle split effects 422
  Experimental system 413
  Hysteresis in 420
  Mechanism of improvement through modulation 420, 422, 424
  Mechanism studies 413
  Modulation strategies 421
  Period effects 418, 422, 424
  Periodic pulsing studies 413
  Process considerations for composition modulation 420
  Product distribution 419
  Redox catalysts 413, 420, 422
  Resonance 424
  Selectivity ratio 424
  Surface restructuring 420
  Water pulsing 422
Pulse mode 6
Pulsed combustors 26
Pulsed operation
  Hydrogen pulsing in ammonia synthesis 225

Quasi periodic behavior 49, 94
Quasi steady state 6

Rate controlling step(s)
  CO_2 gasification of low rank coal 334
Reactor development for composition modulation
  Butane partial oxidation 442
  Butane dehydrogenation 431
Recirculating solids reactors - see Circulating fluidized beds
Redox catalysts
  Benzene partial oxidation 448, 450

Butane dehydrogenation 429
Butane partial oxidation 442
Propane dehydrogenation 426
Propene partial oxidation 413, 420, 422
Xylene partial oxidation 454
Relaxation time
  SO_2 oxidation over vanadium catalysts 271
Relaxed steady state
  Definition of 6
  Models for, in ammonia synthesis 242, 246
Reproducible cycling state, attainment of 92, 118
Resonance
  Ammonia synthesis 214, 218
  Automotive engine exhaust 150, 155, 157
  CO oxidation 39, 46, 48, 51, 73
  Ethene epoxidation 412
  Fischer-Tropsch synthesis 477, 488, 491
  N_2O reduction 124
  NO_x reduction 103, 110, 114
  NO reduction in automotive engine exhaust 150
  Multiple resonant frequencies 93, 287, 480
  Propene partial oxidation 419, 424
  Simulated automotive engine exhaust 155, 157
  SO_2 oxidation 269, 287
Resonance frequency, use of for study of mechanism 531
Ruthenium catalysts
  Ammonia synthesis 226
  Challenges for, with composition modulation 266
  CO_2 methanation 532
  Exposure duration effects 227
  Fischer-Tropsch synthesis 467, 471

SUBJECT INDEX

Mechanism for performance
improvement in modulation
227, 260, 265
Methane homologation
371, 374
Stoichiometric ratio effects 227
Temperature effects 227

Selectivity
Conversion relationship
444, 456
Definition 336
Dependence on conversion
444, 456
Diethanolamine formation 343
Enhancement factor for 337
Fischer-Tropsch synthesis
475, 478
Instantaneous 336
Selectivity ratio 336
Sequential reactions - see
Consecutive reactions
Serial reactions - see Consecutive
reactions
Silver catalysts
CO oxidation 90
Ethene epoxidation
402, 407, 412
Propene epoxidation 409
Similarity of product response
curves, use of in CO_2
methanation 534
Simulated automotive engine
exhaust
Light off with three-way
catalysts 160
Modulation of air-fuel ratio
148, 153
N_2O formation 165
Resonance in frequency
155, 157
Three-way catalysts 153
Simulation - see also Modeling
Ammonia synthesis in a

non-isothermal reactor 246
Bang-bang cycling 525
Convolution method for
polymerization models 515
Emulsion co-polymerization
of styrene and butadiene 527
Imperfect mixing 525
Impurities effects 513
Kinetic models for 281, 502,
513, 522
Modulation cycle 525
Molecular weight distribution
control in polymerization
512, 528
Moment methods for
polymerization reactions
500, 503, 506
Multi-variable modulation 513
Non isothermal operation of
a polymer reactor 504
Olefin polymerization 524
Optimal control policy through
modulation 513, 515
Polydispersity control in
polymerization 512, 516,
524, 528
Polymerization reactor models
for 503, 523
Relaxed steady state 504
SO_2 converter with periodic air
flushing 283
Step growth polymerization
522
Styrene homo and
co-polymerization 504, 505
Transforms to moment form of
models 503
Tubular polymerization reactor
512
Two variable modulation
504, 512
Validity of the Schultz-Flory
model for chain growth 515
Simultaneous reactions - see Dual

oxidation reactions and
Coupled oxidation reduction
reactions
Single crystal studies
CO oxidation on platinum 68
NO_x reduction on platinum
121
SO_2 converter
Periodic air flushing of the final
stage of 272
Smelter off gas 273
Sulfur burning 273
SO_3 breakthrough with air
flushing 276
Sodium carbonate catalysts for
methanol dehydrogenation 305
Sohio research on propene
ammoxidation 413
Space velocity effects 53
Stabilizing an unstable steady
state 29
Start-up behavior in composition
modulation 531
Step growth polymerization
Future challenges for
composition modulation
with 529
Gelation in 524
Kinetic model 502, 522
Polydispersity in 524
Reactor model 523
Stoichiometric number
Definition for CO/NO system
142
Effect on CO partitioning 145
Modulation in CO oxidation
74, 86
Modulation in coupled
oxidation-reduction reactions
145
Three-way catalysts 158, 170
Stoichiometric ratio effects
Acetylene hydrogenation 353
Ammonia synthesis 215, 221,
227, 231
Claus reaction 303
Ethyl adipate saponification
340
Fischer-Tropsch synthesis 471
SO_2 oxidation 269
Stop effect
Dynamic models for 550
Discrimination among rival
models 552, 554
Experimental 306
Exposure duration effect 307
Failure of step change response
method 551
Global enhancement 310
Mechanism of improvement
through modulation 308
Modeling 309
Temperature effects 306
Strategies in composition
modulation
General 9
Dehydrogenation and partial
oxidation reactions 393, 410,
421, 435
Fischer-Tropsch synthesis
464, 471, 479, 482, 484
Multiple reactions
346, 359, 362, 370, 373, 383
Optimal 484
Polymerization 505, 507, 508,
509, 517, 525
SO_2 oxidation 269, 288
Successive pulsing method for
mechanistic studies
Ammoxidation reactions 417
Benzene partial oxidation 450
Epoxidation reactions 413
Ethene partial oxidation 406
Sulfur dioxide oxidation
Air flushing 272, 283, 290
Amplitude effects 269
Experimental system 268, 273
High conversion studies 271

SUBJECT INDEX

Introduction to for composition
   modulation 267
Isothermal CSTR, comparison of
   data and model 289
Low conversion studies 268
Mechanism on vanadium
   catalysts 280
Modeling of 280, 283, 287
Periodic reversal of flow direction
   291
Relaxation time in 271
Resonance in 269
Simulation of SO_2 converters
   under modulation 273
SO_2 converters under
   modulation 272
Stoichiometric ratio effects 269
Sulfur trioxide reduction with CO
   Aperiodic behavior 293
   Autonomous oscillation 294
   Mechanism for improvement
      through modulation 292
   Platinum catalysts 292
Surface restructuring
   Propene partial oxidation 420

Temperature distribution in packed
   beds of catalyst
   Ethene epoxidation 403
Temperature effects
   Butane partial oxidation
      441, 444
   CO_2 gasification of low rank
      coals 329, 334
   Ethene oligomerization 383
   Fischer-Tropsch synthesis 479
   Methane homologation
      374, 376, 381
   Xylene partial oxidation 454
Thermodynamic barrier, bypassing
   through periodic operation 369
Three way catalysts
   Aging effect on light off with
      automotive engine exhaust
      171

Air-fuel ratio effects
   150, 158, 168
Automotive engine exhausts
   - experimental 150, 168, 171
Cerium oxide with CO oxidation
   139
Cerium oxide with oxidation
   -reduction reactions 150
Cerium oxide with sulfur
   poisoning 177
Composition 135
CO oxidation - experimental
   33, 139
CO oxidation - mechanism
   140
Coupled oxidation-reduction
   reactions (CO/NO) 143
Dual oxidation reactions
   (CO/propene) 141
Future challenges for
   composition modulation 182
Improvement in air-fuel ratio
   control 181
Light off behavior with
   automotive engine exhaust
   172
Light off behavior with simulated
   engine exhaust 160
Monoliths as support structures
   163, 173
NO_x reduction 103
Overview of composition
   modulation effects 179
Periodic air exposure effects
   163
Simulated automotive engine
   exhaust 153
Single reaction studies 139
Stoichiometric number effects
   158, 170
Sulfur poisoning 175
Without noble metals 177
Transition Metal catalysts
   Methane homologation 373

SUBJECT INDEX

Two zone fluidized beds
  Decomposition of calcium sulfate in 316
  Experimental system 316

Union Carbide study of oxidative coupling of methane 388
University of Waterloo research
  Butadiene partial oxidation 435
  Oxidative coupling of methane 393
Unstable steady state 29

Vanadium catalysts
  Benzene partial oxidation 559
  SO_2 oxidation 268

Water pulsing in propene partial oxidation 422
Wave front analysis applied to benzene partial oxidation 450
What is composition modulation 2
Wrong-way behavior 95

Yield
  Definition 336
  Diethanolamine formation 343
  Enhancement factor for 337
Yield barrier in the oxidative coupling of methane 386, 390, 398, 401

Zeolite catalysts
  Methane homologation 371
Ziegler-Natta polymerization
  Bang-bang cycling 525
  Frequency effects 525
  Future challenges for composition modulation 529
  Imperfect mixing 525
  Modeling 524
  Modulation cycle 525
  Modulation strategy 525
  Propylene polymerization 525